PRINCIPLES OF SOIL PHYSICS

Mass Spectrometry of Soils, edited by Thomas W. Boutton and Shinichi Yamasaki

Handbook of Photosynthesis, edited by Mohammad Pessarakli

Chemical and Isotopic Groundwater Hydrology: The Applied Approach, Second Edition, Revised and Expanded, Emanuel Mazor

Fauna in Soil Ecosystems: Recycling Processes, Nutrient Fluxes, and Agricultural Production, edited by Gero Benckiser

Soil and Plant Analysis in Sustainable Agriculture and Environment, edited by Teresa Hood and J. Benton Jones, Jr.

Seeds Handbook: Biology, Production, Processing, and Storage, B. B. Desai, P. M. Kotecha, and D. K. Salunkhe

Modern Soil Microbiology, edited by J. D. van Elsas, J. T. Trevors, and E. M. H. Wellington

Growth and Mineral Nutrition of Field Crops: Second Edition, N. K. Fageria, V. C. Baligar, and Charles Allan Jones

Fungal Pathogenesis in Plants and Crops: Molecular Biology and Host Defense Mechanisms, P. Vidhyasekaran

Plant Pathogen Detection and Disease Diagnosis, P. Narayanasamy

Agricultural Systems Modeling and Simulation, edited by Robert M. Peart and R. Bruce Curry

Agricultural Biotechnology, edited by Arie Altman

Plant–Microbe Interactions and Biological Control, edited by Greg J. Boland and L. David Kuykendall

Handbook of Soil Conditioners: Substances That Enhance the Physical Properties of Soil, edited by Arthur Wallace and Richard E. Terry

Environmental Chemistry of Selenium, edited by William T. Frankenberger, Jr., and Richard A. Engberg

Principles of Soil Chemistry: Third Edition, Revised and Expanded, Kim H. Tan

Sulfur in the Environment, edited by Douglas G. Maynard

Soil–Machine Interactions: A Finite Element Perspective, edited by Jie Shen and Radhey Lal Kushwaha

Mycotoxins in Agriculture and Food Safety, edited by Kaushal K. Sinha and Deepak Bhatnagar

Plant Amino Acids: Biochemistry and Biotechnology, edited by Bijay K. Singh

Handbook of Functional Plant Ecology, edited by Francisco I. Pugnaire and Fernando Valladares

Handbook of Plant and Crop Stress: Second Edition, Revised and Expanded, edited by Mohammad Pessarakli

Plant Responses to Environmental Stresses: From Phytohormones to Genome Reorganization, edited by H. R. Lerner

Handbook of Pest Management, edited by John R. Ruberson

Environmental Soil Science: Second Edition, Revised and Expanded, Kim H. Tan

Microbial Endophytes, edited by Charles W. Bacon and James F. White, Jr.

Plant–Environment Interactions: Second Edition, edited by Robert E. Wilkinson

Microbial Pest Control, Sushil K. Khetan

Soil and Environmental Analysis: Physical Methods, Second Edition, Revised and Expanded, edited by Keith A. Smith and Chris E. Mullins

The Rhizosphere: Biochemistry and Organic Substances at the Soil–Plant Interface, Roberto Pinton, Zeno Varanini, and Paolo Nannipieri

Woody Plants and Woody Plant Management: Ecology, Safety, and Environmental Impact, Rodney W. Bovey

Metals in the Environment: Analysis by Biodiversity, M. N. V. Prasad

Plant Pathogen Detection and Disease Diagnosis: Second Edition, Revised and Expanded, P. Narayanasamy

Handbook of Plant and Crop Physiology: Second Edition, Revised and Expanded, edited by Mohammad Pessarakli

Environmental Chemistry of Arsenic, edited by William T. Frankenberger, Jr.

Enzymes in the Environment: Activity, Ecology, and Applications, edited by Richard G. Burns and Richard P. Dick

Plant Roots: The Hidden Half, Third Edition, Revised and Expanded, edited by Yoav Waisel, Amram Eshel, and Uzi Kafkafi

Handbook of Plant Growth: pH as the Master Variable, edited by Zdenko Rengel

Biological Control of Crop Diseases, edited by Samuel S. Gnanamanickam

Pesticides in Agriculture and the Environment, edited by Willis B. Wheeler

Mathematical Models of Crop Growth and Yield, Allen R. Overman and Richard V. Scholtz III

Plant Biotechnology and Transgenic Plants, edited by Kirsi-Marja Oksman-Caldentey and Wolfgang H. Barz

Handbook of Postharvest Technology: Cereals, Fruits, Vegetables, Tea, and Spices, edited by Amalendu Chakraverty, Arun S. Mujumdar, G. S. Vijaya Raghavan, and Hosahalli S. Ramaswamy

Handbook of Soil Acidity, edited by Zdenko Rengel

Humic Matter in Soil and the Environment: Principles and Controversies, Kim H. Tan

Molecular Host Resistance to Pests, S. Sadasivam and B. Thayumanavan

Soil and Environmental Analysis: Modern Instrumental Techniques, Third Edition, edited by Keith A. Smith and Malcolm S. Cresser

Chemical and Isotopic Groundwater Hydrology: Third Edition, Emanuel Mazor

Agricultural Systems Management: Optimizing Efficiency and Performance, Robert M. Peart and W. David Shoup

Physiology and Biotechnology Integration for Plant Breeding, edited by Henry T. Nguyen and Abraham Blum

Global Water Dynamics: Shallow and Deep Groundwater, Petroleum Hydrology, Hydrothermal Fluids, and Landscaping, Emanuel Mazor

Principles of Soil Physics, Rattan Lal and Manoj K. Shukla

Seeds Handbook: Biology, Production, Processing, and Storage, Second Edition, Revised and Expanded, Babasaheb B. Desai

Field Sampling: Principles and Practices in Environmental Analysis, Alfred R. Conklin, Jr.

Sustainable Agriculture and the International Rice–Wheat System, edited by Rattan Lal, Peter R. Hobbs, Norman Uphoff, and David O. Hansen

Plant Toxicology: Fourth Edition, Revised and Expanded, edited by Bertold Hock and Erich F. Elstner

Additional Volumes in Preparation

PRINCIPLES OF SOIL PHYSICS

RATTAN LAL
MANOJ K. SHUKLA

The Ohio State University
Columbus, Ohio, U.S.A.

CRC Press
Taylor & Francis Group
Boca Raton London New York

CRC Press is an imprint of the
Taylor & Francis Group, an **informa** business

CRC Press
Taylor & Francis Group
6000 Broken Sound Parkway NW, Suite 300
Boca Raton, FL 33487-2742

First issued in paperback 2019

ISBN-13: 978-0-8247-5324-5 (hbk)
ISBN-13: 978-0-367-39421-9 (pbk)

Library of Congress Cataloging-in-Publication Data
A catalog record for this book is available from the Library of Congress.

Visit the Taylor & Francis Web site at
http://www.taylorandfrancis.com

and the CRC Press Web site at
http://www.crcpress.com

Preface

This book addresses the topic of soil's physical properties and processes with particular reference to agricultural, hydrological, and environmental applications. The book is written to enable undergraduate and graduate students to understand soil's physical, mechanical, and hydrological properties, and develop theoretical and practical skills to address issues related to sustainable management of soil and water resources. Sustainable use of soil and water resources cannot be achieved unless soil's physical conditions or quality is maintained at a satisfactory level. Fertilizer alone or in conjunction with improved crop varieties and measures to control pests and diseases will not preserve productivity if soil's physical conditions are not above the threshold level, or if significant deterioration of physical conditions occur. Yet, assessment of physical properties and processes of soil is not as commonly done as that of chemical or nutritional properties, and their importance receives insufficient attention. Even when information on soil's physical properties is collected, it is not done in sufficient detail and rarely beyond the routine measurement of soil texture and bulk density.

Sustainability is jeopardized when soil's physical quality is degraded, which has a variety of consequences. The process of decline in soil's physical quality is set in motion by deterioration of soil structure: an increase in bulk density, a decline in the percentage and strength of aggregates, a decrease in macroporosity and pore continuity, or both. An important ramification of decline in soil structural stability is formation of a surface seal or crust with an attendant decrease in the water infiltration rate and an increase in surface runoff and erosion. An increase in soil bulk density leads to inhibited root development, poor gaseous exchange, and anaerobiosis. Excessive runoff lowers the availability of water stored in the root zone, and suboptimal or supraoptimal soil temperatures and poor aeration exacerbate the problem of reduced water uptake.

Above and beyond the effects on plant growth, soil's physical properties and processes also have a strong impact on the environment. Non-point source pollution is caused by surface runoff, erosion, and drainage effluent from agricultural fields. Wind erosion has a drastic adverse impact on air quality. An accelerated greenhouse effect is caused by emission of trace or greenhouse gases from the soil into the atmosphere. Important greenhouse gases emitted from soil are CO_2, CH_4, N_2O, and NO_x. The rate and amount of their emission depend on soil's physical properties (e.g., texture and temperature) and processes (e.g., aeration and anaerobiosis).

The emphasis in this textbook is placed on understanding the impact of the physical properties and processes of soil on agricultural and forestry production, sustainable use of soil and water resources for a range of functions of interest to humans, and the environment with special attention to water quality and the greenhouse effect. Sustainable use of natural resources is the basic, underlying theme throughout the book.

This book is divided into 20 chapters and 5 parts. Part I is an introduction to soil physics and contains two chapters describing the importance of soil physics, defining basic terms and principal concepts. Part II contains six chapters dealing with soil mechanics. Chapter 3 describes soil solids and textural properties, including particle size distribution, surface area, and packing arrangements. Chapter 4 addresses theoretical and practical aspects of soil structure and its measurement. There being a close relationship between structure and porosity, Chapter 5 deals with pore size distribution, including factors affecting it and assessment methods. Manifestations of soil structure (e.g., crusting and cracking) and soil strength and compaction are described in Chapters 6 and 7, respectively. Management of soil compaction is a topic of special emphasis in these chapters. Atterberg's limits and plasticity characteristics in terms of their impact on soil tilth are discussed in Chapter 8.

Part III, comprising eight chapters, deals with an important topic of soil hydrology. Global water resources, principal water bodies, and components of the hydrologic cycle are discussed in Chapter 9. Soil's moisture content and methods of its measurement, including merits and demerits of different methods along with their application to specific soil situations, are discussed in Chapter 10. The concept of soil-moisture potential and the energy status of soil water and its measurement are discussed in Chapter 11. Principles of soil-water movement under saturated and unsaturated conditions are described in Chapters 12 and 13, respectively. Water infiltration, measurement, and modeling are presented in Chapter 14. Soil evaporation, factors affecting it, and its management are discussed in Chapter 15. Solute transport principles and processes including Fick's laws of diffusion, physical, and chemical nonequilibruim, its measurement, and modeling are presented in Chapter 16.

Part IV comprises two chapters. Chapter 17 addresses the important topic of soil temperature, including heat flow in soil, impact of soil temperature on crop growth, and methods of managing soil temperature. Soil air and aeration, the topic of Chapter 18, is discussed with emphasis on plant growth and emission of greenhouse gases from soil into the atmosphere. Part V, the last part, contains two chapters dealing with miscellaneous but important topics. Chapter 19 deals with physical properties of gravelly soils. Water movement in frozen, saline, and water-repellent soils and scale issues in hydrology are the themes of Chapter 20. In addition, there are several appendices dealing with units and conversions and properties of water.

This book is of interest to students of soil physics with majors in soil science, agricultural hydrology, agricultural engineering, civil engineering, climatology, and topics of environmental sciences. There are several unique features of this book, which are important in helping students understand the basic concepts. Important among these are the following: (i) each chapter is amply illustrated by graphs, data tables, and easy to follow equations or mathematical functions, (ii) use of mathematical functions is illustrated by practical examples, (iii) some processes and practical techniques are explained by illustrations, (iv) each chapter contains a problem set for students to practice, and (v) the data examples are drawn from world ecoregions, including soils of tropical and temperate climates. This textbook incorporates comments and suggestions of students from around the world.

The book is intended to explain basic concepts of soil physics in a simplified manner rather than an exhaustive treatise on the most current literature available on the topics addressed. It draws heavily on material, data, graphs, and tables from many sources. The authors cite data from numerous colleagues from around the world. Sources of all data and material are duly acknowledged.

We are thankful for valuable contributions made by several colleagues, graduate students, and staff of the soil science section of

The Ohio State University. We especially thank Ms. Brenda Swank for her assistance in typing some of the text and in preparing the material. Help received from Pat Patterson and Jeremy Alder is also appreciated. Thanks are also due to the staff of Marcel Dekker, Inc., Publishers for their timely effort in publishing the book and making it available to the student community.

Rattan Lal
Manoj K. Shukła

Contents

Preface   *iii*

Part I   **Introduction**

1   Importance of Soil Physics   1
2   Basic Definitions and Concepts   15

Part II   **Soil Mechanics**

3   Soil Solids   33
4   Soil Structure   93
5   Porosity   149
6   Manifestations of Soil Structure   165
7   Soil Strength and Compaction   189
8   Soil Rheology and Plasticity   231

Part III   **Soil Hydrology**

9   Water   255
10   Soil's Moisture Content   287

viii

Contents

11	Soil-Moisture Potential	321
12	Water Flow in Saturated Soils	355
13	Water Flow in Unsaturated Soils	379
14	Water Infiltration in Soil	405
15	Soil Water Evaporation	439
16	Solute Transport	465

Part IV   Soil Temperature and Aeration

17	Soil Temperature and Heat Flow in Soils	515
18	Soil Air and Aeration	557

Part V   Miscellaneous Topics

19	Physical Properties of Gravelly Soils	601
20	Special Problems	625

Appendix A	*The Greek Alphabet*	667
Appendix B	*Mathematical Signs and Symbols*	668
Appendix C	*Prefixes for SI Units*	669
Appendix D	*Values of Some Numbers*	670
Appendix E	*SI Derived Units and Their Abbreviations*	671
Appendix F	*Unit Conversion Factors*	672
Appendix G	*Unit Conversions (Equivalents)*	676
Appendix H	*Conversion Factors for Non-SI Units*	678
Appendix I	*Conversion Among Units of Soil-Water Potential*	679
Appendix J	*Surface Tension of Water Against Air*	680
Appendix K	*Density of Water from Form Air*	681
Appendix L	*The Viscosity of Water 0° to 100°C*	682
Appendix M	*Effect of Temperature of Vapor Pressure, Density of Water Vapor in Saturated Air, and Surface Tension of Water*	684
Appendix N	*Osmotic Pressure of Solutions of Sucrose in Water at 20°C*	685
Appendix O	*Constant Humidity*	686
Appendix P	*Some Common Algebraic Functions*	689
Index		*691*

PRINCIPLES OF SOIL PHYSICS

1

Importance of Soil Physics

1.1 SOIL: THE MOST BASIC RESOURCE

Soil is the upper most layer of earth crust, and it supports all terrestrial life. It is the interface between the lithosphere and the atmosphere, and strongly interacts with biosphere and the hydrosphere. It is a major component of all terrestrial ecosystems, and is the most basic of all natural resources. Most living things on earth are directly or indirectly derived from soil. However, soil resources of the world are finite, essentially nonrenewable, unequally distributed in different ecoregions, and fragile to drastic perturbations. Despite inherent resilience, soil is prone to degradation or decline in its quality due to misuse and mismanagement with agricultural uses, contamination with industrial uses, and pollution with disposal of urban wastes. Sustainable use of soil resources, therefore, requires a thorough understanding of properties and processes that govern soil quality to satisfactorily perform its functions of value to humans. It is the understanding of basic theory, leading to description of properties and processes and their spatial and temporal variations, and the knowledge of the impact of natural and anthropogenic perturbations that lead to identification and development of sustainable management systems. Soil science is, therefore, important to management of natural resources and human well-being.

1.2 SOIL SCIENCE AND ECOLOGY

Ecology is the study of plants and animals in their natural environment (*oikes* is a Greek world meaning home). It involves the study of organisms and their interaction with the environment, including transformation and flux of energy and matter. Soil is a habitat for a vast number of diverse organisms, some of which are yet to be identified. Soil is indeed a living entity comprising of diverse flora and fauna. The uppermost layer of the earth ceases to be a living entity or soil, when it is devoid of its biota.

An ecosystem is a biophysical and socioeconomic environment defined by the interaction among climate, vegetation, biota, and soil (Fig. 1.1). Thus, soil is an integral and an important component of

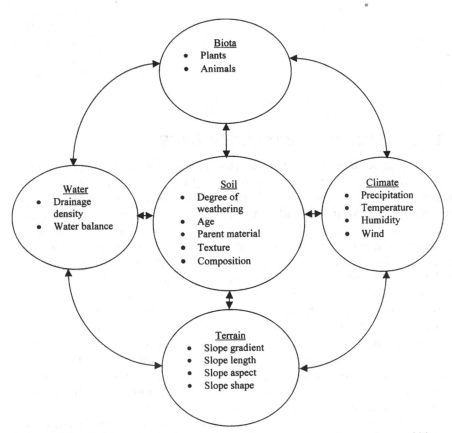

FIGURE 1.1 Soil is an integral component of an ecosystem, also made up of biota, climate, terrain, and water.

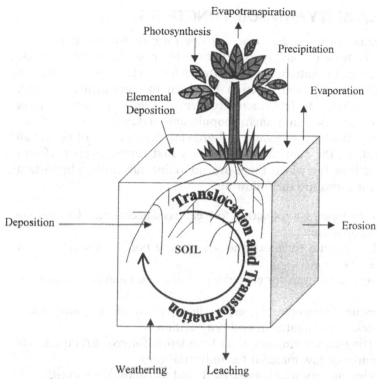

FIGURE 1.2 A pedosphere represents a dynamic interaction of soil with the environment.

any ecosystem. In the context of an ecosystem, soil is referred to as the *pedosphere*. The pedosphere is an open soil system (Buol, 1994). It involves transfer of matter and energy between soil and the atmosphere, hydrosphere, biosphere, and lithosphere (Fig. 1.2). The lithosphere adds to the soil through weathering and new soil formation and receives from the soil through leaching. It receives alluvium and colluvium from soils upslope and transfers sediments to soil downslope. In addition, there are transformations and translocations of mater and energy within the soil. An ecosystem can be natural (e.g., forest, prairie) which retains much of its original structure and functioning, or managed (e.g., agricultural, urban) which has been altered to meet human needs. The productivity of managed and functioning of all (natural and managed) ecosystems depends to a large extent on soil quality and its dynamic nature.

1.3   SOIL QUALITY AND SOIL FUNCTIONS

Soil quality refers to the soil's capacity to perform its functions. In other words, it refers to soil's ability to produce biomass, filter water, cycle elements, store plant nutrients, moderate climate, etc. For an agrarian population, the primary soil function has been the production of food, fodder, timber, fiber, and fuel. Increased demands on soil resources have arisen due to increases in human population, industrialization of the economy, rising standards of living, and growing expectations of people all over the world. In the context of the twenty-first century, soil performs numerous functions for which there are no viable substitutes. Important among these functions are the following:

1. Sustaining biomass production to meet basic necessities of a growing human population
2. Providing habitat for biota and a vast gene pool or a seedbank for biodiversity
3. Creating mechanisms for elemental cycling and biomass transformation
4. Moderating environment, especially quality of air and water resources, waste treatment and remediation
5. Supporting engineering design as foundation for civil structures, and as a source of raw material for industrial uses
6. Preserving archeological, geological, and astronomical records
7. Maintaining aesthetical values of the landscape and ecosystem, and preserving cultural heritage

Soil quality refers to its capacity to perform these functions, and to soils capability for specific functions that it can perform efficiently and on a sustainable or long-term basis (Lal, 1993; 1997; Doran et al., 1994; Doran and Jones, 1996; Gregorich and Carter, 1997; Karlen et al., 1997; Doran et al., 1999). Soil's agronomic capability refers to its specific capacity to grow crops and pasture. In most cases, however, soil cannot perform all functions simultaneously. For example, soil can either be used for crop cultivation or urban use.

Soil degradation refers to decline in soil quality such that it cannot perform one or several of its principal functions. Soil degradation is caused by natural or anthropogenic factors. Natural factors, with some exceptions such as volcanic eruptions and landslides, are usually less drastic than anthropogenic perturbations. Thus, severe degradation is typically caused by anthropogenic perturbations. Soil degradation leads to decline in soil quality causing reduction in its biomass productivity, environmental moderation capacity, ability to support engineering structures, capacity to perform aesthetic and cultural functions, and ability to function as a storehouse of gene pool and archeological/historical records. Thus, a degraded soil cannot perform specific functions of interest/utility to humans.

1.4 SOIL SCIENCE AND AGROECOSYSTEMS

Agroecology is the study of interaction between agronomy (i.e., study of plants and soils) and ecology. It is defined as the study and application of ecological principles to managing agroecosystems. Therefore, an agro-ecosystem is a site of agricultural/agronomic production, such as a farm. In this context, therefore, agriculture is merely an anthropogenic manipulation of the carbon cycle (biomass or energy) through uptake, fixation, emission, and transfer of carbon and energy. Soil quality plays an important role in anthropogenic manipulation of the carbon cycle. More specifically, soil physical quality, which is directly related to soil physical properties and processes, affects agronomic productivity through strong influences on plant growth.

1.5 SOIL PHYSICS

Soil physics is the study of soil physical properties and processes, including measurement and prediction under natural and managed ecosystems. The science of soil physics deals with the forms, interrelations, and changes in soil components and multiple phases. The typical components are: mineral matter, organic matter, liquid, and air. Three phases are solid, solution and gas, and more than one liquid phase may exist in the case of nonaqueous contamination. Physical edaphology is a science dealing with application of soil physics to agricultural land use. The study of the physical phenomena of soil in relation to atmospheric conditions, plant growth, soil properties and anthropogenic activities is called physical edaphology. Study of soil in relation to plant growth is called *edaphology*, whereas study of soil's physical properties and processes in relation to plant growth is called *physical edaphology*. Thus, physical edaphology is a branch of soil physics dealing with plant growth.

Soil physics is a young and emerging branch of pedology, with significant developments occurring during the middle of twentieth century. It draws heavily on the basic principles of physics, physical chemistry, hydrology, engineering and micrometeorology (Fig. 1.3). Soil physics applies these principles to address practical problems of agriculture,

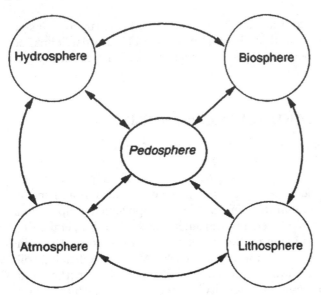

ecology, and engineering. Its interaction with emerging disciplines of geography (geographic information system or GIS), data collection (remote sensing), and analytical techniques (fuzzy logic, fractal analysis, neural network, etc.) has proven beneficial in addressing practical problems in agriculture, ecology, and environments. Indeed, soil physics plays a pivotal role in the human endeavor to sustain agricultural productivity while maintaining environment quality.

1.6 SOIL PHYSICS AND AGRICULTURAL SUSTAINABILITY

Agricultural sustainability implies non-negative trends in productivity while preserving the resource base and maintaining environmental quality. The role of physical edaphology in sustaining agricultural production while preserving the environment cannot be overemphasized. While the economic and environmental risks of soil degradation and desertification are widely recognized (UNEP, 1992; Oldeman, 1994; Pimental et al., 1995; Lal, 1994; 1995; 1998; 2001; Lal et al., 1995; 1998), the underlying processes and mechanisms are hardly understood (Lal, 1997). It is in this connection that the application of soil physics or physical edaphology has an important role

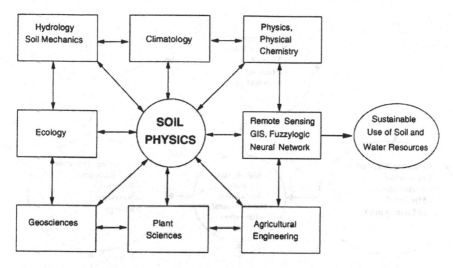

FIGURE 1.4 Interaction of soil physics with basic and applied sciences.

to play in: (i) preserving the resource base, (ii) improving resource use efficiency, (iii) minimizing risks of erosion and soil degradation, and restoring and reclaiming degraded soils and ecosystems, and (iv) enhancing production by alleviation of soil/weather constraints through development and identification of judicious management options (Fig. 1.4). Notable applications of soil physics include control of soil erosion; alleviation of soil compaction; management of soil salinity; moderation of soil, air, and water through drainage and irrigation; and alteration of soil temperature through tillage and residue management. It is a misconception and a myth that agricultural productivity can be sustained by addition of fertilizer and/or water per se. Expensive inputs can be easily wasted if soil physical properties are suboptimal or below the critical level. High soil physical quality (Lal, 1999a; Doran et al., 1999) plays an important role in enhancing soil chemical and biological qualities. Applications of soil physics can play a crucial role in sustainable management of natural resources (Fig. 1.5). Fertilizer, amendments, and pesticides can be leached out, washed away, volatilized, miss the target, and pollute the environment under adverse soil physical conditions. Efficient use of water and nutrient resources depends on an optimum level of soil physical properties and processes. Soil fertility, in its broad sense, depends on a favorable interaction between soil components and phases that optimize soil physical quality. Soil physical properties important to agricultural sustainability are texture, structure, water retention and transmission, heat

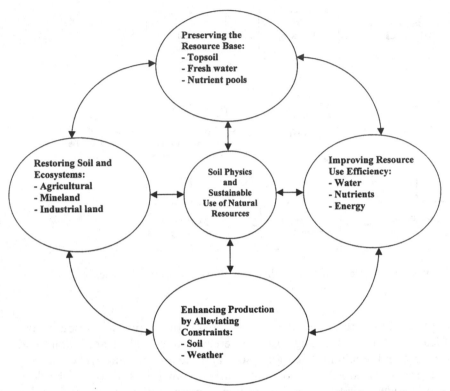

FIGURE 1.5 Applications of soil physics are crucial to sustainable use of natural resources for agricultural and other land uses.

capacity and thermal conductivity, soil strength, etc. These properties affect plant growth and vigor directly and indirectly. Important soil physical properties and processes for specific agronomic, engineering, and environmental functions are outlined in Table 1.1. Soil structure, water retention and transmission properties, and aeration play crucial roles in soil quality.

Soil physical properties are more important now than ever before in sustaining agricultural productivity because of the shrinking global per capita arable land area (Brown, 1991; Engelman and LeRoy, 1995). It was 0.50 ha in 1950, 0.20 ha in 2000, and may be only 0.14 ha in 2050 and 0.10 ha in 2100 (Lal, 2000). Therefore, preserving and restoring world soil resources is crucial to meeting demands of the present population without jeopardizing needs of future generations.

TABLE 1.1 Soil Physical Properties and Processes That Affect Agricultural, Engineering, and Environmental Soil Functions

Process	Properties	Soil functions
Biomass productivity (agricultural functions)		
1. Compaction	Bulk density, porosity, particle size distribution, soil structure	Root growth, water and nutrient uptake by plants
2. Erosion	Structural stability, erodibility, particle size, infiltration and hydraulic conductivity, transportability, tillability	Root growth, water and nutrient uptake, aeration
3. Water movement	Hydraulic conductivity, pore size distribution, tortuosity	Water availability to plants, chemical transport
4. Aeration	Porosity, pore size distribution, soil structure, concentration gradient, diffusion coefficient	Root growth and development, soil and plant respiration
5. Heat transfer	Thermal conductivity, soil moisture content	Root growth, water and nutrient uptake, microbial activity
Engineering functions		
1. Sedimentation	Particle size distribution, dispersibility	Filtration, water quality
2. Subsidence	Soil strength, soil water content, porosity	Bearing capacity, trafficability
3. Water movement	Hydraulic conductivity, porosity	Seepage, waste disposal, drainage
4. Compaction	Soil strength, compactability, texture	Foundation strength
Environmental functions		
1. Absorption/adsorption	Particle size distribution, surface area, charge density	Filtration, water quality regulation, waste disposal
2. Diffusion/aeration	Total and aeration porosity, tortuosity, concentration gradient	Gaseous emission from soil to the atmosphere

1.7   SOIL PHYSICS AND ENVIRONMENT QUALITY

In the context of environment quality, soil is a geomembrane that buffers and filters pollutants out of the environment (Yaalon and Arnold, 2000). It is also a vast reactor that transforms, deactivates, denatures, or detoxifies chemicals. Soil physical properties and processes play an important role in these processes. The environmental purification functions of soil are especially important to managing and moderating the quality of air and water resources (Fig. 1.6). Soil physical properties and processes influence the greenhouse effect through their control on emission of radiatively-active gases (e.g., CO_2, CH_4, N_2O, and NO_x) (Lal et al., 1995; Lal, 1999b; Bouwman, 1990). A considerable part of the 80 ppmv increase in atmospheric CO_2 concentration since the industrial revolution (IPCC, 1995; 2001) has come from C contained in world soils. Soil physical properties and processes determine the rate and magnitude of these gaseous

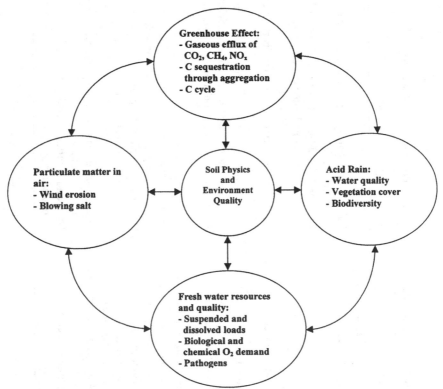

FIGURE 1.6   Applications of soil physics to environment quality.

emissions. Formation and stabilization of soil structure (i.e., development of secondary particles through formation of organomineral complexes), is a prominent consequence of C sequestration in soil. Air quality is also influenced by soil particles and chemicals (salt) airborne by wind currents. Management of soil structure, control of soil erosion, and restoration of depleted soils are important strategies of mitigating the global climate change caused by atmospheric enrichment of CO_2 (Lal, 2001).

Fresh water, although renewable, is also a finite quantity and a scarce resource especially in arid and semiarid regions. Soil, a major reservoir of fresh water, influences the quality of surface and ground waters (Engelman and LeRoy, 1993; Lal and Stewart, 1994). The pedospheric processes (e.g., leaching, erosion, transport of dissolved and suspended loads in water) interact with the biosphere and the atmosphere to influence properties of the hydrosphere. Soil physical properties important to the hydrosphere, in terms of the quality and quantity of fresh water resources, are water retention and transmission properties of the soil, surface area and charge properties, and composition of inorganic and organic constituents.

1.8   SOIL PHYSICS AND THE GRADUATE CURRICULA

Understanding of the soil physical properties and processes is necessary to developing and implementing strategies for sustainable management of soil and water resources for achieving world food security, controlling soil erosion, abating the nonpoint source pollution/contamination of natural waters, developing a strong foundation for stable engineering structures, and mitigating the climate change through sequestration of carbon in soil, biota, and wetlands. Further, understanding soil–climate–vegetation–human interaction is essential to development, utilization, management, and enhancement of natural resources. Therefore, studying soil physics is essential to all curricula in soil science, agronomy/crop–horticultural sciences, plant biology, agricultural engineering, climatology, hydrology, and environmental sciences. This book is specifically aimed to meet the curricula needs of students and researchers interested in these disciplines.

PROBLEMS

1.   Why is soil a nonrenewable resource?
2.   List soil functions of importance to pre- and postindustrial civilization.
3.   Describe soil degradation and its impact.

4. Explain the difference between the terms "property" and "process," and give specific examples in support of your argument.
5. Describe soil quality and factors affecting it.

REFERENCES

Bouwman, A.F. (ed) 1990. Soils and the Greenhouse Effect. J. Wiley and Sons, Chichester, U.K., 574 pp.

Brown, L.R. 1991. The global competition for land. J. Soil and Water Cons. 46: 394–397.

Buol, S. 1994. Soils. In: W.B. Meyer and B.L. Turner II (eds) "Changes in Land Use and Land Cover: A Global Perspective." Cambridge Univ. Press, NY: 211–229.

Doran, J.W., D.C. Coleman, D.F. Bedzicek, and B.A. Stewart (eds) 1994. Defining soil quality for a sustainable environment. Soil Sci. Soc. Amer. Proc., Spec. Publ. 35, Madison, WI.

Doran, J.W. and A.J. Jones (eds) 1996. Methods for Assessing Soil Quality. SSSA Spec. Publ. 49, Madison, WI.

Doran, J.W., A.J. Jones, M.A. Arshad, and J.E. Gilley. 1999. Determinants of soil quality and health. In R. Lal (ed) "Soil Quality and Soil Erosion," CRC/SWCS, Boca Raton, FL: 17–36.

Engelman, R. and P. LeRoy. 1993. Sustaining water: population and the future of renewable water supplies. Population Action International, Washington, D.C., 56 pp.

Engelman, R. and P. LeRoy. 1995. Conserving land: population and sustainable food production. Population Action International, Washington, D.C., 56 pp.

Follett, R.F., J.M. Kimble, and R. Lal (eds) 2000. The Potential of U.S. Grazing Lands to Sequester Carbon and Mitigate the Greenhouse Effect. CRC/Lewis Publishers, Boca Raton, FL, 438 pp.

Gregorich, E.G. and M.R. Carter (eds) 1997. Soil Quality for Crop Production and Ecosystem Health. Developments in Soil Sci., Elsevier, Holland, 448 pp.

IPCC 1995. The Science of Climate Change. Working Group 1. Inter-Governmental Panel on Climate Change, WMO, Geneva, Switzerland, Cambridge Univ. Press, U.K., 572 pp.

IPCC 2001. The Climate Change 2001: The Scientific Basis. WMO, Geneva, Switzerland, Cambridge Univ. Press, U.K., 881 pp.

Karlen, D.L., M.J. Mausbach, J.W. Doran, R.G. Cline, R.F. Harris, and G.E. Schuman. 1997. Soil quality: A concept, definition and framework for evaluation. Soil Sci. Soc. Am. J. 61: 4–10.

Lal, R. 1993. Tillage effects on soil degradation, soil resilience, soil quality and sustainability. Soil & Tillage Res. 27: 1–7.

Lal, R. 1994. Global overview of soil erosion. In "Soil and Water Science: Key to Understanding Our Global Environment", Soil Sci. Soc. Am. Special Publ. 41, Madison, WI: 39–51.

Lal, R. 1995. Erosion-crop productivity relationships for soils of Africa. Soil Sci. Soc. Am. J. 59: 661–667.

Lal, R. 1997. Degradation and resilience of soils. Phil. Trans. R. Soc. London (B) 352: 997-1010.

Lal, R. 1998. Soil erosion impact on agronomic productivity and environment quality. Critical Rev. Plant Sci. 17: 319–464.

Lal, R. 1999a. Soil quality and food security: The global perspective. In R. Lal (ed) "Soil Quality and Soil Erosion," CRC/SWCS, Boca Raton, FL: 3–16.

Lal, R. 1999b. Soil management and restoration for C sequestration to mitigate the greenhouse effect. Prog. Env. Sci. 1: 307–326.

Lal, R. 2000. Soil management in the developing countries. Soil Sci. 165: 57–72.

Lal, R. 2001. World cropland soil as a source or sink for atmospheric carbon. Adv. Agron. 71: 145–191.

Lal, R. and B.A. Stewart (eds) 1994. Soil Processes and Water Quality, Advances in Soil Science, Lewis Publishers, Boca Raton, FL, 398 pp.

Lal, R., J. Kimble, E. Levine, and B.A. Stewart (eds) 1995. "Soils and Global Change," Advances in Soil Science, CRC/Lewis Publishers, Boca Raton, FL, 440 pp.

Lal, R., J.M. Kimble, R.F. Follett, and C.V. Cole. 1998. The Potential of U.S. Cropland to Sequester Carbon and Mitigate the Greenhouse Effect. Ann Arbor Press, Boca Raton, FL, 128 pp.

Oldeman, L.R. 1994. The global extent of soil degradation. In D.J. Greenland and I. Szabolcs (eds) "Soil Resilience and Sustainable Land Use", CAB International, Wallingford, U.K.

Pimmentel, D., C. Harvey, P. Resosudarmo, K. Sinclair, D. Kurz, M. McNair, S. Crist, L. Shpritz, L. Fitton, R. Saffouri, and R. Blair. 1995. Environmental and economic costs of soil erosion and conservation benefits. Science 267: 1117–1123.

UNEP. 1992. World Atlas of Desertification. United Nations Environment Program, Edward Arnold, London.

Yaalon, D.H. and R.W. Arnold. 2000. Attitudes towards soils and their societal relevance: then and now. Soil Sci. 165: 5–12.

2

Basic Definitions and Concepts: Soil Components and Phases

Most soils consist of four components and three phases (Fig. 2.1). The four components include inorganic solids, organic solids, water, and air. Inorganic components are primary and secondary minerals derived from the parent material. Organic components are derived from plants and animals. The liquid component consists of a dilute aqueous solution of inorganic and organic compounds. The gaseous component includes soil air comprising a mixture of some major (e.g., nitrogen, oxygen) and trace gases (e.g., carbon dioxide, methane, nitrous oxide). Under optimal conditions for growth of upland plants, the solid components (inorganic and organic) constitute about 50% of the total volume, while liquid and gases comprise 25% each (Fig. 2.2a). Rice and other aquatic plants are exceptions to this generalization. The organic component for most mineral soils is about 5% or less. Immediately after rain or irrigation, the entire pore space or the voids in between the solids are completely filled with water, and the soil is saturated (Fig. 2.2b). When completely dry, the water in the pores is replaced by air or gases (Fig. 2.2c). General properties of components and phases are listed in Table 2.1. Under optimal conditions for some engineering functions, such as foundation for buildings and roads or runways, the pore space is deliberately minimized by compaction or compression. For such functions, the solid components may compose

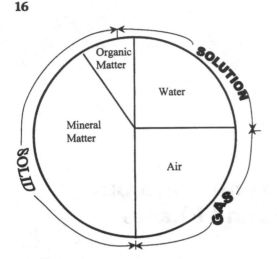

FIGURE 2.1 Soil is made up of four components and three phases.

80–90% of the total volume. There must be little if any liquid component for the foundation to be stable. Some industrial functions (e.g., dehalogenation) may require anaerobic conditions, however.

Anaerobiosis may lead to transformation of organic matter by the attendant methanogenesis and emissions of methane (CH_4) to the atmosphere. In contrast, oxidation and mineralization of organic matter may cause release of carbon dioxide (CO_2) to the atmosphere. Filtration of pollutants and sequestration of carbon (C) in soil as soil organic carbon (SOC), two important environmental functions, also depend on an optimal balance between four components and three phases. The dynamic equilibrium between components and phases can be altered by natural or anthropogenic perturbations.

2.1 DEFINITIONS

Soil physics deal with the study of soil physical properties (e.g., texture, structure, water retention, etc.) and processes (e.g., aeration, diffusion, etc.). It also consists of the study of soil components and phases, their interaction with one another and the environment, and their temporal and spatial variations in relation to natural and anthropogenic or management factors (Fig. 2.3). Soil physics involves the application of principles of physics to understand interrelationship of mass and energy status of components and phases as dynamic entities. All four components are always changing in

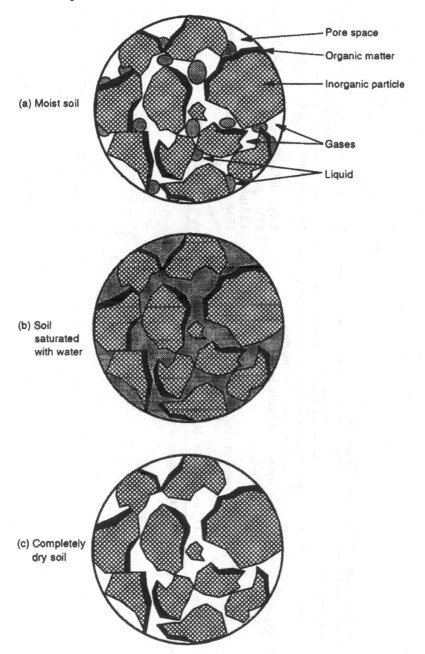

FIGURE 2.2 Interaction among four components and three phases for (a) moist, (b) water-saturated, and (c) completely dry soil.

Table 2.1 Properties and Phases and Components

Phases	Components	Composition	Properties
Solid	Inorganic	Products of weathering; quartz, feldspar, magnetite, garnet, hornblonde, silicates, secondary minerals	Skeleton, matrix $\rho_s = 2.0-2.8 \, Mg/m^3$
	Organic	Remains of plants and animals; living organisms, usually <5%	Large surface area, very active, affects CO_2 in the atmosphere $\rho_s = 1.2-1.5 \, Mg/m^3$
Liquid	Soil solution	Aqueous solution of ions (e.g., Na, K, Ca, Mg, Cl, NO_3, PO_4, SO_4)	Heterogeneous, dynamic, discontinuous $\rho_w = 1.0 \, Mg/m^3$
Gas	Soil air	N_2, O_2, CO_2, CH_4, C_2H_6, H_2S, N_2O, NO	$\rho_a = 1-1.5 \, kg/m^3$ variable, dynamic

ρ_s = particle density, l_w = density of H_2O, l_a = density of air.

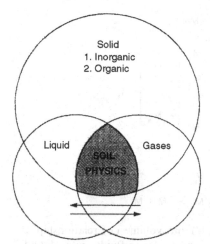

FIGURE 2.3 Soil physics is the study of properties and interaction among four components and three phases. Under optimal conditions for growth of upland plants, the solid phase composes about 50% of the total volume, and liquid and gaseous phases each compose 25% by volume. The volume of liquids increase at the expense of gases and vice versa.

their relative mass, volume, spatial and energy status due both to natural and management factors.

Consider a unit quantity of soil with total mass (M_t) consisting of different components namely solids (M_s, which includes mass of inorganic component M_{in} and organic components M_o), liquids (M_l) and gases (M_g, which is negligible and can be taken as zero for all practical purposes) (Fig. 2.4). Similarly, the total volume (V_t) comprises volume of its different components namely solids (V_s), which includes volume of inorganic components (V_{in}) and organic components (V_o), liquids (V_l) and gases (V_g). Different soil physical properties are defined in the following sections.

2.1.1 Soil Density (ρ)

Density is the ratio of mass and volume. It is commonly expressed in the units of g/cm^3 and Mg/m^3 (lbs/ft^3). Density is defined in four ways as follows:

1. *Particle density* (ρ_s): It is also called the true density, and is the ratio of mass of solid (M_s) divided by the volume of solid (V_s) [Eq. (2.1)].

$$\rho_s = M_s/V_s = (M_{in} + M_o)/(V_{in} + V_o) \tag{2.1}$$

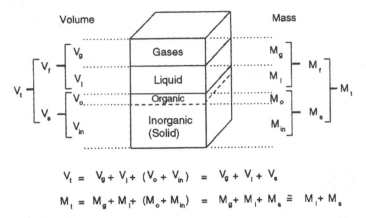

$$V_t = V_g + V_l + (V_o + V_{in}) = V_g + V_l + V_s$$

$$M_t = M_g + M_l + (M_o + M_{in}) = M_g + M_l + M_s \cong M_l + M_s$$

FIGURE 2.4 A schematic showing the mass (M) and volume (V) relationship of four soil components. Subscripts f, g, l, o, in, s, and t refer to fluids, gases, liquid, organic, inorganic, solid, and total, respectively.

Particle density of inorganic soils ranges from 2.6 to 2.8 g/cm³ or Mg/m³, and those of minerals commonly found in soils is shown in Table 2.2. Note that density of organic matter is about half of that of the inorganic mineral. In comparison, the density of water is about 1.0 Mg/m³ and that of the air about 1.0 kg/m³.

2. *Bulk density* (ρ_b):   It is also called the apparent density, and is the ratio of mass of solid (M_s) to the total volume (V_t). Soil bulk density can be defined as wet (ρ'_b) that includes the mass of water [Eq. (2.2)], and dry (ρ_b) which is without water [Eq. (2.3)]. Its units are also that of mass/volume as g/cm³ or Mg/m³.

$$\rho'_b = \frac{M_s + M_w}{V_t} = \frac{(M_{in} + M_o + M_w)}{(V_s + V_w + V_g)} \qquad (2.2)$$

$$\rho_b = \frac{M_s}{V_t} = \frac{M_{in} + M_o}{V_s + V_g} \qquad (2.3)$$

In a dry soil, V_w is zero. Wet soil bulk density is an ever changing entity because of soil evaporation at all times under natural conditions. Therefore, soil bulk density is preferably reported as a dry soil bulk density. A dense soil has more solids per unit volume (Fig. 2.4a) than a porous soil (Fig. 2.5b). Methods of measurement of ρ_b are described by Campbell et al. (2000) and Culley (1993).

Table 2.2 Particle Density of Some Common Soil Minerals, Organic Matter, Water and Air

Mineral	Particle density (Mg/m³)	Other constituents	Particle density (Mg/m³)
Biotite	2.7–3.3	Soil organic matter	1.0–1.4
Brucite	2.38–3.40	Water	1.0
Calcite	2.72–2.94	Air	1.0×10^{-3}
Chlorite	2.60–3.3		
Diamond	3.50–3.53		
Dolomite	2.86		
Gibbsite	2.38–2.42		
Geothite	3.3–4.3		
Gypsum	2.3–2.47		
Hematite	5.26		
Hornblende	3.02–3.45		
Illite	2.60–2.90		
Kaolinite	2.61–2.68		
Magnetite	5.175		
Montmorillonite	2.0–3.0		
Muscovite	2.77–2.88		
Orthoclase	2.55–2.63		
Pyrite	5.018		
Quartz	2.65		
Serpentine	2.55		
Talc	2.58–2.83		
Tourmaline	3.03–3.25		
Vermiculite	2.3		

Source: Adapted from Handbook of Chemistry and Physics (1988).

3. *Relative density or specific gravity (*G_s*)*: Specific gravity is the ratio of particle density of a soil to that of the water. Being a ratio, it is a dimensionless entity, and is expressed as shown in Eq. (2.4).

$$G_s = \rho_s/\rho_w \qquad (2.4)$$

4. *Dry specific volume (*V_b*)*: It is defined as the reciprocal of the dry bulk density [Eq. (2.5)] and has units of volume divided by mass or cm³/g or m³/Mg.

$$V_b = \frac{1}{\rho_b} = \rho_b^{-1} = \frac{V_t}{M_s} = \frac{(V_s + V_f)}{(M_s + M_e)} \qquad (2.5)$$

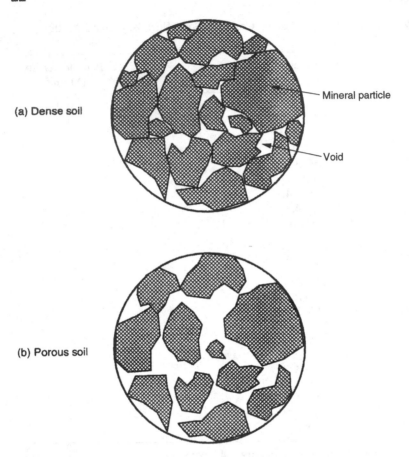

FIGURE 2.5 Dense soils are suitable for engineering functions and porous soils for agricultural land use.

2.1.2 Soil Porosity (f)

Porosity refers to the relative volume of voids or pores, and is therefore expressed as a fraction or percent of the total volume or of the volume of solids. Soil porosity can be expressed in the following four ways:

1. *Total porosity* (f_t): It is the ratio of volume of fluids or water plus air (V_f) to total volume (V_t), as shown in Eq. (2.6).

$$f_t = \frac{V_f}{V_t} = \frac{V_g + V_l}{V_s + V_l + V_g} \tag{2.6}$$

2. *Air-filled porosity* (f_a):   It refers to the relative proportion of air-filled pores [Eq. (2.7)].

$$f_a = \frac{V_g}{V_t} \tag{2.7}$$

In relation to plant growth, the critical limit of air-filled porosity is 0.10 or 10%, below which plant growth is adversely affected due to lack of sufficient quantity of air or anaerobiosis. Air porosity is also equal to total porosity minus the volumetric moisture content (Θ) as computed in Eq. (2.11).

3. *Void ratio* (e):   In relation to engineering functions, where porosity should be usually as low as possible, the relative proportion of voids to that of solids is expressed as void ratio [Eq. (2.8)]. Being a ratio, it is also a dimensionless quantity.

$$e = \frac{V_f}{V_s} = \frac{(V_g + V_l)}{(V_{in} + V_o)} \tag{2.8}$$

4. *Air ratio* (α):   It is defined as the ratio of volume of air to that of the solids [Eq. (2.9)] and has relevance to plant growth and engineering applications.

$$\alpha = \frac{V_g}{V_s} = \frac{V_g}{(V_{in} + V_o)} \tag{2.9}$$

2.1.3 Soil Moisture Content

Soil moisture is the term used to denote water contained in the soil. Soil water is usually not free water, and is, therefore, called soil moisture. Soil moisture content can be expressed in the following four ways:

1. *Gravimetric soil moisture content* (w):   It is the ratio of mass of water (M_w) to that of solids (M_s), and is expressed either as fraction or percent [Eq. (2.10)].

$$w = \frac{M_w}{M_s} = \frac{M_w}{(M_{in} + M_o)} \tag{2.10}$$

2. *Volumetric soil moisture content* (Θ):   In relation to agricultural and engineering functions, it is more relevant to express soil moisture

content on volumetric than on gravimetric basis. Similar to w, Θ is also expressed as a ratio or percent [Eq. (2.11)].

$$\theta = \frac{V_w}{V_t} = \frac{V_l}{(V_s + V_f)} \tag{2.11}$$

3. *Liquid ratio* (θ_p):   Just as in case of void ratio, the liquid ratio has also numerous engineering applications, and is expressed as a ratio [Eq. (2.12)].

$$\theta_p = \frac{V_w}{V_s} = \frac{V_w}{(V_{in} + V_o)} \tag{2.12}$$

The liquid ratio is also a useful property for soils with high swell-shrink properties.

4. *Degree of saturation* (s):   It refers to the relative volume of pore space containing water or liquid in relation to the total porosity [Eq. (2.13)], and is also expressed as a fraction or percentage.

$$s = \frac{V_w}{V_f} = \frac{V_w}{(V_l + V_s)} \tag{2.13}$$

2.1.4   Soil Physical Quality

Thirteen soil physical properties defined above are extremely important in defining soil physical quality in relation to specific soil functions (see Chapter 1; Arshad et al., 1996; Lowery et al., 1996). The objectives of soil management are to optimize these properties for specific soil functions. One or an appropriate combination of these properties is used as an index of soil physical quality. Indicators of soil quality, however, differ among soils and specific functions. The normal range of these indicators is shown in Table 2.3.

General physical properties of three phases and four components are shown in Table 2.4. Solids form the skeleton of the soil or soil matrix in which fluids constitute the plasma. Particle density of the inorganic components is almost twice that of the organic components. The liquid phase is a dilute aqueous solution of numerous salts including nitrates, chlorides, sulphates, carbonates, and phosphate of K, Ca, Mg, Na, and other cations. Soil air or the gaseous phase contains more CO_2 and less O_2 than atmospheric air (see Chapter 18).

TABLE **2.3** Normal Range of Soil Physical Properties in Relation to Plant Growth

Soil physical property	Range	Units
Particle density (ρ_s)	2.6–2.8	g/cm^3, Mg/m^3
Dry bulk density (ρ_b)	0.7–1.8	g/cm^3, Mg/m^3
Porosity (f_t)	0.3–0.7	Fraction, m^3/m^3
Air porosity (f_a)	0–f_t	Fraction, m^3/m^3
Void ratio (e)	0.4–2.2	Fraction
Gravimetric soil moisture content (w)	0–0.3	Fraction, kg/kg
Volumetric soil moisture content (Θ)	0–0.7	Fraction, m^3/m^3
Degree of saturation (s)	0–1	Fraction
Dry specific volume (V_b)	0.5–1	cm^3/g, m^3/Mg
Air ratio (α)	0–1	Dimensionless
Liquid ratio (θ_ρ)	0–1	Dimensionless
Wet bulk density (ρ'_b)	1–2	g/cm^3, Mg/m^3

2.2 INTERRELATIONSHIP AMONG SOIL PROPERTIES

Several of these properties are interrelated and one can be computed from another. Specific examples of these interrelationships are shown below:

$$\theta = w\rho_b/\rho_w \qquad (2.14)$$

$$\theta = \frac{\rho'_b - \rho_b}{\rho_w} \qquad (2.15)$$

$$f_t = (1 - \rho_b/\rho_s) \qquad (2.16)$$

$$e = (\rho_s/\rho_b) - 1 \qquad (2.17)$$

$$\theta_\rho = \Theta(1 + e) \qquad (2.18)$$

$$f_t = f_a + \theta \qquad (2.19)$$

$$\rho_b = \rho_s(1 - f_t) \qquad (2.20)$$

$$w = \frac{\rho'_b - \rho_b}{\rho_b} \qquad (2.21)$$

TABLE 2.4 General Properties of Phases and Components

Phase	Component	Composition	General properties
Solid	Inorganic	Products of weathering of rocks and minerals. Mostly comprise primary and secondary minerals. e.g. quartz, feldspar, magnetite, garnet, hornblende, silicates, and secondary minerals. Usually compose 95% of the dry soil mass.	Skeleton, matrix, ρ_s of 2.6–$2.8\,g/cm^3$. Surface area and charge density depend on size distribution.
	Organic	Remains of plants and animals at various stages of decay and decomposition. Usually comprise <5% of the dry soil mass.	This fraction is highly reactive and dynamic. It has large surface area and high charge density. ρ_s ranges from 1.2 to $1.5\,g/cm^3$.
Liquid	Soil solution	Aqueous and dilute solution of numerous ions. Predominant ions depend on the parent material and land use and may comprise Na, K, Ca, Mg, Cl, NO_3, PO_4, and SO_4.	This is a very heterogenous solution, and is highly variable in time and space. This phase is discontinuous and increases or decreases depending on the degree of wetness and density of soil.
Gas	Soil air	Composition of soil air differs than that of the atmosphere. Soil air comprises a mixture of numerous gases including N_2, O_2, CO_2, CH_4, C_2H_6, H_2S, N_2O, NO, and others.	Composition of soil air is extremely heterogenous, very dynamic, and highly variable over time and space. This is also a discontinuous phase and varies inversely with volume of soil solution. Approximate density of soil air is 1–$1.5\,kg/m^3$.

2.3 ASSESSMENT OF SOIL PARTICLE DENSITY

Methods of assessment of ρ_b, f_t, f_a, w, and Θ are discussed under appropriate sections. There are two common procedures of determining soil particle density. One is based on calculations from the particle density of its constituents [Eq. (2.22)].

$$\frac{1}{\rho_s} = \frac{x_1}{\rho_{s1}} + \frac{x_2}{\rho_{s2}} + \frac{x_3}{\rho_{s3}} \tag{2.22}$$

where x_1, x_2, and x_3 are weight fractions of the constituents, and ρ_{s1}, ρ_{s2}, and ρ_{s3} are the corresponding particle densities of those fractions. The second method of determining the particle density involves the laboratory procedure based on the Archimedes' principle. This procedure involves measurement of the volume displacement of dry soil by a liquid of known density using a pycnometer (Blake and Hartge, 1986). In addition, eletronic pycnometers are also available.

Example 2.1

A soil is sampled by a core measuring 7.6 cm in diameter and 7.6 cm deep. The core weighs 300 g. The total core plus wet soil weight is 1000 g. On oven drying at 105° C the core plus dry soil weighed 860 g. Calculate wet and dry bulk densities and gravimetric moisture contents.

Solution

Total volume of core $= \pi r^2 h = 3.14\ (3.8\ \text{cm}^2) \cdot 7.6\ \text{cm} = 345\ \text{cm}^3$
Core weight $= 300\ \text{g}$
Weight of wet soil $= 1000\ \text{g} - 300\ \text{g} = 700\ \text{g}$
Weight of dry soil $= 860\ \text{g} - 300\ \text{g} = 560\ \text{g}$
Wet bulk density $(M_t/V_t) = 700\ \text{g}/345\ \text{cm}^3 = 2.03\ \text{g/cm}^3$
Dry bulk density $(M_s/V_t) = 560\ \text{g}/345\ \text{cm}^3 = 1.62\ \text{g/cm}^3$
Gravimetric moisture content $(w) = M_w/M_s = (1000\ \text{g} - 860\ \text{g})/560\ \text{g}$
$= 140\ \text{g}/560\ \text{g}$
$= 0.25$ or 25%

Example 2.2

One liter of dry soil sampled from a farm requires 300 g of water to completely saturate it. Calculate: (a) its porosity and (b) volume of water required to saturate the plow layer (20 cm) of 1 hectare of the farmland.

Solution

(a) Porosity $(f_t) = V_w/V_t = 300 \, \text{cm}^3/1000 \, \text{cm}^3 = 0.3 \, \text{m}^3/\text{m}^3$

(b) Depth of water $(Q) = f_t \cdot d$, where d is depth

$$= 0.3 \times 20 \, \text{cm} = 6 \, \text{cm}$$

Total volume of water for one ha $= 6 \times 10^5 \, \text{L}$

Example 2.3

A soil in the greenhouse container has a wet bulk density of $1.7 \, \text{Mg/m}^3$ and dry bulk density of $1.4 \, \text{Mg/m}^3$. Calculate gravimetric and volumetric soil moisture contents, and air-filled porosity.

Solution

$$w = \frac{\rho'_b - \rho_b}{\rho_b} = \frac{1.7 - 1.4}{1.4} = 0.214 \, \text{kg/kg}$$

$$\Theta = \frac{\rho'_b - \rho_b}{\rho_w} = \frac{1.7 - 1.4}{1.0} = 0.3 \, \text{m}^3/\text{m}^3$$

$$f_a = 1 - (V_s + V_w) = 1 - \left(\frac{1.4}{2.65} + 0.3\right) = 0.172 \, \text{m}^3/\text{m}^3$$

An alternative solution is to assume the volume of the container.

Let the pot volume $= 1000 \, \text{cm}^3$

Particle density $= 2.65 \, \text{g/cm}^3$

Wet soil weight $= 1000 \, \text{cm}^3 \times 1.7 \, \text{g/cm}^3 = 1700 \, \text{g}$

Similarly, dry soil weight $= 1400 \, \text{g}$

Mass of water $(M_w) = 1700 \, \text{g} - 1400 \, \text{g} = 300 \, \text{g}$

Volume of water $(V_w) = 300 \, \text{cm}^3$

Gravimetric moisture content $(W) = 300 \, \text{g}/1400 \, \text{g} = 0.214 \, \text{kg/kg}$ or 21.4%

Volumetric moisture content $(\Theta) = 300 \, \text{cm}^3/1000 \, \text{cm}^3 = 0.30 \, \text{m}^3/\text{m}^3$ or 30%

Volume of solids $= \text{mass/density} = 1400 \, \text{g}/2.63 \, \text{g/cm}^3 = 528.3 \, \text{cm}^3$

Air porosity $(f_a) = (1000 \, \text{cm}^3 - 528.3 \, \text{cm}^3 - 300 \, \text{cm}^3)/1000 \, \text{cm}^3$

$$= 171.7 \, \text{cm}^3/1000 \, \text{cm}^3 = 0.172 \, \text{m}^3/\text{m}^3 \text{ or } 17.2\%$$

Example 2.4

One liter of soil has a wet weight of $1500 \, \text{g}$, dry weight of $1200 \, \text{g}$, and volume of soil solids of $450 \, \text{cm}^3$. Compute all 13 soil physical properties.

Solution

1. $\rho_s = M_s/V_s = 1200\,\text{g}/450\,\text{cm}^3 = 2.67\,\text{g/cm}^3$
2. $\rho_b = M_s/V_t = 1200\,\text{g}/1000\,\text{cm}^3 = 1.20\,\text{g/cm}^3$
3. $\rho_b = (M_s + M_w)/V_t = 1500\,\text{g}/1000\,\text{cm}^3 = 1.5\,\text{g/cm}^3$
4. $G_s = \rho_s/\rho_w = 2.67$
5. $V_b = 1/\rho_b = 0.83\,\text{cm}^3/\text{g}$
6. $f_t = (1 - \rho_b/\rho_s) = (1 - 1.2/2.67) = 0.55$ or 55%
7. $f_a = 1 - (V_s + V_w) = 1 - (450/1000 + 300/1000) = 1 - 0.75 = 0.25$ or 25%
8. $e = V_f/V_s = 550\,\text{cm}^3/450\,\text{cm}^3 = 1.22$
9. $\alpha = V_g/V_s = 250\,\text{cm}^3/450\,\text{cm}^3 = 0.56$
10. $w = 300\,\text{g}/1200\,\text{g} = 0.25$ or 25%
11. $\Theta = 300\,\text{cm}^3/1000\,\text{cm}^3 = 0.30 = (w \cdot \rho_b/\rho_w)$
12. $\Theta_p = V_w/V_s = 300\,\text{cm}^3/450\,\text{cm}^3 = 0.67$
13. $S = V_w/V_f = 300\,\text{cm}^3/550\,\text{cm}^3 = 0.55$

Example 2.5

Calculate ρ_s of a mixture containing 48% by weight of quartz, 50% of vermiculite, and 2% by weight of soil organic matter.

Solution

From Table 2.1, ρ_s is 2.65 Mg/m³ for quartz, 2.3 Mg/m³ for vermiculite, and 1.4 Mg/m³ for soil organic matter. The ρ_s is computed by substituting these values in [Eq. (2.22)]:

$$\frac{1}{\rho_s} = \frac{0.48}{2.65} + \frac{0.50}{2.3} + \frac{0.02}{1.4} = 0.181 + 0.217 + 0.014 = 0.412\,\frac{\text{cm}^3}{\text{g}}$$

$$\therefore \rho_s = \frac{1}{0.412}\frac{\text{g}}{\text{cm}^3} = 2.43\,\text{g/cm}^3 \text{ or Mg/m}^3$$

PROBLEMS

1. Calculate particle density of a soil from the following data:

Weight of pycnometer	= 50 g
Weight of the powder dry soil	= 214 g
Mass of soil and deaerated water when pycnometer was filled to capacity + pycnometer	= 352 g
Temperature of water	= 20°C
Volume of pycnometer	= 168 cm³

2. Consider the following data based on field measurements:

    i.   Diameter of the cylindrical core $= 5.05\,\text{cm}$
    ii. Height of the cylindrical core $= 5\,\text{cm}$
    iii. Weight of the core $= 150\,\text{g}$
    iv. Weight of field soil + core $= 312.5\,\text{g}$
    v.   Weight of the oven dried (105°C)
         soil + core $= 282.5\,\text{g}$
    vi. Weight of the oven dried (900°C)
         soil + core $= 276.0$

    Using the particle density calculated in Question 1, calculate W, Θ, ρ'_b, ρ_b, organic fraction, and V_s.

3. Prove or disprove the following:

    i. $f = \frac{e}{1+e}$
    ii. $f = e + 1$
    iii. $e = \frac{f}{1-f}$
    iv. $e = f - 1$
    v. $\Theta = sf$
    vi. $\rho_b = V_s\rho_s + V_w\rho_w + V_a\rho_a$
    vii. $V_w/V_s = \Theta(1+e)$
    viii. $V_s\rho_s = \rho_b(V_s + V_w + V_a)$

4. A soil of one m³ total volume (V_t) has the following properties:
    $V_s = 0.5$
    $V_w = 0.3$
    $V_a = 0.2$
    Assuming $\rho_s = 2.65\,\text{g/cm}^3$, calculate:

    (a) f, f_a, s, e, M_s and ρ_b
    (b) What are the weight and volume of water required to saturate it?

5. In a greenhouse study, a soil is packed in a container at a ρ_b of $1.5\,\text{Mg/m}^3$. The antecedent Θ is 0.2. Assuming the volume of the container is $1000\,\text{cm}^3$, what is the volume of water needed to double the Θ of the entire soil?

6. A sample of moist soil weighed 100 g and had an oven dry moisture content (w) of 0.04. What is the oven dry weight (M_s) of the 100 g sample?

7. 10 mm of rain infiltrated a soil having an initial moisture content by volume (Θ) of $0.1\,\text{m}^3/\text{m}^3$. If the soil absorbed enough of the rainfall to raise its moisture content to $0.2\,\text{m}^3/\text{m}^3$, how many cm would the rainfall penetrate?

8. What are principal soil functions? Briefly describe each function.
9. How does application of soil physics improve environment quality?
10. Describe the term "sustainable use of soil and water resources."
11. How do soil constituents influence environment quality?
12. How do soil constituents influence agricultural sustainability?

REFERENCES

Arshad, M.A., B. Lowery and B. Grossman. 1996. Physical tests for monitoring soil quality. In: J.W. Doran and A.J. Jones (eds) "Methods for Assessing Soil Quality," Soil Sci. Soc. Amer. Spec. Publ. 49, Madison, WI: 123–141.

Blake, G.R. and K.H. Hartge. 1986. Particle density. In A. Klute (ed) "Methods of Soil Analysis", Part 1, Agronomy Monograph 9, American Society of Agronomy, Madison, WI: 377–382.

Campbell, D.J. and K. Henshall. 2000. In: K.A. Smith and C.E. Mullins (eds) "Soil and Environmental Analysis, Physical Methods," Second Edition, Marcel Dekker, Inc., New York: 315–348.

Culley, J.L.B. 1993. Density and compressibility. In M.R. Carter (ed) "Soil Sampling and Methods of Analysis", Lewis Publishers, Boca Raton, FL: 529–539.

Handbook of Chemistry and Physics. 1988. Handbook of Chemistry and Physics. 1st Student Edition. CRC Press, Boca Raton, FL.

Lowery, B., M.A. Arshad and R. Lal. 1996. Soil water parameters and soil quality. In: J.W. Doran and A.J. Jones (eds) "Methods for Assessing Soil Quality," Soil Sci. Soc. Amer. Spec. Publ. 49, Madison, WI: 143–155.

3

Soil Solids

Soil solids, comprising inorganic and organic components, form the matrix or the body of most soils. This matrix, or the visible part of the soil, is the storehouse of water and nutrient elements (e.g., N, P, K, Ca, Mg, Zn, Cu, etc.). It is also the site of most processes that govern soils buffering and filtering capacity, and life support capability. The buffering capacity of the soil refers to its ability to withstand or to adapt to sudden perturbations such as in soil reaction (i.e., pH). The filtering capacity refers to soil's ability to remove pollutants (e.g., pathogens or chemicals including heavy metals) out of the water percolating through the soil by denaturing pollutants or mechanical sieving of suspended particles. Both buffering and filtering capacities depend on soil's reactivity. The latter refers to chemical, physical, and biological reactions in soil and depends on its nature (e.g., relative proportion of the inorganic and organic components, coarse or fine size, small or large surface area, and low or high charge density). Soil quality is determined by these and other properties of soil solids, which in turn moderate the soil's ability to support plant and animal life. Soil's life support capability depends on processes that govern productivity, elemental cycling, and environment quality (see Chapter 1).

3.1 INORGANIC COMPONENTS

The inorganic components comprise more than 95% by weight of total solid fraction for most mineral soils. It is the product of weathering of parent material, and comprises a range of primary and secondary minerals. Important properties of the inorganic components are: (i) size, (ii) shape, (iii) surface area, (iv) clay minerals and charge properties, (v) swelling and shrinkage, (vi) water absorption and heat of wetting, and (vii) packing arrangement.

3.1.1 Particle Size Distribution or Soil Texture

The inorganic component comprises two types of soil particles, primary and secondary. Primary particles are discrete units that cannot be further subdivided, and are also called "soil separates." Secondary particles consist of primary particles and can be subdivided into its "separates" by chemical or mechanical dispersion. Particle size is an important soil physical property. It affects total porosity, pore size, and surface area. Particle size distribution refers to the "quantitative" measure of the particle size that constitutes the solid fraction. In contrast, soil texture refers to a "qualitative" measure of particle sizes based on "feel" of the soil material, which may be coarse, gritty, fine, or smooth.

Size Fractions

Depending on the size distribution, primary particles (textural fractions) or soil separates are usually divided into three classes, e.g., sand, silt, and clay. There are numerous systems of classifying separates into different size classes. Most commonly used systems include: (i) the U.S. Department of Agriculture (USDA), (ii) the International Society of Soil Science (ISSS), (iii) the American Society of Testing Material (ASTM), (iv) the Massachusetts Institute of Technology (MIT), (v) the U.S. Public Road Administration (USPRA), (vi) the British Standard Institute (BSI), and (vii) the German Standards (DIN). There are other local and regional systems as well. The two most commonly used systems by soil scientists and agronomists are the USDA and the ISSS/IUSS (Table 3.1). The ASTM system is widely used by engineers.

Material >2 mm is considered the nonsoil fraction in both USDA and ISSS/IUSS systems. Three principal textural classes of < 2 mm components or the soil fraction are (i) sand, (ii) silt, and (iii) clay. General physical properties of these three fractions are listed in Table 3.2, and are briefly described below.

Sand. This is the coarse fraction, and constitutes the skeleton of the soil body. The sand fraction can be subdivided into coarse, medium,

TABLE 3.1 Two Widely Used Systems of Particle Size Distribution

The USDA System[a]		The ISSS System[b]	
Soil separate	Size range (mm)	Soil separate	Size range (mm)
Very coarse sand	2.00–1.00	coarse sand	2.00–0.20
Coarse sand	1.00–0.50	fine sand	0.20–0.02
Medium sand	0.50–0.25	silt	0.02–0.002
Fine sand	0.25–0.10	clay	<0.002
Very fine sand	0.10–0.05		
Silt	0.05–0.002		
Clay	<0.002		

Note: For both system particles of diameter $(D) > 2$ mm are considered nonsoil (skeletal) fraction.

[a]$D = (ar)^{n-1}$, where $a = 2$, and $r = 1/2$.

[b]$D = ar^{n-1}$, where $a = 2$, and $r = 1/10$.

and fine fractions (USDA system) (Table 3.1). Sand grains comprise mostly quartz but also contain fragments of feldspar and mica, and traces of heavy minerals, e.g., zircon, tourmaline, and hornblende. Sand particles are jagged, hard (hardness of 5 to 7 on mhos scale) (Table 3.2), and can abrade steel as is evident by wearing down of the plow.

Silt. This is an intermediate size fraction, and also constitutes the skeleton of the soil. Properties of coarse silt fraction are similar to that of sand, but that of the fine silt approach that of clay. Mineralogical composition of silt is similar to that of sand, but silt has more surface area (see Section 3.3 in this chapter). Primary minerals present in sand and silt fractions are listed in Table 3.3.

Clay. This is the fine fraction, and constitutes the reactive fraction of the soil. Because of its very fine size, the clay fraction is colloidal, highly reactive, has large surface area, and high charge density. In shape, the clay particles are plate-like or needle-like. In mineralogy, the clay particles comprise a group of clay minerals, called *alumino-silicates*. These are secondary clay minerals, and also contain fine particles of iron oxide (Fe_2O_3), aluminum oxide (Al_2O_3), calcium carbonate ($CaCO_3$), and other salts. Because of its larger surface area, the clay fraction has the most influence on many soil properties. Properties of the clay fraction with a notable influence on soil behavior are listed in Table 3.2 and include: (i) easy hydration because of its high affinity for water, (ii) high swell/shrink capacity because of the expanding nature of the clay lattice, (iii) high

TABLE 3.2 Some Physical Properties of Soil Separates

Property	Soil separates		
	Sand	Silt	Clay
Size	2–0.02 mm	0.02 mm–0.002 mm	< 0.002 mm
Shape	Jagged	Slightly irregular	Platy/tube-like
Feel	Gritty	Smooth, floury	Sticky
Plasticity	Not plastic	Slightly plastic	Plastic
Cohesion	Not cohesive	Slightly cohesive	Cohesive, gelatinous
Surface area	Very low	Moderate	Very high
Mineralogy	Primary	Primary minerals	Secondary clay minerals
Heat of wetting	None	Minimal	High
Secondary particles	None	Few	Forms aggregates
Water holding capacity	None/slight	Moderate	High, hygroscopic
Hardness	5.5–7 (on the mhos scale)	5.5–7.0	—
Ion exchange capacity	None	Very low	High to very high

TABLE 3.3 Common Primary Minerals Found in Sand and Silt Fractions

Mineral	Weatherability
Quartz	Most resistant
Muscovite	↓
Microline	
Orthoclase	↓
Biotite	
Albite	↓
Horneblende	
Augite	↓
Anorthite	
Olivine	Least resistant

Source: Adapted from Brady and Weil 2002.

plasticity because of its ability to retain shape when a moist clay is molded, (iv) sticky when moist and crack because of shrinking, and cake when dry because of the cohesive forces, and (v) high density of negative charge leading to formation of electrostatic double layer when fully hydrated because of the deficit created by ionic substitution or broken bonds/edges. Some of these properties are discussed in detail in this and the following chapters.

Assessment of Particle Size Fractions

The process of determination of particle size distribution is called mechanical analysis. The procedure has two-steps: dispersion and fractionation. Dispersion involves removal of cementing material (compounds or substances which bind the particles together) to break secondary particles into primary particles or soil separates. Dispersion agents used in this determination depend on the nature of the cementing material (Table 3.4). For example, hydrogen peroxide (H_2O_2) is used to remove organic material, dilute acid to remove carbonates/electrolytes, and sodium dithionite to remove sesquioxides. The latter are compounds in which the ratio of metal to oxygen is 2:3 (M_2O_3 where M is a metalic ion such as Fe, Al, Mn, etc.).

Fractionation is the process of physically separating the particles into different size ranges. A wide range of methods of fractionation are used (Table 3.5), and the choice of an appropriate method depends on the particle size, objectives, and the facilities available. Two of the most commonly used procedures in soil physics laboratory are sieving

TABLE **3.4** Dispersive Agents Needed to Remove Binding Agents Prior to Mechanical Analysis

Cementing material	Dispersion agent
Organic matter	Hydrogen peroxide (H_2O_2)
Oxides of Fe and Al	Treatment with oxalic acid, sodium sulfide, sodium dithionite, and sodium citrate
Electrolytes	Dissolution and leaching with dilute acids, electrodialysis, and sodium hexametaphosphate
Cohesion/adhesion	Rehydration by boiling in H_2O, shaking, trituration, stirring, and ultrasound vibration

TABLE **3.5** Approximate Size Range Determined by Different Methods of Particle Size Analysis

Methods of fractionation	Approximate size range (mm)
Sieving	100–0.05
Sedimentation	2–< 0.002
Optical Microscope	1.0–0.001
Gravity sedimentation	0.1–0.0005
Permeability	0.1–0.0001
Gas absorption	0.1–0.0001
Electron microscope	0.005–0.00001
Elutriation	0.05–0.005
Centrifugal sedimentation	0.01–0.00005
Turbidimetry	0.005–0.00005

and sedimentation. Direct sieving involves passing the dispersed soil suspension through a nest of sieves of different sizes (Appendix 3.1). The amount retained on a particular sieve represents the fraction that is larger than the sieve size on which it is retained but smaller than that of the preceding sieve. This method is primarily suited for separating coarse fractions.

The sedimentation procedure is based on the rate of fall of particles through a liquid, which depends on particle size and properties of the liquid. In 1851, G.G. Stokes developed a law that states "The resistance offered by a liquid to the fall of a rigid spherical particle varies with the radius of the particle and not with its surface."

A particle falling freely in a fluid experiences three forces: force of gravity (F_g acting downward), force of friction or resistance (F_r acting upward), and the force due to buoyancy (F_b acting upward).

$$F_g$$
$$\downarrow$$
$$\bullet$$
$$\uparrow$$
$$F_r, F_b$$

When it reaches a constant velocity, called the *terminal velocity*,

$$F_b + F_r = F_g \tag{3.1}$$

Stokes law describes the friction force

$$(F_r) \uparrow = 6\pi r \eta \theta \tag{3.2}$$

where F_r is in dynes, η is viscosity in dynes sec/cm, r is radius of the particles in cm, and θ is the terminal velocity in m/s.

The force of buoyancy (F_b) is equal to the weight of the liquid displaced [Eq. (3.3)].

$$F_b \uparrow = \frac{4}{3}\pi r^3 \rho_l g \tag{3.3}$$

The gravitational force (F_g) = volume × density × g = mg

$$F_g \downarrow = \frac{4}{3}\pi r^3 \rho_s g \tag{3.4}$$

where ρ_s is particle density (Mg/m^3), and g is acceleration due to gravity (9.81 m/s^2).

When particles attain terminal velocity, the sum of the three forces (due to gravity acting downward, buoyancy acting upward and friction acting upward) is equal to zero. The force of gravity is equal to the weight of the particle and the force due to buoyancy is proportional to the volume of water displaced. Adding a positive friction and buoyancy to a negative gravity force equals zero at a steady rate of fall.

$$6\pi r \eta \theta + \frac{4}{3}\pi r^3 \rho_l g = \frac{4}{3}\pi r^3 \rho_s g \tag{3.5}$$

Solve for θ:

$$\theta = \frac{4}{3}\pi r^3(\rho_s - \rho_l)/6\pi\eta r \tag{3.6}$$

$$\theta = \frac{2}{9}\frac{r^2}{\eta}(\rho_s - \rho_l)g = Kr^2 \tag{3.7}$$

where K is a constant, and Eq. (3.6) is referred to as the *settling equation*. Eq. (3.6) states that the velocity of a settling particle is proportional to r^2, $(\theta \alpha r^2)$. If particles differ in their radius by a factor of 10 (2 mm versus 0.2 mm), their settling velocities differ by a factor of 100. If the terminal velocity is attained instantly, then the time needed for a particle to fall a distance h can be calculated as follows:

$$\theta = h/t = \frac{2gr^2}{9\eta}(\rho_s - \rho_l) \tag{3.8}$$

$$\therefore t = 9h\eta/gr^2(\rho_s - \rho_l) \tag{3.9}$$

The same equation can also be solved for r if we know h and t, or for t given h and r.

$$r^2 = \frac{9}{2}\frac{h}{t}\eta/g(\rho_s - \rho_l) \tag{3.10}$$

$$r = \frac{9}{2}\left(\frac{h\eta}{tg(\rho_s - \rho_l)}\right)^{1/2} \tag{3.11}$$

$$r = A/t^{1/2} \tag{3.12}$$

$$t = B/r^2 \tag{3.13}$$

where A and B are constants. If V is in cm/min and d is in mm, then

$$\theta = \frac{1}{30}\frac{g(\rho_s - \rho_l)}{\eta}d^2 \tag{3.14}$$

$$d = \left(\frac{30\eta}{980(\rho_s - \rho_l)} \theta \right)^{1/2} \qquad (3.15)$$

Stokes law and the settling equation are based on several assumptions. If not met, these assumptions are sources of error. Thus, the objective of laboratory experimentation is to create an experimental set-up to meet the protocols as outlined in assumptions described below.

1. The particles are large in comparison to the molecules of the liquid (> 0.0002 mm) so that the Brownian movement (colloids floating in the liquid rather than settling) does not affect their fall.
2. The fall of the particle is unhindered and not affected by the proximity of the wall. If the vessel is less than 10 times the diameter of the particle a correction is necessary:

$$\theta = \frac{2}{9\eta} r^2 \frac{(\rho_s - \rho_l)}{\left(1 = 2.4 \frac{r}{R}\right)\left(1 + 3.1 \frac{r}{L}\right)} \qquad (3.16a)$$

where R is the radius of the vessel, and L is the length of the vessel.
3. The particle is smooth, spherical, and rigid so that there is no *slippage* between the sphere and the medium.
4. The suspension is still and the velocity of particle is small. This means that $\theta < \eta/\rho d$. When $\theta = \eta/\rho d$ then d is called critical diameter. For ρ_s equals 2.65 Mg/m^3, critical diameter is 0.2 mm. In general, particles > 0.2 mm should be fractionated by sieving.
5. Shape of the particles is critical. Rod-shaped particles are not suitable for fractionation by sedimentation. However, most soil particles are not spherical, but their diameters are computed as equivalent cylindrical diameter (e.c.d) or equivalent spherical diameter (e.s.d.).
6. The viscosity must be constant during the experiment. Therefore, temperature control is essential. The velocity of fall is about 12% faster at 30°C than at 25°C.
7. Differences in particle density may cause differences in fall velocity. Particle density can change due to hydration.

Example 3.1

Calculate the settling velocity of 0.2 mm and 0.002 mm size particles in a dilute water suspension at 20°C (units are given in Appendix 3.2 at the end of this chapter).

Solution

Substituting values of η and ρ_w in Eq. (3.7) and assuming ρ_s equals 2.65 Mg m^{-3} leads to the following:

$$\theta = \frac{2}{9}\frac{(1 \times 10^{-4}\,\text{m})^2}{(1 \times 10^{-3}\,\text{Kg/m/s})}\left(2.65 \times 10^3\,\frac{\text{Kg}}{\text{m}^3} - 1.00 \times 10^3\,\frac{\text{Kg}}{\text{m}^3}\right)\frac{9.8\,\text{m}}{\text{s}^2}$$

$$\theta = 0.036\,\text{m s}^{-1}$$

Similarly, the settling velocity of 0.002 mm ($r = 1.0 \times 10^{-6}$ m) can be computed as follows:

$$\theta = \frac{2}{9}\frac{(1 \times 10^{-6}\,\text{m})^2(2.65 \times 10^3\,\frac{\text{Kg}}{\text{m}^3} - 1.00 \times 10^3\,\frac{\text{Kg}}{\text{m}^3} \times 9.8\,\text{m/s}}{1.00 \times 10^{-3}\,\text{Kg/m/s}}$$

$$\theta = 3.6 \times 10^{-6}\,\text{m/s}$$

Two commonly used methods of mechanical analysis by the sedimentation technique are the hydrometer method and the pipet method. For details on these methods, readers are referred to reports by Bouyoucos (1951), Day (1953), Gee and Bauder (1986), Sheldrick and Wang (1992), and Loveland and Whalley (2001).

Expression of Results of Particle Size Analysis

There are numerous methods of expression of results of particle size analyses. The data are commonly expressed as one of the following procedures.

Textural Classes. For agricultural purposes, results of mechanical analysis are expressed into different textural classes. Quantitative information on particle size distribution is used to express the data into textural classes using numerical limits or scale for different systems, textural triangle (Fig. 3.1), and tabular values based on the textural triangle (Table 3.6).

The textural triangle has been appropriately modified for the "feel" method of textural evaluation (Ghildyal, 1988). The feel method is based on feeling the texture while rubbing moist soil between thumb and the finger. Expectedly, this is a highly subjective procedure and requires considerable experience. The procedure is, thus, extremely approximate even at its best.

Summation Curve. For engineering purposes, results of mechanical analysis are expressed in the form of a frequency diagram (Fig. 3.2) in which particle size is plotted against the percentage of the soil that falls within a

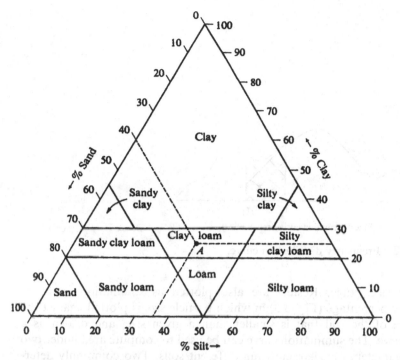

FIGURE 3.1   Textural triangle.

TABLE 3.6   Common Textural Classes Depending on the Relative Distribution of Sand, Silt and Clay

Textural class	Soil separate ranges (%)		
	Sand	Silt	Clay
Sand	85–100	0–15	0–10
Loamy sand	70–90	0–30	0–15
Sandy loam	40–80	0–50	0–20
Loam	23–52	28–50	7–27
Silt loam	0–50	50–88	0–27
Silt	0–20	80–100	0–12
Sandy clay loam	45–80	0–28	20–35
Clay loam	20–45	15–53	27–40
Silty clay loam	0–20	40–73	27–40
Sandy clay	45–65	0–20	35–45
Silty clay	0–20	40–60	40–60
Clay	0–45	0–40	40–100

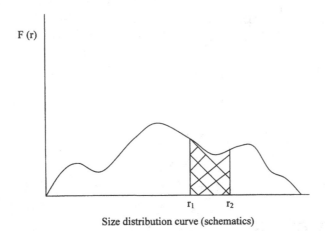

Size distribution curve (schematics)

FIGURE 3.2   Frequency distribution curve.

particular size range. Results are also plotted as summation curve or cumulative percentage (Fig. 3.3) in which particle size is plotted against the percentage of the soil that is smaller than a given size, and drawn as a smooth curve. The summation curve can be used to compute area under two particle diameters for characterizing different soils. Two commonly determined particle diameters are D_{10} and D_{60}, which are used by civil engineers to compute the uniformity coefficient.

Uniformity Coefficient.   For using soil as a construction material, it is appropriate to express the particle size as a coefficient or constant. Two commonly used constants by civil engineers are D_{10} and D_{60} (Table 3.7). The D_{10} refers to the diameter at 10%, which means that 10% of the soil particles are finer than this size. It is also called the *Hazen's coefficient* or the *effective diameter*. Similarly, D_{60} refers to the diameter at 60%, which means that 60% of the soil particles are finer than this size. These two constants are used to compute the *uniformity coefficient*, which is the ratio of $D_{60}:D_{10}$. The uniformity coefficient is an indicator of the uniformity of particle size. A soil with uniform particle size has a uniformity coefficient of about 1, for a soil with a wide range of particle size and $D_{60} > D_{10}$, the *uniformity coefficient* > 1. Soil compactability is strongly related to the *uniformity coefficient*.

3.1.2   Particle Shape

Shape of soil particles varies widely, and often depends on the size, parent material, and degree of weathering. Coarse or large particles (e.g., sand and

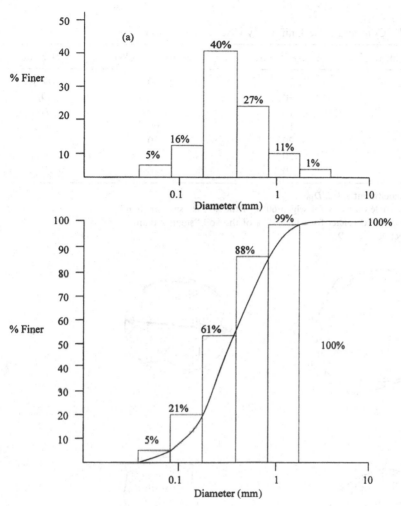

FIGURE 3.3 Summation curve.

silt fractions) are often angular or zigzag in shape. Angularity reflects degree of weathering, highly angular particles, are less weathered and become rounded with progressive weathering by the grinding action of water and wind. In contrast, clay particles are of plate or tubular shape. Particle shape is determined by micrographs, and may be expressed using two

TABLE 3.7 Computing the Uniformity Coefficient of Soil

Diameter (mm)	% by weight	Summation (%)	D value
10–5	20	100	D_{100}
5–2	10	80	D_{80}
2–1	10	70	D_{70}
1–0.5	10	60	D_{60}
0.5–0.2	20	50	D_{50}
0.2–0.1	20	30	D_{30}
< 0.1	10	10	D_{10}

Uniformity coefficient $= D_{60}/D_{10}$
$D_{60} =$ that particle diameter for which 60% of the soil is "smaller than."
$D_{10} =$ that particle diameter for which 10% of the soil "smaller than."
Hazen's effective size $= D_{10}$

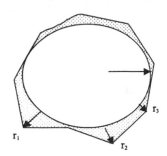

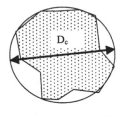

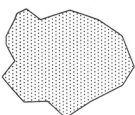

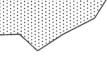

Roundness 0.44   Roundness 0.12
Sphericity 0.78   Sphericity 0.71

FIGURE 3.4 Particle shapes.

indices: roundness and sphericity (Fig. 3.4). Roundness is a measure of the sharpness of the corner, and is computed as per Eq. (3.16b).

$$\text{Roundness} = \sum_{i=1}^{n} \frac{r_i/R}{n} \tag{3.16b}$$

where r_i is the radius of a corner, R is the radius of the maximum circle inscribed within the particle, and n is the number of corners in a particle.

Sphericity is a measure of how closely the particle approaches a sphere, and is computed as per Eq. (3.17).

$$\text{Sphericity} = D_d/D_c \tag{3.17}$$

where D_d is the diameter of a circle with an area equal to that of the particle projection as it rests on its flat side, and D_c is the diameter of the smallest circumscribing circle. Some examples of sphericity and roundness are shown in Fig. 3.5, and other indices of particle shape are listed in Appendix 3.3 at the end of this chapter.

3.1.3 Specific Surface Area

Numerous soil properties are related to specific surface area of particles (a). These properties include cation exchange capacity (CEC), retention and movement of various chemicals, swell-shrink capacity, plasticity, cohesion, and strength. Knowledge of surface area is extremely important for agricultural, industrial, and environmental applications. The specific surface area is expressed using three separate indices: surface area per unit mass (a_m), per unit volume (a_v), and per unit bulk volume (a_b) as expressed by the following equations:

$$a_m = A_s/M_s \, (\text{m}^2/\text{g}) \tag{3.18}$$

$$a_v = A_s/V_s \, (\text{m}^2/\text{m}^3) \tag{3.19}$$

$$a_b = A_s/V_t \, (\text{m}^2/\text{m}^3) \tag{3.20}$$

where A_s is the total surface area, M_s is the mass of soil, V_s is the volume of soil solids, and V_t is the total volume. Surface area depends on particle size and shape. It increases logarithmically with decrease in particle size (Fig. 3.6). Plate, tubular, and chain-shaped particles have more surface area

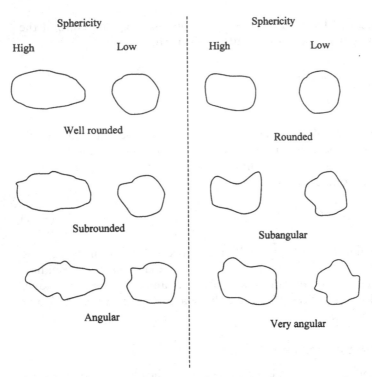

FIGURE 3.5   Soil shapes of particle sizes.

than angular or spherical particles. Surface area can be determined by the following methods.

Particle Geometry

Specific surface area can be computed assuming particle shape as follows:

A Cubic Particle.   A particle of side L has a total surface area of $6L^2$, volume of L^3 and mass of $\rho_s L^3$. Therefore, specific surface area of a cubic particle is given by Eqs. (3.21) and (3.22).

$$a_m = 6L^2/\rho_s L^3 = 6/\rho_s L \tag{3.21}$$

$$a_v = 6L^2/L^3 = 6/L \tag{3.22}$$

Equations (3.21) and (3.22) show that a_m and a_v are inversely proportional to L, the smaller the particle size, the larger the specific surface area. This inverse relationship holds for all geometric shapes.

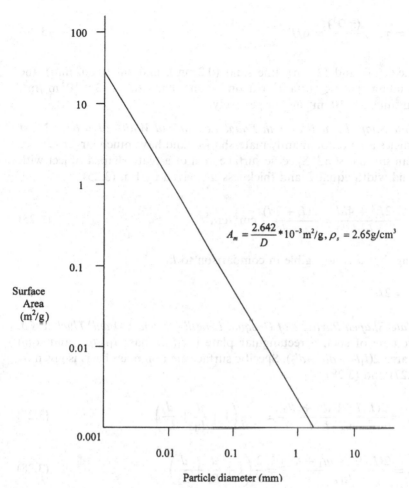

$$A_m = \frac{2.642}{D} * 10^{-3}\,m^2/g, \, \rho_s = 2.65g/cm^3$$

FIGURE 3.6 Surface area on mass basis (A_m) decreases logarithmically with increase in particle diameter.

Spherical Particle. Specific surface area of a spherical particle is similar to that of a cubicle particle. For a spherical particle of diameter D and particle density ρ_s, the total volume is $\pi D^3/6$, mass is $\pi D^3 \cdot \rho_s/6$, and total surface area πD^2. Therefore, the specific surface area is given by Eqs. (3.23) and (3.24).

$$a_m = \pi D^2 / \frac{(\pi D^3 \cdot \rho_s)}{6} = \frac{6}{(D \cdot \rho_s)} \tag{3.23}$$

$$a_v = \pi D^2 / \frac{(\pi D^3)}{6} = 6/D \tag{3.24}$$

Using e.c.d. of sand (2 mm), fine sand (0.2 mm), and silt (0.002 mm), the corresponding specific surface area on volume basis (a_v) is $3 \times 10^3 \, m^2/m^3$, $3 \times 10^4 \, m^2/m^3$, $3 \times 10^5 \, m^2 \, m^3$, respectively.

Plate-Shaped Particles with Equal Length and Width (L=b). Most clay particles are predominantly plate-shaped, and have much larger surface area, than silt and sand. Specific surface area of a plate-shaped object with length and width equal L and thickness d is given by Eq. (3.25).

$$a_v = \frac{2L^2 + 4ld}{L^2 d} = \frac{2(L + 2d)}{Ld} \quad cm^2/cm^3 \tag{3.25}$$

Assuming that d is negligible in comparison to l:

$$a_v = 2/d \tag{3.26}$$

Plate-Shaped Particles of Unequal Length (L_1 and L_2) and Thickness d. Total volume of such a rectangular plate is $l_1 l_2 d$, mass $l_1 l_2 d\rho s$, and total surface area $2(l_1 l_2 + dl_1 + dl_2)$. Specific surface area on mass basis is given by Eqs. (3.27) and (3.28).

$$a_m = \frac{2(L_1 L_2 + dl_1 + dl_2)}{\rho_s L_1 L_2 d} = \frac{2}{\rho_s d}\left(1 + \frac{d}{L_1} + \frac{d}{L_2}\right) \tag{3.27}$$

$$a_v = \frac{2(L_1 L_2 + dl_1 + dl_2)}{L_1 L_2 d} = \frac{2}{d}\left(1 + \frac{d}{L_1} + \frac{d}{L_2}\right) \tag{3.28}$$

Adsorption Isotherms

The relation between the amount of substance adsorbed and the concentration of substance in solution at any given temperature is known as the adsorption isotherm. Specific surface area of soil and other powder substances is determined from such adsorption isotherms using inert or nonreactive materials such as N_2 or ethylene glycol. The shape of the adsorption isotherm may be defined by linear $(y = mx + b)$ or nonlinear $(y = ax^b)$ mathematical function (see Chapter 16). The procedure involves monitoring the amount of gas or liquid needed to form a monomolecular layer over the entire surface. The most commonly used substances include water vapor, inert gas (N_2), or organic liquids (e.g., glycerol and ethylene glycol). A dry soil sample is saturated with ethylene glycol in a vacuum desiccator, and the excess of the polar liquid is removed under vacuum. The surface area is computed from the weight of ethylene glycol retained.

The most common approach to determining the external (non-expanded) surface area of powders, e.g., clays, is based on the work of Brunauer, Emmett, and Teller (1938), commonly referred to as the BET method. The method assumes that nonpolar gas molecules are adsorbed in multilayers on a solid surface, and that the amount of adsorbed gas in the initial monolayer, in contact with the surface, can be determined by constructing an adsorption isotherm and analyzing it mathematically. The BET equation was derived on the assumption that molecules in the initial monolayer, i.e., those directly on the surface, are more energetically adsorbed than molecules in all subsequent layers, and that the heat of adsorption of all layers beyond the first is equal to the latent heat of condensation of the gas. Thus, the equation theoretically differentiates the most energetically held gas molecules, and we assume that these are adsorbed in a regular array over the entire exposed solid surface.

The linear form of the BET equation is Eq. (3.29):

$$\frac{p}{x(p_0 - p)} = \frac{1}{x_m c} + \frac{c-1}{x_m c} \cdot \frac{p}{p_0} \qquad (3.29)$$

where x = weight of gas adsorbed at equilibrium pressure, p = equilibrium gas pressure, p_0 = saturation vapor pressure at temperature T, x_m = weight of gas in a complete monolayer, $c = \exp(E_1 - L)/RT_\mu$, E_1 = heat of adsorption in the first layer, L = latent heat of condensation, R = gas constant/mole (1,336 calories/mole), and T = absolute temperature (cgs units).

The procedure, then, is to conduct an adsorption experiment by varying p and measuring x (or v). The quantity, $p/x(p_0 - p)$ is plotted against p/p_0 and this should yield a straight line with a slope of $c - 1/x_m c$ and an intercept of $1/x_m c$. The amount of gas in a monolayer, x_m, is calculated by solving these two equations (from slope and intercept).

Experimental values of ethylene glycol have been found to deviate from those computed by using the BET equation given above at values of p/p_0 below 0.05 and above 0.35. Hence, useful data for surface area determinations are restricted to this range.

The total surface area of the sample is calculated from the relationship:

$$S_t = \frac{x_m}{M} \times N \times A_m \qquad (3.30)$$

where S_t = total surface area (m^2), x_m = experimentally determined weight of gas in an adsorbed monolayer, M = molecular weight of the adsorbate (28.01 for N_2), N = Avogadro's number 6.02×10^{23}, and A_m = cross-sectional area of gas molecule in the monolayer ($16.2 \times 10^{-20}\,m^2$ for N_2).

The specific surface area, a_m, is obtained by dividing the total surface area by the sample weight.

An adsorption experiment must be conducted at or below the temperature of condensation of the gas in order for significant adsorption to occur. Hence, for N_2 adsorption, the sample cell is immersed in liquid nitrogen ($-195.8°C$). The BET equation is used to calculate surface area from adsorption of nitrogen at liquid nitrogen temperatures on soil (Adamson, 1967; Greg and Sing, 1967; Shaw, 1970).

Fine-textured soils and those with high soil organic matter content have large surface areas. For further details on absorption processes with reference to Boer's law, Langmuir's equation, or BET equation refers to Sposito (1989) and Chapter 16.

3.1.4 Clay Minerals

The inorganic component consists of a wide range of minerals including crystalline and non-crystalline (Uehara and Gillman, 1981). The clay fraction primarily consists of Si, Al, Fe, H and O along with variable concentrations of Ti, Ca, Mg, Mn, K, Na, and P elements. The clay fraction is colloidal, and clay minerals are secondary minerals with significant influence on soil properties, e.g., surface area, cation exchange capacity (CEC), nutrient and water holding capacities, buffering and filtering capacity, swell–shrink properties, plasticity, compactability, and trafficability (or ability to withstand vehicular traffic). The clay minerals are hydrous aluminum silicates, with Mg^{+2} or Fe^{+3} proxying wholly or in part for the Al^{+3} in some minerals and with alkalies or alkaline earth present as essential constituents in others (Grim, 1968). Most commonly observed secondary minerals found in soil are listed in Table 3.8.

Two basic structural units are involved as building blocks in most clay minerals. The first is silicon tetrahedron, which comprises a silicon atom placed equidistant from four oxygen or hydroxyls. The silicon tetrahedral groups are arranged to form a hexagonal network, which is repeated indefinitely to form a sheet of composition $Si_4O_6(OH)_4$. The second unit comprises two sheets of closely packed oxygens or hydroxyls in which Al, Fe, or Mg atoms are embedded in octahedral condition, so that they are equidistant from six oxygens or hydroxyls. These two basic structures are joined together in 1:1 or 2:1 configuration to form a range of clay minerals. The lattice structure may be rigid or expanding type, and has two types of

TABLE 3.8 Commonly Observed Secondary Minerals Found in the Soil Clay Fraction

Secondary minerals	Weatherability
Geothite	Most resistant
Hematite	↓
Gibbsite	
Clay minerals	↓
Dolomite	
Calcite	
Gypsum	Least resistant

Source: Adapted from Brady and Weil 2001.

surfaces, i.e., internal and external. The total specific surface area of clay minerals, therefore, comprises internal and external surface areas. Different types of clay minerals, classified on the basis of number and arrangements of two structures, are listed in Table 3.9. There are nine principal silicate clay minerals of importance in soils. These are chloritic, glauconitic, halloysitic, illitic, kaolinitic, micaceous, montmorillonitic, sepentinitic, and vermiculitic. Predominant clay minerals present in soil affect soil physical properties, and have a profound influence on agricultural sustainability, soil degradation, and environmental quality.

The composition of clay minerals shows that their ultimate constituents are atoms which share electrons. The atoms and their oxidation state commonly found in clay minerals along with their radii are given in Table 3.10. Atoms with similar radii can replace one another within the crystal lattice. Such type of substitution is known as isomorphic substitution. This is a commonly observed process within clay minerals found in the soil.

In fact, it is this "isomorphic substitution" which leads to the formation of different types of clay minerals, and to deficit of positive or negative charge on the crystal. For example, Al^{+3} ($r = 0.57\,\text{Å}$) may substitute for Si^{+4} ($r = 0.39\,\text{Å}$) in the silicon tetrahedron unit causing a strain on the crystal structure because of the large size and producing a net negative charge deficit by one unit [Eq. (3.31)].

$$O^-Si^{++++}O^- \rightarrow O^-Al^{+++}O^- \tag{3.31}$$

Similarly, Mg^{+2} ($r = 0.78\,\text{Å}$), Fe^{+2} ($r = 0.83\,\text{Å}$), and Fe^{+3} ($r = 0.67\,\text{Å}$) may substitute for Al^{+3} in the aluminum octahedron sheet leading to charge

TABLE **3.9**   Classification of the Clay Minerals

I. Amorphous
    Allophane group
II. Crystalline
    A. Two-layer type (sheet structures composed of units of one layer of silica
        tetrahedrons and one layer of alumina octahedrons)
        1. Equidimensional
            Kaolinite group
                Kaolinite, nacrite, etc.
        2. Elongate
            Halloysite group
    B. Three-layer types (sheet structures composed of two layers of silica
        tetrahedrons and one central dioctahedral or trioctahedral layer)
        1. Expanding lattice
            a. Equidimensional
                Montmorillonite group
                    Montmorillonite, sauconite, etc.
                Vermiculite
            b. Elongate
                Montmorillonite group
                    Nontronite, saponite, hectorite
        2. Nonexpanding lattice
            Illite group
    C. Regular mixed-layer types (ordered stacking of alternate layers of different
        types) Chlorite group
    D. Chain-structure types (horneblende-like chains of silica tetrahedrons linked
        together by octahedral groups of oxygens and hydroxyls containing Al and
        Mg atoms)
            Attapulgite
            Sepiolite
            Palygorskite

Source: Adapted from Grim, 1968.

deficit in that sheet. In addition to isomophic substitution, broken bonds on
the edges of the crystals, and ionization of hydroxyl groups attached to
silicon of broken tetrahedron planes in the case of silicic acid, is also a
source of charge [Eq. (3.32)].

$$Si - OH + H_2O = SiO^- + H_3O \qquad (3.32)$$

Broken bonds and shared edges are other sources of charge on the clay
particles. Consequently, clay particles have negative and positive charge on

TABLE 3.10 Radii of Ions Abundant in Common Minerals

Ion species	Symbol	Radius (Å)
Silicon	Si^{4+}	0.39
Aluminum	Al^{3+}	0.57
Ferrous iron	Fe^{+2}	0.83
Ferric iron	Fe^{3+}	0.67
Magnesium	Mg^{2+}	0.78
Calcium	Ca^{2+}	0.99
Cesium	Cs^{+}	1.69
Potassium	K^{+}	1.33
Sodium	Na^{+}	0.95
Lithium	Li^{+}	0.60
Hydroxyl	OH^{-}	1.40
Oxygen	O^{2-}	1.40
Chlorine	Cl^{-}	1.81
Fluorine	F^{-}	1.36

$1 \text{ Å} = 10^{-10} \text{ m}$.

their surfaces, and the magnitude of charge and charge density depends on the type of clay mineral, the degree of substitution, and weathering. The positive or negative charge deficit is balanced by the absorption of anions or cations on the surface of the crystal structure. These ions are also called counter ions or gegen ions, which may be exchanged with those in the soil solution leading to anion exchange capacity (AEC) and cation exchange capacity (CEC).

Ionic bonds can be grouped into two broad categories: (i) primary or high-energy bonds, and (ii) secondary or low-energy bonds.

Primary Bonds

These are high-energy bonds and include ionic and covalent bonds.

Ionic or Electrostatic Bonds. These join two elements with incomplete outer electron shells (Fig. 3.7). These bonds involve the attraction of the unlike electrostatic charges. The atom of one element loses the electron or electrons in its outermost shell to an atom of the second element. In NaCl molecules for example, the Na atom has only one electron in its outermost shell and the Cl atom has seven. The Na atom loses its outermost electron to Cl, which completes its outermost shell. Several cations (Na^{+}, Ca^{+2}, Fe^{+3}, Th^{+4}, P^{+5}) and anions (Cl^{-1}, Br^{-1}, Fe^{-1}, I^{-1}, O^{-2}, S^{-2}, Se^{-2}) form ionic bonds.

Coulomb's law states that between any pairs of oppositely charged ions, there exists an attractive electrostatic force directly proportional to the

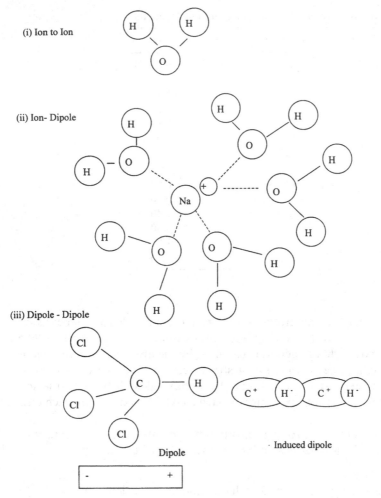

(i) Ion to Ion

(ii) Ion- Dipole

(iii) Dipole - Dipole

Dipole

Induced dipole

FIGURE 3.7 Ionic bonds (i) ion to ion, (ii) ion to dipole and (iii) dipole to dipole.

product of their charges (e_1, e_2) and inversely proportional to the square of the distance between their centers (D). The strength of the ionic bond depends on two factors: (i) the center to center spacing (interionic distance or band length) and (ii) their total charge:

1. Two ions:

$$\text{Force of attraction} = \frac{e_1 e_2}{D^2} \tag{3.33}$$

2. Two dipoles:

$$\text{Dipole moment } M = ed \tag{3.34}$$

$$\text{Force of attraction} = \frac{2M^2}{D^2} \cdot \frac{3D^2 - d^2}{(D^2 - d^2)^2} \tag{3.35}$$

where d is the distance between two equal and opposite point charges (e) of a dipole. The ionic bond or electrostatic attraction may exist for the following combinations: (i) ion-to-ion, (ii) ion-to-dipole, and (iii) dipole-to-dipole (Fig. 3.7).

Covalent Bonds. Covalent bonds develop when two atoms are lacking one or more electrons in their outermost shell. This bond develops when one electron is shared between two adjacent atoms. These two atoms then combine by sharing the electrons in the outermost shell, i.e., the combination of two oxygen atoms forms O_2 molecule (Fig. 3.8). A single covalent bond is the sharing of two electrons between the two bonded atoms (example, H_2). A double-covalent bond is two pairs of electrons being shared (example, O_2). A triple-covalent bond is the sharing of three pairs of electrons. Examples of a triple bond include those between two nitrogen atoms (N_2) or two carbon atoms (C_2H_2).

Two atoms with the same electronegativity share the bonding electron pairs equally. As a result, the bonding electrons are evenly distributed between the bonded atoms. There is no accumulation of bonding electrons on any one atom and the bond dipole moment is zero. Such a covalent bond is called a "nonpolar" bond. The bond between two hydrogen as in H_2, two oxygen as in O_2, or two nitrogen like N_2 or are all nonpolar bonds.

On the other hand, if the two bonded atoms have a different electronegativity, then the bonding pairs of electrons are shared unequally. The atom with the higher electronegativity attracts the bonding electrons closer to itself. As a result, the electron distribution is unequal and a bond dipole moment is formed. For example, the single bond between hydrogen and chlorine as in HCl has the bonding pair closer to the higher electronegative atom (chlorine). As a result, the chlorine end is partially negative since the electrons are closer to the chlorine. The hydrogen end is partially positive since the bonding pair is farther from the hydrogen. This two-pole condition is called a dipole, and it generates a dipole moment that is a vector force directed toward the higher electronegative atom in the bond. Such a bond is referred to as a *polar bond*. The greater the difference in the electronegativity between the two

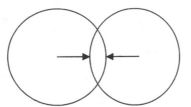

(i) A covalent bond is formed when the electron clouds of two atoms overlap.

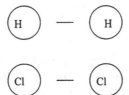

(ii) Single covalent bond. The dash is symbolic of the bonding pair.

(iii) Double covalent bond.

FIGURE 3.8 (i) Schematic of a covalent bond. A covalent bond is formed when the electron clouds of two atoms overlap. (ii) A single covalent bond. The dash is symbolic of the bonding pair. (iii) A double covalent bond.

bonded atoms, the more polar the bond. Elaborate descriptions of a variety of inter atomic bonds can be found in Gruenwald (1993).

Secondary Bonds

These are weak bonds, which include the following:

Hydrogen Bonds. A hydrogen-bond is formed when H in a H_2O molecule is attracted to the O of the neighboring molecule (Fig. 3.9). The hydrogen bond connects cation H^+ to an anion O^-, and links two H_2O molecules. This bond is weak compared with ionic and covalent bonds. In addition to water, such bonds also exist in other molecules such as NH_3. The hydrogen bond has a significant influence on soil physical properties such as

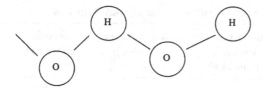

FIGURE 3.9 A hydrogen bond is formed when H in H_2O is attracted to the O of a neighboring molecule.

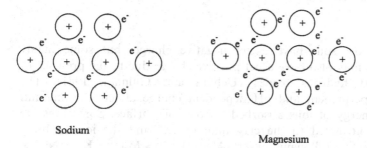

FIGURE 3.10 The strength of metallic bonds increases as the number of outermost electrons increase.

heat of vaporization, dielectric constant, and infrared and ultraviolet absorption. It is because of the hydrogen bond, that the water has high boiling point and heat of vaporization.

Metallic Bonds. Metals conduct electricity because some electrons owe no allegiance to any particular nucleus and are free to drift from one nucleus to another. This type of bond is called a *metallic-bond* (Fig. 3.10).

Charge Properties of Clay

Total charge on the mineral surfaces, due to structural properties including isomorphic substitution and other alterations, is called intrinsic charge density or permanent charge. This charge is independent of soil reaction or pH. There is another variable charge, which is pH or proton-dependent, and is due to the imbalance of complexed proton and hydroxyl charges on

TABLE **3.11**   Charge Properties and Specific Surface Area of Clay Minerals

Clay mineral	Cation exchange capacity (cmol/kg)	Anion exchange capacity (cmol/kg)	Charge density [cmol(+)/ $m^2 \times 10^{-3}$]	Specific surface area (m^2/g)
Kaolinite	3–15	—	60–75	5–20
Illite	10–40	—	10–20	100–200
Vermiculite	100–150	5–10	30–33	300–500
Smectite	80–150		11–19	700–800
Allophane	20–30	10–20	—	> 600

See Appendix 3.1 for units.

the surface. Most soils have a net negative charge, but some highly weathered soils may also have a net positive charge due to the presence of allophanes and hydrous oxides (Uehara and Gillman, 1981). The magnitude of permanent and pH dependent charge affects the amount, activity, and energy of ions absorbed on the soil surface. Some ions are more strongly attracted to the clay than others, and the ionic affinity usually follows the following order: $Al^{+3} > Ca^{+2} > Mg^{+2} > K^+ > Na^+ > Li^+$. The cation and anion exchange capacity differs among clay minerals (Table 3.11).

Electrical Double Layer and Zeta Potential

When clay particles are fully hydrated, the negative charge is balanced by the cations in the soil solution attracted by the Coulomb forces (Fig. 3.11). This negative charge on the clay surface and positive charge of the balancing cations create an electrical double layer around the clay particle (Fig. 3.12a). Three models have been proposed to explain the distribution of ion in the water layer adjacent to the clay minerals. The Helmholtz model assumes that all balancing cations are held in a fixed layer between the clay surface and the bulk solution, which is a condition of minimum energy. In contrast, the Gouy-Chapman model proposes a diffused double layer because cations possess thermal energy that causes a dynamic concentration gradient creating a diffuse double layer, which is a condition of maximum entropy (Fig. 3.12b). The third model by Stern is a combination of the two concepts, and it is a condition of minimum free energy. The double layer comprises a rigid region next to the mineral surface and a diffuse layer joining with the bulk solution. According to Stern's model, the concentration gradients are less steep in the diffuse double layer because the rigid layer lowers the surface

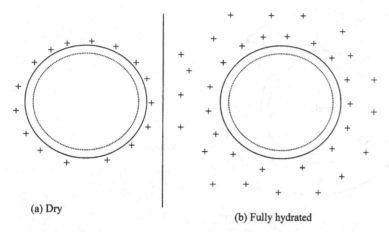

(a) Dry

(b) Fully hydrated

FIGURE 3.11 Negative charge on clay particles: (a) dry; (b) fully hydrated.

charge (Fig. 3.12b). The cations present in the solution neutralize the negative charge on the clay particle and the anions present in the solution. Addition of electrolytes to the system decreases the thickness of the double layer (Fig. 3.12b).

The Stern's double layer, therefore, comprises two parts: (i) a single ion thick layer fixed to the solid surface and (ii) the second diffused layer, which extends to some distance into the liquid phase. There is a potential gradient across these layers, which comprises two components (Zeta and Nernst). The potential difference between the fixed and freely mobile diffuse layer (or the electric potential across the double layer) is called the zeta potential (ζ), or the electrokinetic potential (Fig. 3.12c). It is the potential difference created at the interface upon the mutual relative movement of two phases. The difference in the cross potentials at the interface of two phases when there is no mutual relative motion is called the *Nernst's potential* (also called thermodynamic or the reversible potential). The Nernst's potential does not change with addition of electrolytes to the system, while the ζ is drastically influenced by addition of electrolytes (Fig. 3.12c). The ζ potential can be computed as per Eq. (3.36), and the thickness of the double layer by Eq. (3.37). Thickness of the double layer (U) is defined as the distance from the clay surface at which the cation concentration reaches a uniform or a minimum value. It is the distance over which the electrical influence of the clay platelet on its surroundings vanishes.

$$\xi = \frac{u\pi ed}{\varepsilon}$$

(3.36)

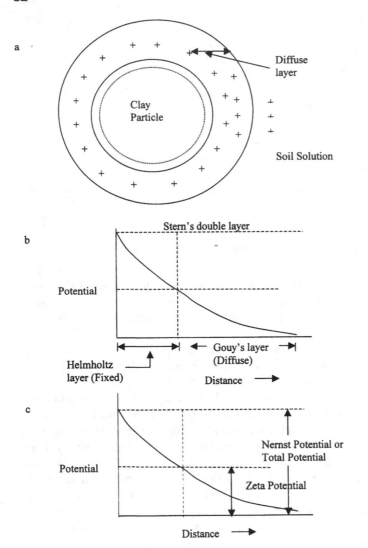

FIGURE 3.12 Electric double layer and the zeta potential.

where e (esu) is charge per cm^2, d is distance in cm within the double layer, ε is the dielectric constant of the media or permittivity $(esu^2/dynes \cdot cm^2)$.

$$U = \left(\frac{\varepsilon K_B T}{8\pi C e^2 V^2}\right)^{1/2}$$ (3.37)

where U is double layer thickness, ε is dielectric constant, K_B is the Boltzmann constant, T is absolute temperature in K, C is counter ion concentration, e is charge per cm^2, and V is counter ion valency. U is inversely proportional to V and C. The Boltzmann constant is given by Eq. (3.38).

$$K_B = \frac{R}{A} = \frac{8.31 \text{ J/K/mol}}{6.022 \times 10^{23}/\text{mol}}$$ (3.38)

where R is the gas constant and A is the Avogadro's number.

Stability of Clay Suspension

The colloidal system involves dispersion in H_2O. A dispersed system involves suspension of soil particles or separates in a dilute mixture of soil in water (Fig. 3.13). Flocculation or coagulation is sticking together of colloidal soil particles in the form of loose and irregular clusters called floccules (Van Olphen, 1963; Hunter, 1987; Gregory, 1989). The process of flocculation or condensation occurs when charged colloidal particles collide with one another and adhere after the collision as a result of favorable conditions in the electrical double layer. Floccules are loose combinations of clay colloids where the original particles can be recognized. The reverse of flocculation is called deflocculation, dispersion, or peptization. The dispersion can be achieved chemically (e.g., addition of sodium hexametaphosphate to soil), or mechanically, by stirring or ultrasound vibration. The dispersity (or ability of a cation to break down the floccules and bring colloids into suspension) of the system follows the lyotropic series, which is based in part on valency of the cations [Eq. (3.39)].

$$\text{Dispersity} = \text{Li}^+ > \text{Na}^+ > \text{K}^+ > \text{Rb}^+ > \text{Cs}^+$$ (3.39)

The DLVO (Derjaguin and Landau, 1941, and Ver Wey and Overbeek, 1948) theory of colloid stability states that dispersion or flocculation depends on the net effect of van der Waals forces of attraction and electrical double layer forces of repulsion. The collision efficiency, the probability of agglomeration when two particles collide, is also important to

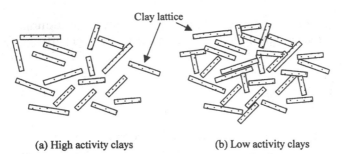

(a) High activity clays   (b) Low activity clays

FIGURE 3.13 Fully hydrated clay particles are completely dispersed. The distance between charged particles may be greater for (a) high activity clays (montmorillonite, vermiculite) than (b) low activity clays (kaolinite).

stability of the colloidal system. Lowering the ζ and decreasing the thickness of the double layer (U) to a critical level by addition of electrolytes causes flocculation. A colloidal suspension is stable as long as ζ exceeds the critical limit. When ζ falls below the critical level, the stability of the suspension is lost and it flocculates. The flocculation may be reversible or irreversible depending on charge properties of the system and of the electrolytes added. Adding electrolytes in excess of a certain amount can result in a system with ζ greater than the critical level and of the opposite sign, thereby reversing the flocculation and restabilizing the colloidal system. The effectiveness of the cation in causing flocculation depends on their valency. The higher the valency of the cation, the lower the concentration of the solution is required to reduce the ζ to the critical level. The effectiveness of monovalent, bivalent, and polyvalent cations is shown in Eqs. (3.40)–(3.42).

Monovalent cations:

$$H^+ > C_s^+ > K^+ > Na^+ > Li^+ \tag{3.40}$$

Bivalent cations:

$$Ba^{+2} > Ca^{+2} > Mg^{+2} \tag{3.41}$$

Polyvalent cations:

$$Th^{+4} > Al^{+3} > Ca^{+2} > Mg^{+2} \tag{3.42}$$

Dispersion agents (e.g., sodium hexametaphosphate) are added during the mechanical analysis to increase ζ so that the colloidal suspension is stable and does not flocculate. In contrast, addition of lime to alkaline soil

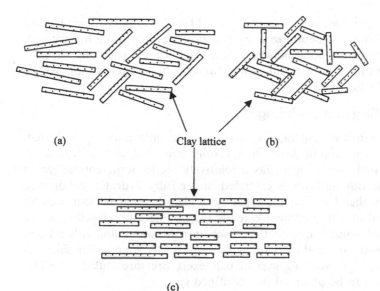

(a)   Clay lattice   (b)

(c)

FIGURE 3.14 Decrease in zeta potential leads to flocculation of clay with different geometric arrangements: (a) partial flocculation, (b) complete flocculation with a card-house structure, and (c) complete flocculation with a plate condensation structure.

lowers the ζ so that soil can flocculate and enhance formation of aggregates. Aggregation, formation of stable soil structure, is flocculation plus cementation by different cementing agents, typically inorganic plus organic matter (see Chapter 4).

Floccules are formed by a decrease in ζ potential because of the presence of ions in the solution. There are different types of flocculation (Fig. 3.14). Fully dispersed clay particles are farther apart in case of high activity (e.g., montmorillonite) than low activity (e.g., kaolinite) clays.

Incomplete Flocculation. Presence of monovalent cations (e.g., K^+) or dilute solution of bivalent cations (e.g., Mg^{+2}) can cause either weak or incomplete flocculation. Further, floccules are unstable and may set in suspension with a minor perturbation.

Random Flocculation. Rather than the plate condensation, flocculation may involve contact at the edges in a random fashion. This "card-house" or "brush-heap" structure of floccules is less stable (see Chapter 4).

Plate Condensation. The cations or ions added to the system are forced/aligned between the two clay crystals, and the distance between the adjoining clay particles is drastically reduced (see Chapter 4). The negative charge on the clay is neutralized by the positive charge of the cations, creating a very strong bond between them. The bond is generally stronger with polyvalent than monovalent cations, and the bond strength follows the order shown in Eq. (3.42).

3.1.5 Swelling and Shrinkage

At low soil moisture content, clay particles are only partially hydrated. Consequently, the double layer is not fully extended and is truncated. Such a truncated double layer has a relatively higher ionic concentration than when the double layer is extended under fully hydrated conditions. Such a system, therefore, has the capacity to absorb water (a polar liquid). Increase in soil moisture content extends the double layer. Swelling is the increase in soil volume due to the absorption of water and other polar liquids. The ratio of swelling caused by a polar to a nonpolar liquid is "swelling index." A swelling system can exert pressure called "swelling pressure," and can be observed in a confined system.

The rate of water absorption and other polar liquids by clay depends on the nature of clay and the exchangeable cations. It is generally rapid at first, then becomes slower with time, and may continue for several days. In comparison, the system of wetting by nonpolar liquids (benzene or carbon tetrachloride) is very rapid and may take only a few minutes. Nonpolar substances do not cause swelling and can be used to measure soil porosity and pore size distribution (see Chapter 5).

The swelling capacity depends on the type of clay mineral and the nature of cations on the exchange complex (Table 3.12). The expanding lattice clay minerals swell more than the nonexpanding clay minerals, suggesting two types of swelling: (i) interlattice swelling, and (ii) interparticle swelling. The interlattice swelling is more in expanding lattice than the nonexpanding clay minerals:

Vermiculite > montmorillonite > beidellite > illite

> Kaolinite > halloysite   (3.43)

With regard to the exchangeable cations, swelling follows the order shown in Eq. (3.39). However, the order may vary with the clay mineral.

$$Li^+ > Na^+ > K^+ > Ca^{+2} = Ba^{+2} > H^+ \tag{3.44}$$

This is the lyotropic series. However, H^+ does not follow the series with real soils. The specific effect of exchangeable cations on swelling depends on: (i) the number of exchangeable ions, (ii) the degree of dissociation or the

TABLE 3.12 The Relation of Swelling to the Type of Clay Mineral and Nature of Exchangeable Cations

Clay mineral	CEC (cmol/kg)	Swelling (cm³/g colloid)					
		H^+	Li^+	Na^+	K^+	Ca^{++}	Ba^{++}
Montmorillonite (Bentonite)	95	2.20	10.77	11.08	8.55	2.50	2.50
Beidellite	65	0.81	4.97	4.02	0.50	0.91	0.85
		Swelling (cm³/mmol cation)					
Montmorillonite	95	2.44	11.3	11.6	9.0	2.63	2.63
Beidellite	65	1.24	7.6	6.2	0.77	1.4	1.3
Ratio: Montmorillonite: Beidellite		1.97	1.49	1.87	11.68	1.88	2.02

Source: Adapted from Baver, Gardner and Gardner, 1972.

energy with which they are held, and (iii) the hydration energy of each ion determined by its hydrated radius and charge density. Both osmotic pressure and swelling increase with ionic hydration of monovalent cations.

There are two types of colloidal hydration or mechanisms involved in the swelling process: (i) water sorption and orientation on the clay surface due to the electrical properties of clay–cation–water system, and (ii) effect of cations. The former or short-range process depends on the cations, and involves van der Waals London forces, electrostatic forces, and hydration energy. The hydration energy plays an important role in the swelling process, and it overcomes the electrostatic attraction forces. During the process, the cation spacing increases significantly. These short-range forces act within the Stern layer from a distance of 10 Å to about 120 Å, and cause a considerable swelling pressure that may exceed 1 MPa. The swelling pressure is the force being exerted by expansion of the diffused double layer. This topic is discussed again in Chapter 8 on soil rheology. The swelling continues until the double-layer repulsive forces are balanced by attractive forces between the layer of particles, e.g., van der Waals force, positive edge–negative force attractions giving a cross-linking force [Eq. (3.45)]. It takes only a few nonparallel cross-linking particles to limit the swelling.

$$\text{Hydration energy } (0\text{--}10\text{Å}) + \text{repulsion due to diffused double}$$
$$\text{layer } (10\text{--}120\text{Å}) = \text{van der Waals forces} + \text{coulombic forces}$$
$$+ \text{cross-linking} \qquad (3.45)$$

Swelling due to diffused double-layer repulsion can be curtailed by strong adsorptive forces of polyvalent cations, e.g., the Coulombic attraction forces hold the two clay particles together against the double-layer repulsion.

In addition to the diffused double-layer concept, there is also a "clay domain" mechanism of swelling of clay colloids. In the dry state, clay particles are organized on a domain basis. A clay domain involves the parallel alignment of individual crystals involving a smaller volume of oriented particles. This alignment and orientation decreases the pore volume. On rewetting, domains swell as an entity, and pore volume increases proportionally to the overall volume.

3.1.6   Water Absorption on Soil Colloids

Soil's capacity to absorb water depends on its affinity for water and the antecedent temperature. The affinity for water is a function of the surface area, charge density, nature of the cations on the exchange complex, and pore size as determined by the packing arrangement. An examination of the water absorption isotherm on soil, a graphic relationship between the amount of water absorbed to the relative humidity or the vapor pressure at a constant temperature, gives information on the relative affinity of soil for water. Soils with high clay content of expanding-lattice clay minerals and higher specific surface area have a higher affinity for water and release more heat upon wetting than soils containing low clay content and nonexpanding type clay minerals.

Two generalized water absorption isotherms are shown in Fig. 3.15. These curves can be divided into three distinct regions. Region 1 shows absorption of H_2O on exchange sites and exchangeable cations, and includes water of hydration of cations. Somewhere at the boundary between regions I and II, the monomolecular layer is complete. Soil water content corresponding to the completion of the monomolecular layer is called the *hygroscopic coefficient*. This is also the amount of soil water content at which the release of the heat of wetting is the maximum. As the vapor pressure increases, the thickness of the water film increases further and the diffuse double layer is completely expanded in the vicinity of the boundary between regions II and III. Thickness of the absorbed water film increases drastically at the relative pressure between 0.9 and 1.0, and the capillary condensation begins.

The interaction of the charges of the clay with the polar water molecules imparts to the first few adsorbed layers of water a distinct and a rigid structure. Here the water dipole assumes the orientation dictated by the charge sites on the solids. This adsorbed water may have a quasi

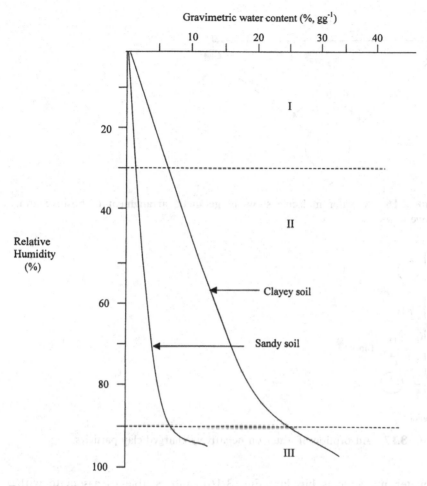

FIGURE 3.15 A schematic of water absorption on sandy and clayey soil equilibrated at different relative humidity. Three stages (I, II, III) correspond with degree of soil wetness and condensation of water in the pore.

crystalline or icelike structure, and can have a thickness of 10–20 Å or 3–7 thick layers of H_2O molecules.

3.1.7 Water Adsorption on Clay Surfaces and Heat of Wetting

There are several mechanisms of adsorption of water on clay surfaces (Low and Lovell, 1959). While the clay particles have a net negative charge,

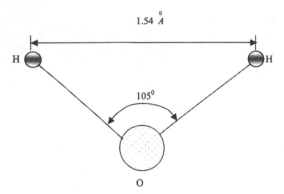

FIGURE 3.16 A water molecule showing geometic arrangment of positive and negative poles.

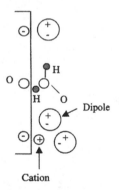

FIGURE 3.17 Adsorption of water on negatively charged clay particles.

the water molecule is bipolar (Fig. 3.16), and is able to associate with charged ions on the clay particles and in the electric double layer, and with the charge on the clay surfaces (Fig. 3.17). Water molecules associated with the cations are held as hydrated water or water of hydration (Fig. 3.18). A water molecule that attaches itself to the oxygen on clay surfaces may be held by hydrogen bonding. The H in H_2O may attach itself to the negative charge on the clay particle through electrostatic forces in which the dipole is attracted and oriented toward the negative charge on the clay surface (Fig. 3.17). The water molecule thus held to clay is called "adsorbed water," and has properties different than that of the "free water." This water is "structured" water because of the bonding to the clay surface. In comparison with the free water, the structured water: (i) has crystalline structure, (ii) is less dense, (iii) is more viscous, (iv) is less mobile,

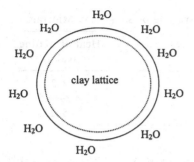

FIGURE 3.18 Water of hydration and formation of a monomolecular layer around a clay particle with moisture content equivalent to hygroscopic coefficient.

(v) has lower energy level, and (vi) has a lower freezing point. The degree of attachment of water decreases with increasing distance from the clay surface. The first layer is rather immobile, and the mobility increases in the bulk volume. The thickness of the absorbed layer differs among clay minerals, and ranges from about 8 Å in kaolinite to about 68 Å in montmorillonite.

The fixed or structured water has less energy than the free water, because the work must be done to remove the bond water. The amount of work that must be done to remove the bond water may be 3–4 Kcal per mol more than the energy released to condense vapor into the liquid state. Therefore, the energy of adsorption also differs among clay minerals.

Water adsorption on clay surfaces leads to release of energy, called "heat of wetting." The heat is also released when other liquids are adsorbed on a dry clay surface, e.g., alcohol. The heat of wetting is generally more for polar than nonpolar liquids. The heat of wetting is related to surface area. Kaolinite, with no internal surface, has a lower surface area and thus a lower heat of wetting than montmorillonite, which has both internal and external surfaces. The range of heat of wetting for some clay minerals is shown in Table 3.13.

The heat of wetting decreases with increase in water content of the clay, and varies with the nature of cations on the exchange complex. All other factors remaining the same, the heat released is generally more for divalent than monovalent cation [Eq. (3.46)]. The heat of wetting also increases with decrease in particle size, increase in surface area, and increase in CEC (Table 3.14).

$$Ca^{+2} > Mg^{+2} > H^+ > Na^+ > K^+ \qquad (3.46)$$

TABLE **3.13** Specific Surface Area and Heat of Wetting of Some Clay Minerals

Mineral	Specific surface area (m^2/g)	Heat of wetting (cal/g)
Kaolinite	11.0–25.0	1.4–2.1
Illite (Hydrous mica)	110–250	4.8–16.5
Montmorillonite	600–800	16.5–22.2

Source: Adapted from Jury, Gardner and Gardner, 1991.

TABLE **3.14** Effect of Particle Size and CEC of Kaolinite on Heat of Wetting

Particle size (μm)	10–20	0.5–10	0.2–4	0.1–0.5	0.5–0.25	0.25–0.10	0.10–0.05
CEC (cmol/kg)	2.4	2.6	3.58	3.76	3.88	5.43	9.50
Heat of wetting (cal/g)	0.95	0.99	1.15	1.38	1.42	1.87	—

Source: Adapted from Grim, 1968.

Heat of wetting is caused by three factors:

1. Change in state of water due to adsorption on the clay particles, or "structured water"
2. Hydration of adsorbed ions
3. Heat due to electric charge on the colloids

The orientation of adsorbed or structured water may be the cause of release of heat of wetting. The structured water is formed due to intermolecular forces. The intermolecular potential decreases as the distance from the surface decreases. If the water molecule does not react with soil colloids, the intermolecular potential energy possessed by H_2O molecules is all converted into heat. The amount of heat for adsorption of H_2O on soil can be calculated by using Eq. (3.47) (Iwata and Tabuchi, 1988).

$$\phi_m = \mu = \frac{-RT}{M} \frac{2.3 \times 0.88}{n^{1.8}} \tag{3.47}$$

where ϕ and μ are the chemical potentials of water in soil expressed in units of energy (ergs or Joules), R is gas constant (1.97 cal/degree/mol), M is molecular weight of water (18 g/mole) and n is statistical number of layers of water molecule adsorbed.

The heat of hydration of ions is very large and differs among ions, being more for trivalent than bivalent, which in turn is greater than for monovalent ions. The heat of hydration is 86.0 Kcal/mol for K^+, 106.0 Kcal/mol for Na^+, 399 Kcal/mol for Ca^{+2}, 477 Kcal/mol for Mg^{+2}, and 1141 Kcal/mol for Al^{+3}. The heat of wetting of clayey soils is in large part due to the heat of hydration of cations.

The hydration of adsorbed ions is usually not complete, because these ions are bonded to the surface and not free. The partial hydration leads to only a partial release of heat of hydration. The electric charge on the soil colloids reduces the internal energy of water molecules adsorbed by the colloid. Therefore, the heat is released when H_2O is adsorbed on the clay surfaces.

The heat of wetting can be measured by using a calorimeter or calculated from the surface tension relation as shown in Eq. (3.48).

$$\frac{U}{A} = \gamma - T \frac{\partial \gamma}{\partial T} \tag{3.48}$$

in which U/A is the energy per unit area, γ is the surface tension, T is the temperature in K (additional information on surface tension will be given in the section on capillarity in Chapter 11). Although heat of wetting is related to surface area, it is difficult to compute surface area of the soil from its heat of wetting because of the confounding effects of exchangeable cations and external and internal surfaces of clay minerals.

3.1.8 Packing Arrangement of Particles

Soil is a heterogenous mixture of solid particles of different sizes and shapes. It is a dynamic mixture, under continuous change due to natural (e.g., climate, biota, gravity) and anthropogenic factors (e.g., plowing, vehicular traffic). The packing arrangement of soil solids influences soil bulk density, pore size distribution and pore continuity, retention and movement of fluids, and substances contained in them (total porosity may not be affected by the packing arrangement). These properties are extremely relevant to agricultural, industrial, urban, and other land uses. Understanding the impact of packing arrangements is, therefore, important to developing and identifying systems of soil manipulation to achieve the desired configuration.

Porosity

Let's assume that a soil comprises spheres of uniform size of radius R. These spheres can be arranged into different forms of packing (Fig. 3.19). For details on different packing arrangements readers are referred to

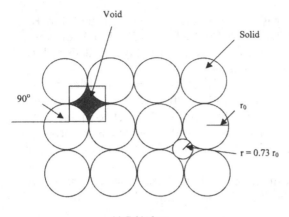

(a) Cubic form

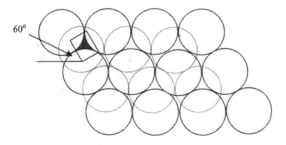

(b) Orthorhombic form

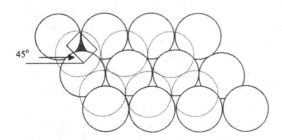

(c) Rhombohedral form

FIGURE 3.19 Different forms of packing of spheres of a uniform size. Within the pore space created by the sphere of radius r in cubic packing, a sphere of radius $r = 0.73\, r_0$ can be inscribed, but the radius of the interconnected passage is $r = 0.41\, r_0$. (a) Cubic form; (b) orthorhombic form; (c) rhombohedral form.

a review by Deresiewicz (1958), Yong and Warkentin (1966) and Childs (1969).

Cubic Form. This is the most open form of packing, with the maximum possible porosity of 47.64% or 48%. The porosity can be computed from simple geometric relationships including the volume of the sphere $(4/3\ \pi R^3)$, total volume of the cube with $2R$ sides $(8R^3)$, and volume of solids in the cube $(4/3\ \pi r^3)$. Therefore, the pore volume in the cube is computed as follows:

Volume of pore space = total volume − volume of solids

$$\text{Porosity} = \frac{\text{volume of pore space}}{\text{total volume}} = \frac{8R^3 - \frac{4}{3}\pi R^3}{8R^3} = 0.48 \qquad (3.49)$$

The pore diameter (d) equals the diagonal of the cube minus the diameter of the sphere or $0.41\,D$ where D is the diameter of the sphere. Foster (1932) computed the radius of pores inscribed by uniform spheres of radius r (Figs. 3.19 and 3.20). The radius of the inscribing circle is $0.73\ r_o$, but that of the interconnected passage is $0.41\ R_o$.

Orthorhombic Configuration. This geometric form involves 3 axes perpendicular to one another. Porosity of such a configuration can be computed as follows:

Total volume of orthorhombal with $2R$ sides $= 2R \cdot 2R \cdot 2R \sin 60°$
$\sin 60° = 0.866$

$$\therefore\ \text{Porosity} = \frac{\text{volume of pore space}}{\text{total volume}} = \frac{4\sqrt{3}R^3 - \frac{4}{3}\pi R^3}{4\sqrt{3}R^3} = 0.40 \qquad (3.50)$$

Rhombohedral Configuration. Rhombohedral is a six-sided prism, whose faces form parallelograms.

Total volume of rhombohedral with $2R$ sides $= 2R \cdot 2R \cdot 2R \sin 45°$
$\sin 45° = 0.785$

$$\therefore\ \text{Porosity} = \frac{\text{volume of pore space}}{\text{total volume of rhombohedral}} = \frac{4\sqrt{2}R^3 - \frac{4}{3}\pi R^3}{4\sqrt{2}R^3} = 0.26$$

$$(3.51)$$

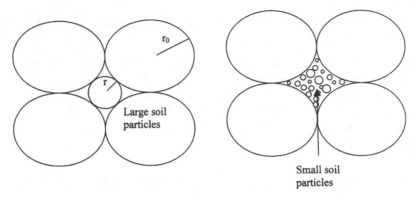

Figure 3.20 (a) Open packing; (b) closed packing ($r = 0.73\,r_0$ in open/cubic packing).

Composite Form. Uniform spheres can also be arranged into composite packing involving cubic and rhombohedral configuration. This situation may happen if soil aggregates or secondary particles were spheres of uniform size. In such a scenario, total porosity of uniform spherical particles within the aggregates in a rhombohedral configuration will be simply the sum of porosity of each configuration.

$$\text{Total porosity} = 0.48 + 0.26(1.00 - 0.48) = 0.62 \qquad (3.52)$$

These simple geometric arrangements lead to the following conclusions:

1. For identical form of packing, total porosity is independent of particle size of uniform spheres. However, the maximum pore diameter is proportional to particle size, and hence the permeability varies as a square function of the particle size. This is discussed under Poiseuille's law in Chapter 6.
2. The particles all have the same diameter, the most open packing or cubic form yields a total porosity of 0.48 and the most dense packing or rhombohedral form yields a total porosity of 0.26.
3. If all soil separates or primary particles are aggregated into secondary particles, the total porosity is much greater than when unaggregated.

Close Versus Open Packing

The packing of soil particles is influenced by particle shape and size distribution. For some engineering applications (e.g., dam construction, embankment, foundation, etc.), a high density is required.

Close rather than open packing is normally observed under natural conditions. For this topic readers are referred to the detailed description of packing arrangements by Yong and Warkentin (1966) and Childs (1969). In this regard, the geometry of "close packed" spheres is important to understand. In close packing, the smaller particles are packed within the pore space of larger particles (Fig. 3.20). The close packing is achieved by arranging the small grain sizes to fill voids created by large particles. Achieving a high density based on close packing necessitates having a material containing a diverse range of particle sizes. The other end of the scale involving open packing is based on a material containing particles of a uniform size. Thus, maximum porosity is achieved with open packing and the least with close packing.

Well-Graded Versus Poorly Graded Material

Packing arrangement of soil material is of relevance to soil compaction and surface seal formation in agricultural soils. It is also of interest to civil engineers concerned with stable foundations. The "well-graded" soil consists mostly of sand and gravel but also contains a small amount of silt and clay to facilitate close packing. "Poorly-graded" soils are those with uniform size fraction, e.g., fine or coarse sand only with little material of other size fractions (Fig. 3.21). Such materials are difficult to manipulate into close packing arrangements, do not compact into a dense mass, and are "poorly graded" soils. Clayey soils, with high swell–shrink capacity and ability to adsorb a large volume of water, are also poor-grade material for construction purposes.

3.2 ORGANIC COMPONENTS

Organic solids form only a small fraction of the total solids (about 5% in surface horizon of many humid-region soils) but play an important role in numerous important soil processes that determine soil quality, its productivity, and environment moderation capacity. Soil organic matter is a complex mixture of living and dead substances of plants and animal origin. Remains of dead plants and animals may be partially or fully decomposed into humic and biochemical substances. There are two principal types of humic substances: (i) insoluble humic acids, and (ii) alkali soluble humic acids and fulvic acids. The latter acids often have high molecular weight.

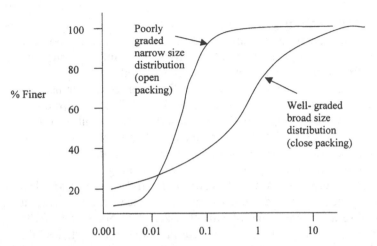

FIGURE 3.21 Particle size distribution for well-graded and poorly graded material.

Humus is dark-colored and amorphous (non-crystalline), and has a low particle density ($0.9–1.5\,Mg/m^3$), high surface area, high charge density, high ion exchange capacity, high buffering capacity, and high affinity for water (hygroscopic). In addition to C, humus contains essential plant nutrients including N, P, S, and micronutrients. Because of its high cation exchange capacity ($300–1500\,cmol/kg$), soil organic matter plays an important role in soil fertility management, buffering capacity and ability to filter contaminants from water passing through the soil. It is particularly effective in retaining heavy metals, e.g., Pb, Cd, Cu. Soil organic matter has a high water retention capacity—it can hold 20 times its weight in water. Being highly reactive, humus and other biochemical products are principal ingredients in formation of organomineral complexes, soil aggregates, or secondary particles. Humus forms stable complexes with several elements, e.g., Cu^{+2}, Mn^{+2}, Zn^{+2}, Al^{+3}, Fe^{+3}. Oxidation or mineralization of soil organic matter can lead to decline in soil structure, and emissions of radiatively active gases into the atmosphere, e.g., CO_2, CH_4, CO, NO, and NO_2.

Depending upon the composition, soil organic matter is classified into several pools. Four principal categories of these pools along with their mean residence time are described in Table 3.15. The easily decomposable fraction is called the "labile or active" pool. The fraction with a long mean residence time is called "recalcitrant or passive" pool. The passive pool may have mean residence time of centuries to millennia. The active fraction has a strong influence on elemental cycling (N, P, S,

TABLE 3.15 Different Pools of Soil Organic Matter

Pool	Constituents	Mean residence time (years)
Labile pool		
(i) Metabolic litter	Plant and animal residues, cellulose	<0.5
(ii) Structural litter	Plant residues, lignin, polyphenol	0.5–2
Active labile pool	Microbial biomass, simple carbohydrates, enzymes	0.2–1.5
Intermediate pool	Particulate organic matter	2–50
Recalcitrant pool	Humic and fulvic acids, organo-mineral complexes	500–2000

Turnover time is calculated by dividing the total pool by flux. For example, if the total soil C pool is 100 Mg and the flux is 50 Mg/ha/yr, then the mean residence time (MRT) is $100/50 = 2$ yrs.
Source: Modified from Parton et al., 1987; Jenkinson and Raynor, 1977; Jenkinson, 1990; Woomer et al., 1994.

Ca, Mg), and on activity of soil fauna and flora. The passive pool influences stability of soil structure through formation of organomineral complexes.

Laboratory determination of soil organic carbon (SOC) is based on methods involving one of the three following principles:

1. Wet oxidation of SOC in acid dichromate solution (Walkley and Black, 1934).
2. Wet oxidation of SOC in acid dichromate solution and measurement of CO_2 evolved (Allison, 1960).
3. Dry combustion of SOC with or without measurement of CO_2 evolved (McKeague, 1976).

Based on these three principles, there is a wide range of methods available for determination of SOC concentration (Nelson and Sommers, 1982; Tiessen and Moir, 1993; Lal et al., 2001). Results obtained are technique dependent, and may vary widely among methods. There is an urgent need to improve upon and standardize the methods of determination of SOC content (Lal et al., 2001).

The soil organic matter pool has a strong impact on the global carbon cycle, and on the atmospheric pool of carbon, especially with regard to the concentration of CO_2. Therefore, assessment of SOC pool, with regards to land use change and soil management, is very important.

Example 3.2

Compute the rate of change in SOC pool upon conversion from natural to agricultural ecosystem if the SOC concentration in 0 to 50 cm depth of a forested soil changed from 2.5% with a bulk density of $0.9\,Mg/m^3$ to 1.2% with a bulk density of $1.2\,Mg/m^3$ over a 10-year period.

Solution

$$\text{SOC pool in a forest ecosystem} = 10^4\,\frac{m^2}{ha} \times 0.5\,m \times 0.9\,\frac{Mg}{m^3} \times \frac{2.5}{10^2}$$
$$= 112.5\,Mg/ha$$

$$\text{SOC pool in an agricultural ecosystem} = 10^4\,\frac{m^2}{ha} \times 0.5\,m \times 1.2\,\frac{Mg}{m^3} \times \frac{1.2}{10^2}$$
$$= 72.0\,Mg/ha$$

$$\text{Rate of change of SOC pool} = (112.5\,Mg - 72.0\,Mg)/10\,yrs$$
$$= 0.41\,Mg/ha/yr$$

3.3 IMPORTANCE OF SOIL SOLIDS

Knowledge of soil solids is important to sustainable use of soil resources for different soil functions and land uses. Properties and processes relevant to inorganic solids and their effects are outlined in Table 3.16. Soil solids have an important effect on agricultural and industrial/engineering land uses, and environments. Agriculturally, soil solids are important to soil tillage and trafficability, plant available soil water, leaching losses of fertilizers and chemicals, formation of soil structure, swell–shrink properties, and physical condition of the soil or soil tilth. In terms of engineering and industrial uses, soil solids are important to foundation strength and stability, water sorption properties, and transmission of fluids in relation to waste disposal. Environmental applications of soil solids are those related to water and air qualities, buffering capacity, and ability to filter contaminants.

There are numerous functions of organic components. The organic components moderate soil and environment qualities. The soil quality effects of organic constituents are due to: (i) improved soil structure, (ii) increased water holding capacity, (iii) increased nutrient availability, and (iv) high soil biodiversity. Environmental effects of soil organic matter are attributed to: (i) high buffering capacity, (ii) chelation with heavy

TABLE 3.16 Importance of Soil Solids to Agriculture, Engineering, and Environments

Property	Agriculture	Engineering	Environments
Texture	Soil tillage and draft power, trafficability, soil compaction, plant available soil moisture	Foundation stability, sedimentation	Water quality and air quality effects of sediments
Surface area	Chemical sorption and buffering capacity, leaching of fertilizer	Strength and stability of material	Filtration of pollutants, contaminants, and pathogens
Diffused double layer	Soil structure formation, swell-shrink properties	Water sorption, and foundation stability	Transport of chemicals in water
Packing arrangements	Soil compaction, porosity	Strength and stability of engineering structures, transmission of fluids in relation to waste disposal	Filtration of chemicals

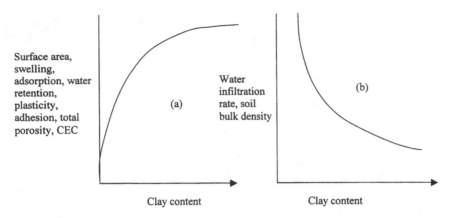

Surface area, swelling, adsorption, water retention, plasticity, adhesion, total porosity, CEC

(a)

Water infiltration rate, soil bulk density

(b)

Clay content   Clay content

FIGURE 3.22   Relation between clay content and soil propertries and processes.

metals and filtration of pollutants and environmental contaminants, and (iii) a large global carbon pool. Soil solids affect numerous properties and processes.

3.3.1   Texture and Soil Processes

Relative proportions of sand, silt and clay affect numerous soil properties. Being the most reactive fraction, increase in clay content increases surface area, swell–shrink capacity, absorption, water retention, plasticity, adhesion, and total porosity (Fig. 3.22a). In contrast, however, increase in clay content decreases water infiltration rate and soil bulk density (Fig. 3.22b). The nature of specific relation depends on other soil parameters (e.g., clay minerals, organic matter content, etc.).

The impact of texture on soil is manifested through its effect on other properties and related processes (Table 3.17). Texture influences soil compaction through its effect on aggregation and porosity, absorption of water and other organic/inorganic compounds by altering surface area, water and nutrient storage through charge properties, transport of solute and gaseous exchange through porosity, etc. In addition to particle size per se, clay minerals also affect surface area, charge density, and in turn, several processes related to these characteristics (Table 3.18).

Textural properties affect agronomic operations and water management. Tillage and traction are strongly influenced by textural properties as well as water content. Soil drainage is strongly influenced by clay content and the nature of clay minerals. There are also numerous engineering applications of textural properties (Table 3.19). Compaction, strength, slope

TABLE **3.17** Soil Properties and Processes Affected by Texture and Inorganic Components

Soil properties	Processes
Bulk density	Compaction, bearing capacity
Surface area	Adsorption, aggregation
Water affinity[a]	Water and nutrient uptake, aeration
Pore size distribution	Transport of solute and solids, leaching, erosion, diffusion
Swelling potential	Cracking, deformation
Plasticity	Moulding, aggregation
Adhesion, cohesion	Formation of soil tilth
Surface charge	Adsorption, absorption, diffusion, chelation
Packing	Compaction

[a]Not retention.

TABLE **3.18** Soil Properties and Processes Influenced by Clay Minerals

Property	Processes
Surface area	Absorption, filtration
Charge density	Ion exchange, leaching
Lattice expansion	Swell–shrink capacity
Shape	Plasticity

TABLE **3.19** Engineering Applications

Property	Application
Size distribution	Compaction, strength, trafficability, foundation stability, filtration
Clay content	Absorption, liquid waste disposal
Clay content	Seepage below drain, drainage, ceramic industry
Clay minerals	Slope stability, ceramic

stability, and seepage are strongly influenced by particle size distribution and the nature of clay minerals.

3.3.2 Organic Fraction and Soil Processes

Similar to clay, soil organic matter is also highly reactive. It has high surface area, charge density, and affinity for water. Thus, it has a strong influence on numerous soil properties and processes. The organic fraction influences

TABLE 3.20 Soil Properties and Processes Affected by Soil Organic Component

Soil properties	Processes
Color	Heat absorption, warming
Surface area	Adsorption, aggregation
Charge density	Cation exchange, chelation, aggregation, buffering capacity
Porosity and pore size distribution	Transport of solute and solids, leaching
Bulk density, particle density	Compaction, erosion, bearing capacity
Gaseous composition of soil air	Soil respiration, gaseous emission to the atmosphere
Microbial biomass and activity	Mineralization, aggregation, soil respiration, nutrient immobilization
Plasticity	Moulding, soil tilth formation
Swelling potential	Cracking deformation
Adhesion, cohesion	Soil tilth, soil structure

TABLE 3.21 Agricultural Applications of Soil Texture and Organic Components

Activity	Applications
Tillage	Timing, type, frequency and intensity of tillage
Fertilizer use	Rate, mode, timing, formulation of fertilizer use (precision farming)
Pesticides	Rate and mode of application
Water management	Rate and frequency of irrigation, and intensity of drainage
Accessibility	Timing of farm operations due to warming and trafficability

thermal properties through alteration of soil color, aggregation through charge properties and surface area, nutrient retention through charge density, and soil tilth through aggregation (Table 3.20). Consequently, the organic fraction affects timing and nature of tillage, rate and type of fertilizers to be used, fate of pesticides, and transport of water and pollutants into the soil (Table 3.21). The generic relationship between soil properties and soil organic matter content is shown in Fig. 3.23. Increase in organic fraction increases aggregation, porosity and available water capacity (Fig. 3.23a), and decreases adhesion, cohesion, and shrinkage (Fig. 3.23b). It is because of these improvements in soil characteristics that increase in soil organic content often leads to increase in crop yields (Fig. 3.24). The magnitude of increase in yield, however, depends on soil type and its organic matter content. Such beneficial effects on

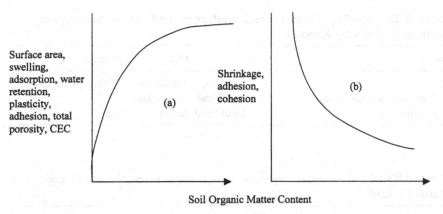

FIGURE 3.23 Relation of soil organic matter content with soil properties.

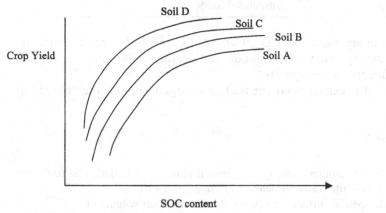

FIGURE 3.24 A generalized relationship between soil organic carbon (SOC) content and agronomic yield.

agronomic yield are especially apparent in subsistence agriculture with low off-farm input.

Beneficial effects of organic fraction on plant growth and yield are also related to improvement in soil quality and decrease in susceptibility to degradative processes. With a strong interaction with texture and clay minerals, the organic fraction affects soil's susceptibility to erosion, compaction, and other degradative processes (Table 3.22). The effects on soil quality are manifested in the overall impact of soil solids (inorganic and organic components) on the environment (Table 3.23). As will be discussed in the chapter on gaseous exchange (Chapter 16), the organic fraction affects flux of several greenhouse gases from soil into the atmosphere.

TABLE 3.22   Soil Degradative Processes Influenced by Inorganic and Organic Components and Clay Minerals

Property	Degradative processes
Texture	Erosion, compaction, leaching, acidification
Soil organic matter content	Acidification, leaching
Clay minerals	Structural decline, crusting

TABLE 3.23   Environmental Applications of Textural Properties and Organic Matter Content

Air quality	Suspended load and particulate matter, smog, soot
	Gaseous emissions (e.g., CH_4, CO_2, NO_x, H_2S)
Water quality	Suspended load

Important among these are CO_2, CH_4, N_2O, and H_2S, etc. Through its buffering capacity and ability to retain and degrade pollutants, the organic fraction influences water quality.

Sand is the skeleton, clay the flesh, and organic matter the "blood" of the soil.

PROBLEMS

1. Calculate the terminal velocity of spherical particles of 2, 0.02, and 0.002 mm diameter in dilute water suspension at 20, 30 and 40°C.
2. Calculate specific surface area per unit mass and unit volume of:
   (a) Spherical quartz particles ($\rho_s = 2.65\,\text{Mg/m}^3$) of radii 2 mm, 0.02 mm and 0.002 mm.
   (b) Plate-shaped particles of length 0.002 mm, width 0.001 mm and thickness 0.00001 mm, and particle density of $2.65\,\text{Mg/m}^3$.
3. Compute specific surface area of the A horizon of a Crosby soil at the Kenny Road Farm with the following characteristics:
   (a) 60% sand with an average e.c.d. of 0.1 mm, $\rho_s = 2.65\,\text{Mg/m}^3$.
   (b) 30% silt with an average e.c.d. of 10 μm, $\rho_s = 2.65\,\text{Mg/m}^3$.
   (c) 10% clay with platy structure of length = 200 nm, width = 100 nm and $d = 5$ nm, $\rho_s = 2.8\,\text{Mg/m}^3$.
      (i) Calculate the relative contribution of each particle size class to the specific surface area.
      (ii) What is the textural classification of this soil? Compare it with that containing 10% sand, 30% silt and 60% clay.
      (iii) What may be possible management problems of these two soils?

4. Assume that a soil has a w equal 0.3 and ρ_b equal 1.3 Mg/m^3. If the soil dries to $w = 0.1$ and shrinks by an amount equal to water loss, calculate ρ_b when $w = 0.1$.

5. A quantity of oven dry soil having a particle density of 2.65 Mg/m^3 and weighing 135.8 g is uniformly disturbed in water to form a total volume of 1000 cm^3 of suspension. After standing for 3 minutes, 10 cm^3 of the suspension removed was 0.437 g. Assuming the temperature of the suspension was 20°C, determine the percentage of particles finer than a specific size fraction.

6. Increase in volume of the suspension was 20 cm^3 when 50 g of dry soil was mixed in a known volume of water. Calculate particle density if soil bulk density is 1.2 Mg/m^3.

7. How do charge properties of soil relate to water quality and filtration attributes of soil?

8. What are the agronomic impacts of soil texture and surface area?

9. Briefly describe some applications of Stokes law in natural and altered ecosystems.

10. Describe effects of soil texture on other soil properties and processes relevant to (a) agronomic, (b) engineering, and (c) industrial uses.

11. Using the data in the table below: (a) plot the frequency and summation curves for three soils, and (b) calculate D_{10}, D_{60}, U.C., and gradation coefficients for three soils.

	% retained		
Particle size	A	B	C
4.0	0	20	1
2.4	1	5	2
2.0	2	2	4
1.2	1	6	5
0.6	11	10	8
0.3	7	8	10
0.15	20	12	12
0.075	16	10	10
0.04	5	4	5
0.01	10	10	8
0.002	10	7	5
0.001	17	6	30

12. What are the sources of charge on clay particles?

13. Describe distribution of charge in a fully hydrated clay particle.

14. What are the factors affecting zeta potential? Describe the process of flocculation.

15. (a) Write a brief essay on methods of measuring specific surface area of soil solids.
    (b) How much is a net charge on a dried out soil?

16. A farmer in Ohio has shifted from conventional plowing to no till farming. By doing so, SOC concentration in the top 1-m depth is increasing at the rate of 0.01%/yr. Assuming mean soil bulk density of $1.5\,Mg/m^3$, calculate the rate of soil carbon sequestration in this 500-hectare farm.

17. Consider a cubic/open packing of spheres of uniform radius of 1 mm. What is the radius of the pore inscribed by four spheres?

APPENDIX 3.1   STANDARD SIEVE SIZES

U.S. Sieve No.	Tyler Mesh No.	Millimeters	Inches
4	4	4.7	0.185
6	6	3.33	0.131
8	8	2.36	0.093
10	9	2.0	0.078
13	10	1.65	0.065
16	14	1.17	0.046
20	20	0.833	0.033
30	28	0.589	0.023
40	35	0.417	0.016
50	48	0.295	0.012
60	60	0.25	0.01
70	65	0.208	0.008
80	70	0.177	0.007
100	100	0.149	0.006
130	150	0.104	0.004
140	170	0.088	0.0035
200	200	0.074	0.0029
400	400	0.038	0.0015

APPENDIX 3.2   COMMON UNITS

Units

$$1\ dyne = g \cdot cm/s^2 \qquad 1\ dyne/cm = g/s^2$$
$$1\ Newton = 1\ kg \cdot m/s^2 = 10^5\ dynes$$
$$1\ Pascal = 1\ N/m^2 = 10^5\ dynes/m^2 = 10\ dynes/cm^2 = 1\ kg/m \cdot s^2$$
$$1\ bar = 10^6\ dynes/cm^2$$
$$1\ atm = 0.101\ MPa$$
$$1\ J = 1\ N \cdot m = 10^7\ erg$$
$$1\ erg = 1\ dyne \cdot cm$$
$$1\ cal = 4.186\ J$$

Poise $= g/cm \cdot s$
$1\ W = 1\ J/s$
Radius of H_2O molecule $= 138$ Å
1 mole of $H_2O = 18\,cm^3 = 6.02 \times 10^{23}$ molecules
$\therefore\ 1\,cm^3$ of $H_2O = 3.34 \times 10^{22}$ molecules

APPENDIX 3.3   INDICES OF PARTICLE SHAPE

Dimensional expression of shape	Index	Formula
2-D	Cailleuxis roundness (R)	$R = (2r/a) \times 1{,}000$
	Powers' scale	visual comparison chart
3-D	Zingg's classification	based on ratios of b/a and c/b
	Krumbein's sphericity (S)	$S = 3\sqrt{(bc/a^2)}$
	Cailleux's flatness (F)	$F = ((a + b)/2c) \times 100$

Note: $a =$ long axis; $b =$ intermediate axis; $c =$ short axis; $r =$ minimum radius of curvature at the end of the longest axis of the particle in its plane of maximum projection (measured by comparison with a set of standard concentric semi-circles).

REFERENCES

Adamson, A.W. 1967. Physical Chemistry of Surfaces, Second Edition, Interscience, New York.

Allison, L.E. 1960. Wet combustion apparatus and procedure for organic and inorganic carbon in soil. Soil Sci. Soc. Am. Proc. 24: 36–40.

Baver, L.D., W.H. Gardner and W.R. Gardner. 1972. Soil Physics, Fourth Edition, J. Wiley and Sons, New York, 498 pp.

Bower, C.A. and J.O. Goertzen. 1959. Surface area of soils and clays by an equilibrium ethylene glycol method. Soil Sci. 87: 289–292.

Brady, N.C. and R. Weil. 2002. The Nature and Properties of Soils. Thirteenth Edition, Prentice Hall, Upper Saddle River, NJ, 960 pp.

Brunauer, S., P.H. Emmett and E. Teller. 1938. Absorption of gases in multi-molecular layers. J. Am. Chem. Soc. 60: 309–319.

Buol, S.W., F.D. Hole and R.J. McCracken. 1989. Soil Genesis and Classification. Third Edition, Iowa State University Press, Ames, IA.

Carter, D.L., M.D. Heilman and C.L. Gonzalez. 1965. Ethylene glycol monoethyl ether (EGME) determining surface area of silicate minerals. Soil Sci. 100: 356–360.

Childs, E.C. 1969. The physical basis of soil water phenomena. J. Wiley & Sons, U.K., 491 pp.

Cihacek, L.J. and J.M. Bremner. 1979. A simplified ethylene glycol monoethyl ether procedure for assessment of soil surface area. Soil Sci. Soc. Am. J. 43: 821–822.

Day, P.R. 1965. Particle fractionation and particle-size analysis. In C.A. Black (ed) "Methods of Soil Analysis," Part 1, American Society of Agronomy, Monograph 9, Madison, WI: 545–567.

Deresiewicz, H. 1958. Mechanics of granular matter. Adv. Applied Mech. 5: 233–306.

Derjaguin, B.V. and L. Landau. 1941. Acta Physiochim. U.R.S.S. 14: 633 pp.

Foster, A.G. 1932. The sorption of condensable vapors by porous solids. Part 1. The applicability of the capillary theory. Trans. Faraday Soc. 28: 645–657.

Gee, G.W. and J.W. Bauder. 1986. Particle size analysis. In A. Klute (ed) "Methods of Soil Analysis," Part 1, American Society of Agronomy, Monograph 9, Madison, WI: 383–411.

Ghildyal, B.P. and R.P. Tripathi. 1987. Soil Physics. J. Wiley & Sons, New Delhi, India, 656 pp.

Gregg, S.J. and K.S.W. Sing. 1967. Adsorption Surface Area and Porosity. Academic Press, New York, 371 pp.

Gregory, J. 1989. Fundamentals of flocculation. Critical reviews in environmental control 19: 185–230.

Grim, R.E. 1968. Clay Mineralogy, Second Edition, McGraw Hill Book Co., 596 pp.

Gruenwald, G. (1993). Plastics: how structure determines properties. Hanser Publishers, New York.

Hunter, R.J. 1987. Foundations of Colloid Science. Vol 1. Clarendon Press, Oxford, 673 pp.

Iwata, S. and T. Tabuchi. 1988. Soil–water interactions: mechanisms and applications. Marcel Dekker, Inc., New York, 380 pp.

Jenkinson, D.S. and J.H. Raynor. 1977. The turnover of soil organic matter in some of the Rothamsted Classical experiments. Soil Science. 123: 298–305.

Jenkinson, D.S. 1990. The turnover of organic carbon and nitrogen in soil. Phil. Trans. Royal Soc. London, 329: 361–368.

Jury, W.A., W.R. Gardner and W.H. Gardner. 1991. Soil Physics, Fifth Edition, J. Wiley and Sons, New York, 328 pp.

Karathanasis, A.D. and B.F. Hajek. 1982. Quantitative evaluation of water adsorption on soil clays. Soil Sci. Soc. Am. J. 46: 1321–1325.

Krumbein, W.C. and F.J. Pettijohn. 1938. Manual of Sedimentary Petrography. Appleton-Century, New York.

Lal, R., J.M. Kimble, R.F. Follett and B.A. Stewart (eds) 2001. Assessment Methods for Soil Carbon. CRC/Lewis Publishers, Boca Raton, FL, 676 pp.

Loveland, P.J. and W.R. Whalley. 2001. Particle size analysis. In K.A. Smith and C.E. Mullins (eds) "Soil and Environmental Analysis: Physical Methods," Marcel Dekker, Inc. New York: 281–314.

Low, P.F. and C.W. Lovell. 1959. The factor of moisture in the frost action. Highway Res. Bd. Bull. 225.

MacCarthy, P., C.E. Clapp, R.L. Malcolm and P.R. Bloom. 1990. Humic substances in soil and crop sciences: selected readings, ASA/SSSA, Madison, WI, 281 pp.

McKeague, J.A. (ed) 1976. Manual on soil sampling and methods of analysis. Soil Research Institute of Canada.

Nelson, D.W. and L.E. Sommers. 1982. Total carbon, organic carbon and organic matter. In G.R. Blake (ed) "Methods of Soil Analysis," Part 2, American Society of Agronomy, Monograph 9, Madison, WI.

Orchiston, H.D. 1953, Absorption of water vapor: 1. Soils at 25°C. Soil Sci. 76: 453–465.

Parton, W.J., D.S. Schimel, C.V. Cole and D.S. Ojima. 1987. Analysis of factors controlling soil organic matter in Great Plains grasslands. Soil Sci. Soc. Am. J. 51: 1173–1179.

Shaw, D.J. 1970. Introduction to Colloid and Surface Chemistry. Butterworths, London, U.K., 236 pp.

Sheldrick, B.H. and C. Wang. 1993. Particle size distribution. In M.R. Carter (ed) "Soil Sampling and Methods of Analysis," Lewis Publishers, Boca Raton, FL: 499–511.

Sposito, G. 1989. The Chemistry of Soils. Oxford University Press, New York, 277 pp.

Stevenson, F.J. 1982. Humus Chemistry: genesis, composition, reaction. J. Wiley & Sons, New York, 443 pp.

Tiessen, H. and J.O. Moir. 1993. Total and organic carbon. In M.R. Carter (ed) "Soil Sampling and Methods of Analysis," Lewis Publishers, Boca Raton, FL: 187–199.

Verwey, E.J.W. and J.T.G. Overbeek. 1948. Theory of the stability of lyophobic colloids. Elsevier, New York.

Uehara, G. and G. Gillman. 1981. The Mineralogy, Chemistry and Physics of Tropical Soils with Variable Charge Clays. Westview Press, Boulder, CO, 170 pp.

Walkley, A. and I.A. Black. 1934. An examination of the Degtjareff method for determining soil organic matter, and a proposed modification of the chromic acid titration method. Soil Sci. 34: 29–38.

Woomer, P.L., A. Martin, A. Albrecht, D.V.S. Resck and H.W. Scharpenseel. 1994. The importance and management of soil organic matter in the tropics. In: P.J. Woomer and M.J. Swift (eds) "The Biological Management of Tropical Soil Fertility." John Wiley & Sons, Chichester, U.K.: 47–80.

Wu, T.H. 1982. Soil Mechanics, The Ohio State University, Columbus, OH 440 pp.

Yong, R.N. and B.P. Warkentin. 1966. Introduction to Soil Behavior. The MacMillan Co., New York, 451 pp.

4

Soil Structure

4.1 DEFINITION AND BASIC CONCEPTS

The arrangement and placement of soil particles determines the response of soil to exogenous stresses such as tillage, traffic, and raindrop impact. This arrangement of soil particles is called "soil structure." The arrangement is dynamic, complex, and is not very well understood. That is why Jacks (1963) stated that "the union of mineral and organic matter to form the organomineral complexes is a synthesis as vital to the continuance of life as, and less understood than, photosynthesis." Numerous advances in clay mineralogy, colloidal science, and sedimentology have since led to better understanding of genesis, characterization, and management of soil structure (Yong and Warkentin, 1966; Baver et al., 1972; Revut and Rode, 1981; Larionov, 1982; Burke et al., 1986; Hartge and Stewart, 1995; Carter and Stewart, 1996). Yet, soil structure remains to be the most complex, the least understood, and among the most important soil physical properties.

One of the reasons for the complexity of soil structure is the range of scales it expresses. Structural processes occur at a scale ranging from a few Å to several cm. Another cause of complexity is the dynamic nature of soil structure. Structural attributes vary in time and space, and the attributes observed at any given time reflect the net effect of numerous interacting factors which may change at any moment. It is truly a moving target. Consequently, it is hard to define soil structure, and the literature is replete with numerous and often confusing terminology, definitions, and approaches. Several terms are used to express easily identifiable structural units including structural form, fabric, aggregate, ped, granule, crumb, tilth, and so on used by different disciplines of soil science.

4.1.1 Different Approaches to Describing Soil Structure

There are at least four related but distinct approaches to describing soil structure. These include pedological, edaphological, engineering, and ecological approaches.

The Pedological Approach

This approach of defining soil structure is based on a mechanistic view with regard to the properties of its components. Therefore, soil structure refers to size, shape, arrangement, and packing of particles into identifiable units called aggregates or peds. In contrast to the synthesis of its components into aggregates, soil structure has also been defined as "the very fragments or clods into which the soil breaks up" (Zakhrov, 1927). In pedological terms, soil structure is a "three-dimensional arrangement of individual mineral grains and organic constituents."

The Edaphological Approach

This approach is based on its functional attributes with regards to plant growth. Functional attributes of soil structure are those related to pores or voids that govern root growth and development, retention and transmission of water, and gaseous diffusion. It is the soil–pore system that is the most important aspect of soil structure, which includes two types of pores (Fig. 4.1): (i) those within an aggregate are determined by textural characteristics and packing state of elementary particles and are called textural pores or intraaggregate pores, and (ii) those between aggregates and which result from arrangement of structural elements and aggregate characteristics are called interaggregate or structural pores (Stengel, 1990). The most important aspect is the number, dimensions, and continuity of pores between primary and secondary particles. Consequently, soil structure has also been defined as the "assemblage of aggregates (peds) and voids, including voids between and within aggregates" (Thomasson, 1978), or "the

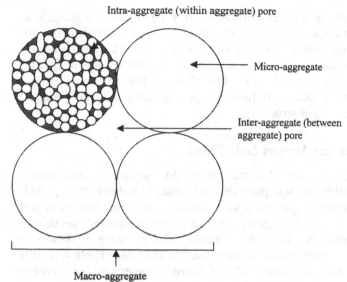

FIGURE 4.1 Interaggregate and intraaggregate pores.

arrangement of solid phase of the soil and of the pore space located between its constituent particles" (Marshall and Holmes, 1979).

The Engineering Approach

It is also appropriate to consider the engineering viewpoint of soil structure. Important among engineering functions of soil structure are bearing capacity, shear strength, slope stability, compressibility, and water permeability (see Chapter 7). With engineering perspective, soil structure is the "strength and stability of aggregates and voids in terms of their compressibility, bearing capacity, and permeability." Another related term used in engineering is sensitivity, which is the ratio of the strength of an undisturbed soil to that of a soil completely remolded at constant volume. Sensitivity refers to the loss in strength of a soil when its original structure is destroyed by remolding (Wu, 1981).

The Ecological Approach

Perhaps the most complete definition of soil structure is the one that combines pedological and edaphological views and takes a holistic or an ecological approach to soil structure. By so doing, soil structure refers to "size, shape, and strength of aggregates and pores, capacity of pores to retain and transmit fluids and dissolved and suspended materials, and ability to support vigorous root growth and development" (Lal, 1991). In other words, soil structure refers to three aspects: (i) degree of aggregation, their size distribution and stability, (ii) porosity, pore size distribution, shape, tortuosity, continuity, and stability, and (iii) spatial and temporal alteration in aggregates and pores in relation to natural (pedogenesis) and anthropogenic (management) factors. Therefore, in this chapter, aggregation is used to denote pedological (form, shape, size, etc.) and edaphological (functional) aspects of soil structure.

4.1.2 Soil Structure Versus Soil Fabric

Pedologists' use of the term "fabric" refers to "geometric and spatial arrangement of individual soil particles and voids" (Bullock et al., 1985; Brewer, 1976). In contrast, soil structure includes "the organization of soil constituents into larger aggregates or secondary/compound particles." Drees (1992) compiled the literature regarding the meaning of these two terms to minimize confusion and inconsistency in their use (Table 4.1). It is important to note that assessment of soil fabric is necessary for a proper evaluation of soil structure (Yong and Warkentin, 1975).

The term "fabric" implies two principal components: the skeleton or the individual mineral grains and the plasma or the soil material that floats

TABLE 4.1 Comparison Between Soil Structure and Soil Fabric

Soil structure	Soil fabric	Reference
1. The organization of soil constituents into larger aggregates or compound particles.	The geometric or spatial arrangement of soil particles or voids.	Drees (1992)
2. The spatial arrangement and total organization of the soil system as expressed by the degree and type of aggregation and the nature and distribution of the pores and pore space.	The arrangement, size, shape, and frequency of the individual solid soil components within the soil as a whole and within features themselves.	Fitzpatrick (1993)
3. The gradation and arrangement of soil particles, porosity, and pore size distribution, bonding agents, and the specific interactions developed between particles through associated electrical forces.	The geometric arrangement of the constituents mineral particles, including the void space which can be observed visually or directly using optical and electron microscopic techniques.	Yong and Warkentin (1975)

in between the skeletal particles (Kubiena, 1938). The term "structural form" refers to the heterogeneous arrangement of solid and void space that exists in a soil at a given time. This term is used to describe arrangement of primary soil particles into hierarchial structural states (Kay, 1990). In contrast, an aggregate is a naturally occurring cluster of soil particles in which the forces holding the particles together are much stronger than the forces between adjacent aggregates. Soil structure, however, is much more than a fabric or an aggregate. It is indeed hard to describe.

4.2 FORCES INVOLVED IN FLOCCULATION

There are several ionic forces involved in formation of floccules, domains, and aggregates (Fig. 4.2). Principal among these are inter and intramolecular forces, electrostatic, and gravitational forces.

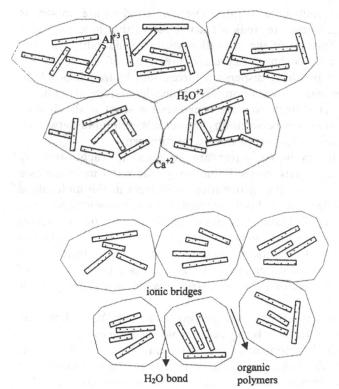

FIGURE 4.2 Floccules of clay particles cemented together lead to granulation. Aggregation is flocculation + cementation.

4.2.1 Intermolecular and Intramolecular Forces

Intermolecular attractions occur between one molecule with a neighboring molecule. The forces of attraction, which hold an individual molecule together (for example, the covalent bonds), are known as intramolecular attractions (see Chapter 3). All molecules experience intermolecular attractions, however, in some cases these attractions are weak. Even in a gas like H_2, cooling slows the H_2 molecules down, and the attractions become large enough for the molecules to stick together to form a liquid and then a solid. For hydrogen, the attractions are so weak that the molecules have to be cooled to 21 K ($-252°C$) before the attractions are enough to condense the hydrogen as a liquid. One type of intermolecular force is the van der Waals forces.

4.2.2 van der Waals Forces: Dispersion Forces

The weak forces that contribute to intermolecular bonding are known as van der Waals forces. These are weak attractive forces that hold nonpolar molecules together. The size of the attraction varies considerably with the size of the molecule and its shape. There are three types of van der Waals forces: intermolecular bonding, dispersion forces, and hydrogen bonding. The dispersion forces exist between nonpolar molecules and are also known as "London forces." Hydrogen bonding is exactly the same as the dipole–dipole interaction that occurs between any molecule with a bond between a hydrogen atom and any of oxygen, fluorine, or nitrogen.

Water molecules in liquid water are attracted to each other by electrostatic or van der Waals forces. Even though the water molecule as a whole is electrically neutral, the distribution of charges in the molecule is not symmetrical and leads to a dipole moment—a microscopic separation of the positive and negative charge centers. This leads to a net attraction between such polar molecules, which finds expression in the cohesion of water molecules and contributes to viscosity and surface. The dipolar interaction between water molecules represents a large amount of internal energy and is a factor in water's large specific heat (1 cal/g/°C or 1 cal/cm^3/°C).

Nonpolar molecules also experience some van der Waals bonding, which can be attributed to their being polarizable. These molecules do not have permanent dipole moments, yet they can have instantaneous dipole moments, which change or oscillate with time. These fluctuations of molecular dipole moments lead to a net attraction between molecules, which allow nonpolar substances like carbon tetrachloride (CCl_4) to form liquids.

4.2.3 Electrostatic Forces

Electrostatic forces work in much the same way as magnetic forces, i.e., like forces repel and unlike forces attracts. Water molecules bond by an oxygen atom joining to one of the hydrogen atoms by means of a covalent bond where electrons are shared (refer to Fig. 3.16). The electrical force (F) is directly proportional to the product of the charges (q_1 and q_2) and inversely proportional to the square of the distance (r) between them.

$$F = \frac{1}{4\pi\varepsilon_o} \frac{q_1 q_2}{r^2} \tag{4.1}$$

where ε_o is the permittivity constant and is equal to 8.854×10^{-12} Coulomb/(newton-m^2).

The lack of electrostatic forces in everyday life reflects that matter consists of almost exactly equal numbers of positively charged protons and negatively charged electrons thoroughly intermingled with one another, mainly in the form of atoms. Electrons move around positively charged nuclei consisting of protons and neutrons. Electrons and protons have equal but opposite charges ($q = 1.602 \times 10^{10}$ Coulomb), and neutrons have zero charge. There is a perfect balance between the number of electrons and protons in ordinary matter, and the net charge is zero. Consequently, two separate objects near each other hardly exert any electrostatic force at all.

4.2.4 Gravitational Forces

Gravitational forces are always attractive, and 10^{36} times smaller than electrostatic repulsion between two protons. Gravitational forces involving massive objects can be strong enough to move Earth, and keep it in a nearly circular orbit around the Sun.

4.3 MECHANISMS OF AGGREGATION

The mechanism of aggregation involves exogenous driving forces and the endogenous interactive forces arising from the soil–water interaction. Consequently, the specific arrangement of soil particles as observed in the field is dictated by the nature of exogenous and endogenous forces involved. Advances in colloid chemistry have facilitated and improved our understanding of the mechanisms and processes of aggregation. The importance of clay and humus colloids in forming aggregates was recognized as early as 1874 by Schloesing. Dumount (1909) also pointed out the importance of amorphous colloidal material in aggregation. For details on earlier literature readers are referred to the review by Harris et al. (1966). Numerous theories have been proposed since the 1930s. For details on interparticle forces in relation to aggregation readers are referred to reviews by Murray and Quirk (1990), Oades (1990), Emerson and Greenland (1990), Tisdall (1996), and others.

4.3.1 Russell's Theory of Crumb Formation

Russell (1934) proposed that clay particles are bonded together into aggregates through ionic bonds (Fig. 4.3). The mechanism of crumb

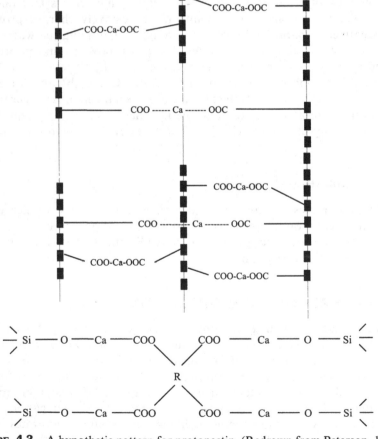

FIGURE 4.3   A hypothetic pattern for protopectin. (Redrawn from Peterson, 1947.)

formation according to this theory is as follows: (i) clay particles have a charge when hydrated, (ii) the charged particle is surrounded by an electric double layer of cations, (iii) polar water molecules are oriented along the lines of force radiating from each ion, and from each free charge of the clay particle, (iv) every clay particle is thus surrounded by an envelope of water, and (v) as the soil moisture content is reduced, the thickness of the envelope is reduced, and each ion shares its envelope with two clay particles thus holding the particles together. Russell observed that crumb formation according to this concept should meet the following requirements: (i) particles must have high cation exchange and large surface area, (ii) particles must be smaller than a certain size (1 µm) because sand and silt fractions are not essential to crumb formation and make a crumb weaker, (iii) the liquid must have an appreciable dipole moment, and (iv) polyvalent cations must be present. Clay particles are absorbed on sand and silt fractions, and the strength of bond between the clay and the sand increases with decreasing particle size of the clay. The process is reversible, because crumbs may disintegrate unless stabilized by appropriate cementing agents, because granulation is flocculation plus cementation.

4.3.2 The Calcium-Linkage Theory

Williams (1935) and Peterson (1947) proposed Ca-linkage as a mechanism in the formation of water-stable aggregates. The linkage was more effective in the presence of polyuronides, a component of soil organic matter, than without it. Negatively charged organic materials such as polysaccharides are absorbed onto the surface of clay by Ca^{+2} or other polyvalent cations (Fe^{+3}, Al^{+3}). This model is schematically presented in Eq. (4.2) for different polyvalent cations, and Eq. (4.3) for Ca^{+2}, and schematically presented in Fig. 4.4.

$$\text{clay–Mg–OH, clay–Be–OH, clay–Fe–(OH)}_2\text{, clay–Fe–OH} \qquad (4.2)$$

$$\text{clay–Ca–OOC–R–COO–Ca–OOC–R–COO–Ca–clay} \qquad (4.3)$$

4.3.3 Clay–Water Structure

Rosenquist (1959) proposed a concept of "clay–water structure." Rosenquist suggested that adhesion between clay particles is based upon the difference in surface energy of the adsorbed water and the liquid pore water. Therefore, creation of interfacial tension between the two types of water may be the cause of cohesion observed in saturated clays. The concept of

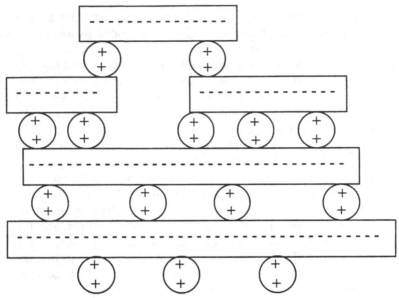

FIGURE 4.4   Plate-condensation of Ca–clay.

clay–water structure was also supported by the work of Lambe (1960), Michaels (1959), and Mitchell (1956).

4.3.4   Edge–Surface Proximity Concept

Schofield and Samson (1954) and Trollope and Chan (1959) proposed a model based on the interparticle forces of attraction and repulsion. Their proposal of a card-house structure is based on the establishment of equilibrium between adjacent particles due to the edge–surface proximity establishing a link bond (Fig. 4.5). Flocculation occurs as a result of electrostatic attraction between the positive edges and negative faces of clay lattices. The link bond is established if the particles are sufficiently close to exceed the potential energy barrier. This model is essentially based on the forces of adhesion between the clay particles. This edge-to-face type of flocculation produces a much more stable system than flocculation caused by lowering of zeta potential due to addition of salt.

4.3.5   Emerson's Model

Emerson (1959) proposed that crumbs are formed by cementation of card-house or brush-heap type of floccules by positive edge-negative face attraction (Fig. 4.6). According to this model, both quartz and clay form the

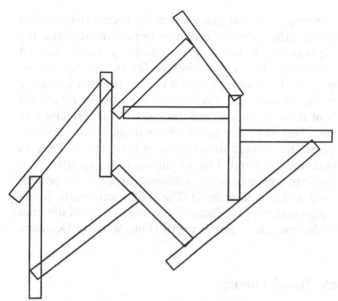

FIGURE 4.5 Card-house structure of floccules.

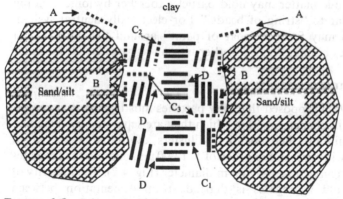

FIGURE 4.6 Schematic of the arrangements of quartz, clay domains, and organic matter in aggregate. Type of bond: A, quartz–organic matter–quartz; B, quartz–organic matter–domain; C, domain–organic matter–domain (C_1, face–face; C_2, edge–face; C_3, edge–edge); D, domain edge–domain face. (Redrawn from Emerson, 1959.)

main components of an aggregate or crumb. However, this structure disappears when soil is dried and 2:1 type clay minerals show an orientation with flat sides parallel. This crumb structure is generally stable when the exchange complex is dominated by Ca^{+2} and other polyvalent cations.

Emerson proposed four types of bonds prevalent in the crumb structure: (i) hydrogen bonding between the carboxyl group in organic matter and the clay, (ii) ionic bonding between the carboxyl group of organic matter and the clay, (iii) interaction of the electric double layers leading to the formation of domains, and (iv) bonding between the organic and inorganic colloids and between the colloids and the large soil particles. Emerson's model is an extension of Russell's model and incorporates the principles of the diffuse double layer. Clusters of clay crystals form domains as a result of orientation and electrostatic attraction to each other. These domains function as a single unit, and are bonded to the surface of the quartz grains and to each other to form aggregates. In addition, organic compounds increase the strength of the clay–quartz bond (Fig. 4.6). Electrostatic forces between the positive edges and negative faces of clay minerals, and presence of polyvalent cations also increase bond strength (Emerson and Dettman, 1960).

4.3.6 The Organic Bond Theory

Greenland (1965a;b) advanced Emerson's model by showing the importance of soil organic matter in strengthening the bond between adjacent clay particles. Soil organic matter may hold particles together by ionic bonding in a manner similar to "string of beads." For electrically neutral system, organic molecules may form a "coat of paint" around the outside of a number of particles binding them together into an aggregate.

4.3.7 Clay-Domain Theory

Williams et al. (1967) proposed that clay particles mostly exist in domains, up to about 5 μm in diameter, within which they are separated by "bonding pores" which maintain their identity. Clusters of domains are called microaggregates, with sizes in the order of 5–1,000 μm, and microaggregates are clustered into aggregates, 1–5 mm in diameter (Fig. 4.7). The integrity of microaggregates and aggregates is dependent on cementation between domains or microaggregates by inorganic precipitates, or on organic materials acting as a lining spread over the surfaces of domains or microaggregates. Oriented clay films and microbial films may also bind microaggregates and aggregates.

4.3.8 Quasi Crystal Theory

Aylmore and Quirk (1971) extended Williams et al. (1967) domain model by introducing the concept of quasi crystals or packets. The latter involves parallel clay crystals (about 5 μm in diameter) which are clustered together

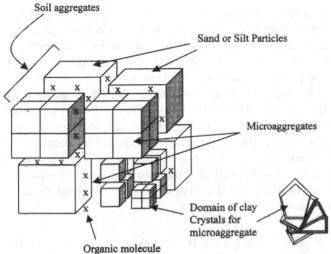

FIGURE 4.7 A hypothetical model of a soil aggregate. (Redrawn from Williams et al., 1967.)

closely enough (0.01–1.3 μm apart) to form domains. Rather than using domains, Quirk and Aylmore proposed the term "quasi crystals" to describe the regions of parallel alignment of individual lamellae of aluminosilicates in swelling type clay minerals which exhibit the intracrystalline swelling (e.g., montmorillonite). In comparison, they used the term domain to describe the regions of parallel alignment of crystals with fixed lattice and which exhibit intercrystalline swelling only (e.g., illite). The quasi crystal model has been verified and supported by Oades and Waters (1991), who argued that clay particles are aggregated into quasi crystals or stable packets. Oades and Waters proposed three distinct size fractions: (i) binding of clay particles into stable packets < 20 μm, (ii) binding of clay packets into stable microaggregates 20–250 μm, and (iii) the binding of microaggregates into stable macroaggregates > 250 μm.

4.3.9 Microaggregate Theory

Edwards and Bremner (1967) proposed that soil consists of microaggregates (< 250 μm) bound into macroaggregates (> 250 μm), and bonds within microaggregates are stronger than those between microaggregates. Micro-aggregates are represented by the structure shown in Eq. (4.4).

$$\text{Microaggreate} = [(Cl\text{–}P\text{–}OM_x]_y \tag{4.4}$$

where Cl is clay, P is polyvalent cation (Ca^{+2}, Al^{+3}, Fe^{+3}), and OM is organometallic complex including humified organic matter complexed with polyvalent metals. There may be more than one polyvalent metal bridge between clay (Cl) and OM in the Cl–P–OM units (Fig. 4.8). $(Cl–P–OM)_x$ and $(Cl–P–OM)_y$ represent compound particles of clay size ($< 2\,\mu m$ in diameter) and x and y are finite whole numbers with limits dictated by the size of the primary clay particles. The bonds linking the Cl–P–OM clusters into the larger $(Cl–P–OM)_x$ and $[(Cl–P–OM)_x]_y$ units can be ruptured by chemical or mechanical treatments. Interparticle bonds are weakened by substitution of polyvalent cations by Na^+ (treatment with sodium hexametaphosphate) and by mechanical shaking (stirring) and ultrasound vibrations. However, reversal of the dispersion process can lead to the formation of stable microaggregates [(Eq. (4.5)].

$$[(Cl–P–OM)_x]_y \rightleftharpoons y(Cl–P–OM)_x \rightleftharpoons xy(Cl–P–OM) \qquad (4.5)$$

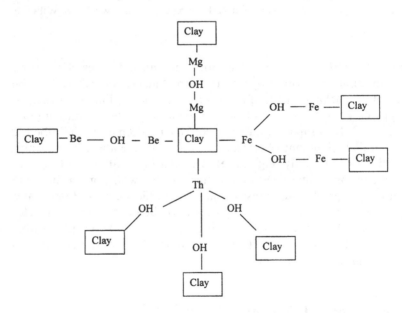

(a)

$$Clay – ca – OOC – R – COO – Ca – OOC – R – COO – Ca – clay$$

(b)

FIGURE 4.8 (a) Bridge between clay and polyvalent cations. (b) The calcium linkage between clay and organic polymers. (For details see Peterson, 1947.)

where D represent dispersion and A aggregation processes. This model has been verified by several researchers for Alfisols and Mollisols (Tisdall and Oades, 1982; Oades and Waters, 1991). Tisdall and Oades proposed that microaggregates themselves are built up in stages with different types of bonds at each stage (Tisdall, 1996; Table 4.1). Stages of aggregation are shown in Eq. (4.6)

$$< 0.2\,\mu m \rightarrow 0.2 - 2\mu m \rightarrow 2 - 20\,\mu m \rightarrow 20 - 250\,\mu m$$
$$\rightarrow > 2000\,\mu m\ \text{diameter} \tag{4.6}$$

4.3.10 The Aggregate Hierarchy Model

Oades and Waters (1991) modified the stages proposed by Tisdall and Oades (1982) especially for soils whose aggregates are mainly stabilized by organic materials. The modification was necessitated by the fact that it was not possible to distinguish steps of aggregation within aggregates less than 20 μm. They proposed that aggregates within the size range of 20–250 μm could be divided into aggregates 20–90 μm and 90–250 μm. Therefore, according to this model, the stages of aggregation or aggregation hierarchy are shown in Eq. (4.7):

$$< 0.2\,\mu m \rightarrow 20 - 90\,\mu m \rightarrow 90 - 250\,\mu m \rightarrow 250\,\mu m \tag{4.7}$$

These aggregation hierarchies (Table 4.2) are developed over many years, and are, therefore, observed only in mature rather than young soils. Binding mechanisms for different size fractions are shown in Fig. 4.9.

4.3.11 The POM Nucleus Model

The hierarchy model presupposes different bonding mechanisms for different aggregate sizes, or spatial distribution and persistence of aggregating agents within the soil matrix. These bonding mechanisms include: (i) bonding of clay into quasi crystals or packets is governed by pedological processes through precipitates of sesquioxes as in Oxisols, and (ii) bonding of packets into microaggregates and aggregates is governed by various organic materials. The particulate organic materials (POM) form a nucleus or core around which clay packets and small microaggregates are bound into larger microaggregates (Elliot, 1996; Golchin et al., 1994) (Fig. 4.9). The POM is colonized by microbial population, and the microflora and its by-products have strong adhesive properties which bind the particles together (Lynch and Bragg, 1985). The plant fragments from

Table 4.2 Models of Aggregation and Major Stabilizing Agents

Soil type	Stabilizing agent	Stage of aggregation (μm)	Reference
Alfisol	Inorgainc materials, organic polymers, electrostatic bonds, coagulation	< 0.2	Tisdall and Oades, 1982
	Microbial and fungal debris	0.2–2 → 2–20	
	Plant and fungal debris	2–20 → 20–250	
	Roots and hyphae[a]	20–250 → >2000	
	Ploysaccharides[b]	20–250 → >2000	
Alfisol, mollisol	Microbial debris, inorganic materials	< 20	Oades and Waters,
	Plant debris	< 20 → 20–90	
	Plant fragments	20–90 → 90–250	
	Roots and hyphae	20–250 → >2000	
Oxisol	Oxides/sesquioxides	< 20 → >250	Oades and Waters,
Oxisol	Oxides/sesquioxides	< 2 → 100–500	Robert and Chenu,
Vertisol	Organic matter	20–35 → >250	Collis-George and Lal,
Andosols	Allophanes and amorphous aluminosilicates	0.001–0.01→0.1–1	Robert and Chenu, 1992

[a]Soil with total organic carbon >2%.
[b]Soil with total organic carbon <1%.
Source: Adapted from Tisdall, 1996.

incorporation of crop residues, therefore, become the center of water stable aggregates (Buyanovsky et al., 1994; Angers and Chenu, 1997).

4.4 AGGREGATION AND STRUCTURAL FORMATION

Bradfield (1936) described that "granulation is flocculation plus." He drew a sharp distinction between flocculation (see Chapter 3) and aggregation. The process of formation of soil aggregates or organomineral complexes, from primary particles and humic and other bonding substances, is called aggregation. It is the first step in the development of soil structure. The process of aggregation is closely linked with the behavior of the diffuse

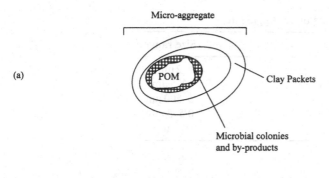

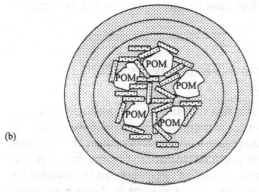

Figure 4.9 Microaggregates are formed around the particulate organic matter (POM) as a nucleus. (a) Microaggregate; (b) cluster of microaggregates forming a macroaggregate.

double layer and its response to ionic composition in the bulk solution (refer to Chapter 3). Aggregation is flocculation plus cementation with numerous forces, agents that stabilize and bind floccules [(Eq. (4.8)]:

$$\text{Aggregation} = \text{flocculation} + \text{cementation} \tag{4.8}$$

Most common cementing agents include soil organic matter, silicate clays, lime, and sesquioxide (FeO_3, Al_2O_3, Mn_2O_3) (Fig. 4.2). Humified organic matter, with its long polymer chains and electric charge balanced by polyvalent cations, is a very effective cementing agent. Fungal hyphae and microbial by-products also serve as cementing agents. In summary, there are four types of binding agents including: (i) oriented clay films, (ii) microbial by-products, fungal filaments, and hyphaes, (iii) inorganic precipitates such

TABLE **4.3** Components of an Aggregate

Component	Size range
Clay	$2\,\mu m$
Domain, quasi crystal, or Packets	$2–5\,\mu m$
Microaggregate	$5–500\,\mu m$
Aggregate	$0.5–5\,mm$
Compound structure	$> 5\,mm$

as oxides of Fe and Al, and (iv) humic substances including organic polymers.

4.4.1 Bonding Agents Responsible for Aggregation and Structural Stability

Structural stability is the ability of a soil to retain its arrangement of solids and void space when external forces are applied. External forces may be natural or anthropogenic. The aggregate stability depends on the bonding agents involved in cementing the particles together. On the basis of the numerous models presented, components of an aggregate can be summed up as those shown in Table 4.3. The smallest component is domain or quasi crystal or packets. These are essentially floccules cemented together by different agents. The largest component is an aggregate that is $< 5\,mm$. Anything larger than $5\,mm$ may be a compound structure or a clod. Mechanisms of aggregation presented in the previous section can be summarized by Eq. (4.9) in which A denotes aggregation and D is dispersion.

$$\text{Clay particle} \underset{D}{\overset{A}{\rightleftharpoons}} \text{floccule} \underset{D}{\overset{A}{\rightleftharpoons}} \text{domains/quasi-crystals/}$$

$$\text{packets} \underset{D}{\overset{A}{\rightleftharpoons}} \text{microaggregates} \underset{D}{\overset{A}{\rightleftharpoons}} \text{macroaggregates} \qquad (4.9)$$

There are different binding agents at each step going from clay particle to macroaggregates.

It has been argued that the reaction shown in Eq. (4.9) is as important as the photosynthesis reaction ($6CO_2 + 6H_2O \rightarrow C_6H_{12}O_6 + 3O_2$). Therefore, understanding the reaction in Eq. (4.9), and developing management strategies that push this reaction forward to the right-hand side are extremely important to crop production, and global food security.

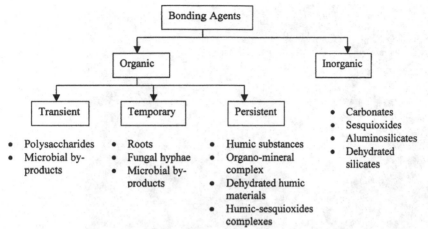

FIGURE 4.10   Different types of binding agents. (For details see Harris et al., 1966.)

The binding agents involved at each stage of aggregation can be grouped into three main categories described below and outlined in Fig. 4.10. For detail discussion on different binding agents, readers are referred to reviews by Harris et al. (1966) and Hamblin (1985).

Transient Binding Agents

These are organic materials that are decomposed very rapidly by microorganisms. These materials include: (i) microbial polysaccharides produced when various organic materials are added to the soil, (ii) and some of the polysaccharides associated with roots and microbial biomass in the rhizosphere. These polysaccharides or glues are associated with large ($> 250 \mu m$ diameter) transiently stable aggregates, and are decomposed readily. Cellulose contributes to only a small fraction of aggregation but is more persistent. The transient polysaccharides (produced by bacteria, fungi, and plant roots) bind clay-sized particles into aggregates which are of the order of $10 \mu m$ diameter. Polysaccharides stabilize aggregates with diameter $< 50 \mu m$.

Temporary Binding Agents

These agents are roots and mycorrhizal hyphae (Tisdall, 1991). Such binding agents are built up in the soil within a few weeks or months as the root system and associated hyphae grow. They persist for months or perhaps years, and are affected by management of the soil.

Roots.   Roots supply decomposable organic residues to soil and support large microbial population in the rhizosphere. Roots of some

FIGURE 4.11 Earthworm casts in a pasture enhance crumb structure and stable aggregates.

plants, e.g., grasses, themselves act as binding agents. Residues released into the soil by roots are: (i) fine lateral roots, (ii) root hairs, (iii) cells from the root cap, (iv) dead cells, and (v) mucilages. The amount of organic carbon released by roots is proportional to the length of root. It can be 20–49 g of organic material per 100 g harvested root. The root system and associated hyphae of pasture plants, especially grasses, are extensive. The upper layer of the soil under pasture is probably all rhizosphere. Water stable aggregates are also formed due to localized drying around roots. Electron micrographs or a drying root show that particles of clay close to root tend to be oriented almost parallel to the axis of the root. Roots also provide food for soil animals, e.g., earthworms and the mesofauna. Population of earthworms in pastures may exceed 1.5×10^6/ha (Fig. 4.11).

Hyphae. Hyphae are sticky and encrusted with fine particles of clay. Stabilization of aggregates by fungi in the field is limited to periods when readily decomposable material is available. Fungal hyphaes are relatively large and usually bind microaggregates greater than 250 μm.

Saprophytic Fungi. This group of fungi includes dark colored fungi that tend to persist in soil.

Vesicular-Arbuscular (VA) Mycorrhizal Fungi. These are abundant in soils and are obligate symbionts. The VA mycorrhizal fungi tend to be most abundant in soils with low or unbalanced level of nutrients. Some plants are, however, mycorrhizal even in fertile soils. Mycorrhizal fungi bind particles into aggregates, and micro- into macroaggregates.

Other Temporary Binding Agents

Fungi constitute more than 50% of the microbial biomass in some soils and contribute more than bacteria to the organic matter in soil. Organic bonds also develop from degraded bacterial cells. In desert soils, filaments of blue-green algae are important. Algae and lichens form crust in desert soils.

Persistent Binding Agents

Persistent bonds include strongly sorbed polymers such as some polysaccharides and organic materials stabilized by association with metals. Degraded, aromatic humic materials associated with amorphous iron, aluminium, and aluminosilicates form the large organomineral fraction of soil that constitutes 52 to 98% of the total organic matter in soils. The persistent binding agents probably include complexes of clay-polyvalent metal—OM, C–P–OM, and (C–P–OM)$_x$ both of which are < 250 μm in diameter. Persistent binding agents are probably derived from the resistant fragments of roots, hyphae, bacterial cells, and colonies developed in the rhizosphere. The organic matter is in the center of the aggregate with particles of fine clay sorbed onto it, as opposed to the Emerson's concept of organic matter sorbed on the clay surface. Persistent bonding agents have not been defined chemically, just as the formula of humic acid cannot be defined. Some of these bonds resist ultrasonic vibrations.

The bonding forces in the formation of clay–organic complexes are summarized by Greenland (1965a). These forces are the same as those involved when atoms and molecules are in proximity. However, the situation is particularly complex when large organic molecules are involved. As is apparent from the discussion of various models of aggregation, the soil organic matter plays an important role in aggregation and structural stability of soils. It is not surprising, therefore, that numerous studies from around the world have demonstrated a high correlation coefficient between aggregation and soil organic matter content (Fig. 4.12). In contrast, there are also numerous studies indicating low or no correlation between soil organic matter content and aggregation. The lack of correlation, however, does not necessarily mean that soil organic matter content is not important to aggregation. The low or no correlation of aggregation with soil organic matter content may be due to several factors: (i) only part of the soil organic present is responsible for aggregation as is the case in soils of high organic matter content, (ii) there is a critical limit or threshold value of soil

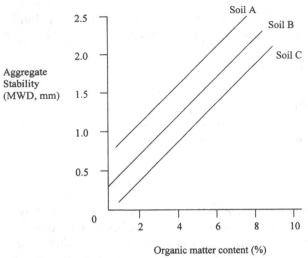

FIGURE 4.12 Schematic of the relationship between aggregate stability (mean weight diameter) and soil organic matter content for a group of soils.

organic matter content above which it has no effect on aggregation, (iii) aggregation is affected by specific organic constituents rather than the bulk soil organic matter, (iv) there are other bonding mechanisms which are as good or more effective than soil organic matter, and (v) aggregation and aggregate stability are affected by other pedological or anthropological factors.

In summary, there are different binding mechanisms for microaggregates and macroaggregates against rapid wetting and disruptive forces of cultivation and other natural or anthropogenic disturbances. Microaggregates are predominately stabilized by organo-mineral complexes. These bonds are relatively stable and not easily disrupted by changes in soil organic matter content brought about by land use and cultivation. In contrast, stability of macroaggregates depends on root hair and fungal hyphae. Therefore, the proportion of stable macroaggregates changes with change in soil organic matter content by land use and cultivation, and with changes in population of root hair and fungal hyphae. The stabilization of macroaggregates depends on management. It increases under fallow and pasture, and decreases with row cropping and plow-based tillage methods.

4.5   PROPERTIES OF AGGREGATES

An aggregate or ped thus formed is a distinct physical entity with quantifiable attributes, and exterior and interior properties. The exterior of an aggregate may be coated with: (i) clay film or "clay skins," (ii) inorganic precipitates and sesquioxides, and (iii) organic matter. The exterior may have distinct shape (angular, subangular, prismatic, columnar, platy), size (coarse, medium, or fine) and strength or grade, and compactness. Similarly, the interior of an aggregate may be compact or loose, anaerobic or aerobic, hygroscopic or hydrophobic, slow to dry when wet, or slow to wet when dry. Single aggregates are more dense compared to bulk soil (Horn, 1990; Kay, 1990). Bulk density generally increases with decrease in size of an aggregate (Becher, 1995). Two principal properties of an aggregate are strength and hydrophobicity.

4.5.1 Strength of Soil Aggregates

Strength refers to the ability of aggregates to withstand disruptive forces (e.g., vehicular traffic, raindrop impact, plowing, root pressure). The knowledge of magnitude and distribution of aggregate strength is key to understanding soil's response to tillage or traffic. Aggregated soils are stronger than nonaggregated or homogenized materials. Strength increases either by an increase in the total number of contact points between floccules and domains, or by increase in shear resistance per contact point (Hartge and Horn, 1984; Horn and Dexter, 1989; Horn et al., 1995). Factors affecting strength of soil aggregates are water content, texture, clay minerals, organic matter content and size of aggregates.

4.5.2 Hydrophobicity of Aggregates

Some coatings on aggregate surfaces impact their hydrophobic properties. Consequently, aggregates do not wet easily. Hydrophobic properties are attributed to some microbial by-products and other organic substances. In some soils, coverage of aggregates by such films is so extensive that water infiltration in soil is severely curtailed (see Chapter 14).

4.6 FACTORS AFFECTING AGGREGATION

There are numerous factors that affect aggregation (Hamblin, 1985; Kay, 1997) most of which can be grouped into two broad categories: endogenous and exogenous factors. The endogenous factors are those that are due to inherent soil properties. These factors include soil characteristics such as texture, clay mineralogy, nature of exchangeable cations, quantity, and quality of the humus fraction. The exogenous factors that affect soil structure include weather, biological processes, land use, and management.

The impact of seasonality, due to wetting and drying and freezing and thawing, on aggregation cannot be overemphasized (Bower et al., 1972). Biological processes, especially the activity and species diversity of soil fauna notably earthworms and termites, are extremely important to soil aggregation (Lal, 1987). Root growth is another important biological process affecting aggregation. Both of these exogenous and endogenous factors interact with one another, vary in both space and time, operate at different scales, and cannot be considered in isolation. Based on these and numerous interacting factors, there is a wide range of possible mechanisms and processes that lead to aggregation.

The literature is replete with analyses of factors affecting soil structure and strategies for its management (Bower et al., 1972; Kay, 1980; Hamblin, 1985; Carter and Stewart, 1996). Therefore, this section provides a brief outline of the salient features of the factors affecting aggregation under field conditions.

4.6.1 Drying and Wetting

Repeated cycles of drying and wetting play a major role in aggregation through shrinking and swelling that lead to formation of aggregates. Swelling or rewetting leads to reorientation of particles. Shrinking or drying leads to formation of cracks and increase in formation of link bonds through cementation. The mechanisms involved, especially the opposing forces, are not clearly understood. Non-uniform drying can lead to unequal strains throughout the soil mass. Consequently, large clods can break down into small aggregates by drying (Figs. 4.13a;b). Similar to rapid drying, rapid wetting also breaks large clods into aggregates because of the effect of entrapped air. That is why slow wetting, wetting by capillarity or wetting in vacuum is suggested for minimizing risks of soil slaking or rapid dispersion (Yoder, 1936; Henin, 1938). There is no slaking of aggregates if air in the soil is replaced by CO_2 (Emerson and Grundy, 1954; Robinson and Page, 1950). Other causes of slaking by rapid wetting include differential swelling (Panabokke and Quirk, 1957), and swelling of the oriented clay coatings or streaks (Brewer and Blackmore, 1956). However, the relative effectiveness of wetting and drying depends on the texture and cohesive properties of the soil (Grant and Dexter, 1989). In heavy-textured soils, desiccation cracks lead to formation of ped faces (White, 1966; 1967). Rewetting of the shrunken soil causes swelling and development of shearing forces between the wet/dry boundary layer. Repeated shrinkage and swelling leads to formation of prismatic, blocky, parallelopiped, or platy peds in subsurface layers of heavy-textured soils.

(a)

(b)

FIGURE 4.13 (a) A freshly plowed field creates cloddy structure. (b) A weak structure creates surface seal that reduces infiltration rate. However, repeated wetting and drying cycles can improve aggregation.

4.6.2 Freezing and Thawing

Water expands on freezing, and its impact on aggregation depends on the size, distribution, and duration (or persistence) of ice crystals (Kay and Perfect, 1988). The in situ freezing of water in pores may lead to a fracturing of the soil. Local redistribution of water may also occur due to freezing

FIGURE 4.14   Repeated cycles of freezing and thawing also improve soil structure.

leading to accumulation of ice in large pores and shrinkage in adjacent areas. Large ice lenses are formed when large quantities of water move from the unfrozen zone up into the frozen zone in response to freeze-induced gradients in soil–water potential. Ice lenses may cause formation of a laminar structure in a silt loam but a distinctly reticular or polygonal structure in a clay loam soil (Ceratzki, 1956; Kay et al., 1985). The most important effect on aggregation is of the cyclic freezing and thawing (Pawluk, 1988) (Fig. 4.14). Fabric changes occur in plastic clays by freezing and thawing (Czurda et al., 1995). Despite numerous observations on the positive effects, Slater and Hopp (1949) and others have reported negative effects of freezing on structural attributes. An important factor determining the effect is the degree of soil wetness at the time of freezing (Logsdail and Webber, 1959), and number of freeze-thaw cycles. There appears to be a maximum in the positive effects of freeze-thaw cycles.

4.6.3 Biotic Factors

Soil biota plays an important role in aggregation and soil structure development (Fig. 4.15). In addition to the significant effects of plant roots, soil fauna drastically alters soil structure (Lal and Akinremi, 1983; Lal et al., 1980; Lee, 1985; Lal, 1991; Lavelle and Pashanasi, 1989; Lee and Foster, 1991; Schrader et al., 1995). The role of root hairs, fungal hyphae, and other mineral by-products of soil biota have been discussed in the previous section. Enhancing microbial activity in soil is an important strategy of improving soil structure. Products of microbial decomposition facilitate clay–organic complex formation.

4.6.4 Soil Tillage

Shearing, compressive, and tensile stresses during seedbed preparation drastically alter porosity and pore size distribution due to change in soil

FIGURE 4.15 Termite activity is more predominant in tropical than temperate region soils, and their activity creates aggregates and channels.

FIGURE 4.16 A wheel rut causes soil compaction.

volume (Spoor, 1988). Wheel traffic has a significant effect on soil structure (Fig. 4.16; Hakansson et al., 1988). Conservation tillage (Fig. 4.17), and use of crop residue mulch (Fig. 4.18), is an important strategy to maintain a favorable structure of some soils.

FIGURE 4.17 No-till farming with residue mulch enhances activity of soil microfauna (e.g., earthworms, termites) and improves soil structure.

FIGURE 4.18 Crop residue mulch, in situ or brought in, also improves soils structure by eliminating the raindrop impact and enhancing activity of soil macrofauna. (The pen points to earthworm casts beneath the mulch layer.)

4.6.5 Soil Amendments

Addition of organic matter (e.g., compost, manure, sludge) has beneficial effects on soil structure through formation of clay–organic complexes (Greenland, 1965a;b; Glass, 1995). Similarly, application of gypsum

($CaSO_4$) leads to improved aggregation of dispersed alkaline soils ($2Na +1-clay + CaSO_4 \rightarrow Ca-clay + Na_2SO_4$) (Gupta and Abrol, 1990). There are also synthetic organic polymers or soil conditioners or soil stabilizers (Levy, 1996). In fact, interest in organic polymers as soil conditioners dates back to the 1950s when the Monsanto company developed Krilium, a trade name comprising several polymers such as vinyl acetate, malic acid, and hydrolyzed polyacrylonitrile (Chepil, 1954; De Boodt and De Leenheer, 1958; Emerson et al., 1978). Polymers are small repeating units or monomers coupled together to form extended chains. Their chain length in solution ranges between a few thousand and 3×10^5 µm with an average diameter of 0.5–1.0 µm. Commonly used polymers are polysaccharides (PSD) and polyacrylamide (PAM). Clay–polymer complexes lead to formation of stable aggregates (De Boodt, 1972; Gabriels et al., 1973; SSSA, 1975). The cost-effectiveness and the persistence of the effect need to be carefully assessed under soil/site specific situations.

4.7 ASSESSMENT OF AGGREGATION AND SOIL STRUCTURE

Soil structure is a dynamic property with numerous aspects, and is difficult to characterize (Coughlan et al., 1991). Methods of aggregation assessment outlined in Fig. 4.19 show two principal techniques: field and laboratory.

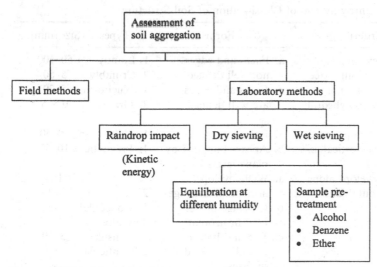

FIGURE 4.19 Methods of assessment of aggregation.

Field methods are primarily used by pedologists in routine soil surveys (Soil Survey Division Staff, 1993).

4.7.1 Pedological Methods

Soils are classified into structureless and structured soils. Structureless soils may either be single-grained such as sand or massive such as large clods without distinctive peds. No peds or units are observed in structureless soils. Structured soils may have simple or compound structure. Simple structure comprises distinct aggregates, which are an entity unto themselves without components or smaller units separated by persistent planes of weakness. Zakhrov (1931) described soil structure based upon the size, shape, and visual appearance of the surface of soil aggregates, fragments, or clods. Soil type refers to shape, size to class, and grade to durability of peds (Table 4.4). The Soil Survey Division of the Soil Conservation Service (now called Natural Resource Conservation Service) of the USDA revised the Zakhrov system. The revised version of the field/pedological method is shown in Fig. 4.20 and Table 4.4. Morphologic features of structural units are also classified based on soil fabric involving petrographic studies (Kubiena, 1938; Brewer and Sleeman, 1960; Brewer, 1964; Ringrose-Voase, 1991).

Table 4.4   Zakhrov System of Classification of Soil Structure

Type	Criteria	Form	Types	Size (mm)
1	Structure develops uniformly along three mutually perpendicular axes (polyhedral or round)	a. Faces and edges not well defined	1. Lumpy	50–100
			2. Crumbly	5–50
		b. Faces and edges are well defined	1. Nuciform	5–20
			2. Grainy	0.5–5
2	Structure develops more toward the vertical axis (prismatic)	a. Rounded apexes	1. Columnar	30–50
		b. Apexes bounded by plane facets	1. Prismatic	10–50
3	Structure develops along the horizontal axis (platy)	a. Well-developed horizontal cleavage	1. Platy	1–5
			2. Leafy	<1
		b. Cleavage planes bent horizontally	1. Concoidal	>3
			2. Flaky	1–3
		c. Top and bottom bound by round surfaces	1. Lenslike	3–10
			2. Lenticular	< 3

Type   Shape

Granular

Crumb

Platy

Prismatic

Columnar

Booky

Single grain

Massive

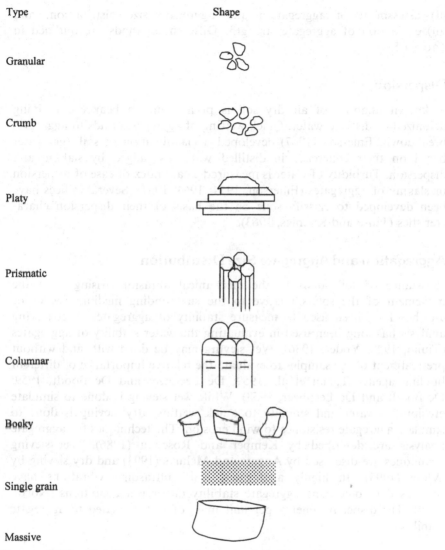

FIGURE 4.20 Classification of soil structure according to shape.

4.7.2 Laboratory Methods

Several books have been written in describing laboratory techniques of soil structure evaluation (Lorinov, 1982; Revut and Rode, 1981; Burke et al., 1986; Hartge and Stewart, 1995). Laboratory methods of aggregate analyses can be broadly grouped into three categories: (i) ease of dispersion,

(ii) assessment of aggregation and aggregate size distribution, and (iii) evaluation of aggregate strength. Different methods are outlined in Table 4.5.

Dispersion

A known quantity of air dry soil is poured into a beaker containing deionized or distilled water. Quick wetting of aggregates leads to aggregate breakdown. Emerson (1967) developed a classification of soil aggregates based on their coherence in distilled water as judged by slaking and dispersion. Turbidity of water is measured as an index of ease of dispersion or slaking of aggregates (Emerson, 1954; 1964; 1967). Several indices have been developed to classify soils on the basis of their dispersion characteristics (Janse and Koenigs, 1963).

Aggregation and Aggregate Size Distribution

Resistance of soil solids to the mechanical abrasion arising from the movement of the solids relative to the surrounding medium (water or air) has long been used to measure stability of aggregates. Wet sieving analysis has long been used in evaluating the water stability of aggregates (Tiulin, 1928; Yoder, 1936). Wet sieving may be done with and without pretreatment of the samples to evaluate the relative importance of different binding agents (Henin et al., 1959; De Leenheer and De Boodt, 1959; De Boodt and De Leenheer, 1958). While wet sieving is done to simulate erosion by water and stability to quick wetting, dry sieving is done to simulate aggregate resistance to wind erosion. The techniques for aggregate analysis are described by Kemper and Rosenau (1986). Wet sieving techniques are discussed by Angers and Mehuys (1993) and dry sieving by White (1993). In highly aggregated soils, ultrasonic vibrations have been used to determine aggregate stability under wet conditions (North, 1979). The dispersive energy per unit mass of soil is related to aggregate stability.

Aggregate Strength

Aggregate strength may be determined by the raindrop technique (McCalla, 1944; Bruce-Okine and Lal, 1975) by evaluating the kinetic energy required to disrupt an aggregate. Dry soil aggregate strength may be evaluated by a procedure that evaluates crushing strength (Skidmore and Powers, 1982; Perfect and Kay, 1994). A soil energy-crushing meter has been developed (Boyd et al., 1983).

Table 4.5 Shapes and Size Classes of Soil Structure

	Shape of structure					
	Units are flat and platelike. They are generally oriented horizontally and faces are mostly horizontal	Units are prismlike and bounded by flat to rounded vertical faces. Units are distinctly longer vertically than horizontally; vertices angular		Units are blocklike or polyhedral with flat or slightly rounded surfaces that are casts of the faces of surrounding peds; nearly equidimensional		Units are approximately spherical or polyhedral and are bounded by curved or very irregular faces that are not casts of adjoining peds
		Tops of units are indistinct and normally flat	Tops of units are very distinct and normally rounded	Faces intersect at relatively sharp angles	Mixture of rounded and plane faces and the vertices are mostly rounded	
Size class	Platy (mm)	Prismatic (mm)	Columnar (mm)	Angular blocky (mm)	Subangular blocky (mm)	Granular (mm)
Very fine or very thin	<1	<10	<10	<5	<5	<1
Fine or thin	1–2	10–20	10–20	5–10	5–10	1–2
Medium	2–5	20–50	20–50	10–20	10–20	2–5
Coarse or thick	5–10	50–100	50–100	20–50	20–50	5–10
Very coarse or very thick[a]	>10	>100	>100	>50	>50	>10

[a] In describing plates, *thin* is used instead of *fine* and *thick* is used instead of *coarse*.
Source: Soil Survey Staff, 1951; 1993.

4.7.3 Expression of Results of Aggregate Analysis

Numerous methods are used to express the results of structural analysis (Table 4.6), and there are different methods to express results of aggregate analysis (Table 4.7). Commonly used methods to express results include percent water stable aggregation (%WSA) and mean weight diameter (MWD) of aggregates (Van Bavel, 1949; Youker and McGuinness, 1956). It is important that the MWD is corrected for the primary particles of the same size to avoid over-estimation of the MWD. The correction in MWD for sand is done as per Eq. (4.10).

$$\% \text{ Stable aggregates on each sieve}$$
$$= \frac{(\text{weight retained}) - (\text{weight of the sand fraction})}{(\text{total sample weight}) - (\text{weight of sand})} \times 100 \qquad (4.10)$$

Results of aggregate analysis are also expressed as geometric mean diameter or GMD (Table 4.7). In general, GMD is lower numerically than MWD.

4.7.4 Indices of Soil Structure

There are also several other indices of soil structure based on soil properties other than aggregation. Important among these are those based on porosity, soil strength, plant available water capacity, and water transmission properties. These indices are outlined in Table 4.8 but discussed in detail in appropriate chapters.

Rather than doing the direct evaluation for total aggregation, and their size distribution and strength, there are numerous indirect indices of soil structure assessment. These indices are based on other soil properties related to soil structure, and have been described by Bryan (1968). Some of these indices include the following:

Dispersion Ratio (Middleton, 1930)

This index is a measure of the clay fraction in dispersed rather than aggregated condition. The dispersion ratio (*DR*) index is given by Eq. (4.11).

$$DR = \frac{a}{b} \times 100 \qquad (4.11)$$

where a is percent (silt + clay) when 50 g of oven dry equivalent sample is mixed end over end without dispersion agent in one liter of distilled water

TABLE 4.6 Methods of Determining Structural Stability

Method	Formula/technique		Reference
1. Slump test	$\dfrac{Z-Y}{Z-X} \times 100$	Z = Initial volume of soil column Y = Final volume of soil column X = Absolute volume of solids	William and Cook (1961)
2. Turbidity/slaking test	Turbidity classes		Panabokke and Quirk (1957) Quirk and Panabokke (1962) Emerson (1967) Janse and Koenigs (1963) Molope et al. (1985) Pajasok and Kay (1990)
3. Thorburn subsoiling test	Amount of cold dispersed in water		Thorburn (cited by Emerson, 1967)
4. Stability against water or wind	Wet and dry sieving		Yoder (1936)
5. Aggregate strength	(i) Kinetic energy (ii) Crushing strength (iii) Rupture energy		Bruce-Okine and Lal (1975) Skidmore et al. (1982) Perfect and Kay (1994)

TABLE 4.7 Some Commonly Used Indices to Express Results of Aggregate Analysis by Wet or Dry Sieving

Mean weight diameter	$MWD \ (mm) = \sum_{i=1}^{n} x_i w_i$, where x_i is mean diameter of each size fraction (mm) and w_i is proportion of the total sample weight occurring in the corresponding size fraction, and n is the number of size fractions.
Geometric mean diameter	$GMD = \exp \left[\dfrac{\sum_{i=1}^{n} w_i \log x_i}{\sum_{i=1}^{n} w_i} \right]$, where x_i is mean diameter of each size fraction (mm), w_i is weight of aggregate in a size class with an average diameter $x_i(g)$, and $\sum_{i=1}^{n} w_i$ is the total weight of the sample.
Distribution percent by	$DPW = S_c / S_o \times 100$, where S_c is oven dry weight of soil remaining on weight.
Percent silt plus clay aggregated	$PSC = \dfrac{W_{ag} - W_p}{W_s} \times 100$, where W_{ag} weight of aggregated soil, W_p is weight of soil particles retained on 0.02 mm sieve, and W_s is weight of original oven dry soil.
Percent clay aggregated	$PC = \dfrac{W_d - W_{nd}}{W_d} \times 100$, % clay with dispersion, W_{nd} is % clay without dispersion
Summation curve	Cumulative % is plotted as a function of the aggregate size.
Log normal statistical distribution	The DPW is plotted on y-axis on a linear scale and the aggregate size on x axis in the log scale.

TABLE 4.8 Some Indices of Soil Structure Based on Properties Other Than
Aggregates

Soil property	Index of soil structure
Porosity	(i) Total porosity (f_t)
	(ii) Pore size distribution (D_{50})
	(iii) Aeration porosity (f_a)
Soil strength	(i) Penetration resistance
	(ii) Modulus of rupture
	(iii) Relative density
Water retention	(i) Plant available water capacity
	(ii) Least limiting water range
Water transmission	(i) Infiltration capacity
	(ii) Profile hydraulic conductivity
	(iii) Soil drainage
Aeration	(i) Oxygen diffusion rate
	(ii) Diffusion coefficient

contained in a cylinder 5 cm diameter and 40 cm deep, and b is actual
silt + clay content determined by routine mechanical analysis with dispersion agent.

Aggregated (Silt + Clay) (Middleton, 1930)

This index is computed from the analysis done for the dispersion ratio, and
is the difference between actual (silt + clay) and the percent suspension
determined without dispersion.

Clay Ratio (Bouyoucos, 1935)

This refers to the ratio between sand and silt + clay. It is a measure of the
amount of binding material and has also been called "mechanical ratio"
(Boyd, 1922).

Colloid Content–Moisture Equivalent Ratio
 (Middleton, 1930)

This ratio is used as an index of soil erodibility. Soil colloid content
comprises clay plus organic matter expressed in percent, and moisture
equivalent is the soil moisture content when soil is subjected to a centrifugal
force equivalent to 1000 G. Nonerodible soils usually have a ratio > 1.5 and
erodible soils < 1.5.

Erosion Ratio (Middleton, 1930)

The erosion ratio is calculated as per Eq. (4.12).

$$\text{Erosion ratio} = \frac{\text{Dispersion ratio}}{\text{Colloid content/moisture equivalent}} \qquad (4.12)$$

For this ratio a value of 10 is thought to be a boundary between erodible and nonerodible soils.

Silica: Sesquioxide Ratio

This ratio is based on the relative proportion of cementing agents (R_2O_3) in comparison with the material to be cemented (SiO_2). This ratio is also an index of soil erodibility and may range from < 1 for nonerodible soils to as high as 9 for erodible soils.

Surface Aggregation Ratio

Anderson (1954) proposed the ratio between the total surface area of particles larger than 0.05 mm diameter and the quantity of aggregated silt + clay content.

Index of Resistance (I_r)

Chorley (1959) proposed an index of resistance against erosion by water as per Eq. (4.13).

$$I_r = \frac{\rho_b \times D_r}{w} \qquad (4.13)$$

where ρ_b is soil bulk density, D_r is the range of particle size, and w is soil moisture content.

Index of Erodibility (I_e)

Chorley combined I_r with permeability to obtain I_e [(Eq. 4.14)].

$$I_e = (I_r \times k)^{-1} \qquad (4.14)$$

where k is soil permeability (see Chapter 12 on soil water movement).

Index of Structural Stability (I_s)

Kay et al (1988) proposed an index of structural stability based on the rate of change in the level of stabilizing material [Eq. (4.15)].

$$C_i/C_o = (1 - e^{-k_1 T_i}) \qquad (4.15)$$

where C is the stabilizing constituent (humic fraction or organo-mineral complexes) representing original (C_o) and final (C_i) concentration, T_i is the time (yr) and k_1 is the rate constant.

Index Based on Texture and Cementing Agents

Henin et al. (1958) proposed an instability index (I_s) based on cementing agents involved in aggregation of tropical soils [Eq. (4.16)].

$$I_s = \frac{(A + LF)_{\text{max}}}{\frac{1}{3}Ag - 0.9SG} \tag{4.16}$$

where $(A + LF)_{\text{max}}$ is the maximum amount of dispersed 0–20 mm fraction obtained after three treatments of the initial soil sample: (i) without any pretreatment (air dry), (ii) following immersion in alcohol, and (iii) following immersion in benzene; and Ag refers to the >200 mm aggregates (air, alcohol, and benzene) obtained after shaking (30 manual turnings and wet sieving of the 3 pretreated samples), SG represents the contents of coarse mineral sand (>200 µm), and $(1/3Ag - 0.9SG)$ represents mean stable aggregates.

Index of Crusting

FAO (1979) proposed an index of crusting (I_c) based on textural composition and soil organic matter content [Eq. (4.17)].

$$I_c = \frac{1.5S_f + 0.75S_c}{Cl + (10 \times SOM)} \tag{4.17}$$

where S_f is % fine silt, S_c is % coarse silt, Cl is % clay, and SOM is % soil organic matter content. Obviously, I_c is inversely related to clay and soil organic matter content, and directly to fine and coarse silt content.

Critical Soil Organic Matter Content

Soil organic matter concentration plays a major role in forming and stabilizing aggregates (Dutartre et al., 1993). Pieri (1991) proposed the concept of critical level of soil organic matter concentration for structural stability of tropical soils [Eq. (4.18)].

$$S_t = \frac{(SOM)}{(Clay + silt\ content)} \tag{4.18}$$

Based on the analysis of about 500 samples from semiarid regions of West Africa, Pierie (1991) proposed the following limits of soil organic matter concentration for characterizing soil structure:

$S_t = <5\%$, loss of soil structure and high susceptibility to erosion
$S_t = 5$ to 7%, unstable structure and risk of soil degradation
$S_t = >9\%$, stable soil structure

Plant Available Water Capacity

Plant available water capacity of the soil (see Chapters 10 and 11) has been used as an index of soil structure. Thomasson (1971) related soil structure to the range of moisture content in which crop growth is optimum. Letey (1985) proposed the "non-limiting water range" or the range of soil water content in which neither O_2 nor water nor soil strength limit crop growth. The concept was further developed by Emerson et al. (1994), da Silva et al. (1994), and da Silva and Kay (1996; 1997) into "least limiting water range" as a characteristic of structural form in relation to plant growth. These methods are rarely used because of the complexity of the procedure and a wide range of parameters involved.

4.7.5 Aggregation and Structural Resiliency

Because of its importance, rather than evaluating aggregation properties per se, it may be prudent and more relevant to assess structural resilience (Kay, 1997). It refers to the ability of soil structure to recover following a major disruption in the aggregation process outlined in Eq. (4.5). The disruption may be caused by alterations in land use, cultivation, or soil management practices that change the composition of cations on the exchange complex, decrease quantity and quality of the humus fraction, and reduce effectiveness of the biotic factors. Numerous soils exhibit self-mulching properties (Fig. 4.21; Blackmore, 1981; Grant and Blackmore, 1991). In other soils, aggregation is restored only when taken out of cultivation and put under a restorative fallow (Lal, 1994). Inevitably, soils with structural resiliency are better suited for intensive management under different land uses than those that do not possess these characteristics. Structural resiliency depends on numerous factors including soil organic matter content, clay mineralogy, wettability characteristics, and biotic factors. It may be important to evaluate soils according to numerous indices outlined in Tables 4.6 and 4.7, and develop a comprehensive index of structural resiliency.

FIGURE 4.21 Surface layer of some vertisols and andisols have self-mulching characteristics with fine- to medium-crumb structure.

4.7.6 Fractal Analyses and Soil Structure

Fractals may describe spatial and temporal systems that may be generated by applying scaling theories using an iterative algorithm (Federer, 1988). These are complex systems at any given scale and therefore, useful for modelling structure in heterogeneous soil. The scaling factors can be unique in a self-similar system and different for each coordinate axis for a self-affine system (Federer, 1988). Spatial fractals are constructed by repeatedly copying a pattern on to the initiator or starting system, or algorithm, which can be accretive, reductive, or mass conserving. For different soil operations, different fractal dimensions and different algorithms are used. Pore size distribution is described by reductive algorithm, whereas, fragmentation and surface irregularity are mostly described by mass-conserving and accretive algorithms (Perfect and Kay, 1995).

The fractal techniques can be used for modelling the structure of heterogeneous soils by quantifying the changes in aggregate size, density and outlines of aggregates, ped shapes, bulk density and pore size distribution. Not all the parameters can be easily assessed, however. The fractal analysis uses the aggregate number–size distribution instead of mass–size distribution determined normally by wet sieving technique. From the known values of aggregate mass–size distribution, bulk density and shape of aggregate in each size fraction, the number–size distribution can be determined by the following equation:

$$N\left(\frac{1}{b^i}\right) = k\left(\frac{1}{b^i}\right)^{-D} ; i = 0,1,2,3\dots\infty \qquad (4.19)$$

where $N(1/b^i)$ is the number of elements of length $1/b^i$, k is the number of initiators of unit length, b is a scaling factor greater than 1, and D is the fractal dimension and can be defined as a fractional dimension (noninteger), which determines the space filling capability of generator in the limit $i \to \infty$.

4.8 IMPACT OF DECLINE IN AGGREGATION AND STRUCTURAL DEGRADATION

Reduction or reversal of the aggregation process has far-reaching local, regional, and global impacts on agriculture (Fig. 4.22). Crusting and surface seal formation (local impacts) (Passioura, 1991) are the precursors to surface compaction, low infiltration, and high soil evaporation. Soil slaking and dispersion lead to exposure of C otherwise tied or locked within the aggregate, which accentuates its microbial decomposition and oxidation. These local processes are determined by biophysical factors and processes, e.g., ion exchange, organomineral complexes, wetting–drying, and freeze–thaw cycles. Local processes of runoff and accelerated erosion are combined at regional scale. Runoff and erosional processes on a watershed scale lead to disruption in cycles of H_2O, and exacerbation of aridization and desertification processes with severe global implications. Disruptions in cycles of C and N also lead to emissions of radiatively-active gases

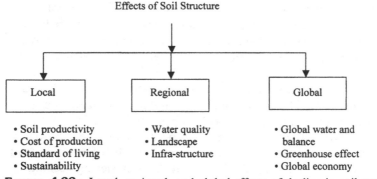

FIGURE 4.22 Local, regional, and global effects of decline in soil structure.

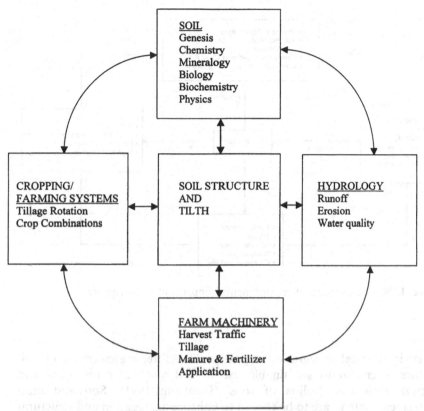

FIGURE 4.23 A multidisciplinary approach to soil structure.

(CO_2, CH_4, CO, N_2O, NO_x) from soil to the atmosphere with attendant risks of the accelerated greenhouse effect (Lal, 1995; 1999; 2001; 2003). At regional and global scales, these processes are driven by socioeconomic and political causes, and policy issues are major considerations. It is because of these interactive effects with numerous impacts that the structure and tilth constitute a central theme of multidisciplinary importance involving soil science, agronomy/plant physiology, engineering, hydrology, and climatology (Fig. 4.23).

4.9 MANAGEMENT OF SOIL STRUCTURE

There are numerous economic and environmental impacts of soil structure (Fig. 4.24), especially those that affect soil quality in relation to productivity

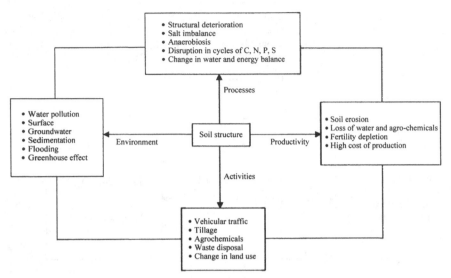

FIGURE 4.24 Economic and environmental impacts of soil structure.

and environmental moderation capacity. Therefore, management of soil structure is crucial to sustainable use of soil and water resources and minimize structural decline of soils (Emerson, 1991). Soil and crop management systems are to be chosen to enhance aggregation and structural stability. For additional readings on this topic, readers are referred to reviews by Baver et al. (1972), Hamblin (1985; 1991), Kay (1990), and Carter and Stewart (1995).

4.9.1 Cropping and Farming Systems

Root systems and canopy cover have an important influence of soil structure. Grasses with their dense and fibrous root system and legumes with their deep tap roots have a profound effect on aggregation characteristics. It is because of these and other differences in legumes and cereals that crop rotations and farming systems have a profound effect on soil structure (Kay et al., 1988). Crops affect structural properties through their impacts on root biomass, amount and rate of water extraction from different depths, total biomass produced, and C:N ratio of the biomass that affects its persistence. From a long-term study in Ohio, Lal et al. (1990) observed that relative aggregation for different rotations was 1.00:1.66:2.1 for corn–oats–meadow, continuous corn, and corn–soybean. The MWD was 1.34 mm for corn–soybean, 1.0 mm for continuous corn, and 0.7 mm for corn–oats–meadow rotation. Perennial forages, both legumes and grasses, improve soil structure (Wilson et al., 1947; Low, 1972; Lal et al., 1979; Lal, 1991). Through their beneficial effects on soil organic carbon (Wilson and Hargrove, 1986; Wilson et al., 1982) and total soil nitrogen contents (Blevins et al., 1990; Camberdella and Corak, 1992). In Ohio, Lal et al., (1997) observed that growing tall fescue and smooth bromegrass for five years increased soil organic carbon content by 18.5%, and total soil nitrogen by 12.5% for 0 to 3 cm depth. Management of the crops and cropping system, use of pasture within a crop rotation, soil surface, and fertility management practices are all important to structural management.

4.9.2 Tillage

Structural effects of tillage depend on the type, frequency, and timing of tillage operation. The antecedent soil moisture content is an important parameter that affects structural properties, because it influences dispersibility of clay. Conservation tillage and mulch farming techniques are beneficial to aggregation and soil structure formation (Lal, 1989; Carter, 1994). Lal et al., (1994) reported that in Ohio, tillage effects on total aggregation and MWD were in the order of no tillage > chisel plowing > moldboard plowing.

4.9.3 Water Management

Drainage of excessively wet soil and irrigation of dry soil may alter aggregation (Collis-George, 1991). The nature and magnitude of effect may depend on soil and environments. In Ohio, Lal and Fausey (1993) observed that the MWD was 2.94 mm for undrained compared with 2.49 mm for drained soil because of decrease in soil organic matter content with drainage. Supplemental irrigation may improve aggregation with good quality water and decrease aggregation with poor quality water containing high proportions of sodium.

4.9.4 Soil Fertility Management and Soil Amendments

Agricultural practices that enhance biomass production have also favorable effects on aggregation and soil structural development. Use of organic manures, compost, and mulches improve aggregation more than chemical fertilizers (Tisdall et al., 1978). Decrease in soil pH due to chemical fertilizers may adversely affect aggregation, especially in soils of low activity clays. Otherwise, use of chemical fertilizers has beneficial effects on aggregation (Emmond, 1971; Hamblin, 1985).

4.9.5 Soil Conditioners

Soil conditioners are synthetic polymers which can be adsorbed by the surface of the clay particles, and alter its relation to water and ions in the solution (see Sec. 4.6.5). One polymer molecule can also link several clay particles through formation of interparticle bonds that facilitate flocculation of a dispersed system or stabilize an existing unstable arrangement of particles. The adsorption of a polymer on clay particles leads to entropy and enthalpy changes due to the change in the state of the molecule in the solution phase to its state in the adsorbed phase, and due to interaction energy involved in the change in the association of soil particle with the polymer molecule. These adsorptive mechanisms have been described by Greenland (1965a;b) and Mortland (1970). The adsorption process is significant when a large net release of enthalpy (ΔH) occurs or the interaction is exothermic. There are two levels of interaction energy that determine the adsorption of polymers on clay surfaces: (i) the net interaction energy E and, (ii) the critical energy E_c. The adsorption process is complete when $E > E_c$. In addition to enthalpy changes, entropy changes may also occur. Restriction of the polymer by interface causes some loss in entropy (ΔS). Gain in entropy may be due to: (i) liberation of water from the clay surface, (ii) movement of water molecules from or to the polymer, as well as from or to the surface phase, and (iii) changes in configuration of the polymer. There is a wide range of polymers that have been used as soil conditioners. Their effectiveness, however, depends on soil properties, management and climate.

PROBLEMS

1. How does soil structure affect: (a) crop growth, (b) quality of ground water, and (c) air quality?
2. Describe the role of aggregation in soil carbon sequestration, and highlight the mechanism involved.
3. A farmer in Ohio has shifted from conventional tillage to no-till farming. By so doing, soil organic carbon content in the top 1-m depth is increasing at the rate of 0.01% per year. Assuming mean soil bulk density of $1.5\,Mg/m^3$, calculate the rate of carbon sequestration in this 1000 ha farm.

4. Dry and wet-sieving analyses were done on 100 g weight of two soils to get the following results:

Sieve size (mm)	Dry sieving (g)		Wet sieving (g)	
	No-till	Plow till	No-till	Plow till
5–8	10	5	8	4
2–5	15	8	12	10
1–2	15	8	10	8
0.5–1	12	10	10	7
0.25–0.5	8	7	6	5
0.1–0.25	8	6	6	4

Calculate and plot summation curve, percent aggregation > 1 mm, MWD, and GMD. Which soil is prone to wind or water erosion, and why?

Sieve No.	8	10	14	20	28	35	48	65	100	150	200
Opening in mm	2.36	1.65	1.17	0.83	0.59	0.41	0.30	0.21	0.15	0.10	0.075
Soil weight (g) A	28.5	25.0	14.8	12.1	6.3	2.0	3.1	2.0	2.1	1.8	2.3
B		2.3	1.8	2.1	2.0	3.1	2.0	6.3	12.1	14.8	25.0	28.5

5. Calculate "mean weight diameter" and "geometric mean diameter" from the following data. The equivalent oven dry weight $= 100$ g.
6. Plot the above data as a summation curve.
7. A soil has 10% of fine silt, 15% of coarse silt, 40% of clay, 35% sand, and 2.5% soil organic matter content. Compute I_c, S_t, and clay ratio.
8. What is the importance of soil structure to plant growth?
9. Jack (1963) stated that soil structure is as important as photosynthesis. List reasons in justification of this statement.
10. In what ways may the projected global climate change affect soil structure in (a) temperate and (b) tropical climates?

APPENDIX 4.1   SPECIFICATION FOR SIEVE SERIES
(SEE ALSO APPENDIX 3.1)

Size of sieve, μ	Sieve number, mesh per inch	Sieve opening, mm	Nominal wire diameter, mm
4000	5	4.000	1.370
2000	10	2.000	0.900
1190	16	1.190	0.650
1000	18	1.000	0.525
840	20	0.840	0.510
500	35	0.500	0.315
250	60	0.250	0.180
210	70	0.210	0.152
177	80	0.177	0.131
149	100	0.149	0.110
74	200	0.074	0.053
53	270	0.053	0.037
37	400	0.037	0.025

REFERENCES

Anderson, H.W. 1954. Suspended sediment discharge as related to stream flow topography, soil and land use. Trans. Am. Geophys. Union 35: 268–281.

Angers, D.A. and G.R. Mehuys. 1993. Aggregate stability to water. In M.R. Carter (ed) "Soil Sampling and Methods of Analysis", Lewis Publishers, Boca Raton, FL: 651–657.

Angers, D.A. and C. Chenu. 1997. Dynamics of soil aggregation and C sequestration. In R. Lal, J.M. Kimble, R. Follett and B.A. Stewart (eds) "Soils and the Carbon Cycle", CRC, Boca Raton, FL (In press).

Aylmore, L.A.G. and J.P. Quirk. 1971. Domains and quasi-crystalline regions in clay systems. Soil Sci. Soc. Amer. Proc. 35: 652–654.

Baver, L.D., W.H. Gardner, and W.R. Gardner. 1972. Soil Physics. Fourth edition. John Wiley & Sons, New York, 498 pp.

Becher, H.H. 1995. Strength distribution in soil aggregates. In: K.H. Hartge and B.A. Stewart (eds) "Soil Structure: Its Development and Functions." CRC/ Lewis Publishers, Boca Raton, FL: 53–70.

Blackmore, A.V. 1981. Self-mulching soils. In "Soils News", Aust. Soc. Soil Sci. (July, 1981).

Blevins, R.L., J.H. Herbek, and W.W. Frye. 1990. Legume cover crops as a nitrogen source for no-till corn and grain sorghum. Agron. J. 82: 769–772.

Bouyoucos, G.J. 1935. The clay ratio as a criterion of susceptibility of soils to erosion. J. Am. Soc. Agron. 27: 738–741.

Boyd, J.R. 1922. Physical properties of sub-grade materials. Am. Soc. Testing Materials Proc. 22: 337–355.

Boyd, E.L. Skidmore, and Thompson. 1983. A soil aggregate crushing energy meter. Soil Sci. Soc. Am. J. 47: 313–316.

Bradfield, R. 1936. The value and limitation of calcium in soil structure. Am. Soil Survey Assoc. Bull XVII: 31–32.

Brewer, R. 1964. Fabric and Mineral Analysis of Soils. John Wiley & Sons, New York.

Brewer, R. 1976. Fabric and mineral analysis of soils. Krieger Publishing Co., Huntington, N.Y.

Brewer, R. and J.R. Sleeman. 1960. Soil structure and fabric. J. Soil Sci. 11: 172–185.

Brewer, R. and A.V. Blackmore. 1956. The effects of entrapped air and optically oriented clay on aggregate breakdown and soil consistence. Aust. J. Appl. Sci. 7: 59–68.

Bruce-Okine, E. and R. Lal. 1975. Soil erodibility as determined by the raindrop technique. Soil Sci. 119: 149–157.

Bryan, R.K. 1968. The development, use and efficiency of indices of soil erodibility. Geoderma 2: 5–26.

Bullock, P., N. Fedoroff, A. Jongerius, G. Stoops, T. Tursina and U. Babel. 1985. Handbook for soil thin section description. Waine Research Publications, Albrighton, Wolverhampton, U.K.

Burke, W., D. Gabriels and J. Buma. 1986. Soil Structure Assessment. A.A. Balkema, Rotterdam, Holland, 92 pp.

Buyanovsky, G.A., M. Aslam, and G.H. Wagoner. 1994. Carbon turnover in soil physical fractions. Soil Sci. Soc. Am. J. 58: 1167–1173.

Cambardella, C.A. and S.J. Corak. 1992. Seasonal dynamics of soil organic N and NO_3–N with and without a cover crop. Agronomy Abstract, ASA, Madison, WI, p 251.

Carter, M.R. and B.A. Stewart. 1996. Structure and Organic Matter Storage in Agricultural Soils. Advances in Soil Science, CRC/Lewis Publishers, Boca Raton, FL, 477 pp.

Ceratzki, W. 1956. Zur Wirkung des Frostes auf die Struktur des bodens. Z. Pflanzenernähr, Düngung Bodenk. 72: 15–32.

Chaney, K. and R.S. Swift. 1984. The influence of organic matter on aggregate stability in some British soils. J. Soil Sci. 35: 223–230.

Chepil, W.S. 1954. The effect of synthetic conditioners on some phases of soil structure and erodibility by wind. Soil Sci. Soc. Am Proc. 18: 386–390.

Chorley, R.J. 1959. The geomorphic significance of some Oxford soils. Am. J. Sci. 257: 503–515.

Collis-George, N. 1991. Drainage and soil structure: a review. Aust. J. Soil Res. 29: 923–933.

Collis-George, N. and R. Lal. 1970. Infiltration into columns of swelling soil as studied by high speed photography. Aust. J. Soil Res. 9: 107–116.

Coughlan, K.J., D. McGarry, R.J. Loch, B. Bridge, and G.D. Smith. 1991. Measurement of soil structure: Some practical initiatives. Aust. J. Soil Res. 29: 869–889.

Czurda, K.A., S. Ludwig, and R. Schababerle. 1995. Fabric changes in plastic clays by freezing and thawing. In K.H. Hartge and B.A. Stewart (eds) "Soil Structure:

Its Development and Functions." CRC/Lewis Publishers, Boca Raton, FL: 71–91.

da Silva, A.P., B.D. Kay, and E. Perfect. 1994. Characterization of the least limiting water range of soils. Soil Sci. Soc. Amer. J. 58: 1775–1781.

da Silva, A.P. and B.D. Kay. 1996. Estimating the least limiting water range of soils from properties and management. Soil Sci. Soc. Am. J. 61: 877–883.

da Silva, A.P. and B.D. Kay. 1997. Effects of soil, water content variation on the least limiting water range. Soil Sci. Soc. Amer. J. 61: 884–888.

de Boodt, M. (ed) 1972. Fundamentals of soil conditioning. Meded., Fak. Landbouwwet. Rijksuniversiteit, Gent.

De Boodt, M. and L. De Leenheer. 1958. Proposition pour l' evaluation de la stabilite des aggregates sur le terrain. Proc. International Symp. Soil Structure, Ghent, Belgium: 234–241.

De Leenheer, L. and M. De Boodt. 1959. Determination of aggregate stability by the change in mean weight diameter. Meded. Landb. Gent. 24: 290–351.

Douglas, J.T. and M.J. Goss. 1982. Stability and organic matter content of surface soil aggregates under different methods of cultivation and in grasslands. Soil & Till. Res. 2: 155–172.

Drees, R. 1992. Optical microscopy and micromorphological terms for thin section descriptions and interpretationsentation. Soil Sci. Soc. Am., Soil Micromorphology Workshop, Oct. 31, 1992, Madison, WI: 6.62.

Dumont, I. (1909). Compt. Rend. Acad. Sci. France 149:1087–1089.

Dutartre, P., F. Bartoli, F. Andreaux, J.M. Portal, and A. Ange. 1993. Influence of content and nature of organic matter on the structure of some sandy soils of West Africa. Geoderma 56: 459–478.

Edwards, A.P. and J.M. Bremner. 1967. Microaggregates in soils. J. Soil Sci. 18: 64–73.

Ehlers, W. 1975. Observations on earthworm channels and infiltration on tilled and untilled loesssoil. Soil Sci. 119: 242–249.

Elliott, E.T. 1986. Aggregate structure and carbon, nitrogen and phosphorus in native and cultivated soils. Soil Sci. Soc. Am. J. 50: 627–633.

Emerson, E.E., R.C. Foster, J.M. Tisdall, and D. Weissmann. 1994. Carbon content and bulk density of an irrigated Natrixeralf in relation to tree growth and orchard management. Aust. J. Soil Res. 32: 939–951.

Emerson, W.W. 1954. The determination of the stability of soil crumbs. J. Soil Sci. 5: 235–250.

Emerson, W.W. 1959. The structure of soil crumbs. J. Soil Sci. 10: 235–244.

Emerson, W.W. 1964. The slaking of soil crumbs as influenced by clay mineral composition. Aust. J. Soil Res. 2: 211–217.

Emerson, W.W. 1967. A classification of soil aggregates based on their coherence in water. Aust. J. Soil Res. 5: 47–57.

Emerson, W.W. 1991. Structural decline of soils, assessment and prevention. Aust. J. Soil Res. 29: 905–921.

Emerson, W.W. and G.M.F. Grundy. 1954. The effect of rate of wetting on water uptake and cohesion of soil crumbs. J. Agr. Sci. 44: 249–253.

Emerson, W.W. and M.G. Dettman. 1960. The effect of pH on the wet strength of the soil crumb. J. Soil Sci. 11: 149–158.

Emerson, W.W., R.D. Bond, and A.R. Dexter. 1978. Modification of soil structure. J. Wiley & Sons, Chichester, U.K.

Emerson, W.W. and D.J. Greenland. 1990. Soil aggregates: formation and stability. In M.F. De Boodt, M.H.B. Hayes and A. Herbillon (eds) "Soil Colloid and Their Associations in Aggregates", Series B, Physics Vol. 25, Plenum Press, New York: 485–511.

Federer, J. 1988. Fractals. Plenum Press, New York.

Fitzpatrick, E.A. 1993. Soil microscopy and micromorphology. Wiley, New York.

Fitzpatrick, E.A., L.A. Mackie, and C.E. Mullins. 1985. The use of plaster of paris in the study of soil structure. Soil Use and Management 1: 70–72.

Gabriels, D.M., W.C. Moldenhauer, and D. Kirkham. 1973. Infiltration, hydraulic conductivity and resistance to water drop impact of clod beds as affected by chemical treatment. Soil Sci. Soc. Am. Proc. 37: 634–637.

Glass, D.J. 1995. Biotic effects of soil microbial amendments. In: J.E. Rechcigl (ed) "Soil Amendments: Impacts on Biotic Systems," Lewis Publishers, Boca Raton, FL: 251–302.

Golchin, A., J.M. Oades, J.O. Skjemstad, and P. Clarke. 1994. Soil structure and C cycling. Aust. J. Soil Res. 32: 1043–1068.

Golchin, A., J.A. Baldock, and J.M. Oades. 1997. A model linking organic matter decomposition, chemistry and aggregate dynamics. In R. Lal, J.M. Kimble, R. Follett and B.A. Stewart (eds) "Soils and the Carbon Cycle", CRC, Boca Raton, FL (In press).

Grant, C.D. and A.R. Dexter. 1989. Generation of microcracks in moulded soils by rapid wetting. Aust. J. Soil Res. 27: 169–182.

Grant, C.D. and A.V. Blackmore. 1991. Self-mulching behavior in clay soils: Its definition and measurement. Aust. J. Soil Res. 29: 155–173.

Greenland, D.J. 1965a. Interaction between clays and organic compounds in soils. Part 1. Mechanisms of interaction between clays and defined organic compounds. Soils and Fertilizers 28: 415–425.

Greenland, D.J. 1965b. Interaction between clays and organic compounds in soils. Part 2. Adsorption of soil organic compounds and its effects on soil properties. Soils and Fertilizers 28: 521–532.

Gupta, R.K. and I.P. Abrol. 1990. Salt-affected soils: their reclamation and management for crop production. In: R. Lal and B.A. Stewart (eds) "Soil Degradation." Adv. Soil Sci. 11: 223–287.

Hakansson, I., W.B. Voorhees, and H. Riley. 1988. Vehicle and wheel factors influencing soil compaction and crop response in different traffic regimes. Soil & Tillage Res. 11: 239–282.

Hamblin, A. 1985. The influence of soil structure on water movement, crop root growth, and water uptake. Adv. Agron. 38: 95–158.

Hamblin, A. 1991. Sustainable agricultural systems: what are the appropriate measures for soil structure. Aust. J. Soil Res. 29: 709–715.

Hargrove, W.L. 1986. Winter legumes as a nitrogen source for no-till grain sorghum. Agron. J. 78: 70–74.

Harris, R.F.G., G. Chesters, and O.N. Allen. 1966. Dynamics of soil aggregation. Adv. Agron. 18: 107–169.

Hartge, K.H. and R. Horn. 1984. Untersuchungen Zur Gültigkeit des Hookeschen Gesetzes bei der Setzung von Boden bei Wiederholter Belastung. Z. Acker-Pflanzenbau 153: 200–207.

Hartge, K.H. and B.A. Stewart. 1995. Soil Structure: Its Development and Function. Advances in Soil Science, CRC/Lewis Publishers, Boca Raton, FL, 424 pp.

Haynes, R.J. and M.H. Beare. 1996. Aggregation and organic matter storage in meso-thermal humid soils. In: M.R. Carter and B.A. Stewart (eds) "Structure and Organic Matter Storage in Agricultural Soils." CRC/Lewis Publishers, Boca Raton, FL.

Hénin, S. 1938. Etude physico-chimique de la stabilite structurale des terres. Monograph National Center of Agronomic Research. Paris, France: 52–54.

Hénin, S., G. Monnier, and A. Combeau. 1958. Methode pour l'etude de la stabilite structural des sols. Ann. Agron. 9: 73–92.

Horn, R, 1990. Aggregates characterization as compared to soil bulk properties. Soil & Tillage Res. 17: 265–289.

Horn, R. and A.R. Dexter. 1989. Dynamics of soil aggregation in a homogenized desert loess. Soil & Till. Res. 13: 254–266.

Horn, R., T. Baumgartl, R. Kayser, and S. Baasch. 1995. Effect of aggregate strength and stress distribution in structured soils. In: K.H. Hartge and B.A. Stewart (eds) "Soil Structure: Its Development and Functions." CRC/Lewis Publishers, Boca Raton, FL: 31–52.

Jacks, G.V. 1963. The biological nature of soil productivity. Soils & Fert. 26: 147–150.

Janse, A.R.P. and F.F.R. Koenigs. 1963. Structural changes of soils on wetting. Boor en Spade 13: 168–177.

Kay, B.D. 1990. Rates of change of soil structure under different cropping systems. Adv. Soil Sci. 12: 1–52.

Kay, B.D. 1997. Soil structure and organic carbon: a review. In R. Lal, J.M. Kimble, R. Follett and B.A. Stewart (eds) "Soils and the Carbon Cycle", CRC, Boca Raton, FL (In press).

Kay, B.D., C.D. Grant, and P.H. Groenevelt. 1985. Significance of ground freezing on soil bulk density under zero tillage. Soil Sci. Soc. Am. J. 49: 973–978.

Kay, B.D., D.A. Angers, P.H. Groenovelt, and J.A. Baldock. 1988. Quantifying the influence of cropping history on soil structure. Can. J. Soil Sci. 68: 359–368.

Kay, B.D. and E. Perfect. 1988. State of the art: heat and mass transfer in freezing soils. Proc. Int'l Symp. on Ground Freezing, Nottingham, U.K. Vol. 1: 3–22.

Kay, B.D., D.A. Angers, P.H. Groenevelt, and J.A. Baldock. 1988. Quantifying the influence of cropping history on soil structure. Can. J. Soil Sci. 68: 359–368.

Kember, W.D. and R.C. Rosenau. 1986. Aggregate stability and size distribution. In A. Klute (ed) "Methods of Soil Analysis", Part 1, American Society of Agronomy Monograph 9, Madison, WI: 425–441.

Kubiena, W.L. 1938. Micropedology. Collegiate Press, Ames, Iowa.

Lal, R. 1991. Soil structure and sustainability. J. Sustainable Agric. 1: 67–92.

Lal, R. 1991. Soil conservation and biodiversity. In D.L. Hawksworth (ed) "The Biodiversity of Microorganisms and Invertebrates: Its Role in Sustainable Agriculture", CAB International, Wallingford, U.K.: 89–104.

Lal, R. 1994. Sustainable land use systems and soil resilience. In D.J. Greenland and R. Lal (eds) "Soil Resilience and Sustainable Land Use", CAB International, Wallingford, U.K.: 41–67.

Lal, R. 1995. Global soil erosion by water and carbon dynamics. In: R. Lal, J.M. Kimble, E. Levine and B.A. Stewart (eds) "Soils and Global Change." CRC/Lewis Publishers, Boca, Raton, FL: 131–141.

Lal, R. 1999. Soil management and restoration for carbon sequestration to mitigate the accelerated greenhouse effect. Prog. Env. Sci. 1: 307–326.

Lal, R. 2000. World cropland soils as source or sink for atmospheric carbon. Adv. Agron. 71: 145–191.

Lal, R. 2003. Soil erosion and the global carbon budget. Env. Intl.: 1–14.

Lal, R., G.F. Wilson and B.W. Okigbo. 1979. Changes in properties of an Alfisol by various cover crops. Soil Sci. 127: 377–382.

Lal, R., D. de Vleeschauwer, and R.M. Nganje. 1980. Changes in properties of a newly cleared Alfisol as affected by mulching. Soil Sci. Soc. Am. J. 66: 827–832.

Lal, R. and O.O. Akinremi. 1983. Physical properties of earthworm casts and surface soil as influenced by management. Soil Sci. 135: 116–122.

Lal, R. and N.R. Fausey. 1993. Drainage and tillage effects on a Crosby-Kokomo soil association in Ohio. IV. Soil Physical Properties. Soil Tech. 6: 123–135.

Lal, R., A.A. Mahboubi, and N.R. Fausey. 1994. Long-term tillage and rotation effects on properties of a central Ohio soil. Soil Sci. Soc. Am. J. 58: 517–522.

Lal, R., P. Henderlong, and M. Flowers. 1997. Forages and row cropping effects on soil organic carbon and nitrogen contents. In R. Lal, J. Kimble, R. Follett and B.A. Stewart (eds) "Management of C Sequestration In Soil", CRC, Boca Raton, FL (In press).

Lambe, T.W. 1960. The structure of compacted clays. Trans. Am. Soc. Civil Engrs. Reprint Paper 3041, 125 pp.

Larionov, A.K. 1982. Methods of Studying Soil Structure. USDA/NSF, Amerind Publishing Co., New Delhi, 193 pp.

Lavelle, P. and B. Pashanasi. 1989. Soil macrofauna and land management in Peruvian Amazonia. Pedobiologia 33: 283–291.

Lee, K.E. 1985. Earthworms: Their ecology and relationships with soils and land use. Academia Press, Sydney, Australia.

Lee, K.E. and R.C. Foster. 1991. Soil fauna and soil structure. Aust. J. Soil Res. 29: 745–775.

Letey, J. 1985. Relationship between soil physical properties and crop production. Adv. Soil Sci. 1: 277–294.

Levy, G.L. 1996. Soil stabilizers. In: M. Agassi (ed) "Soil Erosion, Conservation and Rehabilitation." M. Dekker, New York: 267–299.

Logsdale, D.E. and L.R. Webber. 1959. Effect of frost action on structure of Haldimand clay. Can. J. Soil Sci. 39: 103–106.

Low, A.J. 1972. Improvements in structural state of soils under leys. J. Soil Sci. 6: 179–199.

Lynch, J.M. and E. Bragg. 1985. Microorganisms and soil aggregate stability. Adv. Soil Sci. 2: 133–171.

McCalla, T.M. 1944. Water drop method of determining stability of soil structure. Soil Sci. 58: 117–121.

Michaels, A.S. 1959. Physico-chemical properties of soils: soil water systems. Proc. Am. Soc. Civil Engrs. J. Soil Mech. Found. Div. 85: 91–102.

Middleton, H.E. 1930. Properties of soils which influence soil erosion. USDA Tech. Bull. 178, Washington, D.C., 16 pp.

Mitchell, J.K. 1956. The fabric of natural clays and its relation to engineering properties. Proc. Highway Res. Board 35: 693–713.

Molope, M.B., E.R. Page, and I.C. Grieve. 1985. A comparison of soil aggregate stability tests using soils with contrasting cultivation histories. Comm. Soil Sci. Plant Anal. 16: 315–322.

Mortland, M.M. 1970. Clay-organic complexes and interactions. Adv. Agron. 22: 75–117.

Murray, R.S. and J.P. Quirk. 1990. Interparticle forces in relation to the stability of soil aggregates. In M.F. De Boodt, M.H.B. Hayes and A. Herbillon (eds) "Soil Colloid and Their Associations in Aggregates", Series B, Physics Vol. 25, Plenum Press, New York: 439–461.

Oades, J.M. 1990. Associations of colloids in soil aggregates. In M.F. De Boodt, M.H.B. Hayes and A. Herbillon (eds) "Soil Colloid and Their Associations in Aggregates", Series B, Physics Vol. 25, Plenum Press, New York: 463–483.

Oades, J.M. and A.G. Waters. 1991. Aggregate hierarchy in soils. Aust. J. Soil Res. 29: 815–828.

Panabokke, C.R. and J.P. Quirk. 1957. Effects of initial water content on stability of soil aggregates in water. Soil Sci. 83: 185–195.

Passioura, J.B. 1991. Soil structure and plant growth. Aust. J. Soil Res. 29: 717–727.

Pawluk, S. 1988. Freeze-thaw effect on granular structure reorganization for soil materials of varying texture and moisture content. Can. J. Soil Sci. 68: 485–494.

Perfect, E. and B.D. Kay. 1994. Statistical characterization of dry aggregate strength using rupture energy. Soil Sci. Soc. Am. J. 58: 1804–1809.

Perfect, E. and B.D. Kay. 1995. Application of fractals in soil and tillage research: a review. Soil & Till Res. 36: 1–20.

Peterson, J.B. 1947. Calcium linkage, a mechanism in soil granulation. Soil Sci. Soc. Am. Proc. 12: 29–34.

Pieri, C. 1991. Fertility of soils: A future for farming in the West African savannah. Springer-Verlag, Berlin.

Pojasok, T. and B.D. Kay. 1990. Assessment of a combination of wet sieving and turbidimetry to characterize the structural stability of moist aggregates. Can. J. Soil Sci. 70: 30–42. Quirk, J.P. and C.R. Panabokke. 1962. Incipient failure of soil aggregates. J. Soil Sci. 13: 60–70.

Revut, I.B. and A.A. Rode. 1981. Experimental Methods of Studying Soil Structure. USDA/NSF, Amerind Publishing Co., New Delhi, 530 pp.

Ringrosa-Voase, A.J. 1991. Micromorphology of soil structure: description, quantification, application. Aust. J. Soil Res. 29: 777–813.

Robert, M. and C. Chenu. 1992. Interactions between soil minerals and microorganisms. Soil Biochem. 7: 307–404.

Robinson, D.O. and J.B. Page. 1950. Soil aggregate stability. Soil Sci. Soc. Am. Proc. 15: 25–29.

Rosenquist, I. Th. 1959. Physico-chemical properties of soils: soil water systems. Proc. Am. Soc. Civil Engs., J. Soil Mech. Found. Div. 85: 31–53.

Russell, E.W. 1934. The interaction of clay with water and organic liquid as measured by specific volume changes and its relation to the phenomenon of crumb formation in soils. Phil. Trans. Roy. Soc. Ser. A 233: 361–390.

Russell, E.W. 1971. Soil structure: its maintenance and improvement. J. Sol Sci. 22: 137–151.

Schloessing, T. 1874. Ann. Chim. Phys. [2] 15:514–546.

Schrader, S., M. Joschko, H. Kula, and O. Larink. 1995. Earthworm effects on soil fabric with emphasis on soil stability and soil water movement. In: K.H. Hartge and B.A. Stewart (eds) "Soil Structure: Its Development and Functions." CRC/Lewis Publishers, Boca Raton, FL: 109–133.

Singh, P., R.S. Kanwar, and M.L. Thompson. 1991. Measurement and characterization of macropores by using AUTO-CAD and automatic image analysis. J. Env. Quality 20: 289–294.

Skidmore, E.L. and D.H. Powers. 1982. Dry soil-aggregate stability: energy-based index. Soil Sci. Soc. Am. J. 46: 1274–1279.

Slater, C.S. and H. Hopp. 1949. The action of frost on the water stability of soils. J. Agr. Res. 78: 341–346.

Soil Science Society of America. 1975. Soil Conditioners, SSSA, Madison, WI.

Soil Survey Division Staff. 1951. Soil Survey Manual, USDA Handbook No. 18, Washington, D.C.

Soil Survey Division Staff. 1993. Soil Survey Manual. USDA-NRCS Handbook No. 18, Washington, D.C., 437 pp.

Spoor, G. 1988. Improving the effectiveness of tillage operations. In Proc. Soil Management 88. Darling Downs Ins. Adv. Educ., Toowoomba, Qld. Australia.

Stengel, P. 1990. Characterization of soil structure: objectives and methods. In J. Boiffin and A. Martin la Fliche (eds) "Soil Structure and its Evolution: Agricultural Consequences and its Management". Coll. INRA, France: 15–36.

Thomasson, A.J. 1978. Towards an objective classification of soil structure. J. Soil Sci. 29:38–46.

Tisdall, J.M. 1991. Fungal hyphae and structural stability of soil. Aust. J. Soil Res. 29: 729–743.

Tisdall, J.M. 1996. Formation of soil aggregates and accumulation of soil organic matter. In M.R. Carter and B.A. Stewart (eds) "Structure and Organic Matter Storage in Agricultural Soils", Advances in Soil Science, CRC/Lewis Publishers, Boca Raton, FL: 57–96.

Tisdall, J.M. and J.M. Oades. 1982. Organic matter and water-stable aggregates in soil. J. Soil Sci. 33: 141–163.

Tiulin, A.F. 1928. Questions on soil structure. II. Aggregate analysis as a method for determining soil structure. Perm. Agr. Exp. Sta. Div. Agr. Chem. Rep. 2: 77–112.

Tiulin, A.F. 1933. Considerations on the genesis of soil structure and on methods of its determination. Trans. Ist Com. Int. Soc. Soil Sci., Moscow, Vol. A: 111–132.

Trollope, D.H. and C.K. Chan. 1959. Soil structure and the Step-strain Phenomena. J. Soil Mech. Found 86: 1–39.

Van Bavel, C.H.M. 1949. Mean weight diameter of soil aggregates as a statistical index of aggregation. Soil Sci. Soc. Am. Proc. 14: 20–23.

White, E.M. 1966. Subsoil structure genesis: theoretical consideration. Soil Sci. 101: 135–141.

White, E.M. 1967. Soil age and texture in sub-soil structure genesis. Soil Sci. 103: 288–298.

White, W.M. 1993. Dry aggregate distribution. In M.R. Carter (ed) "Soil Sampling and Methods of Analysis." Lewis Publishers, Boca Raton, FL: 659–662.

Williams, W.R. 1935. Thesis of tenacity and cohesion in soil structure. Pedology 30: 755–762.

Williams, B.G., D.J. Greenland, and J.P. Quirk. 1967. The effect of poly (vinyl alcohol) on the nitrogen surface area and pore structure of soils. Aust. J. Soil Res. 5: 77–83.

Wilson, G.F., R. Lal, and B.N. Okigbo. 1982. Effects of cover crops on soil structure and on yield of subsequent arable crops under strip tillage on eroded Alfisols. Soil & Tillage Res. 2: 233–250.

Wilson, D.O. and W.L. Hargrove. 1986. Release of nitrogen from crimson clover residue under two tillage systems. Soil Sci. Soc. Am. J. 50: 1251–1254.

Yoder, R.E. 1936. A direct method of aggregate analyses and a study of the physical nature of erosion losses. J. Am. Soc. Agron. 28: 337–351.

Yong, R.M. and B.P. Warkentin. 1966. Introduction to Soil Behavior. The Macmillan Co., New York, 451 pp.

Yong, R.M. and B.P. Warkentin. 1975. Soil properties and behavior. Elsevier, Amsterdam.

Youker, R.E. and J.L. McGuiness. 1956. A short method of obtaining mean weight diameter values of aggregate analysis of soils. Soil Sci. 83: 291–294.

Zakhrov, S.A. 1927. Achievements of Russian science in morphology of soils. Russ. Pedolog. Investigations. LL Acad. Sci., USSR.

Zakhrov, S.A. 1931. Kurs Pochvovedeniya (course in soil science). Izd. ANSSSR, Moscow.

5

Porosity

5.1 GENERAL DESCRIPTION

An aggregate is analogous to a building. The functional space of a building includes rooms, interconnecting corridors, and exit and entrance doors that facilitate communication with the exterior. Stability of the exterior and interior walls is important to maintaining functions of all rooms and interconnecting corridors. Continuity of corridors is extremely important for the building to remain functional. Similar to the walls of a building, skeleton structure of microaggregates and aggregates is important to maintaining size, stability, and continuity of pores within and between aggregates. The porosity, or soil architecture, is the functional entity of soil structure. Soil, similar to a building, becomes dysfunctional as soon as it loses its pores and their continuity within the soil profile and to the atmosphere. Therefore, soil structural characterization cannot be complete without assessment of its porosity, pore size distribution, and continuity. Because aggregates are highly dynamic and transient, varying in time and space and ranging in scale from Å to a few cm, so are pores. Porosity is a complex and a moving target, that governs the essence of biological processes that supports life and biochemical and physical processes that determine environment quality. It is this complexity which leads to a wide range of terminology, e.g., porosity, pore, pore space, pore size distribution, voids, channels, biochannels and biopore or macropores, cracks, fissures, fractures, and so on. Therefore, understanding this complexity is important to understanding soil structure.

5.2 TERMINOLOGY

Porosity is a general term used to designate all voids in the soil. There are several systems to designate porosity on the basis of their origin or location within the soil body.

5.2.1 Textural and Structural Porosity

Textural porosity refers to the pores and their size distribution in relation to the particle size distribution. Importance of pores rather than of the size of particles was recognized by Green and Ampt (1911) by stating that "the relations of the soil to the movements of air and water through it...are much less obscure if we direct our attention to the number and dimensions of the spaces between the particles rather than to the sizes of the particles themselves." Soils of coarse texture and single-grain structure have textural pores in between the large particles. Textural pores are also the intra-aggregates pores (see Fig. 4.1). Therefore, the porosity defined by the spatial distribution of soil separates or primary particles is referred to as the "textural porosity."

Primary particles are bonded together to form secondary particles or aggregates, so that in well-aggregated soils the binding between primary particles within an aggregate is stronger than the binding between aggregates. Although these aggregates are transient and vary drastically in temporal and spatial scales, they maintain their integrity at any point in time. Integrity is defined by aggregate size, stability, position, and orientation with respect to one another. Just as primary particles define textural porosity, aggregates define structural porosity (Childs, 1968; Derdour et al., 1993) or inter-aggregate porosity (refer to Fig. 4.1). Structural porosity, total pore volume, and its size distribution and continuity, are extremely important in well-structured soils. Similar to aggregates, structural porosity is a dynamic entity. In addition to endogenous factors that govern aggregation and aggregate size distribution, exogenous factors that affect structural porosity include climate through its effect on wet–dry and freeze–thaw cycles, cropping systems through their effects on root system and other biotic factors, and soil management through tillage and crop residues disposal. In some soils, there are distinct groups of textural and structural pores. In other soils, such a distinction is difficult to make.

5.2.2 Matrix and Non-Matrix Pores

In soil survey terminology, pores are distinguished into three classes: matrix pores, non-matrix pores, and interstructural pores. Matrix pores are formed by the packing of primary soil particles. These are also the textural pores, which are generally small in size. The total volume of matrix pores may change with the soil wetness. Non-matrix pores are large voids created by roots, burrowing animals, action of compressed air, and other agents. The volume of non-matrix pores does not change drastically with change in soil wetness, and is not affected by soil texture. Interstructural pores are defined or delimited by structural units. These are crevices between structural units, and are generally planar.

5.3 METHODS OF EXPRESSION OF SOIL POROSITY

Soil porosity is expressed in numerous ways including total porosity (f_t), aeration porosity (f_a), air ratio (α) and, void ratio (e) (see Chapter 2). Porosity may be expressed in terms of number, size, shape, and vertical/horizontal continuity of pores.

5.3.1 Number

This visual description is particularly useful for describing the non-matrix pores formed by roots, animals, etc. The number of such pores is expressed per unit area that may be $1\,cm^2$ for very fine and fine pores, $1\,dm^2$ for medium and coarse pores, and $1\,m^2$ for very coarse pores. The classification used by the Soil Survey Division Staff (1990) to describe non-matrix pores is as follows:

Few:	< 1 per unit area
Common:	1–5 per unit area
Many:	≥ 5 per unit area

5.3.2 Pore Size Distribution

Rather than the total pore volume, it is its size and distribution that are important to retention and conduction of fluids in and through the soil. Pores in soils range widely from $0.003\,\mu m$ plate separation in clay particles to biopores, cracks, and tunnels tens of centimeters in diameter (Hamblin, 1985). In addition to structural pores of pedological origin, a wide range of pores exists of biological origin (Table 5.1). These pores are extremely important in transmission of water and gaseous exchange.

TABLE 5.1 Pore Dimensions of Biological Origin or Significance

Average pore size (μm)	Biological significance
1500–50,000	Ant nests and channels
500–11,000	Wormholes
300–10,000	Tap roots of dicotyledons
500–10,000	Nodal roots of cereals
100–1,000	Seminal roots of cereals
50–100	Lateral roots of cereals
20–50	1st- and 2nd-order laterals
5–10	Root hairs
1,000	Root plus root hair cylinder in clover
30	"Field capacity" (-10 k Pa)
0.5–2	Fungal hyphae
0.2–2	Bacteria
0.1	Permanent wilting point (-1500 k Pa)

1 kPa = 10 cm of water column at STP
Source: Adapted from Hamblin, 1985.

Non-matrix or macropores are described in terms of the specified diameter size. Five size classes commonly used in soil survey are:

1. Very fine: < 0.5 mm
2. Fine: 0.5–2 mm
3. Medium: 2–5 mm
4. Coarse: 5–10 mm
5. Very coarse: > 10 mm

Complementary to the visual classification used in soil surveys, numerous other systems have been devised for describing pores of different sizes. These systems may be conveniently grouped into two categories based on size (Table 5.2) and pore functions (Table 5.3). There is evidently a wide discrepancy in the nomenclature, and there exists a strong need for standardization of the terminology. Toward an attempt to standardize, it is suggested that Kay's (1990) classification for size and Greenland's (1977) classification for function be used in pore characterization. In terms of their size, pores of equivalent cylindrical diameter (ECD) > 30 μm are defined as macropores, between 0.2 and 30 μm as mesopores, and < 0.2 μm as micropores. In terms of their functions in relation to plant growth, pores of ECD > 50 μm are described as transmission pores, those between 0.5 and 50 μm as storage pores, and those < 0.5 μm as residual pores. Functions of these pores in relation to plant growth are listed in Table 5.4. Pores > 500 μm, especially the biopores, are called fissures, and those < 0.005 μm

TABLE 5.2   Some Classification Systems of Soil Pores Based on Their Size
Distribution

Reference	Equivalent cylindrical diameter (ECD, μm)	Pore category
Manegold (1957)	100–5000	Voids
	30–100	Capillaries
	0.002–30	Force spaces
Jongerius (1957)	100–5000	Macropores
	30–100	Mesopores
	0.002–30	Micropores
Johnson, et al. (1960)	> 5000	Coarse
	2000–5000	Medium
	1000–2000	Fine
	75–1000	Very fine
	< 75	Micropores
Brewer (1964)	> 5000	Coarse macropores
	2000–5000	Medium macropores
	1000–2000	Fine macropores
	75–1000	Very fine macropores
	30–75	Mesopores
	5–30	Micropores
	0.1–5	Ultramicropores
	< 0.1	Cryptopores
IUPAC[a] (1972)	0.1–5000	Macropores
	0.005–0.1	Mesopores
	< 0.005	Micropores
McIntyre (1974)	500–5000	Superpores
	50–500	Macropores
	0.1–50	Minipores
	< 0.1	Micropores
Smart (1975)	100–5000	Minipores
	30–100	Macropores
	< 30	Micropores
Kay (1997)	> 30	Macropores
	0.2–30	Mesopores
	< 0.2	Micropores
Soil Survey Division (1990)	≥ 10 mm	Very coarse
	5–10 mm	Coarse
	2–5 mm	Medium
	1–2 mm	Fine
	< 0.5 mm	Very fine

[a]International Union of Pure and Applied Chemistry.

TABLE 5.3 Some Classification Systems of Soil Pores Based on Functional Characteristics

Reference	Equivalent cylindrical diameter (μm)	Classification
Greenland (1977)	< 0.005	Bonding pores
	< 0.5	Residual pore
	0.5–50	Storage pore
	50–500	Transmission pore
	> 500	Fissures
Luxmoore (1981)	< 10	Pressure gradient pore
	10–1000	Gravitational pore
	> 1000	Channel-flow pore

TABLE 5.4 Pore Classification in Relation to Pore Function

Name	Equivalent cylindrical diameter (μm)	Function
Transmission pores	> 50	Air movement and drainage of excess water.
Storage pores	0.5–50	Retention of water against gravity and release.
Residual pores	0.5–0.005	Retention and diffusion ions in solutions.
Bonding pores	< 0.005	Support major forces between soil particles.

are called *bonding pores*. These are the pores that separate clay particles to form quasi crystals or domains (refer to Chapter 4). Readers are referred to a review by Kay (1990; 1998) for conceptual interrelationship among size distribution of aggregates and pores.

5.3.3 Shape and Continuity

Pore shape and geometry are assessed to describe non-matrix pores, most of which are either vesicular (e.g., spherical or elliptical) or tubular (e.g., cylindrical or elongated). Some pores may also be irregular, as is the case in gravelly soils. Continuity and tortuosity of pores are also important to fluid transmission and transport processes in soil, and root growth. Vertical continuity through the horizon is relevant to transport of water across it and gaseous exchange with the atmosphere. The vertical continuity is expressed by assessing the average distance through which the mean pore diameter exceeds 0.5 mm (> fine pores) when soil is moist. Three classes of pores are recognized: low, < 1 cm; moderate, 1 to 10 cm; and high, > 10 cm (Soil Survey Division Staff, 1990).

5.4   ORIGIN AND FORMATION OF PORES

A classification system may also be based on the origin or genesis of soil pores. Macropores or transmission pores are formed by biotic activity, development of shrinkage cracks, formation of ice lenses, activity of soil animals, and tillage operations. Soil organic matter content and clay minerals also play an important role in formation and stabilization of macropores. Further, macropores are strongly influenced by anthropogenic activities, and thus altered by land use and soil management. Mesopores or retention pores are important to plant growth. Mesopores are created by creation of microcracks through shrinkage, freeze–thaw cycles, collapse or plugging of macropores by sedimentation or precipitation, and development of root hair, fungal hyphae, and mycorrhizae. These pores comprise textural porosity and are influenced by particle size distribution, organic matter content, and clay mineralogy, and are only slightly influenced by management. Micropores are created by shrinkage of the soil matrix and collapse of mesopores. Micropores or residual pores are least impacted by soil management and are biologically inactive. These pores are essentially always filled with water, inaccessible to microorganisms, and can be strategically helpful in soil carbon sequestration.

5.5   ASSESSMENT OF POROSITY AND PORE SIZE DISTRIBUTION

There are numerous methods of characterizing porosity, some of which are briefly described in this section.

5.5.1   Total Porosity and Void Ratio

Total porosity (f_t) is usually determined from the bulk density and particle density relationship ($f_t = 1 - \rho_b/\rho_s$). The f_t can also be determined from the saturation moisture content (Θ_s), provided that there is no entrapped air. These relationships hold for non-swelling soils. Void ratio (e) is another indirect measure of porosity, and can also be determined from the bulk density and particle density analysis ($e = \rho_s/\rho_b - 1$) (refer to Chapter 2). In swelling soils, however, in which both the pore volume and bulk volume change substantially with change in w or Θ, it is more appropriate to compute e than f_t. The e value may range from 0.25 to 0.8 for subsoils and 0.8 to 1.4 for surface soil.

5.5.2   Air-Filled Porosity (f_a)

The air-filled porosity is a measure of the macropores, and is generally measured at field capacity or 60 cm water suction ($f_a = f_t - \Theta_{60\,cm}$). Some of these concepts will be explained in Chapter 10 dealing with soil moisture retention. The critical limit of f_a in relation to plant growth is 0.1 for sensitive upland plants (apparently not for the hydromorphic plants such as rice).

5.5.3   Pore Size Distribution

Assessment of the pore size distribution is a principal goal of characterization of soil structure. Similar to the nomenclature, there are also numerous methods of determining the pore size distribution.

Field Methods

Visual Methods.   Macropores, comprising cracks and fissures and biochannels, are often determined in the field using visual methods. Fissures and channels are easily visible and can be counted and measured as such (Douglas, 1986). Small pores can be impregnated with a substance that enhances their visibility. A commonly used procedure involves using a super saturated solution of gypsum ($CaSO_4$), which is poured over the soil. The soil is then removed layer by layer horizontally to assess pore continuity as indicated by transport of gypsum by the pores (Ehlers, 1975). Pores can also be lined with a fluorescent dye (e.g., rhodamine-B dye) to improve their visibility. In a field setting the dye solution (3 g of 45 mm brilliant blue FCF dye dissolved in one liter of deionized water) is uniformly applied on a soil surface (1 × 1.5 m) for 6 hours using a field sprinkler (Flury and Fluehler, 1995a;b). One day after dye application, a trench of 12 m depth is opened at a distance of 0.3 m from the border of sprinkled area to prepare a vertical profile of 1 × 1 m. The dye coverage is estimated from the photograph of the stained area. The blue stained areas represent macropores or preferential flow paths (Fig. 5.1). The continuous stained pores can be traced on an acetate sheet. The dye method is usually visible in soils of neutral color. Pictures of impregnated or dye-lined pores can be taken, magnified and pore dimensions assessed in the laboratory using micrometer, planimeters, image analyzer, and other devices (Anderson et al., 1990; Grevers and deJong, 1990). An alternative to staining is the direct measurement

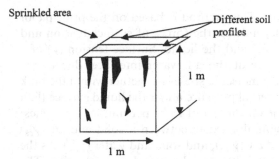

Sprinkled area

Different soil profiles

1 m

1 m

FIGURE 5.1 Schematic of an experimental setup to assess macropores using Brilliant Blue dye-tracer.

technique by x-ray computed tomography (Anderson et al., 1990; Carter and Ball, 1993).

Fractal Analyses. Field assessment of pore size distribution can be described using fractals in three different ways. In the first method, the number–size distribution of voids is obtained in two-dimension by image analysis and is fitted to equation (refer to Chapter 4). These results are then extrapolated to three-dimensions using the relation $D_{R3} = D_{R2} + 1$. The parameter b in Eq. (4.19) is related to the air-entry value and provides the measure of length of the largest pore. The parameter k is linked to representative elementary volume, using the equation $a = k^{-D_{R3}}$, where a is the minimum sample length to represent pore size distribution by using soil water retention curve discussed in detail in Chapter 11. In this method, D_{R3} is related to the pore size distribution index (λ) as $D_{R3} = 3 - 1/\lambda$. The third method uses the modified Campbell's function (Ross et al., 1991) to predict zero-relative saturation at a finite tension. The fractal pore space between tensions at air entry and dryness can be given by $D_{R3} = c + 3$, where c is a constant (Perfect and Kay, 1995).

Laboratory Methods of Determining Pore Size Distribution

Microscopic Measurements. Thin sections made from appropriately impregnated soil clods are examined under the microscope to determine the size and number of different pores (Burke et al., 1986). Different types of microscopes are used depending on the pore size to be assessed. For example, an optical microscope is used for determining pores of 250 nm, scanning electron microscope for pore size of 10 nm, and transmission electron microscope for size range of 1 nm (Burke et al., 1986).

Water Desorption Method. This method is based on the principle of capillarity. The capillary rise depends on the forces of surface tension and the contact angle between the solid and the liquid. Surface tension (γ) of a water is the difference in pressure at the air–water interface, due to the cohesive forces created by the like molecule sticking together within the bulk volume and creating a greater internal pressure under the liquid surface than above it. Surface tension has the dimension of force per unit length (dynes/cm). The force of surface tension also exists between a solid and air (γ_{ra}) compared with that of water and air (γ_{wa}), and solid and water (γ_{sw}). As the solid is immersed in water, there are interfacial forces due to adhesion. The work (W_{sw} in ergs or joules) to separate the solid from water depends on the surface tensions and the interfacial area (A_s) and is given by Eq. (5.1).

$$\frac{W_{sw}}{A_s} = \gamma_{sa} + \gamma_{wa} - \gamma_{sw} \tag{5.1}$$

The interface between the solid and water forms a definite angle, or the angle of contact [Eq. (5.2)].

$$\cos\alpha = \frac{\gamma_{sa} - \gamma_{sw}}{\gamma_{wa}} \tag{5.2}$$

This method is based on the assumption that pores in a soil are a bundle of rigid capillaries. The height of rise of water in a capillary tube depends on the surface tension of the wetting liquid with the surface, and the diameter of the tube. Assume that a liquid has risen to height h in the capillary tube shown in Fig. 5.2. At steady state, when the liquid has stopped rising, the net force acting on the meniscus is zero. The downward force ($F\downarrow$) is the gravitational pull [Eq. (5.3)].

$$F\downarrow = \pi r^2 h \rho_1 g \tag{5.3}$$

where r is the radius of the capillary, h is the height of rise of liquid, ρ_1 is the density of the liquid, and g is the acceleration due to gravity. The upward force ($F\uparrow$) is due to the surface tension [Eq. (5.4)].

$$F\uparrow = 2\pi r \gamma \cos\alpha \tag{5.4}$$

where γ is the surface tension of the liquid against the wetting surface (in this case glass) and α is the contact angle for units of surface tension of H_2O and Hg at different temperatures and against a range of solid surfaces.

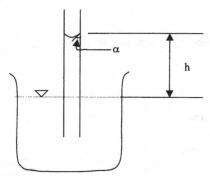

FIGURE 5.2   Capillary rise of water to height h in a glass tube.

At steady state, $F\!\downarrow = F\!\uparrow$

$$\therefore \; \pi r^2 h \rho_1 g = 2\pi r \gamma \cos \alpha \tag{5.5}$$

$$h = \frac{2\gamma \cos \alpha}{r \rho_1 g} \tag{5.6}$$

$$r = \frac{2\gamma \cos \Theta}{h \rho_1 s} \tag{5.7}$$

Assuming that the wetting liquid is H_2O at 20°C, then γ is 72.75 dynes/cm or g/s², l_w is 0.9982 g/cm³, g is 980 cm/s², and α is 0 and cos 0 is one. Substituting these values and rearranging Eq. (5.4) to solve for r leads to:

$$r = \frac{2(72.75)}{h} \frac{g}{s^2} \cdot \frac{cm^3}{0.9982\,g} \frac{s^2}{980\,cm} = \frac{0.15}{h} cm, \quad \text{and} \quad h = 0.15/r\,cm$$

Being a polar liquid, water reacts with soil and a nonreactive substance is used instead, i.e., Hg.

Example 5.1

Calculate size of the pores corresponding to a capillary height of water of 10 cm, 100 cm, 1000 cm, and 10,000 cm at 20°C.

Solution

Using Eq. (5.7) at 20°C, pore radius for corresponding capillary height is:

$$10\,\text{cm} = \frac{0.15}{10}\,\text{cm} = 1.5 \times 10^{-2}\,\text{cm}$$

$$100\,\text{cm} = 0.15 \times 10^{-2}\,\text{cm} = 1.5 \times 10^{-3}\,\text{cm}$$

$$1000\,\text{cm} = 0.15 \times 10^{-3}\,\text{cm} = 1.5 \times 10^{-4}\,\text{cm}$$

$$10,000\,\text{cm} = 0.15 \times 10^{-4}\,\text{cm} = 1.5 \times 10^{-5}\,\text{cm}$$

Mercury Intrusion Method. The mercury intrusion technique is similar to the water desorption method based on the capillary rise. This method is often used for fine pores ranging in size from 10 nm to 100 μm (Danielson and Sutherland, 1986). Because Hg does not wet the soil (and the contact angle is 140°), positive pressure has to be used to inject Hg into the soil pores. A principal advantage of the mercury injection technique lies in its non-wettability. Therefore, pore size does not shrink due to swelling.

In this method, the soil sample is dried, evacuated, and inundated in Hg and pressure is applied at discrete steps. The volume of pores at each pressure step is related to the diminution of Hg. Hg is a non-wetting fluid, therefore, the contact angle is >90°. The pressure required to force Hg into soil pores is a function of contact angle, size, and geometry of pore and surface tension. The equivalent radii of smallest pores (r_p) can be calculated by the following equation:

$$r_p = \frac{-2\gamma \cos\theta}{P} \tag{5.8}$$

where γ is surface tension of Hg (J/m^2), θ is the contact angle of Hg on soil, and P is absolute pressure (N/m^2). The negative sign used in the above equation cancels the negative value of cos θ and provides a positive value of r_p. The r_p values calculated by this method for each pressure steps are consistently lower than the actual and, therefore, multiplied by a correction factor of 1.31. For a detailed description on the above methods, readers are referred to Sills et al. (1973a;b), and Danielson and Sutherland (1986).

Nitrogen Sorption. Similar to Hg, other nonpolar liquids also do not react with clay. Soil sample must be dried, however, prior to using any nonpolar liquid. Freeze-drying is preferred because it does not cause shrinkage. The N-sorption is done on freeze-dried soil cooled to a low temperature of 78 K when a liquid–gas interface is formed for N. Equations (5.5) to (5.7) can be used for N for computing r (Aylmore and Quirk, 1967). A comparison between mercury injection and nitrogen sorption for evaluating pore size distribution is shown by Sills et al. (1973a).

PROBLEMS

1. Calculate the height of capillary rise in a soil pore of $50\,\mu$m inner diameter in winter (5°C), spring (20°C), summer (30°C), and the tropics (40°C).
2. Compute the pressure difference at the air–water interface in Question 1 above.
3. Consider the following equation of the height of capillary rise $r = 2\gamma/(\rho gh)$, where γ and ρ refer to the surface tension and density of the fluid, respectively. Calculate the difference in the height of the capillary rise in $20\,\mu$m diameter pore for (a) water and (b) alcohol at 20°C.
4. Compute the maximum size of the pores that will retain water in soil corresponding to suction (capillary height) of 330 cm and 15,000 cm of water.
5. A soil has a perched water table at 1-m below the surface. Predominant soil capillary pores have an ECD of 0.05 mm. If corn roots penetrate to 30 cm depth, can corn survive a prolonged drought without severe decline in yield?
6. What is the principal of mercury-injection porosity meter? Why is mercury injected under pressure?
7. Determine ECD corresponding to Hg injection pressure of 10,000 and 1,000 cm.

REFERENCES

Anderson, S.H., R.L. Peyton and C.J. Gantzer. 1990. Evaluation of constructed and natural soil macropores using x-ray computed tomography. Geoderma 46: 13–29.

Aylmore, L.A.G. and J.P. Quirk. 1967. The micropore size distribution of clay mineral systems. J. Soil Sci. 18: 1–17.

Brewer, R. 1964. Fabric and Mineral Analysis of Soils, Wiley, New York.

Carter, M.R. and B.C. Ball. 1993. Soil porosity. In M.R. Carter (ed) "Soil Sampling and Methods of Analysis", Lewis Publishers, Boca Raton, FL: 581–588.

Childs, E.C. 1968. The Physical basis of Soil Water Phenomena, Wiley, London.

Danielson, R.E. and P.L. Sutherland. 1986. Porosity. In A. Klute (ed) "Methods of Soil Analysis" Part 1, American Society of Agronomy, Monograph 9, Madison, WIP 443–461.

Derdour, H. D.A. Angers and M.R. Laverdiere. 1993. Caracterisation de l'espace poral d'un sol argileux: effets de res constituants et du travail du sol. Can. J. Soil Sci. 73: 299–307.

Douglas, J.T. 1986. Macroporosity and permeability of some cores from England and France. Geoderma 37: 221–231.

Ehlers, W. 1975. Observations on earthworm channels and information on tilled and untilled loess soils. Soil Sci. 119: 242–249.

Flury, M. and H. Fluehler. 1995a. Modeling solute transport in soils by diffusion-limited aggregation. Basic concepts and application to conservative solutes. Water Resources Res. 31: 2443–2452.

Flury, M. and H. Fluehler. 1995b. Tracer characteristics of Brilliant Blue FCF. Soil Sci. Soc. Am. J. 59: 22–27.

Greenland, D.J. 1977. Soil damage by intensive arable cultivation: temporary or permanent? Phil. Trans. Roy Soc. London, B, 281: 193–208.

Grevers, M.C.J. and E. De Jong. 1990. The characterization of soil macroporosity of a clay soil under ten grasses using image analysis. Can. J. Soil Sci. 70: 93–103.

Hamblin, A. 1985. The influence of soil structure on water movement, crop root growth, and water uptake. Adv. Agron. 38: 95–158.

IUPAC, 1972. Manual of Symbols and Terminology, International Union of Pure and Applied Chemistry, London.

Johnson, W.M., J.E. McClelland, S.B. McCaleb, R. Ulrich, W.G. Harper, and T.B. Hutchings. 1960. Classification and description of soil pores. Soil Sci. 89: 319–321.

Jongerius, A. 1957. Morphological Investigations of Soil Structure, Bodemkundige studiea, No.2, Meded. van der Stickting Bodemkartiering, Wageningen.

Kay, B.D. 1990. Rates of change in soil structure under difficult cropping systems. Adv. Soil Sci. 12: 1–52.

Kay, B.D. 1998. Soil structure and organic carbon: A review. In: R. Lal, J.M. Kimble, R.F. Follett and B.A. Stewart (eds) "Soil Processes and the Carbon Cycle." CRC/Lewis Publishers, Boca Raton, FL: 169–197.

Lepiec, J., R. Hatano, A. Slowinskia-Jurkiewicz. 1998. The fractal dimension of pore distribution patterns in variously-compacted soils. Soil & Tillage Res. 47: 61–66.

Luxmoore, R. 1981. Micro-, meso-, and macroporosity of soil. Soil Sci. Soc. Am. J. 45: 671–672.

Manegold, 1957. Quoted by A.E. Scheidegger (1957) "The Physics of Flow Through Porous Media." Univ. of Toronto Press, Toronto, Canada.

McIntyre, D. 1974. Porespace and aeration determinations. In: J. Loveday (ed) "Methods for Analysis of Irrigated Soils." Commonwealth Agric. Bureau, Farnhan Royal.

Perfect, E. and B.D. Kay. 1995. Application of fractals in soil and tillage research: a review. Soil & Till. Res. 36: 1–20.

Ross, P.J., J. Williams and K.L. Bristow. 1991. Equation for extending water retention curves to dryness. Soil Sci. Soc. Am. J. 55: 923–927.

Sills, I.D., L.A.G. Aylmore and J.P. Quirk. 1973a. A comparison between mercury injection and nitrogen sorption as methods of determining pore size distribution. Soil Sci. Soc. Amer. Proc. 37: 535–537.

Sills, I.D., L.A.G. Aylmore and J.P. Quirk. 1973b. An analysis of pore size in illite-kaolinite mixtures. J. Soil Sci. 24: 480–490.

Smart, P. 1975. Soil microstructure, Soil Sci. 119: 385–393.
Soil Survey Division Staff, 1990. Soil Survey Manual. Natural Resource Conservation Service, Washington, D.C.
Tyler, S.W. and S.W. Wheatcraft. 1990. Fractal processes in soil water retention. Water Res. Res. 26: 1047–1054.

6

Manifestations of Soil Structure

The dynamic of soil structure has numerous agronomic, economic, and ecological implications. Thus, sustainable management of natural resources requires optimization of soil structural characteristics. Structural degradation and decline in aggregation of structured soils lead to soil dispersion, crusting, compaction, formation of pans, accelerated soil erosion, and emission of CO_2 and other greenhouse gases into the atmosphere (Fig. 6.1). These ramifications can have a drastic impact on plant growth and net primary productivity, hydrologic cycle, water quality, elemental cycling, and emission of trace gases (Fig. 6.2). The interactive effects of soil processes, soil properties, plant growth, and environment can adversely impact ecosystem functions (Fig. 6.3). The latter includes biomass production, purification of water, detoxification of natural and anthropogenic pollutants, restoration and resilience of ecosystems, and cycling of elements.

6.1 CRUSTING AND SURFACE SEAL FORMATION

Crusting is a soil surface phenomena caused by susceptibility of aggregates at the soil–air interface to disruptive forces of climatic elements and perturbations caused by agricultural practices (e.g., tillage and traffic). Slaking, deflocculation, or dispersion of aggregates on rapid wetting or

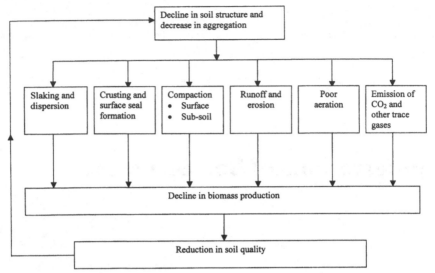

FIGURE 6.1 Impact of decline in soil structure on soil physical quality.

submersion in water, is attributed to numerous factors including the effect of entrapped air, predominance of Na^+ on the exchange complex, and weak aggregate strength caused by low level of soil organic matter content and weak ionic bonds. These factors and processes governing them are discussed by Sumner and Stewart (1992). Dispersion, reorientation of dispersed particles, drying, and desiccation, lead to formation of a thin crust on the soil surface. Soil crust or surface seal, therefore, refers to the thin dense layer on the soil surface characterized by low porosity, high density, and low permeability to air and water.

6.1.1 Types of Crusts

There are three principal categories of crust: chemical crusts, biological crusts, and physical crusts (Figs. 6.4a;b;c). *Chemical crusts* are formed due to salt incrustations on soil surface in arid and semi-arid regions. *Biological* or *microbiotic crusts* are primarily formed by algal growth. Ponded water on surface of slowly permeable soils in arid and semi-arid tropics lead to formation of algal crusts. Such crusts are extremely hydrophobic, and drastically reduce the rate of water infiltration into a soil. *Physical crusts* are formed due to alteration in structural properties of the soil, and may be *structural* or *depositional*.

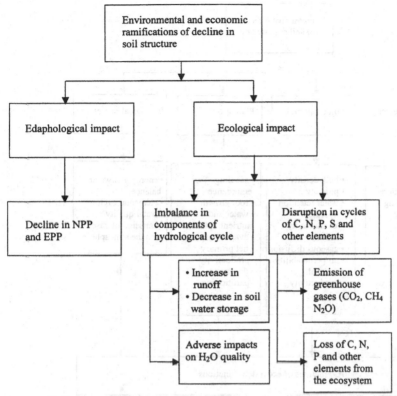

FIGURE 6.2 Economic and environmental ramifications of decline in soil structure (NPP is net primary productivity, and EPP is ecosystem primary productivity).

Structural Crust

Structural crust is formed due to the disruption of aggregates by raindrop impact and physiochemical dispersion of soil clays (McIntyre, 1958a;b). The upper surface of the structural crust, or "skin seal," has low permeability and is about 1–3 mm thick. Sodic soils, those with high percentage of exchangeable Na^+ on the exchange complex, are extremely prone to formation of structural crust.

Depositional Crust

Depositional crust is formed by transport and deposition of fine particles by surface flow (Chen et al., 1980). Depositional crusts are thicker than structural crusts, and are formed wherever suspended fine-textured material in water gets settled. Kinetic energy of raindrops and dispersional properties of soil have no effect on formation of depositional crusts.

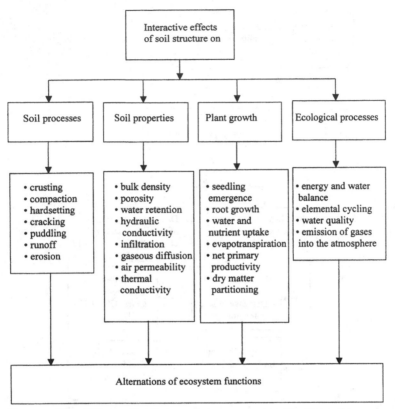

FIGURE 6.3   Effects of soil structure on ecosystem functions.

6.1.2   Factors Affecting Slaking and Deflocculation

There are three principal factors: Kinetic energy of rainfall, soil properties, and anthropogenic factors (for anthropogenic factors, refer to Sec. 6.1.7 in this chapter).

Rainfall Factor

Slaking is principally caused by the kinetic energy of impacting raindrops (McIntyre, 1958b; Shainberg et al., 1989; Bradford and Huang, 1992). The kinetic energy ($E = 1/2\, mv^2$, where m is the mass of rain per unit area and v is the impact velocity of rain drop) and momentum ($M = mv$) are the primary sources of energy that disrupts an aggregate. The rate and intensity of crust formation increases with increase in energy of the raindrop impact. The energy of flowing water may also have indirect impact, probably due to its influence on transport and deposition of sediments.

(a)

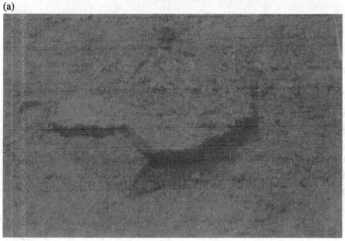

(b)

FIGURE 6.4 (a) Silt loam soils with low organic matter content are prone to formation of surface seal or crust. (b) High strength surface seals inhibit germination and retard seedling growth by limiting gaseous exchange. (c) Seedlings that emerge through hard crust can suffer from drought stress because of low water infiltration into the soil.

Weather

Wetting–drying and freeze–thaw cycles affect aggregation (see Chapter 4). Consequently, these processes also influence formation and strength of crust. Crust strength is more if a heavy rain is followed by dry and hot weather that desiccates the crust.

(c)

FIGURE 6.4 Continued.

Soil Properties

Susceptibility to crust formation also depends on numerous soil properties. Important among these are texture, clay mineralogy, soil organic matter content, and degree and strength of aggregates. Resistance of surface aggregates to raindrop impact, shearing force of overland flow, and to the disruptive force of entrapped air upon quick wetting are important soil factors. The mean weight diameter (MWD) and the median aggregate diameter (see Chapter 4) (D_{50}) are strongly correlated with susceptibility to crusting (Bajracharya, 1995).

Field Moisture Content

The antecedent soil moisture content or soil wetness at the beginning of the rainfall influences aggregate strength, slaking or dispersion, infiltration rate, and the rate of overland flow (le Bissonnais, 1990). Under initial dry soil conditions, the dispersion is caused by slaking. Slaking causes rapid aggregate breakdown, quickly filling the intraaggregate pore space with microaggregates or dispersed primary particles. Under initial dry soil conditions, the aggregate breakdown depends more on rainfall rate than on its kinetic energy or momentum. Under wet soil conditions, aggregates are less prone to slaking but more to the raindrop impact. The surface seal formation is caused by the kinetic energy or momentum of the rain and overland flow. Raindrop impact easily disrupts the aggregate when the aggregate strength is low due to wetness (Farres, 1978).

Microrelief

Microrelief is defined by surface cloddiness, clod size, and geometry. The microrelief is prominent soon after plowing (see Fig. 6.13a). Rough seedbed decreases susceptibility to crust formation (Burwell and Larson, 1969). Microrelief also controls the physical processes occurring at the soil surface, e.g., microrills, surface depressions, infiltration rate, etc.

6.1.3 Mechanisms of Crust Formation

Crust formation involves dispersion of aggregates followed by orientation and hardening by desiccation. Thus, properties of the double layer and stability of the colloidal system are important to crusting (van Olphan, 1963; Young and Warkentin, 1966; Sumner, 1992) (see also Chapter 3). Flocculation (which is caused by attractive forces) and slaking (which is caused by repulsive forces) are both present in the electric double layer. In addition, colloid particles are also subject to Brownian movement. Therefore, dispersion depends on the following factors:

Charge Distribution on Soil Colloids. The charge distribution on soil colloids depends on surfaces with permanent charge (e.g., 2:1 clay minerals, 1:1 clay minerals), surfaces with variable charge (e.g., oxides, amorphous minerals, soil organic matter), and other soil conditions. Soils with low-activity clays are more prone to dispersion than those with high-activity clays. Similarly, soils with low concentration of soil organic matter are more prone to crusting than those with higher concentrations.

Properties of the Electric Double Layer. Effective thickness of the double layer, the surface charge, surface potential, and other properties of the double layer are influenced by relative proportion of the colloidal surfaces with permanent and variable charge, nature of the cations on the exchange complex, and degree of hydration. The thickness of the double layer also depends on the nature of cations on the exchange complex. Predominance of monovalent cations (e.g., Na^+) increases the thickness of the double layer (see Chapter 3).

Surface Charge on Soil Particles. All soils have both permanent and variable charge, and these charges change with soil pH especially in soils with variable charge surfaces. Coulombic interactions are extremely important in dispersion, these interactions depend on variations in surface charges. Under dilute electrolyte conditions, there is a maximum overlap of oppositely charged double layer that results in maximum positive Coulombic interactions and flocculation.

Particle Repulsion. The colloidal stability is determined by the net effect of van der Waals forces of attraction and the electrical double layer repulsion forces. The double layer repulsion is given by Eq. (6.1) (Olphen, 1963; Sumner, 1992).

$$E_r = \left[\frac{64nKT}{k}\right]\left[\frac{\exp(Ze\psi_o/2K_BT) - 1}{\exp(Ze\psi_o/2K_BT) + 1}\right]^2 e^{-2K_Bd} \qquad (6.1)$$

where E_r is the repulsive energy of the double layer, n is the electrolyte concentration in the equilibrium solution, Z is the valency of the counter cations, e is electronic charge K_B is Boltzmann constant, T is temperature. $1/k$ is an expression of the effective thickness of the double layer, and d is the half distance between the plates. The magnitude of repulsive energy between particles suspended in electrolytes of varying counter-ion concentration and valency as computed by Eq. (6.1) is graphically illustrated in Fig. 6.5. The graph shows a rapid increase in repulsive force with reduction in the concentration or valency of the counter ion. For colloidal particles, where the distance between the plates is small compared with the thickness, the attractive energy due to van der Waals forces is given by Eq. (6.2) (Gregory, 1989; Sumner, 1992).

$$E_a = \frac{A}{48\pi d^2} \qquad (6.2)$$

where A is the Hamaker constant, and d is the half distance between the plates. The net energy $(E_n = E_r - E_a)$, which determines the dispersion or flocculation, also depends on the electrolyte concentration. In the case of low electrolyte concentration, the repulsive energy (E_r) dominates the attractive energy (E_a) and the clay particles remain dispersed and the colloidal system is very stable. In case of high concentration, the E_a dominates and rapid flocculation takes place. There exists a critical flocculation concentration (CFC) where the energy barrier just disappears (Gregory, 1989). In addition, there are other numerous repulsive forces, such as hydration repulsive forces. Similarly, some other attractive forces include hydrophobic attractive forces. For additional details, readers are referred to a review by Sumner (1992).

Rearrangement of Particles. Once soil particles are dispersed, the next step in the formation of crust is the reorientation and development of a close packing arrangement of particles. The rearrangement may occur due to electrokinetic processes, and movement of dispersed particles with the infiltrating water. Smaller particles get lodged in

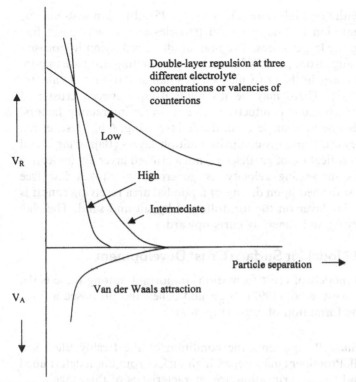

Figure 6.5 Schematic representation of the variation in repulsive and attractive forces between colloidal particles of like charge with distance from the particle surface. (Redrawn from Van Olphen, 1963, and Summer, 1992.)

between the larger particles, clogging the pores and increasing soil bulk density.

Desiccation. Rapid drying and desiccation soon after dispersion and reorientation are crucial to crusting and surface seal formation. Crust formation is weak or it completely breaks down if the weather conditions favor freeze–thaw or wet–dry cycles.

6.1.4 Properties of Crust

The crusted layer is more dense but may be of similar textural makeup than the unaffected soil beneath it. The crust is primarily characterized by reduction in total volume, size, shape, and continuity of pores. Thickness of the crust may range from < 1 mm to 10 mm (Norton, 1987). Very thin crusts are called "skin seal." These microlayers are usually < 0.1 mm thick, extremely dense with no visible pores (McIntyre, 1958a;b). Skin seals may be formed by reorientation of fine dispersed particles and/or washed-in fine material that plug the larger pores. The magnitude of reduction in porosity of the crust may range from 30 to 90%, with corresponding decrease in pore size. The pore diameter in the crust may be as small as 0.075 mm (Valentin and Figueroa, 1987). There may be no relationship between crust and infiltration rate or hydraulic conductivity due to other interacting factors. The crust may also be in a single or multiple layer (Fig. 6.6, West, et al., 1992). Sedimentary crusts usually comprise multiple layers (Bajracharya and Lal, 1999). The stratification of particles within a crusted layer are indicative of the differences in settling velocity as governed by Stokes law (see Chapter 3). A crust formed upon drying of a ponded area receiving runoff is characterized by clay layer on the top followed by silt and sand. The clay skin cracks on drying and generally curls upward.

6.1.5   General Model for Surface Crust Development

There are several models of crust formation. Important among these is the one proposed by West, et al. (1992). West and colleagues proposed a four-stage model of the formation of crust (Fig. 6.7):

Stage 0.   Stage 0 represents the condition of the freshly tilled soil before any rainfall. Prominent microrelief, high surface roughness determined by large clods, and lack of crustation are characteristics of this stage.

Stage 1.   Stage 1 or the initial stage of crust development involves breakdown of aggregates and particle rearrangement due to raindrop impact and slaking. The aggregate disruptions result in formation of a disruptional layer.

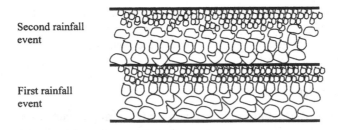

Second rainfall event

First rainfall event

Structural crust or natural soil

Figure 6.6   Multiple layer crust formed due to successive rainfall events. (Redrawn from West et al., 1992.)

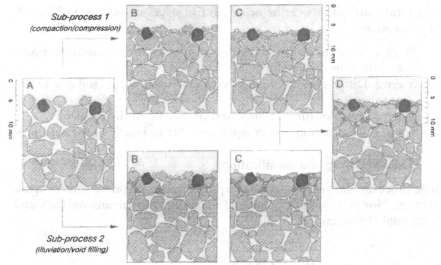

FIGURE 6.7 Conceptual model of crust formation processes and resulting crust. Black polygons represent stones, gray polygons—aggregates, small circles—sand grains, and dark gray shading—oriented fine particles (clay). (From Bajracharya, 1995.)

Stage 2. Stage 2 may involve two pathways. For a soil of high aggregates stability and low susceptibility to dispersion, this stage represents continued development of the disruptional layer. In addition, aggregate coalescence may occur beneath the zone of aggregate disruption and thicken the disruptional layer. For a soil with weak aggregation and high potential for dispersion, the particle disfunction is more extensive, and the released micromass may move downward to form a washed-in layer. The surface layer may become smooth due to removal of the microrelief.

Stage 3. Stage 3 represents the maximum development of the crust, leading to maximal runoff and erosion of the washed-out layer. There may be further thickening of the disruptional layer and formation of a secondary washed-out layer. However, the released micromass may be washed out in the runoff. The microrelief may flatten during this stage, and soil surface may be covered by a sedimentary crust.

Bajracharya and Lal (1999) proposed another model. They observed that there are two parallel subprocesses leading to formation of crust on an Alfisol in central India. These are: (i) physical compaction and compression due to the force of raindrop impact, and (ii) close packing of particles by filling in of pores by aggregate breakdown products. Formation of a "structural crust" of this nature occurs in five stages as outlined in Fig. 6.7. These stages are:

Stage 1: Mechanical breakdown of aggregates due to raindrop impact and the attendant slaking

Stage 2: Differential swelling, slaking and dispersion of soil due to soil wetting

Stage 3: Translocation of dispersed particles into the pores

Stage 4: Compaction and compression due to kinetic and mechanical forces

Stage 5: Drying and densification

These processes are generic and may apply to all crust-prone soils of weak structure. However, specific steps and stages may differ among soils and ecoregional characteristics.

6.1.6 Characterization of Crust

There are several methods to characterize properties of crust (Fig. 6.8). Properties of the crust may be characterized by evaluation of: (i) thickness, (ii) micromorphology by thin section (Norton, 1987), (iii) hydraulic properties by measuring crust conductance (McIntyre, 1958a; Falayi and Bouma, 1975), (iv) strength by penetrometer measurement, and (v) potential adverse effect on seedling emergence by measuring crust strength through the buried nail or buried balloon technique (Arndt, 1965a;b). Crust strength can also be measured by modulus of rupture (see chapter on strength properties). Simple techniques of characterizing soil crust have been developed for use in the field and laboratory conditions (Brossman et al., 1982; Franzmeier et al., 1977; Parker and Taylor, 1965; Taylor, 1962, etc.). A simple device used in the laboratory, described by Sutch, et al. (1983), is shown in Fig. 6.9.

6.1.7 Crust Management

Crusting has adverse impacts on seedling emergence and growth (Arndt, 1965a;b; Parker and Taylor, 1965). Thus, crust management is important to obtaining high yields. There are several technological options for crust management (Fig. 6.10), and the choice of technology also depends on the causes of crust formation. In addition to the impact of raindrops on an unprotected soil, crust may also be caused by the trampling action of livestock or humans, or vehicular traffic of farm operations. Preventative measures are based on strategies of enhancing aggregation,

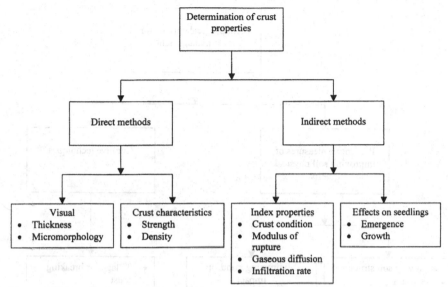

FIGURE 6.8 Methods of determining properties of crust.

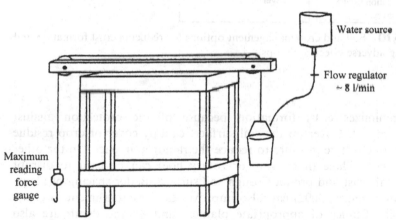

FIGURE 6.9 An apparatus used to measure crust strength. (Sutch et al., 1983.)

improving soil structure, and minimizing the disruptive effects of raindrop impact. The curative measures involve strategies of managing crust once it has been formed. Use of inorganic (gypsum) and organic amendments (compost, farmyard manure) helps to maintain clay in an aggregated or flocculated state. Use of conservation tillage and residue

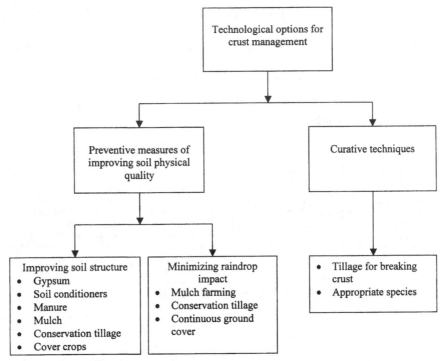

FIGURE 6.10 Soil and crop management options for reducing crust formation and minimizing adverse effects on crops.

mulch minimizes crust formation because of the protection against raindrop impact. Cover on the soil surface, canopy cover or crop residue mulch, is an effective measure to reduce the raindrop impact. On the other hand, tertiary tillage (harrowing or rotary hoe) can be used to disrupt depositional crust and produce rough soil surface. Better spacing of plants in the row (Metzer, 2002) can also improve stand establishment in crust-prone soils. Choice of appropriate planters and sowing depth are also critical to reducing adverse impact of crust on stand establishment (Nabi et al., 2001; Hemmat and Khashoei, 2003). Management and enhancement of soil organic matter content is a useful strategy to increase aggregate strength and stability and minimizes risks of structural crust formation. Soil conditioners and polymers have also been found useful to improve aggregation and minimize crusting (Shainberg et al., 1989). Application of soil conditioners, manure, or mulch on the seed row can reduce the risks of crusting.

FIGURE 6.11   Puddling is a deliberate attempt to break aggregates and destroy structure by plowing when the soil is wet. The objective is to decrease infiltration rate and increase water retention in the puddled layer.

6.2   PUDDLING

Puddling refers to physical manipulation of a wet soil to slake and disrupt structural aggregates and decrease total and macroporosity (Fig. 6.11; see also Chapter 5). Puddling implies reduction in apparent specific volume (ρ_b^{-1} or inverse of bulk density) and void ratio (e) of a soil by mechanical work done on it (Bodman and Rubin, 1948; Ghildyal and Tripathi, 1987). The stress applied when soil is wet ($\Theta = s$), leads to reorientation of clay and reduction in air porosity (f_a). The term puddlability (P) expresses the susceptibility of soil to puddling, and is numerically equal to the change in apparent specific volume of a soil (dv) per unit of work (dw) expended in causing such a change.

$$P = dv/dw \tag{6.3}$$

The change in volume per unit of work is related to the air-filled pore space on drying. Cohesion of a puddled soil increases with progressive decrease in soil moisture content until it reaches the maximum value when soil is dry. Increase in cohesion on drying is due to an increase in interparticle contacts and forces of surface tension as the water film drains into small pores. Puddling of a soil leads to: (i) reduction of macroaggregates, (ii) decrease in total and air-filled porosity, (iii) reduction in hydraulic conductivity,

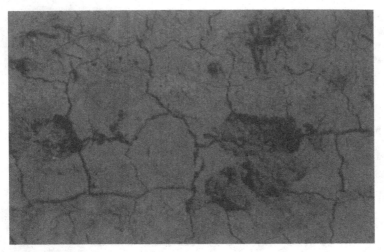

FIGURE 6.12 Drying of the puddle soil leads to formation of cracks, which may have an adverse impact on root growth of rice seedlings.

and (iv) increase in retention pores. Puddling leads to change in soil from a 3-phase (solid, liquid, and gases) to a 2-phase (solid and liquid) system (Fig. 6.11). Drying of a puddled soil, transformation from a 2-phase to 3-phase system, lead to formation of wide cracks (Fig. 6.12).

Mechanical puddling is done for rice cultivation. Being a semiaquatic plant, rice is grown under saturated soil conditions with surface ponding. Therefore, maintaining a ponded water condition is important to rice growth. Such ponding conditions increases losses of water by deep percolation and seepage (see Chapter 9). These losses must be reduced for improving water use efficiency. In order to reduce percolation and seepage losses, soil aggregates are destroyed to reduce transmission pores and increase retention pores. Aggregates are weakest when saturated with water, and the electric double layer of the clay particles is fully expanded. Easy to puddle soils are those that contain high clay content, 2:1 expanding lattice clay minerals, high proportion of Na^+ on the exchange complex, and low concentration of sesquioxides (see Chapter 3). It is difficult to puddle coarse-textured soils with low clay and high organic matter contents.

The process of puddling occurs in two stages. The first stage involves increasing soil water content, the second is the mechanical work done to disrupt the aggregates and reduce soil volume. Increase in soil moisture content decreases cohesion and soil strength. The work done during puddling involves two kinds of deformation stresses: (i) normal stress causing compression, and (ii) tangential stress causing shear (see Chapter 7). The work done during puddling may be computed from these two stresses. The porosity, and therefore the hydraulic conductivity, of a puddled soil decreases rapidly with increase in the degree of puddling.

6.3 HARDSETTING

"Hardsetting" refers to a process in which soils set hard into a structureless mass following drying and ultradesiccation (Mullins et al., 1990). When dry and set hard, these soils have a high bulk density, high penetration resistance, high strength, and are difficult to plow or dig. Hard setting soils have a narrow range of workable soil moisture content. Extreme types of such soils are often called "lunch-time soils." These soils may be too wet to plow before lunch and too hard after lunch. Hardsetting soils have a weekly developed structure characterized by: (i) low aggregation, (ii) aggregates prone to slaking and dispersion, (iii) low infiltration rate, and (iv) high runoff and erosion (Fig. 6.13). (Ley et al., 1989; 1993). The hardsetting process begins with slaking followed by slumping or consolidation, and desiccation. The major difference between hardsetting and compaction is that the densification in hardsetting occurs without the application of an external load (e.g., machinery traffic, trampling by animals or humans) (for definition of compaction, see Chapter 7). The forces leading to hardsetting are generated within the soil itself. Hardsetting is also different than surface seal formation or crusting. Some soils that exhibit crusting may not be hardsetting. A hardsetting soil differs from the one that crusts by the fact that the A horizon is extremely unstable that mere wetting causes the slaking, dispersion, and mobilization of the fine material. The kinetic energy of raindrop or running water and low electrolyte concentration in soil solution, essential to crusting, are not necessary to hardsetting.

There are some soil attributes that make it susceptible to hardsetting. Hardsetting soils have textural properties ranging from loamy sand to sandy clay, low swell-shrink capacity, low soil organic matter content, and predominantly low activity clays. Risks of hardsetting are accentuated by factors and processes that increase susceptibility to slaking, dispersion, and slumping including: (i) cultivation under wet conditions, (ii) mechanical soil disturbance, (iii) low application of compost and organic amendments, and (iv) clean cultivation.

Hardsetting behavior has numerous limitations with regards to timings of cultivation, restricted root growth, high-energy requirement for soil management, low crop stand, and poor yield (Ley et al., 1989; 1993). Management of hardsetting soils involve techniques that improve aggregation and aggregate strength. These techniques include use of residue mulch,

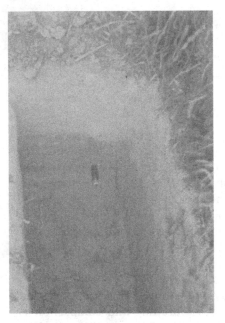

(a)

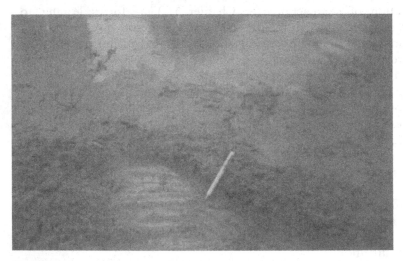

(b)

FIGURE 6.13 (a) Hardsetting soils are characterized by predominantly low activity clays, low organic matter content and structurally inert characteristics. Consequently, they set hard on drying. (b) Hardsetting soils may also have low infiltration rates, especially when combined with a hydrophobic surface crust.

no-till or conservation tillage, cover crops, etc. Application of gypsum and other soil amendments is crucial. Maintenance of soil temperature and moisture regimes in optimal range by avoiding too dry and too hot conditions minimizes risks of hardsetting.

6.4 CRACKING

Heavy textured soils containing high amounts of expanding lattice clays have a high coefficient of expansion and contraction and develop large and deep shrinkage cracks on drying (Fig. 6.14). This process is also discussed in Chapter 20. It is the three-dimensional shrinkage which is accompanied by cracking. A crack is initiated where soil cohesion (strength) is the lowest and the soil moisture content is the highest (Mitchell and Van Genuchten, 1992). Crack initiation occurs where soil is the wettest, i.e., in the middle of two rows in the inter-row zone or in between two plants. The phenomenon of between-row cracking has long been observed by farmers and soil scientists/ agronomists. Johnson and Hill (1944) reported extensive between-row cracking in Houston black clay and Austin clay under corn. In New South Wales, Australia, Fox (1964) proposed a theory of root-anchoring that increases soil strength and reduces cracking. Plant roots provide a skeleton to which soil adheres as it shrinks causing formation of large cracks along the outer boundaries of the rooted volume. Because of additional surface

FIGURE 6.14 Veritsols and other soils containing predominantly high activity clays develop wide and deep cracks on drying.

area exposed to evaporation, cracks accelerate soil drying. If the soil is not disturbed and rows are planted at the same location, as with a no-till system of seedbed preparation, the crack will appear on the same place upon redrying after wetting or in the next season. Cracking intensity and number of cracks per unit area depend on clay mineralogy and structural attributes such as particle arrangement. A large number of cracks are formed in a soil with flocculated clay. In contrast, a few cracks are formed in soils with high cohesive strength. Soils with well-developed crumb structure and self-mulching characteristics usually do not exhibit intensive cracking.

Formation of cracks or soil failure involves energy. Cracking occurs when the release of energy per unit area by the crack is more than the increase of surface energy due to creation of additional surface area (Ghildyal and Tripathi, 1987).

Soil cracking is a special case of soil failure. It occurs when the release of energy per unit area by the crack is greater than the increase of surface energy. There are two separate energy terms involved. First, energy is due to the forces of surface tension (γ_s) which is proportional to the new surface created by cracking [Eq. (6.4)].

$$dU/dA = 2\gamma_s \tag{6.4}$$

where U is the energy of soil surfaces, A is the area of the exposed new crack, and γ_s is the surface tension at the soil–air interface. The second energy involved in cracking is due to the tensile strength of the soil which is released per unit free surface energy due to the new area exposed by cracking [Eq. 6.5)].

$$dU/dA = \frac{\pi\sigma^2 D}{E} \tag{6.5}$$

where σ is tensile stress normal to the plane of the crack, D is major diameter of the crack which is assumed to be elliptical, and E is Young's modulus of soil (see Chapter 7).

Combining Eq. (6.4) and (6.5) lead to the Griffith formula related to the development of crack [Eq. (6.6)].

$$\sigma_s = \left(\frac{2E\gamma_s}{\pi S}\right)^{1/2} \tag{6.6}$$

where σ_s is the limiting stress in dynes/cm^2. Both σ_s and E depend on soil moisture content and ρ_b.

PROBLEMS

Write a brief note to answer the following questions.

1. Why is crusting a more serious problem in soils of loamy rather than sandy or clayey texture?
2. Why does a "clay skin" formed on a dry soil after ponding curl upward?
3. Why is "dense planting" or high seed rate recommended for crust-prone soils?
4. Why does manuring and application of biosolids decrease risks of crusting?
5. List factors affecting thickness of soil crust.
6. In what soil and environmental conditions does plowing increase and decrease the risks of crusting?
7. Complete a matrix listing processes involved in crusting, hardsetting, and cracking.

Number	Crusting	Hardsetting	Cracking
1			
2			
3			

REFERENCES

Arndt, W. 1965a. The nature of the mechanical impedance to seedlings by soil surface seals. Aust. J. Soil Res. 3: 45–54.

Arndt, W. 1965b. The impedance of soil seals and forces of emerging seedlings. Aust. J. Soil Res. 3: 55–68.

Bajracharya, R.M. 1995. Soil crusting and erosion processes on an Alfisol in south-central India. Ph.D. Dissertation, The Ohio State University, Columbus, OH.

Bajracharya, R.M. and R. Lal. 1999. Land use effects on soil crusting and hydraulic response of surface crusts on a tropical Alfisol. Hydrological Processes 13: 59–72.

Bodman, G.B. and J. Rubin. 1948. Soil puddling. Soil Sci. Soc. Amer. Proc. 13: 27–36.

Bradford, J.M. and C. Huang. 1992. Mechanisms of crust formation: physical components. In M.E. Sumner and B.A. Stewart (eds) "Soil Crusting: Chemical and Physical Processes", Lewis Publishers, Boca Raton, FL: 55–72.

Brossman, G.D., J.J. Vorst, and G. Steinhardt. 1982. A technique for measuring soil crust strength. J. Soil Water Conserv. 37: 225–226.

Burwell, R.E. and W.E. Larson. 1969. Infiltration as influenced by tillage-induced random roughness and pore space. Soil Sci. Soc. Am. Proc. 33: 449–452.

Chen, Y., J. Tarchitzky, J. Brouwer, J. Morin, and A. Banin. 1980. Scanning electron microscope observations on soil crusts and their formation. Soil Sci. 130: 45–55.

Falayi, O. and J. Bouma. 1975. Relationships between the hydraulic conductance of surface crusts and soil management in a Typic Hapludalf. Soil Sci. Soc. Am. Proc. 39: 957–963.

Farres, P. 1978. The role of time and aggregate size in crusting process. Earth Surface Processes 3: 243–254.

Fox, W.E. 1964. Cracking characteristics and field capacity in a swelling soil. Soil Sci. 98: 413–415.

Franzmeier, D.P., G.C. Steinhardt, J.R. Crum, and L.D. Norton. 1977. Soil characterization in Indiana. I. Field and laboratory procedures. Res. Bull. #943, Purdue Univ., Indiana Agric. Exp. Sta., pp. 30.

Ghildyal, B.P. and R.P. Tripathi. 1987. Soil Physics. J. Wiley & Sons, New Delhi, India, 656 pp.

Gregory, J. 1989. Fundamentals of flocculation. Crit. Rev. Env. Control 19: 185–230.

Hemmat, A. and A.A. Khashoei. 2003. Emergence of irrigated cotton in flatland planting in relation to furrow opener type and crust-breaking treatments for Cambisols in central Iran. Soil & Till. Res. 70: 153–162.

Johnson, J.R. and H.O. Hill. 1944. A study of the shrinking and swelling properties of rendzina soils. Soil Sci. Soc. Am. Proc. 9: 24–29.

Kruyt, H.R. 1952. Colloid Science vol. 1, Irreversible Systems, Elsevier, Amsterdam.

le Bissonnais, Y. 1990. Experimental studying and modelling of soil surface crusting processes. In R.B. Bryan (ed) "Soil Erosion: Experiments and Models", Catena Supplement 17, Catena Verlog, Cremlingen-Destedt, Germany: 13–28.

Ley, G.J., C.E. Mullins, and R. Lal. 1989. Hard setting behavior of some structurally weak tropical soils. Soil & Till. Res. 13: 365–381.

Ley, G.J., C.E. Mullins, and R. Lal. 1993. Effect of soil properties on the strength of weakly structured tropical soils. Soil & Till. Res. 28: 1–14.

McIntyre, D.S. 1958a. Permeability measurements of soil crusts formed by raindrop impact. Soil Sci. 85: 185–189.

McIntyre, D.S. 1958b. Soil splash and the formation of surface crust by raindrop impact. Soil Sci. 85: 261–266.

Metzer, R.B. 2002. Comparison of sweep and double-disk type planter for stand establishment and lint yield. Texas Agric. Expt. Service. http://entowww.tamu.edu/cotton/26.htm

Mitchell, A.R. and M. Th. Van Genuchten. 1992. Shrinkage of bare and cultivated soil. Soil Sci. Soc. Am. J. 56: 1036–1042.

Mullins, C.E., D.A. MacLeod, K.H. Northcote, J.M. Tisdall, and I.M. Young. 1990. Hardsetting of soils: behaviour, occurrence and management. In R. Lal and B.A. Stewart (eds) "Soil Degradation", Adv. Soil Sci. 11: 37–108.

Nabi, G., C.E. Mullins, M.B. Montemayor, and M.S. Akhtar. 2001. Germination and emergence of irrigated cotton in Pakistan in relation to sowing depth and physical properties of the seedbed. Soil & Tillage Res. 59: 33–44.

Norton, L.D. 1987. Micromorphological study of surface seals developed under simulated rainfall. Geoderma 40: 127–140.

Parker, J.J., Jr. and H.M. Taylor. 1965. Soil strength and seedling emergence relations. I. Soil type, moisture tension, temperature and planting depth effects. Agron. J. 57: 289–291.

Shainberg, I., M.E. Sumner, W.P. Miller, M.P.W. Farina, M.A. Pavan, and F.V. Fey. 1989. Use of gypsum on soils: a review. Adv. Soil Sci. 9: 1–111.

Sumner, M.E. 1992. The electrical double layer and clay dispersion. In M.E. Sumner and B.A. Stewart (eds) "Soil Crusting: Chemical and Physical Processes", Lewis Publishers, Boca Raton, FL: 1–31.

Sumner, M.E. and B.A. Stewart (eds) 1992. Soil Crusting: Chemical and Physical Processes. Adv. Soil Sci., Lewis Publishers, Boca Raton, FL, 371 pp.

Sutch, J.A., G.C. Steinhardt, and D.B. Mengel. 1983. Crust strength of two Indiana soils as influenced by soil properties and tillage. Indiana Academy of Science, Indianapolis, IN.

Taylor, H.M. 1962. Seedling emergence of wheat, grain sorghum and guar as affected by rigidity and thickness of surface crusts. Soil Sci. Soc. Am. Proc. 26: 431–433.

Valentin, C. and J.F. Ruiz Figueroa. 1987. Effects of kinetic energy and water application rate on the development of crusts in a fine sandy loam soil using sprinkling irrigation and rainfall simulation. In N. Fedoroff, L.M. Bresson, and M.A. Courty (eds) "Soil Micromorphology", L'Association Francaise Pour P'Etude du Sol, Plaisir, France: 401–408.

Van Olphen, H. 1963. An introduction to clay colloid chemistry. J. Wiley & Sons Inc., New York.

West, L.T., S.C. Chiang, and L.D. Norton. 1992. The morphology of surface crusts. In M.E. Sumner and B.A. Stewart (eds) "Soil Crusting: Chemical and Physical Processes", Lewis Publishers, Boca Raton, FL: 73–92.

Young, R.N. and B.P. Warkentin. 1966. Introduction to Soil Behavior. The Macmillan Co., New York, 451 pp.

7

Soil Strength and Compaction

Soil strength is an important soil physical property, with numerous applications to agronomy and engineering. Important agronomic applications are those related to impacts of crusting and compaction on plant growth and agronomic yield. Relevant engineering applications are related to trafficability, draft power required to till the soil for alleviating soil compaction, and soil as a foundation for hydraulic and civil structures (e.g., dams, roads, buildings). For detailed discussions on soil strength, readers are referred to textbooks on soil mechanics (Wu, 1982; Whitlow, 1995; Aysen, 2002; Brown, 2001; NAS, 2002).

7.1 BASIC RHEOLOGICAL MODELS

Rheology deals with the study of flow, and the degree and principles of deformation (see Chapter 8). There are several rhelogical models relevant to understanding the soil responses called strain (ε) (or deformation) and stress (σ) (or pressure). Some basic models used to explain stress–strain behavior ae discussed by Yong and Warkentin (1966) and Hillel (1980). Available models can be grouped under three categories: elementary, complex, and compound.

7.1.1 Elementary Models

The stress–strain behavior is explained by three simple models:

Hookean Model. This linear spring model states that strain (ε) is proportional to stress (σ), and that strain occurs instantaneously when stress is applied and it disappears when the stress is removed.

$$\sigma = K\varepsilon \tag{7.1}$$

where σ is expressed in units of pressure or force per unit area (PSI, bars, Pa), K is constant of proportionality (units of pressure), and ε is a dimensionless ratio (L/L). This model applies to perfectly elastic bodies.

Newtonian Model. The stress–strain relationship is characterized by a constant rate of strain (ε) under an applied stress (σ).

$$\sigma = K'\dot{\varepsilon} \tag{7.2}$$

where $\dot{\varepsilon} = d\varepsilon/dt$ and K' is constant of proportionality and has units of stress (bars) x time. When $\varepsilon = 0$ at $t = 0$, Eq. (7.3) can be rewritten as follows:

$$\sigma K' = \dot{\varepsilon} \tag{7.3}$$

Yield Stress Model. There is a threshold stress needed to initiate a strain. Such a type of stress–strain behavior follows a yield–stress model.

$$\sigma > \alpha_0 \quad \text{for } \varepsilon = 0 \text{ where } \sigma_0 = \text{frictional resistance}$$
$$\sigma > \alpha_0 \text{ for finite } \varepsilon \tag{7.4}$$

7.1.2 Complex Models

Soil is a complex mixture of four components and three phases (see Chapter 2). Thus, stress–strain behavior of soils does not follow any of the elementary models. Such models are not sufficient to accurately represent stress–strain–time behavior of soils. Thus, a combination of two or three models is often used to assess the stress–strain behavior of soils. Elementary models, however, comprise essential components of complex models.

St. Vincent Model. This model involves a combination of the Hookean and Yield Stress models in a series. The stress–strain behavior is explained by the condition of an elastic strain up to the yield point.

Kelvin Model. The Kelvin model is a combination of the Hookean and Newtonian models. It involves the parallel coupling of two models. The strain is characterized by elastic deformation delayed by time effects. This behavior is also sometimes called the Voigt model.

$$\sigma = \sigma_{\text{Hookean}} + \sigma_{\text{Newtonian}}$$

$$\sigma = K\varepsilon + K'\dot{\varepsilon} \tag{7.5}$$

$$\frac{\sigma}{K'} = \frac{K}{K'}\varepsilon + \frac{d\varepsilon}{dt}$$

Maxwell Model. This complex model is used to explain the stress–strain behavior using the series coupling of the Hookean and Newtonian models. Thus,

$$\varepsilon_{\text{total}} = \varepsilon_{\text{Hookean}} + \varepsilon_{\text{Newtonian}} \tag{7.6}$$

$$\varepsilon_{\text{total}} = \frac{\sigma}{K} + \frac{\sigma}{K'} \tag{7.7}$$

7.1.3 Compound Models

These models involve a combination of complex and simple models to achieve a higher order of combination for explaining the stress–strain behavior of soils.

Linear Model. A combination of Hookean model and Maxwell model in parallel is called the Linear model. This model is used to explain the stress–strain relationship of a material with skeletal structure.

$$\sigma = \sigma_{\text{H}} + \sigma_{\text{M}} \tag{7.8}$$

Burger Model. This model combines in series the Maxwell and Kelvin models.

Bingham Model. This model combines Newtonian model in a series with the St. Venant model.

Of the three compound models, the Burger model is applicable to simulating the soil behavior.

7.2 STRESS–STRAIN RELATIONSHIP

Soil rheology also involves the study of soil strength or soil's ability to bear or withstand stress without collapsing or deforming excessively. Soil strength is attributed to forces of cohesion and adhesion and varies with soil moisture content. When subjected to external force or stress, soil undergoes different types of deformation or strain. There are different types of stress that result in different types of strain.

7.2.1 Stress (Tension or Compression)

Stress refers to the force per unit area. For a given plane at a point, the resultant stress vector may be divided into two components: normal and tangential stress.

Normal Stress (σ). Normal stress is caused by a force vector perpendicular to the area of action [Eq. (7.9)]

$$\sigma = F_n/A \tag{7.9}$$

where F_n is the force acting normal to the area A. The transmitted normal stress generally decreases with distance from the applied load and with distance from its line of action.

Tangential Stress (τ) or Shearing Stress. This stress is caused by a force vector parallel to the area of action [Eq. (7.10)].

$$\tau = F_t/A \tag{7.10}$$

where F_t is the tangential force acting on area A.

7.2.2 Strain

Strain refers to soil's reaction to stress in the form of deformation that the stress has created. There are two principal types of strain: longitudinal strain and shear strain.

Longitudinal Strain (ε). Longitudinal strain refers to the relative change in length [Eq. (7.11)].

$$\varepsilon = \Delta L/L \tag{7.11}$$

where ΔL is the change in soil length and L is the original length. The soil may be compressed or expanded (swelling).

Shear Strain (γ) or Tangential Strain. This strain refers to the angular deformation [Eq. (7.12)].

$$\gamma = u/h \tag{7.12}$$

where u is lateral or tangential displacement, h is the height of the soil, and the ratio u/h is the tangent of the deformation angle (Fig. 7.1). The strain defined by Eqs. (7.11) and (7.12) refers to a small degree of deformation, usually less than 0.1%.

7.2.3 Time-Dependent Stress and Strain

Time-dependent longitudinal strain (ε') refers to the rate of change in longitudinal strain over time (t). Differentiating Eq. (7.11) with respect to time (t):

$$\varepsilon' = \frac{d\varepsilon}{dt} = \frac{1}{L}\frac{d(\Delta L)}{dt} \tag{7.13}$$

where ε' is the time rate of elongation or contraction, L is length and t is time.

Similarly, time-dependent stress application can be expressed as per Eq. (7.14), which is obtained by differentiating Eq. (7.12) with respect to time (t):

$$\gamma' = \frac{d\gamma}{dt} = \frac{1}{h}\frac{du}{dt} = v/h \tag{7.14}$$

where γ' is the velocity (v) gradient (du/dt) in the direction perpendicular to that of the shearing displacement. The time dependent stress–strain relationship of soil (body) govern several rheological properties such as elasticity and plasticity. Plastic properties are important to soil tilth.

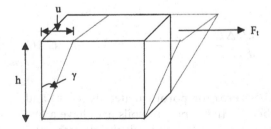

FIGURE 7.1   Shear strain exemplified by angular deformation.

7.3 ELASTICITY

An elastic material deforms under stress instantaneously and retains its new form as long as the stress is maintained. However, it returns to its original form when the stress is released. Soil, similar to other solids, is not a perfectly elastic material. Most natural bodies do not return to their original form, and exhibit some residual deformation after release of stress. The rate and total magnitude of deformation is called "creep," which depends on the "relaxation" characteristic of the material. Relaxation refers to the tendency of a material to relieve stresses gradually through internal structural adjustments. Perfectly elastic bodies exhibit the following characteristics that can be expressed through well-defined laws called "elastic constants":

1. *Young's Modulus*: Based on the college physics experiment relating weights hung from a spring and the length to which it is stretched, Hooke's law states that strain (ε) is proportional to stress (σ). Further, strain (ε) occurs instantly when the stress (σ) is applied and it disappears when the stress is removed. This relationship between normal stress and the attendant strain it produces is expressed in terms of Young's modulus [Eq. (7.15)].

$$\varepsilon = \frac{\sigma}{Y_m} \tag{7.15}$$

where Y_m is Young's modulus.

2. *Poisson's Ratio (v)*: Normal (σ) or tangential stress (γ) may result in change in length (L) as well as thickness of a material (d). Poisson's ratio (P_R) is defined as the "ratio of elongation along one axis to the corresponding contraction of another axis." It is dimensionless and its value ranges from 0 for rigid bodies to 0.49 for rubber. The value of P_R for soils depends on total porosity (f_t) and macroporosity (f_a).

$$P_R = -\frac{\Delta d/d_o}{\Delta L/L_o} = -\frac{\Delta d/\Delta L}{d_o/L_o} \tag{7.16}$$

Poisson's ratio is small (approaches zero) for porous materials (cork) and about 0.5 for elastic material (rubber). Highly porous soils may have a low Poisson's ratio and extremely clayey soils with high swell/shrink properties may have a high Poisson's ratio.

3. *Modulus of Shearing*: Similar to the Hooke's law and Young's modulus in case of the normal stress, the elastic relation for shearing stress is expressed by the modulus of shearing or rigidity [Eq. (7.9)]:

$$\gamma = \tau/M_R \tag{7.17}$$

where M_R is the modulus of rigidity or shearing.

4. *Bulk Modulus*: Rather than decrease in length (in case of normal stress) or thickness (in case of tangential stress), isotropic stress (e.g., immersion of a body in a liquid) can change the total volume. The magnitude of change in volume (ΔV) is proportional to the pressure (P) as per Eq. (7.18).

$$\Delta V \propto P, \quad \Delta V = \frac{P}{B_M} \tag{7.18}$$

The proportionality constant B_M is called the *bulk modulus* and refers to the volume compression or expansion relative to the original volume. Depending on soil structure and layering, it may be isotropic or anisotropic. Isotropism in soil may also depend on soil properties. Soil may be anisotropic in relation to hydraulic conductivity (refer to Chapter 11) but isotropic in relation to texture. These four elastic constants are inter-related [Eqs. (7.19) to (7.22)], and can be verified through solving the algebraic equations:

$$Y_m = 9B_M \cdot M_R/(3B_M + M_R) \tag{7.19}$$

$$P_R = (3B_M - 2M_R)/(6B_M + 2M_R) \tag{7.20}$$

$$M_R = Y_m/2(1 + P_R) \tag{7.21}$$

$$B_M = Y_m/3(1 - 2P_R) \tag{7.22}$$

7.4 PLASTICITY

Plasticity refers to the property of a body to deform progressively under stress and to retain its deformed shape when the stress is removed. Some materials are ideally plastic. In such materials, the behavior is elastic up to a certain magnitude of stress (σ_o) beyond which the deformation exhibits plastic behavior. This threshold or critical stress (σ_o) is called the yield point. Transition from elastic to plastic behavior may be gradual rather than abrupt and is determined by a property of the material called "strain hardening." Strain hardening in metals under stress is caused by deformation, internal structural changes, and recrystallization. A soil under compactive stress may also undergo structural changes and exhibit "strain hardening." A third category of materials is ideally brittle material, which exhibits elastic properties under stress up to the peak stress and all strength is lost upon failure.

For details on the stress–strain relationship of materials, readers are referred to reviews on soil mechanics (Barber, 1965; Hillel, 1980; Ghildyal and Tripathi, 1987).

7.5 STRESS–STRAIN RELATIONSHIP IN SOIL

In soils, strain is often large as is evident from an increase in soil bulk density from $0.8\,Mg/m^3$ under forest to $1.6\,Mg/m^3$ with cultivation (Lal and Cummings, 1979; Lal, 1985; 1996). Further, change in bulk density or strain (ε) may not be uniform. Therefore, stress–strain relationship in soils is difficult to predict, is soil-specific and must be determined experimentally. Soil is neither a perfectly elastic nor an ideally plastic material. Being highly heterogenous, soil deformation in response to stress is a complex process.

Assume a soil is subjected to a known stress. When the stress is small, the soil may deform slightly (low strain) and may recover its original shape when the stress is released (elastic deformation). If the stress is large, it may produce larger strain resulting in permanent deformation from which soil may not recover even when the stress is released (plastic deformation). The strain increases linearly up to the critical or threshold stress (σ_o). This region of linear response represents the "elastic region" and response follows the theory of elasticity. Permanent deformation occurs as the stress is increased beyond the threshold, critical stress or yield. This is also known as the failure stress or the highest stress that the soil can safely withstand. In case of brittle or sensitive soil, it completely loses its strength. Tensile failure of soil is a measure of the cohesive component of the shear strength. In contrast, failure of soil by shear is definable when it is in rigid or brittle state and exhibits a distinct failure plane. This type of failure is observed in relatively dry and cohesive soils (see Sec. 7.1 on basic rheological models). In contrast to elastic, plastic or viscous material, soil may exhibit a combination of these responses as follows:

1. Elastoplastic soils are those that exhibit partial recovery when stress is removed.

2. Viscoelastic soils are those that exhibit time-dependent soil deformation [Eq. (7.5)], as is the case in the creep phenomena.
3. There are numerous ramifications of the stress–strain behavior of the soil including soil compaction and soil strength.

7.6 SOIL STRENGTH

Soil strength is the resistance that has to be overcome to obtain a known soil deformation. It refers to the capacity of a soil to resist, withstand, or endure an applied stress (σ) without experiencing failure (e.g., rupture, fragmentation, or flow). It is soil's resistance that must be overcome to cause physical deformation (ε) of a soil mass. It implies the maximal stress which may be induced in soil without causing it to fail. As stated in the introductory paragraph of this chapter, the concept of soil strength has numerous applications in agriculture and engineering. In agriculture, soil strength has applications to root growth, seedling emergence, aggregate stability, erodibility and erosion, compaction and compactability, and draft requirements for plowing. In engineering, soil strength has applications to soil and slope stability, foundation engineering, and bearing capacity with regard to agricultural application, high soil strength may have both positive and negative effects. Positive effects are those related to soil trafficability and bearing capacity, and resistance to compactive and erosive forces. Negative effects are those due to high draft power requirement, poor root growth, low seedling emergence, and poor crop stand.

Soil strength may be of two types: (i) resistant to volumetric compression, and (ii) resistant to linear deformation or shear strength. The resistance to volumetric compression can be measured by evaluating stress density relationship at different soil moisture content. This may involve measurement of penetration resistance of a soil at different density and different soil moisture content (potential). For a given bulk density, soil strength decreases with increasing soil moisture content. For a given soil moisture content, soil strength increases with increase in soil bulk density. In general, fine-textured soils at low moisture content exhibit high strength. Shear strength of a soil is the resistance to deformation by continuous shear displacement of soil particles due to tangential (shear) stress. Soil's shear strength is due to three separate but interactive forces: (i) the structural resistance to displacement of soil particles, (ii) the frictional resistance to translocation between the individual soil particles due to interparticle contacts, and (iii) forces of cohesion and adhesion.

7.6.1   Mohr Theory of Soil Strength

This theory is based on the functional relationship between normal stress (σ) and tangential or shearing stress (τ). The envelope of the family of circles is used as a criterion of shearing strength of soil. When a series of stress states just sufficient to cause failure is imposed on the same soil material, these states can be plotted as a set or family of Mohr circles. The line tangent of these circles, called the envelope of the family of circles, is used as a criterion of shear strength. When this envelope is a straight line, it can be described mathematically by Eq. (7.23) (Fig. 7.2).

$$\tau = \tau_0 + b\sigma \qquad\qquad (7.23)$$

where the constant τ_0 is the intercept of the envelope line on the τ axis, and constant b is the tangent of angle ϕ which the envelope line makes with the horizontal line. This linear relationship between τ and σ is analogous to the Coulomb's law that states that "the frictional resistance toward a tangential stress tending to slide one planar body over another is proportional to the normal force pressing the bodies together." In view of this analogy for sliding friction between bodies, the angle ϕ is called the angle of internal friction. The intercept (τ_o) is the shear stress needed to cause failure when normal stress (σ) is zero, and is called soil cohesion (C) or cohesiveness. Substituting these terms in Eq. (7.23) yields Eq. (7.24) used to express soil shear strength.

$$\tau = C + \sigma \tan\phi \qquad\qquad (7.24)$$

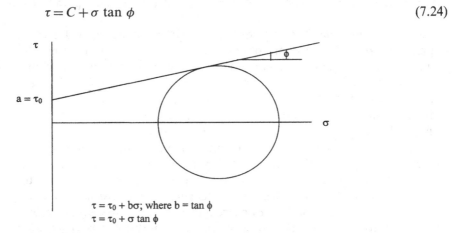

$\tau = \tau_0 + b\sigma$; where $b = \tan\phi$
$\tau = \tau_0 + \sigma \tan\phi$

FIGURE 7.2   The functional relationship between shearing stress (τ) and normal stress (σ) is given by Mohr's circle (a or τ_o is the intercept and constant b is the tangent of angle ϕ).

7.6.2 Factors Affecting Soil Strength

Soil deformation under stress happens when solid constituents (both primary and secondary particles) are able to separate and move with respect to each other. Particle movement under stress is restricted by particle-to-particle friction and interparticle bonds. Frictional forces increase with: (i) increase in soil bulk density, (ii) decrease in soil moisture content, and (iii) increase in over burden pressure. Forces due to interparticle bonds include: (i) cohesion due to surface tension at the air–water interface and soil matric potential or pore water pressure, (ii) link bonds or particle-to-particle contents, e.g., mineral–mineral, mineral–organic–mineral, etc. There are numerous types of cementing agents that bind the particles together (refer to Chapter 4).

Soil properties affecting soil strength are discussed by Guerif (1994) and include the following:

Soil Structure. Aggregate size is an important determinant of soil strength. Stress at fracture decreases exponentially with increase in aggregate (clod) diameter.

Soil Bulk Density. It determines the magnitude of particle-to-particle contacts. Effects of soil bulk density on soil strength are confounded with those of soil moisture content. Because soil bulk density is related to total volume (V_t) and total porosity (f_t), soil strength may be expressed on the basis of strength–volume or strength–porosity relationships. Soil strength decreases with increase in total soil volume [Eq. (7.25); Braunack, et al., 1979].

$$\ln S = - F \ln V + A \qquad\qquad (7.25)$$

where S is soil strength, V is soil volume, A is an adjustment factor, and F is soil constant which is a measure of the ease of breakdown of large clods into smaller aggregates. The factor F, called soil friability (Utomo and Dexter, 1981), is defined as "the tendency of a mass of unconfined soil to break down and crumble under applied stress into a particular size range of smaller fragments." The topic of soil friability is discussed in Chapter 8.

Properties of Soil Solids. Soil constitution (i.e., particle size distribution, clay mineralogy, and soil organic matter concentration) affects soil strength through changes in aggregation, soil bulk density and specific volume, moisture content, and types of pores. Relative proportion of textural versus structural pores can affect soil strength. Soil organic matter influences soil strength through its effects on aggregation and porosity.

Clayey soils have more strength and cohesiveness (*C*) than sandy soils. Dry sand, being non-cohesive, may actually expand during shear, a phenomenon known as dilatancy. Moist sand is apparently cohesive and can withstand traffic (is trafficable) but dry sand cannot. Guerif (1990) observed that tensile strength increases linearly with clay content [Eq. (7.26)].

$$S_T = m\,(\text{clay}) + b \qquad\qquad\qquad (7.26)$$

where S_T is the mean tensile strength of dry spherical aggregates of 2–3 mm diameter, clay content is expressed as a fraction (g/g), and *b* is an empirical constant. The intercept *m* is considered as the mean tensile strength of an ideal clay representative of different soils involved in the regression analysis. Textural tensile strength is an intrinsic property of the soil. The textural strength, defined at the scale of the smallest significant elementary volume of cohesive material, is considered as the upper limit of the strength that a given soil may exhibit following a severe compaction (Guerif, 1994).

Soil Moisture Content. Soil strength increases with decrease in soil moisture content or moisture potential. Soil drying increases strength by increasing capillary cohesion as it increases the effective stress, and compactness by shrinkage.

7.6.3 Measurement of Soil Strength

Tensile strength is a sensitive indicator of the condition of a soil and is a useful measure of strength of individual soil aggregates. Two principal theories describing the strength of porous materials like soils are: (i) Mohr–Coulomb maximum shear strength and (ii) Griffiths's tensile failure theory. The Mohr–Coulomb theory states that shear failure occurs when the maximum resolved shear stress on fracture plane is attained. According to Griffin theory, fracture occurs when the highest local tensile stress in the longest cracks reaches the critical tensile strength of the material (Hadas and Lennard, 1988).

The tensile strength of a spherical particle can be determined by a simple crushing test. A force of magnitude *F* applied across the poles of a particle causes elastic deformation of the particle (Fig. 7.3). This produces a proportional tensile stress in the center of particle perpendicular to the direction of applied force. If the force *F* is increased gradually, the internal tensile stress reaches the tensile strength (*Y*) of the particle, and a slight increase thereafter results in cracking of particle on a plane through the polar diameter.

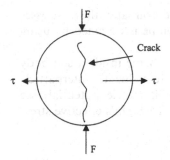

FIGURE 7.3 Schematic of loading (F) of an aggregate at poles and the resultant tensile stress (τ) at right angles to F and development of a crack as τ approaches tensile strength (S_T) of soil.

Measurement of soil strength involves characterization of two parameters of Eq. (7.24): (i) cohesiveness C, and (ii) angle of internal friction ϕ. The cohesiveness factor C represents the adherence or bonding of soil particles which must be broken if the soil is to be sheared. The angle of internal friction ϕ represents the frictional resistance encountered when soil is forced to slide over soil.

There are direct and indirect methods of measuring C and ϕ, the strength properties of soil. These methods are described in details by Sallberg (1965), Wu (1982), Snyder and Miller (1985), Ghildyal and Tripathi (1987), Guerif (1994), and others.

Direct Methods
The direct methods involve a direct application of stress to a soil sample.

Laboratory Techniques. In the direct shear test, the shear strength (or the shearing resistance) and the normal stress are both measured directly at a predetermined plane of a soil. The primary objective of strength measurement is to determine the failure envelope, or the relationship between τ and σ. The values are plotted on $\tau-\sigma$ coordinate system, and the line connecting the points is an envelope from which C and ϕ are computed (Fig. 7.2).

The direct shear test has several limitations: (i) the shearing plane does not remain constant during the test, (ii) stresses vary even though normal and tangential forces remain constant, and (iii) test results are influenced by the size and shape of the container.

The triaxial shearing test is designed to overcome these limitations. In this test, the failure surface is not predetermined, and longitudinal and lateral stresses are applied to a sample of the soil and these stresses determine the plane of failure. The strength envelope is obtained by using different combinations of the applied stresses.

The internal or total stress (σ) acting on any plane inside a soil body consists of two components: (i) the effective stress due to interparticle pressure, and (ii) the pore–water pressure or soil matric potential (see Chapter 10). These relationships are described by Terzaghi's effective stress equation [Eq. (7.27), Terazghi, 1953].

$$\bar{\sigma} = -\Psi + \sigma \tag{7.27}$$

where $\bar{\sigma}$ is the effective stress, σ is the internal or total stress, and Ψ is the hydrostatic pressure. In unsaturated soil, Ψ is negative and increases effective stress. The term effective stress is also called the inter-granular stress, and Ψ the neutral stress, because in saturated soil the hydrostatic pressure acts equally in all directions.

A special case of the cylindrical shearing test is called the "unconfined compression test" in which no lateral pressure is applied. There are other laboratory techniques of measuring tensile strength (Gill, 1961; Vomocil et al., 1961).

Cohesive strength of soil is also measured under laboratory conditions by measuring the modules of rupture (Richards, 1953; Reeve, 1965). Modulus of rupture is defined as the maximum force per unit area that a material can withstand without breaking. It is a measure of the breaking strength of the soil, and is used to assess the physical status of seedbed, especially the crust strength (see Chapter 6). This method involves making a small briquette of the soil of known width (b) and thickness (d). The briquette is prepared to simulate seedbed preparation involving wetting and drying of soil and eventually crust formation. The briquette is loaded on both ends until it fails. The modulus of rupture (σ_b) is computed from Eq. (7.28).

$$\sigma_b = 3Fl/2bd^2 \tag{7.28}$$

where F is the force applied to cause failure, l is the length of the briquette, b is the breadth, and d is depth or thickness. For a cylindrical briquette, σ_b is computed by Eq. (7.29).

$$\sigma_b = Fl/r^3 \tag{7.29}$$

Modulus of rupture is also related to soil crusting (Richard, 1953), and is an indirect method of measuring soil strength. Changes in the dimension of the briquette upon drying are used to compute linear shrinkage [Eq. (7.30)].

$$\% \text{ Linear shrinkage} = \frac{\text{decrease in length}(\Delta L)}{\text{original length of the model}(L)} \times 100 \quad (7.30)$$

Field Techniques. Kirkham, et al. (1959) used the cylindrical specimen to determine the strength required to split a specimen laterally into two longitudinal halves. The modulus of rupture for lateral failure is given by Eq. (7.31).

$$\sigma_b = F/\pi lr \quad (7.31)$$

In situ determination of soil strength under field conditions is done by two methods. A first and simple one is the Vane shear test (ASTM, 1956). A vane is driven into the soil to a known depth and then rotated to measure the torque (T). The torque is related to soil cohesiveness as per Eq. (7.32).

$$T = C\pi(1/2\, d^2 l + 1/3\, d^3) \quad (7.32)$$

where C is soil cohesiveness, d is diameter of the vane, and l is length of the vane. If the length-to-radius ratio is 4:1, soil cohesiveness can be computed from Eq. (7.33).

$$C = 6T/7\pi d^3 \quad (7.33)$$

The second field method is based on the measurement of tensile strength, which is the normal force per unit area required to detach or pull apart one section of soil from another (Sourisseau, 1935).

Indirect Methods

The indirect methods involve indirect failure induced by applying external compressive forces or bending moments that generate tensile or shear stresses within the sample.

Strength of Soil Aggregates. Soil aggregates are highly irregular, they are placed in the most stable position, and force is applied across the minor principal diameter. For a particle of incompressible material with Poisson's ratio (ratio of transverse contraction strain to longitudinal extension strain in the direction of stretching force) of 0.5 and diameter d, the tensile strength for a polar force F at failure is given by Eq. (7.34):

$$Y = R\frac{F}{d^2} \qquad (7.34)$$

where R is the proportionality constant and usually equal to 0.576, although it may be correlated to bulk density and/or pore size distribution. Tensile deformation is considered positive and compressive deformation is considered negative. The definition of Poisson's ratio [Eq. (7.16)] contains a minus sign so that normal materials have a positive ratio. Aggregate diameter needs to be determined before tensile strength can be calculated from above equation. Since aggregates are irregularly shaped, exact determination of an effective spherical diameter is not possible. One method employs sieving of soil aggregates through two sieves of opening sizes as s_1 and s_2 ($s_1 > s_2$). The mean diameter of the aggregates passing through s_1 but retained on s_2 can be calculated as Eq. (7.35).

$$d = \frac{s_1 - s_2}{2} \qquad (7.35)$$

the ratio $(s_1-s_2)/s_2$ is to be kept small. The other method involves measurement of the diameter of each individual aggregate (with calipers) and then calculating the effective mean diameter as the arithmetic or geometric mean or as a weighted mean mass or weighted mean density basis (Dexter and Kroesbergen, 1985).

There are numerous factors that affect tensile strength of aggregates. Analysis of the fracture of air-dry soil aggregates is important for the management of soil structural stability, root growth and tillage operations (Hadas and Lennard, 1988; Causarano H., 1993). The effect of aggregate size on root growth and nutrient uptake is due to the increase in mechanical stress adjacent to the soil–root interface with increasing aggregate size (Mishra et al., 1986). The knowledge of magnitude and distribution of aggregate strengths is key to understanding the amount of aggregate break up during tillage or movement of farm machineries. Factors influencing the tensile strength of soil aggregates are: moisture content, clay content, organic matter content, and size of aggregate. The tensile strength of soil aggregates generally decreases with increasing moisture content and/or aggregate size (Causarano, 1993).

7.7   SOIL COMPACTION

Soil compaction can be conceptually viewed in a dynamic or a static situation, and in practical applications. In a dynamic situation, it is a physical deformation or a volumetric strain. In a static situation, it is the characteristic related to soil resistance to increase its bulk density. In practice, soil compaction is a process leading to compression of a mass of soil into a smaller volume and deformation resulting in decrease in total and macroporosity and reduction in water transmission and gaseous exchange. The degree or severity of soil compaction is expressed in terms of soil bulk density (ρ_b), total porosity (f_t), aeration porosity (f_a), and void ratio (e). The volume decrease is primarily at the cost of soil air, which may be expelled or compressed. The compression of soil solids (i.e., change in ρ_s) and water (i.e., change in ρ_w) is evidently not possible. However, soil solids may be rearranged or deformed as a result of compactive pressure.

Compression of a moist soil due to external load may displace the liquid and increase the contact area between two particles (Fig. 7.4). The magnitude of increase in contact area depends on the degree of rearrangement or deformation of the particles. The menisci formed by the liquid may also change due to differences in the contact area. The shape of the meniscus depends on surface tension forces, which are usually small compared with the external load. The deformation may be elastic and soil particles may regain their original shape when the applied load is released.

The degree of deformation and rearrangement depends on soil structure and aggregation, and on the extent to which soil particles can change position by rolling or sliding. For partly saturated clayey soils, the volume change depends on reorientation of the particles and displacement of water between particles. The particle rearrangement may lead to closed packing (Chapter 3) with attendant decrease in void ratio [Eq. (7.36)].

$$e = e_0 - c \log P/P_0 \tag{7.36}$$

where e_0 is the void ratio at the initial pressure P_0, c is the slope of the curve on semilogrithmic plot, and P is the applied pressure that changed the final void ratio to e. Degree of soil compaction may also be expressed in terms of total porosity in relation to the external load (Soehne, 1958) [Eq. (7.37)].

$$f_t = -A \ln P + f_{10} \tag{7.37}$$

where f_t is total porosity, f_{10} is the porosity obtained by compacting loose soil at a pressure of 10 PSI, A is the slope of the curve, and P is the applied pressure.

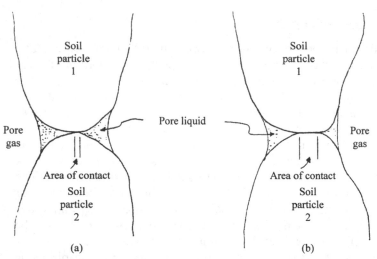

FIGURE 7.4 Two soil particles in contact in a partly saturated condition: (a) no external load; (b) with an external load applied.

Soil compaction is extremely relevant to agriculture because of its usually adverse impact on root development and crop yields (Table 7.1); civil engineering because of its relation to settlement, stability, and groundwater flow; and to environments because of its effects on erosion, anaerobiosis, transport of pollutants in surface and sub-surface flow, and nature and rate of gaseous flow from soil to the atmosphere. From an agricultural perspective especially in relation to plant root growth, there is an optimal range of soil bulk density, which for most soils is $< 1.4\,Mg/m^3$. However, the optimum range of soil bulk density may differ among soils and crops (Kyombo and Lal, 1994). For some soils (e.g., Andisols or soils of volcanic origin) the optimal density may be as low as 1.0. A similar case may be in soils containing a high level of soil organic matter content. It is precisely because of these differences in response to bulk density that effects of compaction on crop yield are highly soil-dependent. An example of variable response is shown by the data in Table 7.1, which indicate severe adverse effects on yield of corn on a clayey soil but slight or more on a loamy soil. Soils of the tropics are easily compacted, and can cause severe reductions in crop yields (Tables 7.2 and 7.3). Thus, the objective of soil management is to maintain soil bulk density within the optimal range that favors root growth, water retention and transmission, and gaseous exchange. In contrast, engineers consider soil bulk density in terms of the strength and stability of the foundation. The desirable goal, therefore, is to

TABLE 7.1 Effects of Axle Load on Corn Grain Yield on Coarse- and Fine-Textured Soils

Compaction level/axle load	Grain yield (Mg/ha)		
	Wooster silt loam soil–corn grain yield		
	1988	1997	1998
Control	5.8	6.1	8.6
7.5 Mg (controlled traffic)	5.0	5.4	8.2
7.5 Mg (entire plot)	5.0	5.4	7.3
LSD (0.05)	0.8	NS	0.9
	Hoytville clay soil–soybean grain yield (1996)		
	No till	Chisel plow	Moldboard plow
Control	2.6	2.6	2.3
10 Mg	2.4	2.2	2.2
20 Mg	2.4	2.1	2.0
LSD (0.05)			
(i) compaction	0.2		
(ii) tillage	0.2		
	Hoytville clay soil–corn grain yield (1990)		
	No till	Chisel plow	Moldboard plow
Control	9.3	7.7	6.5
10 Mg	5.2	3.9	3.6
20 Mg	2.7	3.5	4.1
LSD (0.05)			
(i) compaction (C)	0.6		
(ii) tillage (T)	1.6		
C × T	0.9		

Source: Adapted from Lal and Ahmadi, 2000.

form the densest and tightest possible soil condition. While achieving the highest soil compaction is the goal for civil engineers, it is a major concern for soil scientists and agricultural engineers.

7.7.1 Soil Compactibility

Soil compaction or densification happens due to external load or force applied to the soil. The force applied per unit area is defined as stress, which

TABLE 7.2 Effects of Progressive Decline in Structure of a Tropical Alfisol with Continuous Cultivation on Corn Grain Yield in Southwestern Nigeria

Tillage method	First season corn grain yield (Mg/ha)							
1980	1981	1982	1983	1984	1985	1986	1987	Mean
Plow till	2.7	3.1	3.8	3.7	4.8	2.2	2.0	1.7	3.0
No till without mulch	2.1	2.8	4.2	3.6	4.2	3.6	2.3	1.7	3.1
No till with mulch	2.5	3.6	4.6	4.4	5.1	3.5	2.8	1.6	3.5
LSD (0.05)	NS	NS	NS	NS	0.9[a]	0.8[a]	0.5[a]	NS	

Source: Lal, 1997.
[a]Treatments differ at 5% level of probability.

TABLE 7.3 Decline in Corn Grain Yield on a Tropical Alfisol Due to Soil Compaction Caused by Vehicular Traffic Under Mechanized Farming

Tillage method	Maize grain yield (Mg/ha)					
	1975	1976	1977	1978	1979	1980
No till	2.8	4.5	4.8	5.0	3.8	3.0
Plow till	2.7	4.0	3.9	4.0	2.9	1.0

Source: Lal, 1984.

may be normal stress when it is perpendicular to the soil or shear stress when it has a tangential component. Compression is the process of increase in soil mass per unit volume due to external load. The load may be static or dynamic. The latter is applied in the form of vibration, rolling, or trampling (Fig. 7.5). While compression in unsaturated soils is called "compaction," that in saturated soils is termed "consolidation." Soil compressibility is the "resistance of a soil against volume decrease by external load." In comparison, soil compactability is the difference between the initial bulk density and the maximum bulk density to which a soil can be compacted by a given amount of energy at a defined moisture content. Factors affecting soil compactability include the following:

Soil Wetness

Soil's response to external load depends on soil moisture content (w). There is an optimum range of w at which the soil is most compactable. In general, ρ_b changes nonlinearly in relation to change in w (Fig. 7.6). Beginning with a low moisture content, increase in w serves to render the soil more plastic and workable and facilitate the compaction process (Hogentogler, 1936;

(a)

(b)

FIGURE 7.5 (a) A single-axle grain cart with capacity of 10 or 20 Mg can cause severe compaction during harvest in the fall. (b) A kneading roller is used to create a compact road bed. Spikes cause more compaction than a smooth roller.

Olson, 1962). The dry bulk density increases with an increase in w, and the maximum ρ_b is obtained at an optimum w, beyond which ρ_b drops with further increase in w. The magnitude of the peak ρ_b at a given w depends on soil texture and the load applied. The laboratory evaluation of soil's compactability in relation to w and the load is done according to the Proctor compaction test (Proctor, 1933; Lambe, 1951). The zero-air-void curve

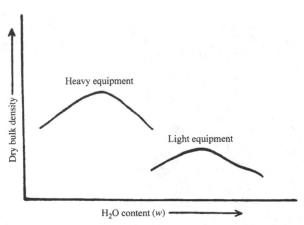

FIGURE 7.6 Relationship between moisture content and soil compaction.

obtained when the Proctor test is done at different moisture content is the w vs. ρ_b curve for a saturated soil. Compaction curves of all soils approach this curve at high w. Well-graded soils can be compacted to higher ρ_b than poorly graded soils, and the effect of w on ρ_b is more pronounced in heavy-textured than coarse-textured or cohesionless soils. Compactability is significantly influenced by soil organic matter content and slip-induced shear. In addition to determining compactability, it is also useful to compute the relative density [Eq. (7.38)].

$$R_d = \frac{e_{max} - e}{e_{max} - e_{min}}$$
(7.38)

where R_d is the relative density, e is the void ratio of the soil in situ, e_{max} is the void ratio of the soil in the loose state that can be attained in the laboratory, and e_{min} is the void ratio of the soil in the densest state. Rather than a simple proctor density, vibratory maximum density test is done for cohesionless or sandy soils (ASTM, 1965).

Soil Compaction and Wheel Traffic

Heavy traffic of agricultural machinery is a major cause of compaction on arable lands (Gill and Vanden Berg, 1967; Harris, 1971; Chancellor, 1976; Soane and Van Ouwerkerk, 1994). The pressure exerted by pneumatic tires of a single-axle load is proportional to the total weight [Eq. (7.39)].

$$W_v = \sum_{i=1}^{n} (P_w \times A_w)$$
(7.39)

where W_v is total weight of the vehicle at rest, P_w is the pressure exerted by the wheel (inflation pressure in the pneumatic tire), and A_w is the area of contact of wheel with the soil. Therefore, an increase in load increases the pneumatic pressure and/or the contact area. For a rigid surface, increase in pneumatic pressure results in an increase in the contact area. For porous media, however, increase in pressure is also accompanied by soil deformation that causes compaction and formation of a wheel rut. Because of the wall rigidity, the shape of the wheel rut is of W shape, because pressure at the edges is more than that at the center (Gill and van den Berg, 1967; Figs. 7.7 and 7.8). Wheel rut depth or shrinkage of the soil under a load is related to the pressure, as per Eq. (7.40) (Bekker, 1961).

$$Z = M_d P^n \tag{7.40}$$

where P is pressure, Z is depth of wheel rut, M_d is modulus of deformation, and n is constant. For most mineral soils, M_d is about 4 and n is about 2. The pneumatic tire behaves like a rigid wheel in case of extremely high pressure and very soft (extremely wet) soil (Chancellor, 1976). Soil compaction by vehicles with crawler tracks is complicated by other additional factors: (i) backward tilt of the vehicle increasing pressure on the rear side two to three times that of the average pressure, (ii) shearing

FIGURE 7.7 The cold method is a useful technique to determine soil bulk density under field conditions. The cold dipped in saran or any other resin can be used to determine total volume by the water displacement method.

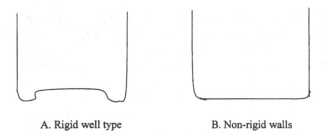

A. Rigid well type   B. Non-rigid walls

FIGURE 7.8   A wheel tire with rigid walls creates a W shaped rut because of high pressure on the edges. A. Rigid well type; B. Nonrigid walls.

force due to tilting and shift of pressure, and (iii) particle displacement due to slippage. Similar compactive effects are observed under moving wheels.

The pressure distribution under wheel can be computed by using the Boussinesq equation, details of which are given by Soehne (1958) and others [Eq. (7.41)].

$$\sigma_z = \frac{3FZ^3}{2\pi(r^2 + Z^2)^{5/2}}\tag{7.41}$$

where σ_z is the stress at a depth Z, F is the total force applied, and r is the radial distance away from the center. When r is 0, directly beneath the wheel, $\sigma_z = 3F/2\pi Z^2$. Soehne (1958) applied the Boussinesq theory to compute the pressure distribution under the tire. A schematic of the pressure distribution under the tire is shown in Fig. 7.9.

7.7.2   Measurement of Soil Compaction

Soil compaction may be measured by assessing bulk density and porosity, and pore size distribution (see Chapter 5). Thus, there are direct and indirect methods of measuring soil compaction (Fig. 7.10).

Soil Bulk Density

There are several methods of measuring soil bulk density. Basic principles, practical applications, and limitations of different methods are described in details by Gardner (1986), Campbell and Henshall (2001), and Campbell (1994). Most methods fall under two categories: (i) measurement of mass and volume, and (ii) assessment of other properties. Because dry bulk density is computed by dividing dry soil mass by its total volume, most direct methods are based on different techniques of measurement of soil volume (Table 7.4).

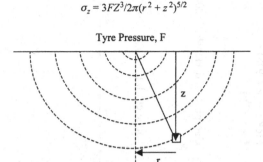

$$\sigma_z = 3FZ^3/2\pi(r^2 + z^2)^{5/2}$$

Tyre Pressure, F

FIGURE 7.9 Pressure distribution under a tire per the Boussinesq equation.

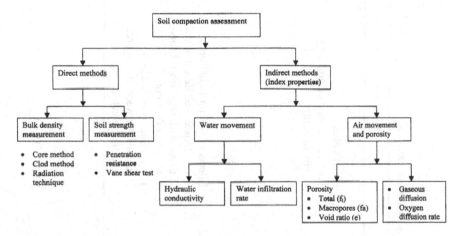

FIGURE 7.10 Assessment methods of soil compaction (properties related to water and air movement are discussed in later chapters).

Radiation methods are based on the principle of measuring the attenuation of γ-rays which is exponentially related to wet bulk density (ρ_b'). Therefore, dry soil bulk density can be determined only if Θ is also determined. Most instruments are equipped with both γ-ray (C_s^{137}) to measure ρ_b and neutron (Am/Be or Ra/Be) sources to measure Θ (see Chapter 10). Radiation techniques for measuring soil bulk density involve γ-rays. The γ-ray photons are emitted with one or more characteristic energies by radioactive nuclei during the decay process. When passed through the soil, γ-rays are either scattered or transmitted.

TABLE 7.4 Methods of Measurement of Soil Bulk Density

Method	Principle	Reference
I. Mass/volume relationship		
1. Core method	Fixed volume	Blake and Hartge (1986)
2. Clod method (Fig. 7.7)	Variable volume enclosed by wax or saran	McKeague (1978); Abrol and Palta (1968); Russell and Balcerek (1944)
3. Sand cone method		Blake (1965); Cernica (1980)
4. Rubber balloon method		McKeague (1978)
5. Excavation method		Lal (1979)
II. Indirect method		
1. Radiation method		
(i) Backscatter gauges	Effects of soil on radiation γ-ray source and detector are fixed without direct transmission of photons	Campbell and Henshall (2001)
(ii) Transmission gauges	The sample to be tested is located between the source and the detector	Soane et al. (1971)

During scatter, the gamma photon is deflected by the electrons within the medium with an attendant loss of energy related to the angle of deflection. The photons interact principally with the electrons, and electron density is related to the bulk density of soil. With backscatter technique, both the source and the detector, usually a Geiger–Mueller (GM) tube, are located within the instrument, thus facilitating a nondistractive evaluation. This technique usually works well for the surface soil horizons. In comparison, the intensity of photons transmission depends on the bulk density of the soil. Attenuation by transmission requires that the source and/ or the detector be lowered down a pre-augured hole. The transmission technique may involve one probe or dual probe. Most density probes need to be calibrated for soil-specific situations. Density probe calibration is influenced by texture, gravel concentration and even soil wetness (Lal, 1974; Fig. 7.11).

There are pros and cons of both direct and the radiation methods (Table 7.5). The choice of methods used depends on objective and the resources available.

Penetration Resistance

Soil compaction is routinely determined by measuring the penetration resistance, which is a measure of soil strength or resistance to deformation. Penetration resistance is "the capacity of the soil in its confined state to resist penetration by a rigid object" (Soil Survey Division Staff, 1993). In addition to soil strength, the penetration resistance also depends on the shape, size, and orientation of the axis of the penetrating object. The penetrating object may be a finger, pencil, stick, nail, root, or a specially designed probe with a specific geometric shape and a device to measure the resistance as it is pushed into the soil. A simple probe is often used to measure soil resistance to penetration to a known depth.

There are two types of penetration tests. In the static penetration test, the penetrometer is pushed steadily into the soil. In a dynamic penetration test, the penetration is driven into the soil by a hammer or falling weight (Davidson, 1965). There are several types of penetrometers including: (i) pocket penetrometer, (ii) proctor penetrometer, (iii) cone penetrometer, and (iv) split-spoon penetrometer. The penetrometer may also have either a flat tip or a conical tip (Carter, 1990). The cone penetrometer is the most commonly used device to measuring soil's mechanical condition. It is an easy device to use (ASAE, 1986). A cone penetrometer is an instrument in the form of a cylindrical rod with a cone-shaped tip designed for penetrating soil and for measuring the end-bearing component of penetration resistance (SSSA, 1997). The resistance to penetration developed by the cone equals the vertical force applied to the cone divided by its horizontally projected area.

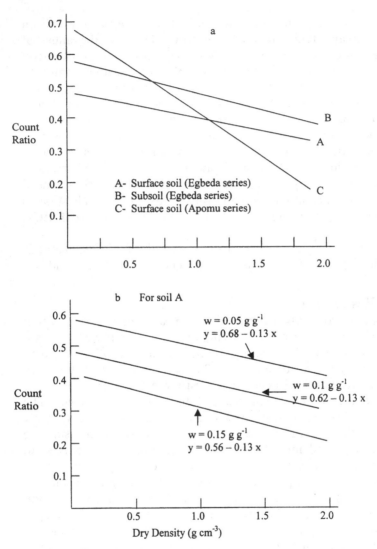

FIGURE 7.11 The effect of (a) soil texture, and (b) gravimetric soil moisture content on density probe calibration. (Redrawn from Lal, 1974.)

Two 30° cone penetrometer tips are specified by the American Society of Agricultural Engineers (1986). One has a base area of $1.3\,\text{cm}^2$, the other $3.2\,\text{cm}^2$. These tips are inserted into the soil to where the base of the cone is flush with the soil surface. The "cone index" is defined as the force per unit

TABLE 7.5 Pros and Cons of Direct and Radiation Techniques of Bulk Density Measurement

Technique	Pros	Cons
1. Direct method	Cheap, simple, safe, routine	Destructive, laborious, small
2. Radiation method	Large sample volume, repeated measurement overtime for the same site, large number of measurements	Expensive, health hazards, require careful calibration, difficulties in transport and repairs

basal area required to push a cone penetrometer through a specified increment of soil (SSSA, 1997). The recommended insertion time is two and four seconds for the smaller and the larger cones, respectively. A penetrometer may be light, easily carried from one site to another, and pushed into the soil by hand. Some hand-held penetrometers are equipped with devices that automatically integrate the penetrometer force over depth (Carter, 1969; Anderson et al., 1980). Other penetrometers are heavy and either tractor-mounted (Wilkerson et al., 1982) or mounted on a frame to which two wheels are fitted for towing the device on the field (Olsen, 1988). Such heavy penetrometers are driven into the soil by an electric motor, and the test data are transferred to a microcomputer equipped with RAM memory.

All cones must be calibrated. The penetrometer measurements are strongly influenced by the antecedent soil moisture content, density, and soil type. Therefore, it is important that penetration resistance measurements are made in conjunction with those of soil moisture measurements. Soil penetration resistance is measured in units of pressure, or the force per unit area (Kg/cm^2, PSI, Kpa, or MPa). The Soil Survey Division Staff has prepared a standard rating table for classifying soils into various resistance classes (Table 7.6).

7.7.3 Management of Soil Compaction in Agricultural Lands

Some soil compaction is inevitable with the use of agricultural machinery and trampling effect of cattle and other traffic (ASAE, 1971; Soane and Van Ouwerkerk, 1994). Soil compaction can cause drastic reductions in crop yields, especially in clayey soils of low permeability and poor internal

TABLE 7.6 Penetration Resistance Classes

Class	Penetration resistance (MPa)
Small	< 0.1
Extremely low	< 0.01
Very low	0.01–0.1
Intermediate	0.1–2
Low	0.1–1
Moderate	1–2
Large	> 2
High	2–4
Very high	4–8
Extremely high	≥ 8

Source: Soil Survey Division Staff, 1990.

drainage (Raghavan et al., 1990; Lal, 1996; Hakansson and Petelkau, 1994; Lindstrom and Voorhees, 1994; Kayombo and Lal, 1994). Yield reduction is caused by mechanical impedance to root growth (Gregory, 1988; Bennie, 1991; Vepraskas, 1994; McKenzie, 1996). Within a textural class, there is a critical limit for root growth, which differs among crops. For example, Taylor and Gardner showed that for a sandy loam soil at field capacity, critical limit for cotton root growth was 3000 KPa, measured with a 5 mm diameter cylindrical tip penetrometer. The topic of critical limit has been reviewed by numerous researchers (e.g., McKenzie, 1996). Compaction management becomes necessary when soil strength exceeds the critical limit. There are two strategies of soil compaction management: (i) minimizing risks of soil compaction or compaction prevention, and (ii) compaction alleviation (Fig. 7.12). Preventive strategies are economic and have less adverse impacts on crop yields and environments than the curative measures of compaction alleviation (Larson et al., 1994). A useful strategy to prevent soil compaction is to minimize the vehicular traffic to the absolutely essential by reducing the number and frequency of operations, and performing farm operations only when the soil moisture content is below the optimal range for the maximum proctor density. Mulch farming and conservation tillage (Lal, 1989; Carter, 1994) reduce the risk of soil compaction for some soils and environments. Guided traffic system, low ground pressure tires (Vermeulen and Perdok, 1994), adoption of dual tires (Fig 7.13), and wide tires (Fig. 7.14) are other innovative ideas of decreasing pressure on soil. The guided traffic system involves confining vehicular traffic to permanent narrow lanes and reducing the fractional area affected by

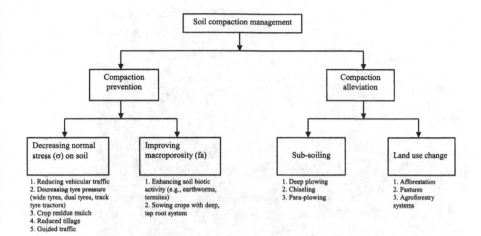

FIGURE **7.12**   Strategies of soil compaction management.

traffic wheels to as little as possible (Taylor, 1994). Wide tires of low inflation pressure cause less soil compaction, minor ruts, low rolling resistance, and high traction. The larger the contact area of the wheel, the less deep are the ruts. In practical terms, Eq. (7.39) can be written as:

$$W_v = T_R \times P \times A \qquad (7.42)$$

where W_v is the weight of the vehicle, P is inflation pressure in Kg/cm^2, and A is the area of contact in cm^2. The constant T_R depends on the rigidity of the tire and its value is usually 1.0 to 1.2 with an average of about 1.1, and is a function of the stiffness of the tire. Total vehicular load remaining the same, the pressure on the soil is inversely proportional to the areas [Eq. (7.43)].

$$A = W_v / T_R \cdot P \qquad (7.43)$$

Compaction alleviation through subsoiling, deep plowing and chiselling (Fig. 7.15) is an expensive strategy. Subsoil alleviation is an extremely difficult task, and usually soil settles back to the original density. Increasing efficiency of subsoiling requires adoption of

FIGURE 7.13 The vehicle load can be distributed over a large area by using dual tires. The soil compaction hazard is less when the load is distributed over a large area.

FIGURE 7.14 Similar to dual tires, wide tires can also distribute the load over a large area. Nonetheless, the strategy to minimize soil compaction is to decrease the frequency of heavy vehicular traffic.

FIGURE 7.15 Subsoiling by chisel plow can decrease bulk density and reduce soil strength temporarily. The long-term goal is to create stable biochannels in the subsoil.

compaction-preventive technologies for subsequent farm operations. Biological measures, which create biopores through worm holes (Fig. 7.16), or root channels (Fig. 7.17), or macrofauna (Fig. 7.18) are better options. Mulch farming techniques (Fig. 7.19) minimize risks of soil compaction.

7.8   SOIL CONSOLIDATION

Soil consolidation refers to the densification process when reduction in volume occurs due to expulsion of water under saturated conditions. Soil may be either initially saturated or compacted to attain saturation. The process of soil consolidation occurs at much slower pace than that of soil compaction because water is several orders of magnitude more viscous than air. Therefore, soil consolidation has more application in foundation engineering than in agriculture. The rate of consolidation is quicker in dense sandy than in porous clayey soils, but the magnitude of consolidation is more in clayey than in sandy soils. Terzaghi (1953) developed the theory of consolidation, which is available in standard texts on soil mechanisms (Wu, 1982).

(a)

(b)

FIGURE 7.16 (a) Earthworm channels or stable biopores enhance infiltration rate and promote root growth and proliferation. (b) The presence of biopores improves gaseous exchange and enhances infiltration even in a compacted subsoil of platy or massive structure. (c) Dr. William Edwards, soil scientist at Coshocton, Ohio, demonstrates earthworm channels in the soil managed with a no-till system and manure application for a long period of time.

(c)

FIGURE 7.16 Continued

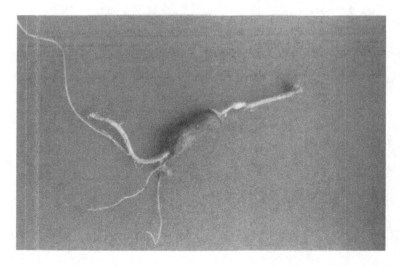

FIGURE 7.17 Biochannels created by tap roots of a tree or a woody perennial can be exploited by fibrous roots of an annual crop such as corn.

FIGURE 7.18 Large macropores of 2 to 5 cm in diameter are created by the burrowing activity of rodents and other animals that inhabit soil.

FIGURE 7.19 Crop residue mulch in a no-till farming system buffers the impact of heavy vehicles and minimizes the risk of soil compaction.

PROBLEMS

1. How many times is the wheel rut deeper in a tire with doubling of the tire pressure?
2. What is soil compactability? List and briefly describe processes, causes, and consequences of soil compaction on agricultural lands (1–2 pages).
3. Consider the following data from a Proctor test.

Gravimetric moisture content (%)	Wet density $(\rho'_b, g/cm^3)$
5	1.4
10	1.6
15	1.8
20	2.0
25	1.8
30	1.4
35	1.2

   Determine the range of moisture for the optimum density.
   Assuming it to be a mineral soil, calculate the degree of saturation corresponding to each moisture content.
4. What are mechanical means of alleviating soil compaction.
5. What are environmental impacts of soil compaction?
6. How does soil compaction impact eutrophication of surface water?
7. Calculate the void ratio (e) of a soil whose total porosity is 0.40.
8. Calculate the modulus of rupture of a briquette 4-cm wide and 1-cm thick, with a distance between two supports of 5 cm when the normal force applied is 200 g.
9. Soil porosity was measured with a mercury injection porosimeter. Calculate pore size corresponding with injection pressure of 100 cm, 1000 cm, and 10,000 cm ($r = 2\gamma \cos \Theta / \rho Pg$), γ for mercury is 430 dynes/cm, ρ of mercury is 13.6 Mg/m^3, and g is 980 cm/s^2).
10. Calculate porosity of a soil sample whose void ratio is 0.7 and particle density is 2.6 Mg/m^3.

REFERENCES

Abrol, I.P. and J.P. Palta. 1968. Bulk density determination of soil clod using rubber solution as a coating material. Soil Sci. 106: 465–468.

Anderson, G., J.D. Pidgeon, H.B. Spencer and R. Parks. 1980. A new hand-held penetrometer for field studies. J. Soil Sci. 31: 279–296.

ASAE 1971. Compaction of Agricultural Soils. American Soc. Agric. Eng., St. Joseph, MI, 471 pp.

ASAE 1986. Soil cone penetrometer. ASAE Standards. American Soc. Agric. Eng.,
S313.2, St. Joseph, MI, 466 pp.

ASTM 1956. Symp. Vane Shear Testing Soils. American Society for Testing
Materials Spec. Tech. Publ. 193, Philadelphia, PA.

ASTM 1965. Compaction of Soils. American Society for Testing and Materials.
Chicago, IL, 135 pp.

Aysen, A. 2002. Soil Mechanics: Basic Concepts and Engineering Applications. A.A.
Balkene, Rotterdam, Holland, 459 pp.

Barber, E. 1965. Stress distributions. In "Methods of Soil Analysis", part 1, ASA
Monograph 9, Madison, WI: 413–430.

Bekker, M.G. 1961. Mechanical properties of soil and problems of compaction.
Trans. of the ASAE 4: 231–234.

Bennie, A.T.P. 1991. Growth and mechanical impedance. In: Y. Waisal, E. Amram
and U. Kafkafi (eds) "Plant Roots," Marcel Dekker, New York: 393–414.

Blake, G.R. 1965. Bulk density. In C.A. Black (ed) "Methods of Soil Analysis," part 1,
Am. Soc. Agron. Monograph 9, Madison, WI: 374–390.

Blake, G.R. and K.H. Hartge. 1986. Bulk density. In: A. Klute (ed) "Methods of Soil
Analysis Part I: Physical and Mineralogical Methods." Second Edition, ASA
Monograph 9, Madison, WI: 33–376.

Braunack, M.V., J.S. Hewitt and A.R. Dexter. 1979. Brittle fracture of soil
aggregates and the compaction of aggregate beds. J. Soil Sci. 30: 653–667.

Brown, R.W. (ed) 2001. Practical Foundation Engineering Handbook. McGraw
Hill, New York.

Campbell, D.J. 1994. Determination and use of soil bulk density in relation to soil
compaction. In B.D. Soane and C. Van Ouverkerk (eds) "Soil Compaction
in Crop Production," Elsevier Science Publishers, Amsterdam, Holland:
113–140.

Campbell, D. and J.K. Henshall. 2001. Bulk density. In K.A. Smith and C.E.
Mullins (eds) "Soil and Environmental Analysis: Physical Methods," Marcel
Dekker, Inc., New York: 315–348.

Carter, L.M. 1969. Integrating penetrometer provides average soil strength. ASAE
Agric. Eng. 50: 618–619.

Carter, M.R. 1990. Relationship of strength properties to bulk density and
macroporosity in cultivated loamy sand to loam soils. Soil Tillage Res. 15:
257–268.

Carter, M.R. (ed) 1994. Conservation tillage in temperate agroecosystems. Lewis
Publishers, Boca Raton, FL: 390 pp.

Causarano H. 1993. Factors affecting the tensile strength of soil aggregates. Soil &
Till. Res. 28: 15–25.

Cernica, J.N. 1980. Proposed new method for the determination of density of soil in
place. Geotech. Testing J. 3: 120–123.

Chancellor, W.J. 1976. Compaction of soil by agricultural equipment. Div. of Agric.
Sci., Univ. of Ca, Richmond, Bull. 1881.

Davidson, D.T. 1965. Penetrometer measurements. In "Method of Soil Analysis,"
Part 1, Am. Soc. Agron. Monograph 9, Madison, WI.

Dexter A.R. and B. Kroesbergen. 1985. Methodology for determination of tensile strength of soil aggregates. J. Agric. Eng. Res. 31: 139–147.

Flowers, M.D. and R. Lal. 1998. Axle load and tillage effects on soil physical properties and soybean grain yield on a Mollic Ochraqualf in northwest Ohio. Soil & Tillage Res. 48: 21–35.

Gardner, W.H. 1986. Water content. In A. Klute (ed) "Methods of Soil Analysis," Part 1, Am. Soc. Agronomy, Monograph 9, Madison, WI: 493–544.

Ghildyal, B.P. and R.P. Tripathi. 1987. Soil Physics. Wileys, Eastern Ltd, New Delhi, India, 656 pp.

Gill, W.R. 1961. Mechanical impedance on plants by compact soils. Trans. ASAE 4: 438–442.

Gill, W.R. and G.E. van den Berg. 1967. Soil dynamics in tillage and traction. USDA-ARS Handbook 316, Washington, D.C.

Gregory, P.J. 1988. Growth and functioning of plant roots. In: A. Wild (ed) "Russel's Soil Conditions and Plant Growth," 11th Edition, Longman, Harlow, U.K.: 113–167.

Guerif, J. 1990. Factors influencing compaction induced stresses in soil strength. Soil Tillage Res. 16: 167–178.

Guerif, J. 1994. Effects of compaction on soil strength parameters. In B.D. Soane and C. Van Ouwerkerk (eds) "Soil Compaction in Crop Production," Elsevier Science Publishers, Amsterdam: 191–214.

Hadas A. and G. Lennard. 1988. Dependence of tensile strength of soil aggregates on soil constituents, density and load history. J. Soil Sci. 39:577–586.

Hakansson, I. and H. Petelkau. 1994. Benefits of limited axle load. In B.D. Soane and C. Van Ouwerkerk (eds) "Soil Compaction in Crop Production," Elsevier Science Publishers, Amsterdam, Holland: 479–500.

Harris, W.L. 1971. The soil compaction process. In "Compaction of Agricultural Soils," ASAE Monograph, St. Joseph, MI: 9–44.

Hillel, D. 1980. Stress–strain relationship and soil strength. In "Fundamentals of Soil Physics," Academic Press, New York: 318–354.

Hogentogler, C.A. 1936. Essentials of soil compaction. Proc. Highway Res. Bd 20, p. 329.

Kayombo, B. and R. Lal. 1994. Responses of tropical crops to soil compaction. In B.D. Soane and C. Van Ouwerkerk (eds) "Soil Compaction in Crop Production," Elsevier Science Publishers, Amsterdam, Holland: 287–315.

Kirkham, D., M.F. de Boodt, and L. De Leenheer. 1959. Modulus of rupture determination on undisturbed soil core samples. Soil Sci. 87: 141–144.

Lal, R. 1979. Concentration and size of gravel in relation to neutron moisture meter and density probe calibration. Soil Sci. 127: 168–173.

Lal, R. 1984. Mechanized tillage systems effects on soil erosion from an Alfisol in watersheds cropped to maize. Soil and Tillage Res. 4: 349–360.

Lal, R. 1985. Mechanized tillage systems effects on properties of a tropical Alfisol in watershed cropped to maize. Soil & Tillage Res. 6: 149–162.

Lal, R. 1989. Conservation tillage for sustainable agriculture. Adv. Agron. 42: 85–197.

Lal, R. 1996. Axle load and tillage effects on crop yields on a Mollic Ochraqualf in northwest Ohio. Soil & Tillage Res. 14: 359–373.

Lal, R. 1996. Axle load and tillage effects on crop yields on a Mollic Ochraqualf in northwest Ohio. Soil & Tillage Res. 37: 143–160.

Lal, R. 1997. Long-term tillage and maize monoculture effects on a tropical Alfisol in western Nigeria. I. Crop yield and soil physical properties. Soil & Tillage Res. 42: 145–160.

Lal, R. and D.J. Cummings. 1979. Changes in soil and microclimate after clearing a tropical forest. Field Crops Res. 2: 91–107.

Lal, R. and M. Ahmadi. 2000. Axle load and tillage effects on crop yields for two soils in central Ohio. Soil & Tillage Res. 54: 111–119.

Lambe, T.W. 1951. Soil testing for engineers. J. Wiley & Sons, New York, 165 pp.

Larson, W.E., A. Eynard, A. Hadas and J. Lipiec. 1994. Control and avoidance of soil compaction in practice. In B.D. Soane and C. Van Ouwerkerk (eds) "Soil Compaction in Crop Production," Elsevier Science Publishers, Amsterdam, Holland: 597–626.

Lindstrom, M.J. and W.B. Voorhees. 1994. Responses of temperate crops in North America to soil compaction. In B.D. Soane and C. Van Ouwerkerk (eds) "Soil Compaction in Crop Production," Elsevier Science Publishers, Amsterdam, Holland: 265–286.

McKeague, J.A. (ed). 1978. Manual on soil sampling and methods of analysis. Can. Soc. Soil Sci., Ottawa, Canada.

McKenzie, D.C. 1996. Measurement and management of compaction damage on Vertisols under irrigated cotton. Ph.D Thesis, Dept. of Agric., Chemistry and Soil Science, Univ. of Sydney, Australia.

Mishra R.K., A.M. Alston and A.R. Dexter (1986). Penetration of soil aggregates of finite size. II. Plant roots. Plant Soil, 94:59–85.

NAS 2002. Soil Mechanics 2002: Soils, Geology and Foundations. Transportation Research Board, National Academy of Sciences, Washington, D.C.

Olsen, H.J. 1988. Technology showcase: electronic penetrometer for field use. J. Terramechanics 25: 287–293.

Olson, R.E. 1962. The shear strength properties of calcium-illite. Geotechnique 12, p. 23.

Proctor, R.R. 1933. Fundamental principles of soil compaction. Description of field and laboratory methods. Engin. News Record 111: 286–289.

Raghavan, G.S.V., P. Alvo and E. McKyes. 1990. Soil compaction in agriculture: a view toward managing the problem. In R. Lal and B.A. Stewart (eds) "Soil Degradation," Adv. Soil Sci. 11: 1–36.

Reeve, R.C. 1965. Modulus of rupture. In "Methods of Soil Analysis," Part 1, ASA Monograph 9, Madison, WI: 466–471.

Richards, L.A. 1953. Modulus of rupture of soils as an index of crusting of soil. Soil Sci. Soc. Am. Proc. 17: 321–323.

Russell, E.W. and W. Balcerek. 1944. The determination of volume and airspace of soil clods. J. Agric. Sci. (Camb.) 34: 123–132.

Sallberg, J.R. 1965. Shear strength. In "Methods of Soil Analysis," part 1, ASA Monograph 9, Madison, WI: 431–447.

Snyder, V.A. and R.D. Miller. 1985. A pneumatic fracture method for measuring the tensile strength of unsaturated soils. Soil Sci. Soc. Am. J. 49: 1369–1374.

Soane, B.D., D.J. Campbell and S.M. Herkes. 1971. Hand-held j-ray transmission equipment for the measurement of bulk density of field soils. J. Agric. Eng. Res. 16: 146–156.

Soane, B.D. and C. Van Ouwerkerk (eds) 1994. Soil Compaction in Crop Production. Elsevier Science Publishers, Amsterdam, Holland, 662 pp.

Soehne, W. 1958. Fundamentals of pressure distribution and soil compaction under tractor tires. Agric. Eng. 39: 276–281, 290.

Soil Survey Division Staff, 1990. Soil Survey Manual. USDA-NRCS, Washington, D.C.

Soil Survey Division Staff, 1993. Soil Survey Manual. USDA-NRCS, Washington, D.C.

Sourisseau, J.H. 1935. Determination and study of physico-mechanical properties of soil. Organ. Raps. II Congs. Int. Genie Rural, Madrid, pp. 159–194.

SSSA 1997. Glossary of Soil Science Terms. Soil Sci. Soc. Am., Madison, WI.

Taylor, J.H. 1994. Development and benefits of vehicle gantries and controlled traffic systems. In B.D. Soane and C. Van Ouwerkerk (eds) "Soil Compaction in Crop Production," Elsevier Science Publishers, Amsterdam, Holland: 521–538.

Terzaghi, K. 1953. Theoretical Soil Mechanics. Wiley, New York.

Utomo, W.H. and A.R. Dexter. 1981. Soil friability. J. Soil Sci. 32: 203–213.

Vepraskas, M.J. 1988. Plant response mechanisms to soil compaction. In: R.E. Wilkerson (ed) "Plant Environment Interactions," Marcel Dekker, New York: 263–287.

Vermenlen, G.D. and U.D. Perdok. 1994. Benefits of low ground pressure tire equipment. In B.D. Soane and C. Van Ouwerkerk (eds) "Soil Compaction in Crop Production," Elsevier Science Publishers, Amsterdam, Holland: 447–477.

Vomocil, J.A., L.J. Waldron and W.J. Chancellor. 1961. Soil tensile strength by centrifugation. Soil Sci. Soc. Am. Proc. 25: 176–180.

Whitlow, R. 1995. Basic Soil Mechanics. Third edition. Longman, Essex, U.K., 559 pp.

Wilkerson, J.B., F.D. Tompkins and L.R. Wilhelm. 1982. Microprocessor-based, tractor-mounted soil cone penetrometer. ASAE Paper #82-5511.

Wu, T.H. 1982. Soil Mechanics. Second Edition. T.H. Wu Publisher, Worthington, OH, 440 pp.

Yong, R.N. and B.P. Warkentin. 1966. Introduction to Soil Behavior. McGill Univ., the MacMillan Co., New York, 451 pp.

8

Soil Rheology and Plasticity

Rheology is a science dealing with the study of deformation-time properties of materials in response to applied stresses. It refers to the study of the change in the form and flow of the soil, embracing elasticity, plasticity, and viscosity. These are the dynamic properties of the soil and are expressed in terms of soil movement as a result of external forces. Soil response to applied stress may be perfectly elastic, perfectly plastic, elastoplastic, viscoelastic, and viscoplastic. Soil has both elastic and plastic behaviors. Soil properties that affect rheological characteristics include texture, structure, the nature of clay minerals, exchangeable cations, properties of the diffused double layer, saturation void ratio, and moisture content.

   Soil plasticity has a strong impact on soil tilth, especially in soils with high plasticity or clayey soils containing 2:1 type clay minerals. Therefore, understanding soil's plastic characteristics is important to identifying strategies for maintaining good soil tilth (refer to Chapter 4).

8.1   SOIL CONSISTENCE

Soil consistence (or consistency) refers to the manifestations of the physical forces of cohesion and adhesion acting within the soil at a range of soil moisture contents. The term consistence is not to be confused with

231

penetration resistance. It specifically refers to "attributes of soil material as expressed in degree of cohesion and adhesion or in resistance to deformation or rupture" (Soil Survey Division Staff, 1993). It is a soil physical property that is manifested by its resistance to flow. In other words, it refers to the resistance offered by the soil against the force that tends to deform it.

It is important to distinguish between the forces of cohesion and adhesion operating in a soil. Adhesion, attraction between dissimilar objects, refers to the attraction of water to the soil solids because water molecules adhere to the soil particles. Cohesion, attraction between similar objects, is bonding between soil particles. Cohesive forces in soil are due to attractive forces between the particles. These forces arise due to physicochemical mechanisms including van der Waals forces, electrostatic attraction between negatively charged clay surfaces and positively charged clay edges, cationic bridges, cementing effects of humic substances and salts, and surface tension of water. These manifestations may include soil behavior to: (i) gravity, pressure, thrust, and pull, (ii) adhesive forces with foreign bodies and substances, and (iii) human feel and sensations experienced. Therefore, soil consistence encompasses several attributes including friability, tilth, plasticity, stickiness, and resistance to compression (Russell, 1928; Baver et al., 1972).

Atterberg (1911; 1912) defined five different forms of soil consistence depending on soil wetness (Fig. 8.1). Atterberg's consistence constants are indicative of the workability of soil at different moisture contents.

Harsh: Dry soil has a harsh consistence to touch. Soil is hard, and the degree of harshness depends on texture and soil organic matter content. Soil is highly cohesive because of clay to clay cementation. When soil is plowed at harsh consistence, it has high-energy requirement and produces a cloddy and rough soil surface.

Friable: A soil has a friable consistence when it easily crumbles into granules or crumbs. Plowing and other tillage operations should be done when soil moisture content is such that soil has a friable consistence. Plowing when soil moisture content is at friable consistency leads to a favorable soil tilth.

Soft: When soil is visibly wet, it has a soft consistence. Soil is not trafficable at this consistence, and is prone to formation of deep ruts. In the dry range, a soft soil may have a friable consistence.

Harsh	Friable	Soft	Plastic	Sticky	Viscous

| Dry | | Moist | Wet | | Saturated | |

Soil wetness (Θ)

FIGURE 8.1 Forms of soil consistency in relation to wetness.

Plastic: Soil is wet enough to be molded into different forms and shapes. Soil particles are orientated in a laminar fashion due to the layer of water between them.

Sticky: The soil adheres to other objects, e.g., farm implements, shoes, etc. Scouring point refers to the soil moisture content at which the soil no longer sticks to the foreign object. Soil moisture content at the sticky point is sufficient to satisfy the attractive power of the soil for water, and there is a free water film between the surface of the foreign object and the soil that prevents the object from sticking to the soil. The water film is connected to the bulk of the soil water at the same tension that exists throughout the soil. Therefore, the sticky point is the moisture content at which maximum adhesion occurs, and at which normal soils will scour during tillage.

Liquid consistence: The soil wetness is near saturation ($\theta = s$), and soil behaves like a viscous liquid. In comparison with five consistence levels defined by Atterberg, Soil Survey Division Staff (1993) defined nine levels of consistence (Table 8.1). These levels are based on rupture resistance of block like specimens of soil, and are measured in terms of the stress (force per unit area) or blows (force through a distance) applied to the soil. A favorable soil tilth is obtained when soil is plowed at soft consistence with rupture resistance of < 8 Pa. A cloddy tilth is obtained when soil is plowed at hard consistence (40–80 Pa stress). It is difficult, if not impossible, to dig or plow a soil at rigid consistence (>800 Pa stress).

8.1.1 Soil Tilth and Consistence

Knowledge of soil consistency is important to preparation of a good "tilth," which is produced when soil is tilled at a moisture content corresponding to friable consistency. Tillage results in clods if soil is plowed when at harsh consistency, in a good tilth when at friable consistency, and in puddling when at plastic or sticky consistency.

Harsh Consistence and Soil Tilth

The soil is dry and represents the lower moisture limit beyond which it does not shrink any further. If plowed, soil produces large clods and massive structure. Knowing the water content at which a friable consistence is achieved is important to producing a good soil tilth.

Friable Consistence and Soil Tilth

During friable state, soil particles are randomly oriented. A friable soil results in good soil tilth. Knowledge of the range of soil moisture content at which it has a friable consistence is important to minimizing risks of

TABLE 8.1 Different Forms of Consistence Based on Rupture Resistance of Block-Like Specimens of Soil

Classes for moisture states			Test description	
Moderately dry and very dry	Slightly dry and wetter	Air dry, submerged	Operation	Stress applied (Pa)
Loose	Loose	Not applicable	Specimen not obtainable	—
Soft	Very friable	Non-cemented	Fails under very slight force applied slowly between thumb and forefinger	< 8
Slightly hard	Friable	Extremely weakly cemented	Fails under slight force applied slowly between thumb and forefinger	8–20
Moderately hard	Firm	Very weakly cemented	Fails under moderate force applied slowly between thumb and forefinger	20–40
Hard	Very firm	Weakly cemented	Fails under strong force applied slowly between thumb and forefinger	40–80

Very hard	Extremely firm	Moderately cemented	Cannot be failed between thumb and forefinger but can be between both hands or by placing on a non-resilient surface and applying gentle force underfoot	80–160
Extremely hard	Slightly rigid	Strongly cemented	Cannot be failed in hands but can be underfoot by full body weight applied slowly	160–800
Rigid	Rigid	Very strongly cemented	Cannot be failed underfoot by full body weight but can be by <300 J blow	800 Pa–300 J blows
Very rigid	Very rigid	Indurated	Cannot be failed by blow of <300 J	≥300 J blows

Force of $1 N$ = Newton = $1 kg \cdot m/s^2$, stress of pressure of $1 N = 1 N/m^2 = Pa$ (pascal), $1 J = 1 N \cdot m$ or application of $1 N$ force through a distance of $1 m$ as in blows applied to a soil during the Proctor test (refer to Chapter 8).

Source: Adapted from Soil Survey Division Staff, 1993.

tillage-induced soil degradation and producing a good tilth. Soil tilth is defined as "the physical condition of a soil as related to its ease of tillage, fitness as a seedbed, and its importance to seedling emergence and root penetration" (SSSA, 1979). Most definitions of soil tilth are vague, subjective and qualitative, because tilth is used as a blanket term describing all soil conditions that relate to seed germination, and seedling growth and crop development (Yoder, 1937). Russell (1961) observed that "soil tilth is a property that a farmer can feel with the kick of his boot and a soil scientist cannot describe it." In addition to inherent soil properties, soil friability and tilth also depend on numerous exogenous factors, e.g., crop rotation, soil fertility management, vehicular traffic, tillage systems, and soil biotic activity. In fact, soil is a complex term and implies combination of soil structure, consistence, and biotic activity. Attempts are being made to develop a tilth index based on soil properties (Karlen et al., 1990; Singh et al., 1990), yet there are numerous research needs to make this soil tilth concept an objective and quantitative criterion. These priorities include establishing relationship between: (i) soil properties (e.g., soil moisture content, aggregation, porosity, water transmission characteristics) and soil tilth, (ii) soil tilth index and plant growth, (iii) soil tilth and fluxes of water, energy, and nutrients, and (iv) soil management and tilth. Establishing relationship between soil moisture content and soil tilth for major soils is a high priority.

8.2 SOIL PLASTICITY

With a progressive increase in soil moisture content, soil consistence changes from friable to soft and plastic. When plastic, soils are cohesive and can be molded like putty. Plasticity refers to "soil's ability to change its shape without cracking when it is subjected to deforming stress." Plasticity enables a soil to be deformed without rupture when a material is subjected to a force in excess of the yield value and maintain the deformed shape even after the stress is removed and water is drained or dried. Soil plasticity depends on the clay content and is the resultant effect of stress and deformation. Sandy or coarse-textured soils are not plastic. Such soils can be molded when wet but fall apart when dried. The stress needed to produce a specific degree of deformation is proportional to the magnitude of cohesive forces that hold the soil particles together. Cohesive forces depend on the properties of the clay and the degree of soil wetness or the thickness of the water film. Soil plasticity is explained by several theories (Kurtay and Reece, 1970).

     1. *Water film theory.* Soil cohesion is attributed to several interparticle forces including those due to van der Waals forces, electrostatic forces, catonic bridges, surface tension at the soil–water interface, and

cementing effects of humic substances and sesquioxides. The magnitude of these cohesive forces determines soil behavior under stress (whether brittle, plastic, or viscous) and depends on soil wetness. As soil wetness and the diffused double layer is extended, the repulsive forces balance the cohesive forces and soil consistency changes from friable or soft to plastic. At this juncture, the interparticle force F is related to the particle size [Eq. (8.1)] (Haines, 1925; Nichols, 1931).

$$F = \frac{4k\pi r \gamma \cos \alpha}{d} \qquad (8.1)$$

where r is particle radius, γ is surface tension, α is the angle of contact, d is the distance between the particle, and k is constant. When soil is sufficiently wet and each particle is surrounded by a water film, particles get oriented in a laminar fashion, and the cohesive forces are overcome under stress and the soil deforms (Baver, 1930). As the soil wetness increases and cohesion decreases, soil becomes capable of a viscous flow. Therefore, soil factors that affect its plasticity are particle size, surface area, nature of clay minerals, and exchangeable cations.

    2. *Critical state theory.* This simple explanation was proposed by Kurtay and Reece (1970). The soil is said to be in critical state when it continues to deform under stress without any change in volume. When a loose soil sample is subjected to a progressively increasing uniaxial stress while the confining stress is kept constant, the soil volume (V_t) progressively decreases due to soil compression. With continued increase in stress, a soil reaches a point at which it cannot be compressed any more. At this point, when the soil cannot be compressed with additional stress but is deformed without change in volume, the soil reaches critical state. When soil is plastic, it is at the critical state. It deforms under stress without changing its volume.

8.3 ATTERBERG CONSTANTS

Atterberg, a Swedish agriculturist, proposed a concept dividing the entire cohesive range of the soil into five stages and six divisions of soil wetness. These limits, corresponding with soil moisture content from harsh consistency to viscous flow, are called Atterberg constants.

    Shrinkage Limit. This represents the soil moisture content corresponding with the lower limit of the volume change at which there is no further decrease in soil volume (V_t) as soil moisture is evaporated. The moisture content below which the soil ceases to shrink is called the shrinkage limit, and represents the lower moisture limit of the semisolid state of consistency. It is a moisture content at which soil transforms from the semisolid state to the solid state, and the volume of soil remains constant with progressive drying. Soil shrinkage is caused by the tension formed at the air–water interfaces at the surface of the soil–water system. The diffused double layer shrinks as water is evaporated causing the soil particles to be drawn closer together (see Section 8.5).

Lower Plastic Limit. This refers to the moisture content corresponding with the lower limit of the plastic range. This is the moisture content at which the soil starts to crumble when rolled into a thread (3 mm diameter) under the palm of the hand. It represents the minimum soil moisture content at which the soil can be puddled. The thickness of the water film is enough to satisfy the need for formation of the bonded water layer plus capillary condensation that lubricates the particles to enable them to slide over one another. Soil moisture is held at a suction of about 500 to 2000 cm of water, and the magnitude of *w* depends on the clay content and nature of clay minerals.

Cohesion Limit. It refers to the soil moisture content at which crumbs of soil cease to adhere when placed together or in contact with one another.

Sticky Limit. This is the soil moisture content above which the mixture of soil and water will adhere or stick to a steel spatula or another object that can be wet by water.

Upper Plastic Limit. This is also called the liquid limit or the lower limit of viscous flow. It signifies the moisture content at which the moisture film becomes thick, cohesion is decreased, and soil–water mixture flows under stress but possesses a small shear strength of about 1 g/cm^2. The water film's coalesce to fill most pores, and the ratio of the bond water to the free or unoriented water is extremely small. The soil is almost saturated with a soil moisture suction of <10 cm of water. In practical terms, this is the moisture content at which the mixture of soil and water flow as a viscous liquid and below which the mixture is plastic.

Upper Limit of Viscous Flow. This is the soil moisture content above which the mixture of soil and water flows like a liquid.

8.3.1 Soil Indices Based on Atterberg's Limits

Soil behaviors in relation to moisture content, expressed in terms of different Atterberg's limits or constants, has important implications to agricultural, engineering, and industrial uses of the soil. Although agricultural uses in relation to soil tillage are extremely important, civil engineers have used these concepts to define soil strength and deformation behavior. Some important indices based on Atterberg's limits are the following:

Plasticity Index (PI) or Plastic Range. This represents the difference in moisture content between the upper and lower plastic limits [Eq. (8.2)].

$$PI = UPL - LPL \tag{8.2}$$

where PI is the plasticity index, and UPL and LPL refer to the moisture content at upper and lower plastic limits. This index is an indirect measure of the force required to mold the soil, and is a measure of the distance d [Eq. (8.1)] between particles that corresponds with the soil moisture content ranging from extremely low suction to the free water present that enables the soil to flow under an applied force. There is generally a good correlation between the lower and upper plastic limits, because the soil moisture content at these levels is affected by similar or same factors. There exists a good relationship between the upper plastic limit and the PI (Casagrande, 1932). When PI is plotted on the y-axis as a function of the UPL on the x-axis, points that represent samples from the same soil stratum or soil of similar mineralogical composition fall along lines that are approximately parallel to the A-line. The PI is also strongly correlated with soil adhesion to metal, e.g., plow. An example of the UPL, LPL, and PI for some soils of Nigeria is shown in Table 8.2. Most coarse-textured soils do not exhibit strong plastic characteristics, as is the case with soils of the surface horizon in Table 8.2. In contrast, however, clayey subsoils exhibit some plastic characteristics. Both UPL and LPL are strongly influenced by clay content (Table 8.3).

Liquidity Index (LI). Similar to PI, the liquidity index also reflects the properties of soil and also depends on the UPL and LPL. The LI is

TABLE 8.2 Plasticity Properties of Surface Horizon of Some Soils from Nigeria

	UPL		LPL	
	Surface	Subsoil	Surface	Subsoil
Soils parent material	%, weight basis			
Precambrian basement complex	24 ± 5	24 ± 7	20 ± 4	18 ± 6
Arenaceous sedimentary rocks	27 ± 6	27 ± 14	20 ± 5	15 ± 7

Source: Lal, 1979.

TABLE **8.3** Relationship Between Soil Constituents and Plastic Properties for Some Nigerian Soils

Soil constituent	UPL			LPL			PI		
	r	m	b	r	m	b	r	m	b
Organic carbon	-0.25^a	-3.01	32.20	NS	NS	NS	-0.32^a	-3.22	12.70
Clay	0.26^a	0.16	24.59	0.29^a	0.14	14.64	NS	NS	NS
Sand	-0.28^a	-0.17	40.41	-0.34^a	-0.17	29.88	NS	NS	NS

[a] =Implies that r value is significant at 5% level of probability.
r = correlation coefficient
m = slope
b = intercept
NS = not significant
Source: Lal, 1979.

related to the percent antecedent soil moisture content (w), UPL and PI [Eq. (8.3)].

$$LI = \frac{w(\%) - UPL}{PI} \tag{8.3}$$

The LI also describes soil's water content range between the upper and lower plastic limits, and is a measure of soil consistency. The LI value is generally 1 for soils of low strength, and 0 for soils of high strength or stiff soil. Soils that are compressed under heavy loads have an LI of about zero.

Activity Ratio (AR). It is the ratio of the plasticity index to the percent clay content (<2 μm).

$$AR = PI/Clay\ Content(\%) \tag{8.4}$$

If PI (%) is plotted on the y-axis as a function of clay content (x-axis), the slope of these lines is the activity ratio. The activity ratio depends on the clay content, clay mineralogy, nature of the exchangeable cations, and concentration of the soil solution.

8.3.2 Factors Affecting Atterberg's Limits

Atterberg's limits are affected by the nature of soil solids. Atterberg's limits are affected by similar factors that affect the thickness and dynamics of the diffused double layer. These include clay, sand, and organic matter content.

Clay Content

Plasticity is a function of the total surface area of the colloidal fraction or fine particles. The amount of water absorbed depends on the surface area, which determines cohesion and plasticity. Therefore, soil plasticity depends on the clay content. The PI increases with an increase in clay content. Soils with low clay content have low upper plastic limit and, therefore, low PI. The PI is an indirect measure of the clay content. The data in Table 8.3 show effects of textural properties, and soil organic matter content on plastic attributes of some Nigerian soils. The degree of correlation depends on soil composition and the nature of clay minerals.

Clay Minerals

The type of clay minerals (i.e., 1:1 or 2:1, expanding or non-expanding lattice) affect soil moisture content of the molded soil. Soil moisture absorption, with all other factors remaining the same, is usually in the order of montmorillonite > illite > kaolinite.

Exchangeable Cation

Soil plasticity and Atterberg's limits are influenced by the exchangeable cations through their effects on hydration, dispersion, flocculation, and characteristics of the diffused double layer. Polyvalent cations hold the expanding lattice together compared to monovalent cations. All other factors remaining the same, the PI follows the order $Na^+ > K^+ > Mg^{+2} > Ca^{+2} > Al^{+3} > Th^{+4}$. However, the order may vary among clay minerals (Baver et al., 1972).

Soil Organic Matter Content

In predominantly inorganic soils with soil organic matter content of less than 5%, increase in organic matter content increases both upper and lower plastic limits. Therefore, organic matter content may have no effect on the PI (Table 8.3). Atterberg's constants of some soils from western Nigeria are shown in Table 8.2. It is apparent that plasticity indices are measurable in clayey soils only.

8.3.3 Measurement of Atterberg's Limits

Standard procedures for measuring Atterberg's limits are described in details by Ghildyal and Tripathi (1987), Campbell (2001), and McBride

(1993). Most common methods include the following:

Casagrande Test.   The upper plastic limit is determined by a standard equipment to determine the moisture content at which the soil on two sides of a groove flows together after the dish which contains the soil has been dropped through a distance of 1 cm 25 times. This test is analogous to the soil strength test, because soil strength at the UPL is about 1 g/cm^2. There have been several modifications in the test including the "one-point method," which involves making the soil paste such that the number of blows required to close the grove is about 25.

The lower plastic limit is determined by measuring the soil moisture content at which the soil crumbles when it is rolled down to a thread about 3 mm in diameter. The soil is described as nonplastic if it cannot be rolled or the lower plastic limit is close to that of the upper plastic limit.

Drop–Cone Test.   The Casagrande test is highly subjective and there is a lot of variation in results due to the personal judgment of the operator. Some soils can slide in the cup, liquefy from shock, rather than flowing plastically. Sherwood and Ryley (1968) proposed that the drop cone test may be more accurate for determining the upper plastic limit than the Casagrande test. A 30° cone mounted on a shaft, with a total weight of about 80 g, is allowed to drop on a cup (50 mm deep and 55 mm in diameter) full of soil for 5 seconds. The linear relationship between soil moisture content (*x*-axis) and the penetration (*y*-axis) is plotted. Soil moisture content (%, *w*) corresponding to a penetration of 20 mm is determined and considered as a cone penetrometer liquid limit or the upper plastic limit (Campbell, 2001). There exists a good correlation between the Casagrande test and the Drop–Cone test for some soils (Campbell, 1975). Similar to Casagrande test, attempts have also been made to develop a one-point Drop–Cone test.

Indirect Methods.   There are several indirect methods of determining Atterberg's limits, most of which are based on correlation with other soil physical properties called the pedotransfer functions.

   1. *Proctor test*: Measurements of the Proctor Density test have been used to estimate the upper and lower plastic limits. For some soils, the moisture content at the maximum density corresponds to the upper plastic limit and that at the lowest bulk density to the lower plastic limit (Faure, 1981). This concept is in accord with the "critical state" theory of plasticity.

   2. *pF curves*: Pedotransfer functions have been developed to relate soil moisture constants determined from pF curves (refer to Chapter 11) to the Atterberg's limits. Within a given textural group, the liquid limit or the upper plastic limit may correspond with a narrow range of soil moisture

potential. The moisture potential corresponding with the lower plastic limit, however, may depend on clay content (Russell and Mickle, 1970). Archer (1975) observed high correlation coefficient between the lower plastic limit and the field capacity.

    3. *Hydraulic conductivity*: Saturated hydraulic conductivity generally increases with increase in the upper plastic limit (refer to Chapter 12). Such pedotransfer functions have been used to design systems to reduce seepage losses from ponds.

    4. *Viscosity*: Viscometers have been used to measure flow behavior of clays (Yasutomi and Sudo, 1967; Hajela and Bhatnagar, 1972). This method may be inaccurate for determining the lower plastic limit.

    5. *Shear strength*: The lower plastic limit in some soils may be estimated by measuring the moisture content remaining when a soil paste has been subjected to a standard stress (Vasilev, 1964). Soil strength at the lower plastic limit may be 100 times that at the upper plastic limit.

8.3.4 Applications of Atterberg's Limits

There are numerous engineering and agricultural applications of the concepts involved in Atterberg's limits. Engineering applications are those relevant to soil strength and stability, and agricultural in relation to soil tilth, compactability, and shrinkage.

Tillage

A complex and interactive relationship between Atterberg's limits, soil tilth, and soil moisture content is shown in Fig. 8.2. Soil produces a good tilth when cultivated at a moisture content corresponding to a friable consistency or in the vicinity of the lower plastic limit. Soil does not produce clod when plowed at this moisture content. Soils are highly susceptible to compaction and puddling when cultivated within the plastic range. Because of high adhesion and frictional forces, the draft power is also high for cultivation within the plastic range. For subsoiling to be effective, it must be done when soil moisture content is just below the lower plastic limit. If the lower plastic limit is smaller than field capacity, soil structure may be adversely affected if soil is cultivated at moisture content between the lower plastic limit and the field capacity. If the lower plastic limit is greater than the field capacity, good soil tilth is produced when it is cultivated at moisture content between the lower plastic limit and field capacity. Hardsetting soils have a very narrow range of workable moisture content.

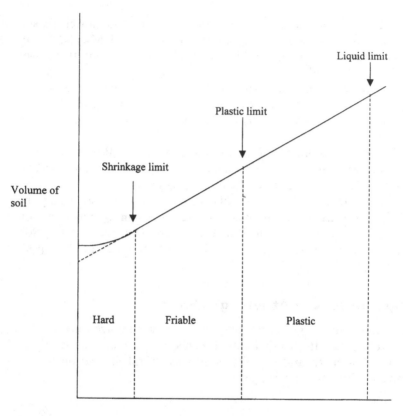

FIGURE 8.2 A schematic showing relationship between soil moisture content, soil volume, shrinkage behavior, and soil consistency.

Properties related to soil–tillage interaction and other dynamic properties of soil during and after tillage operation are closely associated with the Atterberg limits. Soil dynamics refers to the relation between forces applied to the soil and the resultant soil reaction (Gill and van den Berg, 1967). Soil properties that affect soil dynamics include texture, nature and the amount of clay content, and antecedent soil moisture content. A principal dynamic property involved in soil–tillage interaction is the shear strength (comprising soil cohesion and internal friction). Shear strength is the maximum at a soil moisture content in vicinity of the lower plastic limit.

Friction between soil and metal is another important factor that develops in three phases: (i) phase 1 represents true friction between metal and dry soil, (ii) phase 2 is governed by the forces of adhesion between soil and the metal which increase with increase in soil moisture content, and reaches the maximum value near the upper plastic limit. Detail description of soil dynamics and physics of low action and other tillage operations is given by Nichols (1929), Nichols et al. (1958), Baver et al. (1972), and Horn et al. (1994).

Scouring or the self-cleaning flow of soil over the tillage implements and draft power are also related to Atterberg limits. Non-scouring, the process in which the soil mass is pushed away, is due to: (i) the minimum angle that implement makes with the direction of travel, (ii) high cohesion as in dry soil, (iii) high coefficient of soil–soil friction, (iv) low coefficient of soil–metal friction, and (v) low adhesion of soil to metal when soil is below the lower plastic limit or above the sticky point.

The draft power is needed to overcome the forces of cohesion, adhesion, resistance to compression, shear strength, and soil–metal friction. The power needed is usually the maximum when the soil wetness if just above the lower plastic limit and the draft power increases logarithmically with the increase in PI. The draft power is the least when the soil is at friable consistency. Sohne (1956) attributed power requirements to several types of work done during the tillage operation to: (i) cut, (ii) overcome cohesion and shear forces involved in compressing, shearing, and turning the soil, (iii) lift and turn the furrow slice, and (iv) overcome friction between soil and the tool on all sides. The relative magnitude of these forces in relation to different implements and soil condition has been evaluated by Gill and van den Berg (1967) and Soane and Van Overkerk (1994).

Mole Drainage

Knowledge of the plastic behavior can be useful in installing mole drains. Mole drainage channels are stable if established when the soil moisture content at the mole depth is above the lower plastic limit. However, soil above the mole channel must be at the friable consistency. Appropriate soil moisture content most suitable for mole drain establishment may correspond to a specific PI which may vary among soils.

Soil Strength and Compaction

Soil is generally most susceptible to compaction when its moisture content is in the vicinity of the lower plastic limit. In contrast, soil is most susceptible to puddling when soil moisture content exceeds the upper plastic limit. Road and foundation engineers can determine the moisture content corresponding to the maximum Proctor density from the lower plastic limit. There is generally a good correlation between PI and various parameter related to soil strength, e.g., cohesion, angle of internal friction, and shear strength. All soils may have similar strength when soil moisture content is in the vicinity of the upper plastic limit.

8.4   SOIL VISCOSITY

As soil moisture content increases, its consistency changes from plastic, to sticky, to viscous. When viscous, soil flows under stress and the flow is proportional to the force applied. When plastic, a certain amount of force must be applied before any flow is produced. The flow behavior of a soil is explained by the Bingham equation [Eq. (8.5)].

$$V = k\mu(F - F')$$ (8.5)

where V is the volume of flow, μ is the coefficient of mobility, F is the force applied, F' is the force necessary to overcome the cohesive forces (also called the yield value), or F' is zero and the volume of flow is proportional to the force ($V \propto F$). The constant of proportionality k in viscous flow is the coefficient of viscosity of the liquid (Fig. 8.3).

8.5   SOIL SHRINKAGE

Atterberg limits also have an important application to soil shrinkage. Atterberg defined "shrinkage limit" as the soil moisture content below which the soil ceases to shrink, and represents the lower moisture limit of the semisolid state or soft-friable consistency. The process of shrinkage is due to the manifestations of the diffused double layer, and due to the forces of surface tension at the air–water interface. The magnitude of volume change depends of soil structure, aggregate shape, porosity and pore size distribution, nature, and amount of clay. Therefore, the shrinkage process is related to the change in total volume (V_t) in relation to the change in volume of water (θ) in the soil (Fig. 8.4).

A schematic of the shrinkage process shown in Fig. 8.5 shows two distinct types of shrinkage. The normal shrinkage (curve segment labelled AB) refers to the process in which decrease in total soil volume (V_t) is proportional to the volume of water (θ) withdraw from the soil. The slope of the normal line is an important indicator of the kind of shrinkage. If the angle is 45°, the soil displays a normal shrinkage. If the angle is <45°, the soil displays less than normal shrinkage. The angle of the line of normal

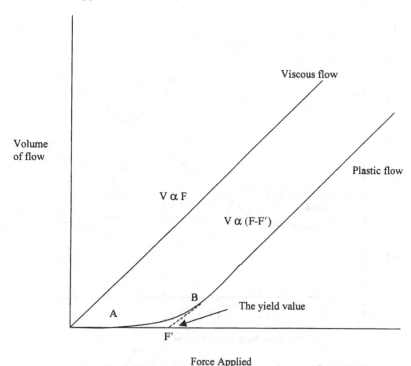

FIGURE **8.3** Viscous versus plastic flow.

shrinkage is an important soil characteristic (Mitchell, 1992) and is influenced by management (Mitchell and Van Genuchten, 1992; Flowers and Lal, 1999) (Fig. 8.5). The normal shrinkage continues until the point when there is a strong interaction between particles, and further shrinkage is caused by compression and orientation of particles rather than due to decrease in V_t. This shrinkage is called the *residual shrinkage* (curve segment labeled BC). At this point, the air enters the soil. There is a change in slope of the curve from 1 for the normal shrinkage line AB to less than 1 for the segment BC. The point B signifies the moisture content at which air enters the soil and corresponds to the shrinkage limit. In practice, the curve ABC is simplified by drawing ABC, and the shrinkage limit is then defined as the moisture content corresponding to the point B. The magnitude of shrinkage beyond this point or the residual shrinkage depends on soil properties. The amount of air that enters the soil during the residual shrinkage (f_a) can be calculated by extending the line AB to point D on the y-axis. The segment DC corresponds to the air-filled porosity during the residual shrinkage. The residual shrinkage is usually more in well-structured than poorly

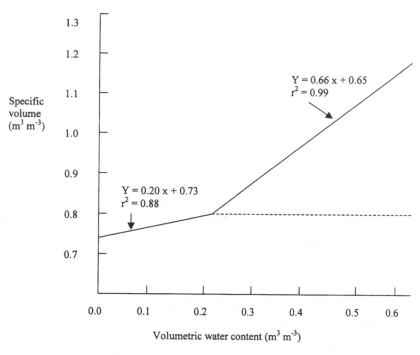

FIGURE 8.4 Normal (upper solid line) and residual (lower solid line) shrinkage curves for a soil bulk density of 1.1 $Mg\,m^{-3}$. (Modified from Flower and Lal, 1999.)

structured or puddled soils and in heavy-textured than light-textured soils. The shrinkage curve is strongly influenced by soil bulk density (Flowers and Lal, 1999).

8.5.1 Methods for Determining Soil Shrinkage

There are several methods of determining soil shrinkage (Holtz, 1965; Warkentin, 1993). The choice of methods depends on the objective. Field assessment of shrinkage involves measurement of the height of soil surface overtime. Assessment of the volume of cracks in the field is considered as an indication of the horizontal shrinkage. These are, however, extremely crude measurements and may be highly subjective. Shrinkage of well-structured or aggregated soils is measured by taking an undisturbed sample (clod) 3–10 cm across, and its volume change is measured as it is dried slowly under high humidity environment. Another method used for measuring shrinkage of structured soil involves determining bulk density of a clod at different moisture contents (Grossman et al., 1968). Shrinkage of remolded samples

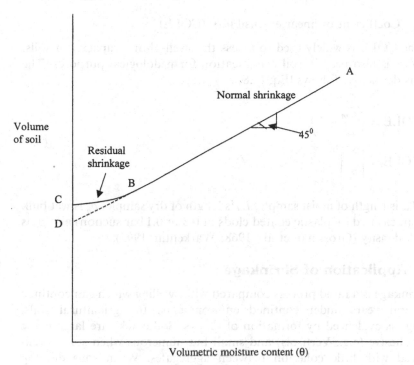

FIGURE 8.5 A schematic of the soil shrinkage characteristic curve (SSCC).

is determined by evaluating change in volume of a saturated paste upon progressive drying (ASTM, 1989) to develop the soil shrinkage characteristic curve or SSCC (Fig. 8.5). Measured shrinkage is usually the maximum for remolded soil samples.

There are numerous indices of expressing the shrinkage behavior of a soil including the following (Warkentin, 1993):

1. Volume decrease per unit weight of soil in units of m^3/Mg [$\Delta V/M_s$, Eq. (8.6)]

$$\text{Shrinkage} = (\rho_b')^{-1} - (\rho_b)^{-1} = \left(\frac{1}{\rho_b'} - \frac{1}{\rho_b}\right) \tag{8.6}$$

2. Decrease in porosity per unit volume of soil [$\Delta f/V_t$, Eq. (8.7)]

$$\Delta f = \frac{\rho_b - \rho_b'}{\rho_s} \tag{8.7}$$

3.   Coefficient of linear extensibility (COLE)

The COLE is widely used to assess the swell–shrink capacity of soils. This index is also used in soil classification for pedological purposes. The COLE is defined as follows [Eq. (8.8)]:

$$COLE = \left(\frac{L_m}{L_d} - 1\right)$$
$$COLE = \left[\frac{\rho_b}{\rho'_b}\right]^{1/3} - 1 \tag{8.8}$$

where L_m is length of moist sample, L_d is length of dry sample, ρ'_b is wet bulk density (measured on plastic coated clods at 0.3 or 0.1 bar suction) and ρ_b is dry bulk density (Grossman et al., 1968; Warkentin, 1993).

8.5.2 Application of Shrinkage

Soil shrinkage is a rapid process compared with swelling which can continue for several years under confined environments. In agricultural soils, shrinkage is evidenced by formation of cracks. Soil cracks are large if the soil is cohesive (e.g., Vertisols) and small but numerous when soil is well-structured with little cohesion between aggregates. When soils develop large cracks, there is a considerable damage to plant roots. Roots in a severely cracked soil are confined to the small and dense soil mass between the cracks, thereby decreasing water and nutrient use efficiencies. Roots also affect soil shrinkage (Mitchell and Van Genuchten, 1992). Soil shrinkage can also be used to estimate soil profile water content and for scheduling irrigation (Yule, 1984; Mitchell, 1991). In engineering applications, soil shrinkage jeopardizes safety of buildings, roads, and dams. There is also interest in soil shrinkage with regards to soil subsidence.

PROBLEMS

1. Describe the agronomic significance of the upper and lower plastic limits, and of the plasticity index.
2. What is the practical significance of the numerical value of liquidity, index, and activity ratio?
3. What are the engineering applications of Atterberg's limits?
4. Distinguish between elasticity and plasticity of soils, and what inherent properties of soil determine these characteristics.
5. How do change in plasticity characteristics influence engineering properties of soils?

6. While using deionized water, plastic properties of a soil are as follows: upper plastic limit = 80%, lower plastic limit = 60%. While using a dilute solution, properties change as follows: upper plastic limit = 70%, lower plastic limit = 40%. Describe the reasons for change in properties, and practical significance of such behavior.

7. Consider the following data in a shrinkage test:

   Volume of saturated wet soil = 15 cm^3

   Weight of saturated wet soil = 25 g

   Volume of dry soil = 6 cm^3

   Weight of dry soil = 15 g

   Calculate shrinkage limit and particle density of soil.

8. The analysis of two soils produced the following data:

Property	Soil A	Soil B
Upper plastic limit	25%	10%
Lower plastic limit	10%	5%
Field moisture content	18%	12%
Soil particle density	2.7 Mg m^{-3}	2.65 Mg m^{-3}

9. Tabulate soil factor's that affect Atterberg's limits, and briefly explain reasons for these effects.

10. How do soil organic matter and clay contents and clay type influence plastic properties?

REFERENCES

Archer, J.R. 1975. Soil consistency. In "Soil Physical Conditions and Crop Production". Tech. Bull. 29, Ministry of Agric. Fisheries and Food, HMSO, London: 289–297.

ASTM. 1989. Standard testing methods for one-dimensional expansion, shrinkage and uplift pressure of soil–lime mixtures. Method D 3877-80. American Society for Testing and Materials, Annual Book of Standards Vol. 04.08.

Atterberg, A. 1911. Die Plastizität der Tone. Int. Mitt. Bodenk. 1: 10–43.

Atterberg, A. 1912. Die Konsisteng und die Bindigkeit der Boden. Int. Mitt. Bodenk. 2: 148–189.

Baver, L.D. 1930. The Atterberg consistency constants: Factors affecting their values and a new concept of their significance. J. Am. Soc. Agron. 22: 935–948.

Baver, L.D., W.H. Gardner and W.R. Gardner. 1972. Soil Physics. Second Edition. John Wiley & Sons, New York, 498 pp.

Campbell, D.J. 1975. Liquid limit determination of arable topsoils using a drop–cone penetrometer. J. Soil Sci. 26: 234–240.

Campbell, D.J. 2001. Liquid and plastic limits. In K.A. Smith and C. Mullins (eds) "Soil and Environmental Analysis: Physical Methods." Marcel Dekker, Inc. New York: 349–376.

Casagrande, A. 1932. Research on Atterberg's limits of soils. Pub. Roads 13: 121–130.

Faure, A. 1981. A new conception of the plastic and liquid limits of clay. Soil & Tillage Res. 1: 97–105.

Flowers, M. and R. Lal. 1999. Axle load and tillage effects on the shrinkage characteristics of a Mollic Ochraqualf in northwest Ohio. Soil & Tillage Res. 50: 251–258.

Ghildyal, B.P. and R.P. Tripathi. 1987. Soil Physics. Wiley Eastern Ltd., New Delhi, India, 656 pp.

Gill, W.R. and G.E. van den Berg 1967. Soil dynamics in tillage and traction. Agric. Handbook 316. U.S. Govt. Printing Office, Washington, D.C.

Grossman, R.B., B.R. Brasher, D.P. Franzmeier and J.L. Walker. 1968. Linear extensibility as calculated from natural clod bulk density measurements. Soil Sci. Soc. Am. Proc. 32: 570–573.

Haines, W.B. 1925. Studies in the physical properties of soils. II. A note on the cohesion developed by capillary forces in an ideal soil. J. Agric. Sci. Camb., 15: 529–535.

Hajela, R.B. and J.M. Bhatnagar. 1972. Application of rheological measurements to determine liquid limit of soils. Soil Sci. 114: 122–130.

Holtz, W.G. 1965. Volume Change. In "Methods of Soil Analysis," ASA Monograph 9, Madison, WI: 448–465.

Horn, R. and M. Lebert. 1994. Soil compactability and compressibility. In: B.D. Soane and C. Van Ouwerkerk (eds) "Soil Compaction in Crop Production," Elsevier, Amsterdam, Holland: 45–70.

Karlen, D.L., D.C. Erbach, T.C. Kaspar, T.S. Colvin, E.C. Berry and D.R. Timmons. 1990. Soil tilth: a review of past perceptions and future needs. Soil Sci. Soc. Am. J. 54: 153–161.

Kezdi, A. 1974. Handbook of Soil Mechanics. Vol. 1 Soil Physics, Elsevier Science Publishers, Holland.

Kurtay, T. and A.R. Reece. 1970. Plasticity theory and critical state soil mechanisms. J. Terramech. 7: 23–56.

Lal, R. 1979. Physical properties and moisture retention characteristics of some Nigerian soils. Geoderma 21: 209–223.

McBride, R.A. 1993. Soil consistency limits. In: M.R. Carter (ed) "Soil Sampling and Methods of Analysis," Lewis Publishers, Boca Raton, FL: 519–529.

Mitchell, A.L. 1992. Shrinkage terminology: escape from "normalcy." Soil Sci. Soc. Am. J. 56: 993–994.

Mitchell, A.R. and M.T. Van Genuchten. 1992. Shrinkage of bare and cultivated soils. Soil Sci. Soc. Am. J. 56: 1036–1042.

Nichols, M.L. 1929. Methods of research in soil dynamics. Alabama Agr. Exp. Sta. Bull. 229.

Nichols, M.L. 1931. The dynamic properties of soil. I. An explanation of the dynamic properties of soils by means of colloidal films. Agric. Eng. 12: 259–264.

Nichols, M.L., I.F. Reed and C.A. Reaves. 1958. Soil reaction: to plow share design. Agric. Eng. 39: 336–339.

Russell, E.J. 1961. Soil Conditions and Plant Growth. Longman, U.K.

Russell, J.C. 1928. Report of Committee on Soil Consistency. Am. Soil Survey Assoc. Bull. 9: 10–22.

Russell, E.R. and J.L. Mickle. 1970. Liquid limit values by soil moisture tension. J. Soil Mech. Found. Eng. Am. Soc. Civ. Eng. 96: 967–989.

Sherwood, P.T. and M.D. Ryley. 1968. An examination of the cone penetrometer methods for determining the liquid limit of soils. Report No. LR233, Transport Road Res. Lab., Crowthorne, U.K.

Singh, K.K., T.S. Colvin, D.C. Erbach, and A.Q. Mughal. 1990. Tilth Index: An approach towards soil condition quantification. Trans. ASAE 35(6): 1777–1785.

Soane, B.D. and C. Van Overkerk (eds) 1994. Soil Compaction in Crop Production. Elsevier, Amsterdam, 662 pp.

Sohne, W. 1956. Einige Grundlagen für eine lanoltechnischa Bodenmechanik. Grundlagen der Landtechnik 7: 11–27

Soil Survey Division Staff. 1990. Soil Survey Manual. USDA Handbook 18, U.S. Govt. Printing Office, Washington, D.C.: 172–183.

Soil Survey Division Staff. 1993. Soil Survey Manual. USDA Handbook 18, U.S. Govt. Printing Office, Washington, D.C.

SSSA 1979. Glossary of soil science terms. Soil Sci. Soc. Am., Madison, WI.

Vasilev, Y.M. 1964. Rapid determination of the limits of rolling out. Pochvovedenie 7: 105–106.

Warkentin, B.P. 1993. Soil shrinkage. In M.R. Carter (ed) "Soil Sampling and Methods of Analysis," Lewis Publishers, Boca Raton, FL: 513–518.

Yasutomi, R. and S. Sudo. 1967. A method of measuring some physical properties of soil with a forced oscillation viscometer. Soil Sci. 104: 336–341.

Yoder, R.E. 1937. The significance of soil structure in relation to the tilth problem. Soil Sci. Soc. Am. Proc. 2: 21–23.

9

Water

The hydrosphere, with a strong influence on the pedosphere, comprises the sum total of all water bodies (oceans, rivers, lakes), groundwater (renewable and fossil), and soil water. Although water is the most abundant of all resources covering 70% of the earth's surface, freshwater is a source resource. The data in Table 9.1 indicate that 97.2% (volume basis) of the world water is in oceans and seas (1370 M Km^3). Freshwater accounts for merely 2.8% of the total volume, of which groundwater is 0.6%, and soil water accounts for less than 0.1% of the total (Fig. 9.1).

Soil is a major reservoir of freshwater, which accounts for about 50 times that in rivers and streams (Table 9.1). Some hydrologists classify the freshwater pools using simple nomenclature that reflects their functional characteristics. For example, *blue water* refers to water in water bodies that is lost from the land as runoff or seepage flow. This is the water that is temporarily lost for use by humans, animals, or plants. Freshwater, usable by primary producers (and comprising soil water, groundwater, and other irrigable sources), can be termed *green water*. The fraction of freshwater that is lost to the atmosphere through direct and soil evaporation may be termed *red water*. Fossil water is difficult to assess, is not renewable, and may be termed *gray water*. While simple and easy to comprehend, such terminology is vague, subjective, and arbitrary.

TABLE 9.1 Global Water Resources

Reservoir/Pool	Quantity (Km^3)	Percent of total
Water bodies		
Oceans	1,370,000,000	97.2
Freshwater lakes	125,000	<0.1
Saline lakes and inland areas	104,000	<0.1
Rivers and streams	1,300	<0.1
Ice sources		
Polar ice cap and glaciers	29,200,000	2.2
Lithosphere		
Soil water	67,000	<0.1
Groundwater	8,350,000	0.6
Atmosphere	13,000	<0.1

Source: Nace, 1971; Edwards et al., 1983; Goldman and Horne, 1983; Van der Leeden et al., 1990; Alley et al., 2002.

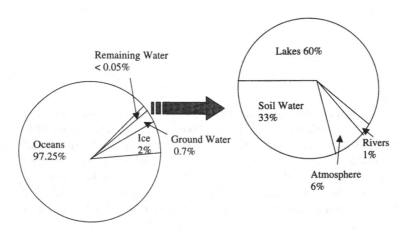

FIGURE 9.1 Different types of natural water.

9.1 PROPERTIES OF WATER

Principal properties of H$_2$O relevant to soil physical properties and processes are listed in Table 9.2. Some specific properties are described below.

9.1.1 Water Molecule

A single water molecule has a radius of 1.38 Å, at the center of which lies the oxygen nucleus. The oxygen and hydrogen protons in the water molecule

TABLE 9.2 Properties of Water Relevant to Soil Physical Properties and Processes

Property	Value
Density at 20°C	998.20 Kg m^{-3}
Density at 3.98°C	1000.0 Kg m^{-3}
Viscosity at 0°C	1.787 centipoise
Viscosity at 20°C	1.002 centipoise
Surface tension against air at 0°C	75.6 dynes cm^{-1}
Surface tension against air at 20°C	72.75 dynes cm^{-1}
Boiling point at NTP	100°C
Freezing point at NTP	0°C
Heat of vaporization	590 cal g^{-1}
Heat of freezing	80 cal g^{-1}

Source: Adapted from Weast, 1987.

are about 0.97 Å apart, and hydrogen protons are about 1.54 Å apart. Two hydrogen atoms are at an angle of about 105° from each other (Fig. 9.2), giving water an electric dipole of about 1.87×10^{-18} esu.

The dipole moment produces electric field in the vicinity of each molecule. The electric field of adjacent water molecules creates an attractive force creating relatively weak intermolecular hydrogen bond between the proton of the hydrogen atom of one molecule and the oxygen atom of the other. Therefore, water molecules are joined together through hydrogen bonding (Fig. 9.3). These bonds are weaker than covalent bonds.

One mole of water, about 18 cm^3 (18 g), contains 6.02×10^{23} molecules. Therefore, 1 cm^3 of water contains 3.3×10^{22} individual molecules. When water crystallizes at 0°C, it develops an open crystalline structure. Therefore, ice is less dense than liquid water at the same temperature because water expands on freezing. It is this expansion of water on freezing that causes changes in soil structure by repeated cycles of freezing and thawing (refer to Chapter 4).

Water has a very high boiling point, a very high melting point, and low density in the liquid phase (Table 9.2; see also Appendix 9.1). The liquid water molecules are freer to move, and they have greater internal energy. About 80 calories (334 joules or 3.34×10^8 ergs) of heat energy per g of water is liberated when water changes from liquid to solid. Therefore, the entropy of water is higher in the liquid than in the more orderly crystalline, or solid state. The heat of vaporization of water, the heat absorbed to change from liquid to vapor state, is about 590 calories (2463 joules, or 2.45×10^9 ergs) per g of water. Therefore, entropy of water is higher in the vapor than in the liquid state (see Chapter 17).

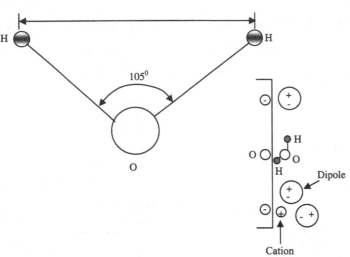

FIGURE 9.2 Water adsorption.

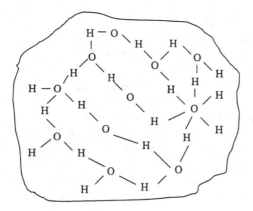

FIGURE 9.3 Water molecules joined together through hydrogen bonding.

9.1.2 Surface Tension

Water molecules at the air–water, solid–water, or another fluid–water interface are subjected to different forces than molecules within the bulk volume of the fluid. Water molecules within the bulk volume are hydrogen bonded to adjacent molecules and the cohesive forces are the same in all directions. At the air–water interface (or solid–water interface), the force pulling into the air is much different than force within the bulk volume. For a molecule on the surface, there is a resultant attraction inward (Fig. 9.4).

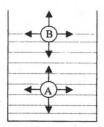

FIGURE 9.4 Forces acting on a molecule resting inside the liquid (A) and on the surface (B). The molecule on the surface has an unbalanced force making the water surface behave like a stretched membrane.

This imbalance in the force has the net effect of pulling the molecules in the one or two molecular layers near the surface into the bulk volume. The result is an orientation of molecules at the surface in such a way that pressure beneath the surface is much greater than above it, the surface behaves as if it were a stretched membrane, and the surface of the liquid always tends to contract to the smallest possible area. That is why the drop of liquid and bubbles of gas in a liquid become spherical. For a sphere, the surface is minimum for the given volume. In order to extend the surface, work has to be done to bring the molecules from the bulk of the liquid into the surface against the inward attractive force. This is called *free surface energy*. The difference in pressure is the cause of surface tension (γ), which is expressed in units of force per unit length or dynes/cm.

If a solid is immersed in water, the interfacial tension is due to the forces of adhesion (e.g., the forces of attraction for the water molecules by the solid and vice versa). These forces are the reasons for the work to be done to separate the solid from the liquid. The amount of work required is given by Eq. (9.1).

$$\frac{W_{sw}}{A_s} = \gamma_{sa} + \gamma_{wa} - \gamma_{sw} \qquad (9.1)$$

The work (W_{sw}) is expressed in units of energy, and γ_{sa}, γ_{wa}, and γ_{sw} represent surface tension at the solid–air, water–air, and solid–water interfaces, and A_s is the area of the solid surface. Eq. (9.1) is called the Dupré equation (1969).

9.1.3 Contact Angle

Soil is a three-phase system: solid, liquid, and gas (see Chapter 2). Assuming that two fluid phases (liquid and gas) are in contact with soil solid, the

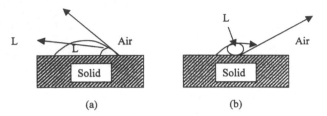

FIGURE 9.5 Angle of contact (a) is acute in a liquid that wets the solid, and (b) is obtuse in a liquid that does not wet the solid (L refers to liquid).

interface between air and water, forms a definite angle called the *contact angle* (Fig. 9.5). This angle is determined by Eq. (9.2) or Young's equation.

$$\cos\alpha = \frac{\gamma_{sa} - \gamma_{sw}}{\gamma_{wa}} \tag{9.2}$$

The contact angle α thus depends on three interfacial tensions. However, whether it is acute ($<90°$) depends on the relative magnitude of γ_{sa} and γ_{sw}. If γ_{sa} exceeds γ_{sw}, then cos α is positive, and α is less than 90°. This is generally the case with most mineral soils and water, because water wets soils. If γ_{sw} exceeds γ_{sa}, then cos α is negative, and α is between 90° and 180°. This is the case of mercury and soil, because mercury does not wet the soil. When the liquid wets the solid (soils) the contact angle is acute and the liquid meniscus is convex, when it does not, the contact angle is obtuse and the meniscus is concave (Fig. 9.5).

Hydrophilic Versus Hydrophobic Soils

If the adhesive forces between the soil and water are greater than the cohesive force inside the water, and greater than the forces of attraction between the air and the soil, then the soil–water contact angle is acute and water will wet the soil. Therefore, hydrophilic soil can be defined as having the following characteristics:

> Adhesive force (water–soil) > cohesive force (water–water)
>
> > adhesive force (soil–air)

A contact angle of zero implies complete flattening of the drop and perfect wetting of the soil surface by the water, and soil has absolute preference for the water over air.

A contact angle of 180° would mean a complete nonwetting or rejection of the water by the air-full soil. The water drop would retain its spherical shape without spreading over the soil surface.

When water is wetting the soil, the contact angle is low or acute. This low angle is called "wetting" or "advancing" angle. When the soil is drying and the water film is receding, the contact angle is different. It is called "receding" or "retreating angle." This difference in wetting and retreating angle is also the cause of soil–water hysteresis, which will be discussed in Chapter 10. Soil hydrophobicity is affected by some organic substances coated on aggregate surfaces, such as in the case of the formation of algal crust. In such cases, the angle of contact can be modified through management. Plowing and physically rupturing the crust can improve wetting. In irrigated soils, wettability can be improved by use of surfactants.

9.1.4 Capillarity

A capillary tube in a body of water forms a meniscus as a result of the contact angle of water with the walls of the tube. The curvature of this meniscus will be greater (i.e., radius of curvature smaller), the narrower the tube (Fig. 9.6a vs. 9.6d). The height of capillary rise depends on the diameter of the section that corresponds with the pressure difference (Fig. 9.6b vs. 9.6c). Because of the difference in the contact angle, the water rises in the glass tube but mercury falls in the glass tube (Fig. 9.7).

A liquid with an acute angle will have less pressure inside meniscus than atmospheric pressure [Eq. (9.3)].

$$P_i < P_o \tag{9.3}$$

For a capillary of uniform radius r, at equilibrium the forces per unit area pulling down.

$$F_{\downarrow_g} = \pi r^2 h \rho_w g = \text{force of gravity} \tag{9.4}$$

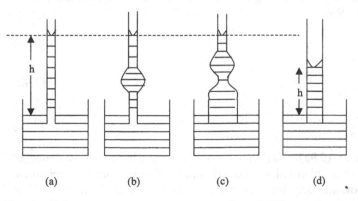

(a)   (b)   (c)   (d)

FIGURE 9.6 Rise of water in capillary tubes of different diameters.

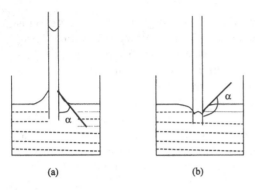

(a)   (b)

FIGURE 9.7   Angle of contact for (a) glass and water is acute (<90°), and (b) glass and mercury is obtuse (>90°).

$$F_{\uparrow_\gamma} = 2\pi r\gamma \cos \alpha = \text{surface tension} \qquad (9.5)$$

At equilibrium $F_{\downarrow_g} = F_{\uparrow_\gamma}$

$$\pi r^2 h \rho_w g = 2\pi r\gamma \cos \alpha \qquad (9.6)$$

$$h = \frac{2\gamma \cos \alpha}{r\rho_w g} \qquad (9.7)$$

$$\frac{2\gamma \cos \alpha}{r} = \rho_w g h = -\Delta p \qquad (9.8)$$

where ΔP is the pressure difference across the interface. Eq. (9.8) is the equation of Young and Laplace for a spherical surface. If the contact angle between solid and liquid is zero (glass, mineral soil particle) than Eq. (9.8) is as simple as Eq. (9.9).

$$\Delta P = \frac{2\gamma}{r} \qquad (9.9a)$$

$$h = \frac{2\gamma}{r\rho_w g} \qquad (9.9b)$$

$$r = \frac{2\gamma}{h\rho_w g} \qquad (9.9c)$$

Equations (9.8) or (9.9a) state that as a consequence of the existence of surface tension at a spherical surface of radius of curvature r, mechanical equilibrium is maintained between two fluids (water and air in soil) at different pressures.

Substituting the appropriate values for H_2O at 20°C in Eq. (9.9c) ($\gamma = 72.75$ dynes/cm or g/s^2, $\rho_w = 0.9982$ g/cm^3, and $g = 980$ cm/s^2), we can solve for r assuming that α is zero [Eq. (9.10)].

$$r = (0.1487/h)\,\text{cm} \tag{9.10}$$

Soil being an extremely heterogeneous mass, there are numerous radii to influence the pressure across the water film. In Figs. 9.6b and c, there are two radii to be considered. The radius (r) is within the liquid phase and forms a convex surface. If the bubble is not spherical and has two principal radii (r_1 and r_2), and assuming that the contact angle is zero, the pressure difference across an interface with two principal radii is given by Eq. (9.11).

$$\Delta P = \gamma \left(\frac{1}{r_1} + \frac{1}{r_2} \right) \tag{9.11}$$

The rise or fall of liquids in capillary tubes is used to calculate the pore size distribution (Chapter 6). Whether a liquid rises in a glass capillary (as water) or is depressed (as mercury) depends on the relative magnitude of the forces of cohesion (between the liquid molecules themselves) and the forces of adhesion (between the liquid and the wall of the tube). These forces determine the contact angle α.

The occurrence of a concave meniscus leads to the capillary rise, whereas a convex meniscus leads to capillary depression. As soon as the concave meniscus is formed, the pressure in the liquid under the curved surface is less than the pressure in the air. The liquid thus rises in the tube until the weight of the liquid column just balances the pressure difference ($\Delta P = 2\gamma/r$) and restores the hydrostatic equilibrium. The liquid column acts as a manometer to register the pressure difference across the meniscus.

Example 9.1

What is ΔP at the surface of a droplet of water and mercury with r of 1 mm?

Solutions

(a) For the air–water interface

$$\Delta P = \frac{2\gamma}{r} = 2 \left(72.75 \, \frac{\text{dynes}}{\text{cm}} \middle/ 0.1 \, \text{cm} \right) = 1455 \, \text{dynes/cm}^2$$

(b)   For the mercury–air interface $= 8600\,\text{dynes/cm}^2$

The difference calculated across the interface does not refer to the difference due to vapor pressure, which may be substantial.

Example 9.2

What vacuum is needed to draw all the water out of a sintered glass funnel if the minimum pore size is 4 µm in diameter?

Solution

As the water is withdrawn from the pores, the maximum pressure is reached when a hemispherical bubble is formed with a radius just equal to that of the pore. Therefore, the

$$\Delta P = (2\gamma/r) = 2(72.75\,\text{dynes/cm})/2 \times 10^{-4}\,\text{cm} = 72.75 \times 10^4\,\text{dynes/cm}$$

9.1.5   Osmotic Pressure

Osmotic pressure is a property of solutions, expressing the decrease of the potential energy of water in solution relative to that of pure water (see Chapter 20).

When an aqueous solution is separated from pure water (or from a solution of lower concentration) by a membrane that is permeable to water alone, water will tend to *diffuse* or *osmose* through the membrane into the more concentrated solution, thus diluting it or reducing the potential energy difference across the membrane. The osmotic pressure is the counter pressure that must be applied to the solution to prevent the osmosis of water into it.

In dilute solutions, the osmotic pressure is generally proportional to the concentration of the solution and to its temperature according to Eq. (9.12).

$$P_s = KTC_s \tag{9.12}$$

where P_s is osmotic pressure, T is absolute temperature, and C_s is concentration of solute.

An increase in the osmotic pressure is usually accompanied by a decrease in the vapor pressure, a rise of the boiling point, and a depression of the freezing point.

9.1.6 Solubility of Gases

The concentration of gases in water generally increases with pressure and decreases with temperature. According to Henry's law, the mass concentration of gas C_m is proportional to the pressure of gas P_i [Eq. (9.13)].

$$C_m = S_c \frac{P_i}{P_o} \tag{9.13}$$

when C_m is mass concentration of gas, S_c is solubility coefficient of the gas in water, P_i is pressure of the gas, P_o is total pressure of the atmosphere, and C_m is mass of dissolved gas relative to the mass of H_2O.

The volume concentration is similarly proportional to Eq. (9.14).

$$C_v = S_v \frac{P_i}{P_o} \tag{9.14}$$

where S_v is solubility expressed in terms at volume ratio, and C_v is volume of dissolved gas relative to the volume of H_2O.

9.1.7 Viscosity

Viscosity of a fluid is its resistance to flow. For example, water has lower viscosity than syrup or honey. When fluid is moved in shear (adjacent layers of fluid are made to slide over each other), the force required is proportional to the velocity of shear. The proportionality factor is called viscosity (η). It is the property of fluids to resist the rate of shearing, and can be visualized as an internal friction.

The coefficient of viscosity η is defined as the force per unit area necessary to maintain a velocity difference of 1 cm/sec between two parallel layers of fluid that are 1 cm apart. The viscosity equation is shown in Eq. (9.15).

$$\tau = \frac{F_s}{A} = \eta \frac{du}{dx} \tag{9.15}$$

where τ is shearing stress, F_s is force, A is area of action for the force, and du/dx is velocity gradient normal to the stressed area.

Kinematic Viscosity (η_K)

The ratio of the viscosity to the density of the fluid is called the *kinematic viscosity* (η_K). It expresses the shearing-rate resistance of a fluid mass independently of the density.

While η of water is about 50 times more than that of air, η_K of water is actually lower. Viscosity has the units of poise or centipose (see Appendix 9.1 and Appendix L).

9.1.8 Newtonian and Non-Newtonian Fluids

Newtonian fluids obey Newton's law of viscosity, which states that shear stress (τ) is proportional to shear rate, with the proportionality constant being the coefficient of viscosity (η) as shown in Eq. (9.15) and Fig. 9.8.

For solids, shear stress divided by shear strain gives an elastic modulus [refer to Eq. (7.17)]. For viscous liquids, since the strain is increasing all the time, shear stress divided by the rate of shear strain gives the viscosity coefficient. Newtonian fluids have a constant viscosity at a given temperature. Examples of Newtonian fluids are water, salt solution, milk, mineral oil, etc. In general, all gases and most liquids with simpler molecular formula and low molecular weight (e.g., water, benzene, ethyl alcohol, CCl_4, hexane, and most solutions of simple molecules) are Newtonian fluids.

Non-Newtonian fluids do not obey Newton's law of viscosity. Such fluids have a variable viscosity at a constant temperature $\eta = f(t)$, and viscosity depends on the force applied (time and temperature).

$$\tau = \eta \frac{\partial v}{\partial x} \tag{9.16}$$

where v is the velocity and x is the distance, η is the apparent viscosity and is not a constant (Fig. 9.8). Examples of non-Newtonian fluids are a syrupy mixture of cornstarch and water, quicksand, slurries, pastes, gels, polymer solutions, etc.

In some non-Newtonian fluids, properties are independent of time under shear. Such fluids include the following:

Bingham Plastic

These fluids resist a small shear stress but flow easily under larger shear stresses e.g., toothpaste, jellies, and some slurry.

Pseudoplastic

Viscosity of the fluids decreases with increasing velocity gradient (e.g., polymer solutions, blood). Pseudoplastic fluids are also known as shear thinning fluids. At low shear rates (dv/dx) the shear thinning fluid is more viscous than the Newtonian fluid, and at high shear rates it is less viscous. Most non-Newtonian fluids fall into this group.

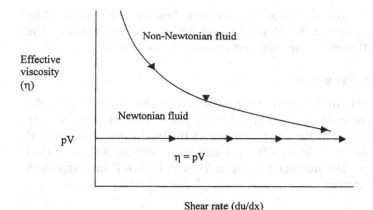

FIGURE 9.8 Newtonian and non-Newtonian fluids.

Dilatant Fluids

Viscosity of these fluids increases with increasing velocity gradient. They are uncommon, but suspensions of starch and sand behave in this way. Dilatant fluids are also called shear thickening fluids.

In other non-Newtonian fluids, properties are dependent upon duration of shear. Such fluids include the following:

Thixotropic Fluids

For thixotropic fluids the dynamic viscosity decreases with the time for which shearing forces are applied (e.g., thixotropic jelly paints).

Rheopectic Fluids

For Rheopectic fluids the dynamic viscosity increases with the time for which shearing forces are applied (e.g., gypsum suspension in water).

Viscoelastic Fluids

Viscoelastic fluids have elastic properties, which allow them to spring back when a shear force is released (e.g., egg white).

9.1.9 Fluidity

Fluidity is the reciprocal of viscosity and has the units of 1/poise or 1/centipose.

$$F = \rho g/\eta = \frac{g}{cm^3} \cdot \frac{cm}{s^2} \cdot \frac{s \cdot cm}{g} = \frac{1}{cms} = \frac{1}{poise} \qquad (9.17)$$

Fluids of lower viscosity flow more readily than those of high viscosity. The fluidity of water increases by about 3% per 1°C rise in temperature. The fluidity is also affected by the type and concentration of solutes.

9.1.10 Vapor Pressure

The change of state of water from liquid to vapor phase is related to the kinetic theory. The molecules in a liquid move past one another in a variety of speeds. A molecule in the upper regions of the liquid with a high speed may leave the liquid momentarily and fall back. Others with a critical speed may escape. The number of molecules with high KE increases with increase in temperature.

Evaporation $E \propto T$

Water evaporation is an endothermic process. When H_2O molecules with high energy escape, the velocity and kinetic energy of those remaining is less, the lesser the velocity the lower is the temperature. The liquid is, therefore, cool.

Saturated Vapor Pressure

The vapor is in equilibrium with a liquid because the rate of molecules escaping and those returning back by condensation is equal.

The equilibrium exists when the space is saturated. The pressure of the vapor when it is saturated is called the "saturated vapor pressure." The saturated vapor pressure does not depend on the size of the container. It depends on:

1. Pressure of water
2. Temperature of water
3. Chemical condition (solutes)

Boiling Point

Water boils at a temperature when vapor pressure becomes equal to atmospheric pressure. The saturated vapor pressure is related to the temperature (T) as per the simplified version of the Clasius–Clapeyron equation [Eq. (9.18)].

$$\ln P_o = a - \frac{b}{T} \tag{9.18}$$

where $\ln P_o$ is logarithm to the base e of the saturation vapor pressure P_o, T is absolute temperature, and a and b are constant.

Pressure of the liquid water also affects vapor pressure. Water in soil is a dilute solution of various electrolytes. The vapor pressure of electrolytes is lower than that of pure water, soil–water also has a lower vapor pressure even when the soil is saturated. In an unsaturated soil, capillary and adsorptive effects further lower the potential and the vapor pressure.

Vapor pressure is expressed in units of pressure, e.g., dynes/cm², bar, mm of Hg, or water. The vapor pressure of atmosphere can also be expressed in the following different ways:

1. $RH = \dfrac{\text{partial pressure of water}}{\text{saturated pressure of water}} \times 100$

2. (ρ_v) Vapor density $= \dfrac{\text{mass of water vapor}}{\text{volume of air}}$

3. Specific humidity $= \dfrac{\text{mass of water vapor}}{\text{mass of air}}$

4. Saturation (or vapor pressure) deficit = the difference between the existing vapor pressure and the saturation vapor pressure

5. Dew point temperature: The temperature at which the existing vapor pressure becomes equal to the saturation vapor pressure, i.e., the temperature at which a cooling body of air with a certain vapor content will begin to condense dew.

9.2 THE HYDROLOGIC CYCLE

Water is a completely renewable resource. It changes from one form to another and from one environment to another. Water transfer or movement from one form and/or one environment to another governs the hydrologic cycle (Fig. 9.9). The hydrologic cycle involves interchange (fluxes) between principal pools. These fluxes are: (i) evaporation, transpiration, or evapotranspiration (red water) by which water enters the atmosphere, (ii) precipitation by which returns to the land and ocean, and (iii) infiltration, percolation, interflow, and runoff or overland flow by which water is returned from land to streams, rivers, lakes, and oceans (blue water). Soil water (green water) is a principal pool of the freshwater reserves. The magnitude of these pools and fluxes is shown in Fig. 9.9.

The data in Fig. 9.9 and Table 9.3 can be used to calculate the mean resident time (T_r) which is equal to the mass/flux. The T_r for water in the atmosphere, streams/rivers, and oceans is given by Eqs. (9.19) to (9.21).

$$T_r \text{ atmosphere} = 13{,}000 \, \text{Km}^3 / 496{,}000 \, \text{Km}^3 \text{ per yr} = 0.026/\text{yrs} \quad (9.19)$$

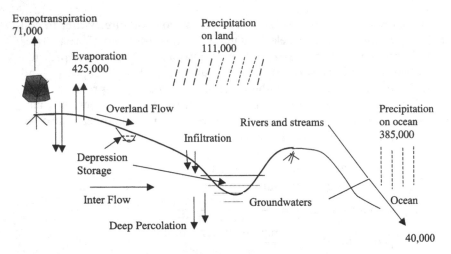

FIGURE 9.9 Schematic of global transfer rates (km³/yr) for water movement in the hydrologic cycle. (Flux rates are from Spiedel and Agnew, 1982 and Alley et al., 2002).

TABLE 9.3 Major Water Pools and the Mean Residence Time (Tr) of Water in Specific Pool

Pool	Capacity (Km³)	Flux (Km³/yr)	Tr (yr)
Oceans	1,370,000,000	425,000	3223.5
Freshwater lakes	125,000		—
Saline lakes and inland seas	104,000		—
Rivers and streams	1,300	40,000	0.0325
Glaciers and ice caps	29,200,000		—
Soil water	67,000	111,000	0.604
Groundwater	8,350,000		—
Atmosphere	13,000	496,000	0.0262

Source: Calculated from Spiedel and Agnew, 1982, and Alley et al., 2002.

$$T_r \text{ streams/rivers} = 1,300 \, Km^3/40,000 \, Km^3 \text{ per yr} = 0.033 \, \text{yrs} \quad (9.20)$$

$$T_o \text{ oceans} = 1,320 \times 10^6 \, Km^3/485 \times 10^3 \text{ per yr} = 3,100 \, \text{yrs} \quad (9.21)$$

Therefore, atmospheric and stream/river pools are highly dynamic and can transfer contaminants or pollutants from one pool to another very rapidly. The T_r of water in soil is highly variable, and depends on soil properties.

9.3   SOIL AS A RESERVOIR OF WATER

The total pool of soil–water is estimated at about 67,000 Km^3. In terms of the freshwater reserves, soil is, in fact, a very efficient storage system. Assume that a one-hectare area of soil has water content of 20% by weight in the top 1-m depth with an average bulk density of 1.25 Mg/m^3. The total amount of water in the soil is 0.25 hectare-meter, 2.5×10^3 Mg, 2.5×10^6 Kg, or 2.5×10^6 L. This is indeed a large quantity of water. If human consumption of water is about 100 L/day, this water is enough for one person for 2.5×10^4 days or 68.5 years or for 25,000 people for one day. If half of this water were available for plant uptake at the consumptive use rate of 0.5 cm/day, it can support plant growth for 25 days.

Rather than the absolute quantity, it is often change in the soil–water pool that is of major interest. The change in the soil–water pool can be computed from the water balance Eq. (9.22).

$$\Delta S = P + I - (R + D + \text{ET}) \tag{9.22}$$

where ΔS is the change in the soil–water pool, P is precipitation, I is irrigation, R is surface runoff, D is deep drainage, and ET is evapotranspiration. Different components listed in Eq. (9.22) are determined by lysimetric evaluation.

9.4   COMPONENTS OF THE HYDROLOGIC CYCLE

Different components of the hydrologic cycle are outlined in Eq. (9.22), which is normally written in a form to solve for ET [Eq. (9.23)].

$$ET = P + I - (R + D) \pm \Delta S \tag{9.23}$$

Therefore, different components of the hydrologic cycle include: (i) precipitation (P) including rain, snow, hail, fog, mist, (ii) irrigation (I) is not a component in natural ecosystem but is an important factor in the hydrologic cycle of managed and especially agricultural ecosystems in arid and semi-arid regions, (iii) R is surface runoff, (iv) D is deep drainage leading to groundwater recharge, (v) ΔS is change in soil water storage, and (vi) ET is evapotranspiration. Methods of measurement and estimation or prediction of evapotranspiration are described in detail by Monteith (1985), and standard methods of measuring precipitation are discussed in texts on climatology or any hydrologic manual (USDA, 1979).

9.4.1 Precipitation

Accurate measurement of precipitation is important for reliable assessment of the water balance. In addition to simple or non-recording and recording rain gauges normally used at the meteorological stations (Figs. 9.10–9.13), rainfall measurement under a vegetation cover involves measurement of: (i) through fall using a spider gauge (Fig. 9.14a), and (ii) stem flow (Fig. 9.14b). Measurement of through fall and stem flow can be highly variable depending on the tree canopy and foliage characteristics.

9.4.2 Runoff

There are numerous methods of measuring surface runoff for different scales. The scale may range from a microplot of a few square meters to a watershed of several Km^2 or more (Table 9.4). Hydrologic parameters that are measured to compute surface runoff include total volume stage or water level, velocity, discharge, and their variation over time. Installation, measurements, and calibration procedures of these devices are described in USDA (1979), and shown in Figs. 9.15–9.18.

9.4.3 Lysimetric Analysis

A lysimeter is a confined volume of soil, in which input, output and change in water storage can be quantified. The size, shape, and material used in

FIGURE 9.10 A meteorological station installed within a rice paddy.

constructing lysimeters vary widely. Lysimeters may be square (Fig. 9.19) or circular (Figs. 9.20–9.22), and made of steel, galvanized material, fiberglass, or plastic. Hydrologic inputs comprise precipitation and supplemental addition of water depending upon the management systems imposed. Hydrologic output comprises deep drainage or percolation water.

FIGURE 9.11 Recording and non-recording rain gauges.

FIGURE 9.12 A snow gauge.

(a)

(b)

FIGURE 9.13   (a) Class A pan evaporemeter; (b) a device to measure evaporation in a lake.

Changes in soil–water storage can be measured by using neutron moisture meter or gypsum blocks.

There are several types of lysimeter depending on the method of construction, and evaluating hydrologic balance. Common types of lysimeters are outlined in Table 9.5. The drainage is facilitated by using about a 5 cm thick layer of gravel, sand, or diatomaceous clay at the base

(a)

(b)

FIGURE 9.14 (a) Spider gauge to measure through-fall and (b) stem flow.

TABLE 9.4 Method of Measuring Surface Runoff

Technique/plot	Size	Equipment
Microplots	1–10 m^2	A drum with a capacity of about 200 liter, or a small flume with water stage recorder
Field runoff plots	0.0025–100 ha	Multidivisor tanks, flume, water stage recorder
Small watersheds	1–10 ha	Flume, water stage recorder, proportional samplers
Large watersheds	>10 ha	Weirs, waterstage recorders

Source: Adapted from Lal, 1990.

FIGURE 9.15 A multidivider tank and a flume with water stage recorder to measure runoff from a plot.

(Fig. 9.23). Lysimeters may be cited or different landscape positions in the field, or constructed at one cite to facilitate specific measurement (Figs. 9.24 and 9.25).

Lysimetric data are used to compute consumptive water use by plants or crops grown. An example of the method to use these data is shown below. Consider the data in Table 9.6 for 30-day period from a lysimetric experiment:

Consumptive use or ET per day = 16 cm/30 days = 0.53 cm/day

FIGURE 9.16 An H-flume and a water stage recorder to measure runoff from a steep agricultural watershed.

FIGURE 9.17 A wier with a slot-pipe to collect runoff sample.

There are numerous uses of lysimetric experiments, with the primary use of measuring the components of hydrologic cycle, especially deep drainage, soil-water storage, and evapotranspiration. In addition, chemical analyses of the deep drainage or percolation water can be extremely useful to study transport of chemicals applied to the soil, e.g., fertilizers and pesticides. Temporal changes in concentration of NO_3–N, PO_4–P, organic P, dissolved organic carbon can provide useful information on the risks of contamination of groundwater. Fate and pathways of pesticides can also be studied by lysimetric analyses.

Lysimetric studies are also useful to evaluate transport of clay from surface to the subsoil by the process of illuviation (Roose, 1977). The

FIGURE 9.18 A Coshocton wheel sampler to obtain runoff sample.

(a)

FIGURE 9.19 A square filled in lysimeter (a) method and (b) with removable cover.

(b)

FIGURE 9.19 Continued.

FIGURE 9.20 Installation of a circular monoleith lysimeter.

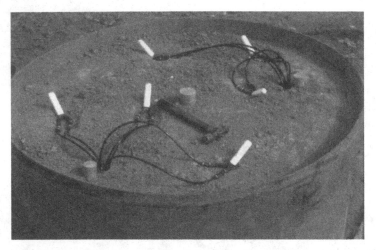

FIGURE 9.21   A suction cup and neutron probe access tube are installed at the base.

FIGURE 9.22   Suction cups are embedded in the diatomaceous clay.

information on solution weathering or rate of new soil formation can also be obtained by chemical analyses (Al^{+3}, Si^{+4}, cations) of the percolating water. For these measurements, lysimeters must be deep enough and include bedrock as a part of the monolith or soil solum being studied.

Table 9.5 Types of Lysimeters Used for Evaluating Components of the Hydrologic Cycle

Basis		Lysimeter types
Soil disturbance	(i)	Filled in, where disturbed soil is packed layer by layer at ρb similar to the field situation
	(ii)	Monolith, where a block of undisturbed soil is encased under natural conditions
Weighing	(i)	Nonweighing or drainage lysimeter in which water balance is obtained by carefully measuring the volume of water drained
	(ii)	Weighing lysimeters monitor changes in total weight on a continuous basis or at regular time intervals. Weighing lysimeters may use a mechanical balance or a hydrologic weighing technique
Drainage	(i)	Gravity drainage
	(ii)	Suction drainage
Location	(i)	In situ, constructed with soil in place
	(ii)	Constructed with soil transported from different regions

Example 9.3

A runoff plot has a dimension of 25 m × 4 m. The runoff collection system involves a Coshocton Wheel Sampler, which collects 1% of the runoff. Total runoff collected after 2.5 cm of rainfall is 10 liters. The sediment load in runoff is 5 g/liter. Calculate runoff and erosion.

Solution

Total runoff volume $= 10$ liters $\times 100 = 1000$ liters

$$\text{Runoff depth} = \frac{\text{volume}}{\text{area}} = 10^3 \, \text{L} \times \frac{10^3 \, \text{cm}^3}{\text{L}} \times \frac{1}{100 \, \text{m}^2} \times \frac{10^4 \, \text{cm}^2 \, \text{m}^2}{10^4 \, \text{cm}^2} = 1 \, \text{cm}$$

$$\text{Runoff C\% of rainfall} = \frac{1 \, \text{cm}}{2.5 \, \text{cm}} \times 100 = 40\%$$

$$\text{Total sediments} = 1000 \, \text{L} \times \frac{5 \, \text{g}}{\text{L}} \times \frac{\text{kg}}{10^3 \, \text{g}} = 5 \, \text{kg}$$

$$\text{Soil erosion} = \frac{5 \, \text{kg}}{100 \, \text{m}^2} \times \frac{10^4 \, \text{m}^2}{\text{ha}} = 500 \, \text{kg/ha} = 0.5 \, \text{Mg/ha}$$

PROBLEMS

1. A lake has a capacity of 1200 Km^3. The steady state evaporation flux is 200 Km^3 y^{-1}. What is the mean residence time of water in the lake?

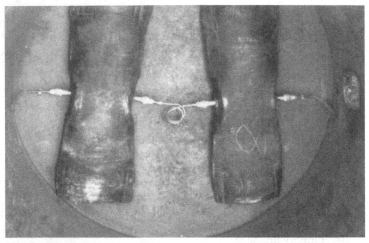

(a)

(b)

Figure 9.23   (a) A hydraulic weighting device may involve water-filled pillows placed beneath the lysimeter, and (b) connected to a pressure gauge.

2. A one hectare field contains $0.2\,\mathrm{g\,g^{-1}}$ of water to $10\,\mathrm{m}$ depth. Assuming a uniform soil bulk density of $1.5\,\mathrm{Mg\,m^{-3}}$, calculate the total water content of soil in liters and equivalent depth.

3. Draw a landscape, and list principle components of the hydrologic cycles.

4. Tabulate methods of monitoring components of a hydrologic cycle along a hill slope.

(a)

(b)

FIGURE 9.24 A battery of drainage lysimeter (a) with a trench to collect seepage; (b) an underground weighing and seepage collection facility.

5. Draw up a table or a nomograph comparing different units of measuring water capacity and flux, and compute conversion factor to change from one unit to another.
6. Calculate the height of capillary rise in a soil pore of 50 μm inner diameter in winter (0°C), spring (10°C), early summer (20°C), and tropics (40°C).
7. Compute the pressure difference at the air–water interface in Question 1 above.

FIGURE 9.25 A series of lysimeters under a plastic shelter.

TABLE 9.6 Lysimetric Measurements

Period (days)	Precipitation	Irrigation	ΔS	Runoff	Deep drainage
0–10	0	5	−1	0	0
10–15	12	0	+4	3	2
15–30	5	0	−2	0	0

Calculate ET:
Solution:
$ET = P + I - (R + D + \Delta S)$
ET For Period $1 = 0 + 5 - (0 + 0 - 1) = 6$ cm
ET For Period $2 = 12 + 0 - (3 + 2 + 4) = 3$ cm
ET For Period $3 = 5 + 0 - (0 + 0 - 2) = 7$ cm
Total ET $= 16$ cm

8. Consider the following equation of the height of capillary rise:

$$r = \frac{2}{h\rho g}$$

where γ and ρ refer to the surface tension and density of the fluid, respectively. What is the difference in the height of capillary rise in 20 μm diameter pore for water and alcohol at 20°C?

9. Write a brief essay on "surface tension." As a diagram, explain interactive forces, and define units.

10. The 0–50 cm layer of a lakebed soil in northwestern Ohio has a field capacity of 30% by weight, soil–water content of 15% by weight, and bulk density of

$1.2\,\text{Mg m}^{-3}$. A rainfall of 4 cm was received of which 75% was lost as runoff. Calculate the following:

1. What is the volume of runoff from a test plot of $25\,\text{m} \times 40\,\text{m}$?
2. What is soil erosion (t/ha) if the runoff contained sediments of 25 g/liter?
3. What is the total NO_3 loss if concentration in runoff is 5 g/liter?

11. Why are some soils more wettable than others? Why does burning crop residue or any biomass make a soil hydrophobic?

APPENDIX 9.1 SOME PHYSICAL PROPERTIES OF WATER AT ATMOSPHERIC PRESSURE

Temperature (°C)	Density ρ (Mg/m³)	Specific weight γ (N/m³ × 10³)	Dynamic viscosity μ (N × s/m² × 10⁻³)	Kinematic viscosity $(\eta\kappa)$ (m²/s × 10⁻⁶)
0	1.0	9.810	1.79	1.79
5	1.0	9.810	1.51	1.51
10	1.0	9.810	1.31	1.31
15	0.999	9.800	1.14	1.14
20	0.998	9.790	1.00	1.00
25	0.997	9.781	0.891	0.894
30	0.996	9.771	0.797	0.800
35	0.994	9.751	0.720	0.725
40	0.992	9.732	0.653	0.658
50	0.988	9.693	0.547	0.553
60	0.983	9.643	0.466	0.474
70	0.978	9.594	0.404	0.413
80	0.972	9.535	0.354	0.364
90	0.965	9.467	0.315	0.326
100	0.958	9.398	0.282	0.294

$0.001\,\text{N} \times \text{s/m}^2 = 0.001\,\text{Pa} \times \text{s} = -0.01\text{P} = 1\,\text{cP} = 1\,\text{centipose}$
Source: Adapted from Weast, 1987; Julien, 1998.

REFERENCES

Alley, W.H., R.W. Healy, J.W. LaBaugh, and T.E. Reilly. 2002. Flow and storage in ground water systems. Science 296: 1985–1990.

Edwards, K.A., G.A. Classen, and E.H.J. Schroeten. 1983. The water resource in tropical Africa and its exploitation. ILCA Res. Rep. No. 6, Addis Ababa, Ethiopia, 103 pp.

Goldman, C.R. and A.J. Horne. 1983. Limnology, McGraw Hill, New York. Hillel, D.R. 1994. Rivers of Eden.

Julien, P.Y. 1998. Erosion and sedimentation. Cambridge Univ. Press, Cambridge, U.K., 280 pp.

Lal, R. 1990. Soil Erosion in the Tropics: Principles and Management. McGraw Hill, New York, 580 pp.

Monteith, J.L. 1985. Evaporation from land surfaces: progress in analysis and prediction since 1948. In "Advances in Evapotranspiration", American Soc. of Agric. Eng., St. Joseph, MI: 4–12.

Nace, N.R. 1971 (ed). Scientific framework of the world water balance. UNESCO Tech. Papers Hydrol. 7, UNESCO, Paris.

Roose, E.J. 1977. Application of the Universal Soil Loss Equation of Wischmeier and Smith in West Africa. In: D.J. Greenland and R. Lal (eds) "Soil Conservation and Management in the Humid Tropics," J. Wiley and Sons, Chichester, U.K.: 177–187.

Spiedel, D.H. and A.F. Agnew. 1982. The natural geochemistry of our environment. Westview Press, Boulder, Co.

USDA 1979. Field manual for research in agricultural hydrology. USDA Agric. Handbook No. 224, Washington, D.C., 545 pp.

Van der Leeden, F., F.L. Troise and D.K. Tod. 1990. The water encyclopedia. Lewis Publishers, Chelsea, MI, 808 pp.

Weast, R.C. (ed.) 1987. Handbook of chemistry and physics. Int. student edition. CRC Press, Boca Raton, FL.

10

Soil's Moisture Content

Soil's moisture content is defined as the water that may be evaporated from soil by heating at 105°C to a constant weight. The choice of the temperature limit is arbitrary, and clayey soils retain a considerable quantity of water at this temperature.

Water in the soil is held by the forces of cohesion and adhesion in which surface tension, capillarity, and osmotic pressure play a significant role. There are two types of forces acting on soil moisture. Positive forces are those that enhance soil's affinity for water (e.g., forces of cohesion and adhesion). In contrast, some negative forces that take water away from soil include gravity, actively growing plant roots, and evaporative demand of the atmosphere. At any given point in time, soil's moisture content is the net result of these positive and negative forces. Considerable advances in our understanding of soil moisture regime were made in the first half of the twentieth century. Historical developments in the science of soil moisture are given in Taylor and Ashcroft (1972), Rode (1969), Rose (1966), Childs (1969), and others.

10.1 SOIL-WATER REGIME

There are three forms of soil moisture. The liquid water is held in the transmission and retention pores. The absorbed water is held by the forces of cohesion and adhesion on the soil particles, mostly colloidal particles such as clay and organic matter. The third form of water is the one held within the lattice structure of clay minerals. Two edaphologically important aspects of the liquid water held within the pores are field moisture capacity and permanent wilting point.

10.1.1 Field Moisture Capacity (FC)

When a fully saturated soil ($s = \Theta = 1.0$) is allowed to drain freely under the force of gravity and there is no loss due to evaporation, after some time the soil's moisture content will approach an equilibrium level (Fig. 10.1). This equilibrium in soil's moisture content is called *field moisture capacity*. It is the moisture content that a given soil reaches and maintains after it has been thoroughly wetted and allowed to drain freely. It is the upper limit of moisture content that a soil can hold. It is the moisture content when all macropores or transmission pores have been drained and water in the macropores has been replaced by air.

Being a highly heterogenous mixture, most natural soils do not have a well-defined field moisture capacity. Clayey soils (curve B in Fig. 10.1) rarely attain a field moisture capacity because they continue to drain for a long period of time. Soils with impeded drainage (curve C in Fig. 10.1) never attain a field moisture capacity.

Free drainage under the force of gravity removes excess water from the upper layer and transmits it to the lower layers (Fig. 10.2). If the water drained from the upper layer is more than that needed for attaining the field moisture capacity of the lower layer, the excess water will be drained and transmitted to the third layer, and so on.

Example 10.1

A soil with a bulk density of 1.2 g/cm^3 has an initial gravimetric moisture content of 0.083. If its field moisture capacity is 0.25 (g/g), how deep will 2 cm of rain penetrate into the soil? Assume density of water (ρ_w) is 1.0 g/cm^3.

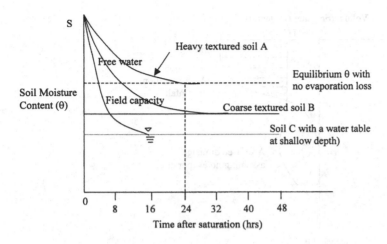

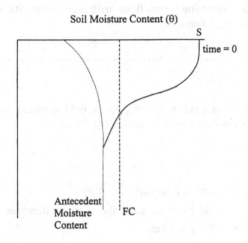

FIGURE 10.1 Field moisture capacity is the moisture held in the soil when free water in macropores is allowed to drain under the force of gravity.

Solution

$$\text{Initial volumetric moisture content } (\Theta_i) = \frac{0.083\,\text{g}}{\text{g}} \times \frac{1.2\,\text{g}}{\text{cm}^3} \times \frac{\text{cm}^3}{1\,\text{g}} = 0.10$$

$$\text{Volumetric field moisture capacity } (\Theta_{fc}) = \frac{0.25\,\text{g}}{\text{g}} \times \frac{1.2\,\text{g}}{\text{cm}^3} \times \frac{\text{cm}^3}{1\,\text{g}} = 0.30$$

$$\therefore \text{ Soil moisture deficit} = \Theta_{fc} - \Theta_i = 0.30 - 0.10 = 0.20$$

$$\text{Depth of rain penetration} = \frac{\text{Rain fall amount}}{\text{Soil moisture deficit}} = \frac{2\,\text{cm}}{0.20} = 10\,\text{cm}$$

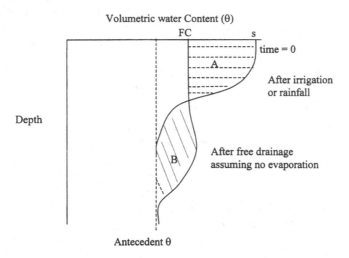

Volumetric water Content (θ)

FIGURE 10.2 Free drainage following rainfall or irrigation transmits water in excess of field capacity to the layer beneath.

Example 10.2

Consider that soil in the above example is to be irrigated to field moisture capacity to 50 cm depth. How much of irrigation water is needed for 10 ha?

Solution

Soil moisture deficit $= \Theta_{fc} - \Theta_i = 0.30 - 0.10 = 0.20$

∴ Water needed to attain field moisture capacity to 50 cm depth =

depth × deficit $= 0.20 \times 50\,\text{cm} = 10\,\text{cm}$

Total water needed to irrigate 10 ha $= 10\,\text{ha} \times 10\,\text{cm}$

$$= 100\,\text{ha\,cm} \times 10^4\,\frac{\text{m}^2}{\text{ha}} \times 10^{-2} \times \frac{\text{m}}{\text{cm}} = 10^4\,\text{m}^3$$

There are numerous soil factors that affect its FC. Important among these are texture and especially the clay content, clay minerals, porosity and pore size distribution, and soil organic matter content. The FC is more for soils with high than low clay content. For the same clay content, soils with 2:1 swelling type clay minerals have more FC than those with 1:1 clay minerals, and those with high % WSA and structural porosity have more FC than those with low % WSA and contain predominantly textural porosity. Soil's organic matter content has a positive effect on FC. All other factors remaining the same, soils with high organic matter content have a higher FC than those with low organic matter content. Effects of these factors on field capacity are shown in Figs. 10.3 and 10.4 (Lal, 1979a).

10.1.2 Permanent Wilting Point (PWP)

This is the lower limit of the moisture content of soil at which forces of cohesion and adhesion holding moisture in soil far exceed the pull that plant roots can exert to extract moisture from the soil. It is a unique moisture content that a soil attains beyond which soil moisture is no longer available to plants. This is the moisture content at which plant leaves wilt permanently and do not regain turgidity even when placed in an atmosphere with a relative humidity of 100%. The PWP is the moisture content at which even the retention pores have been depleted of their moisture content. The residue moisture content in soil at the PWP is of little use to plants.

Similar to field moisture capacity, moisture content at PWP also differs widely among soils. The PWP is higher in soils with higher clay content. It is higher with 2:1 type than 1:1 type clay minerals, and with expanding-lattice and more surface area than those with fixed-lattice and low surface area (Lal, 1979c). In contrast to FC, the PWP is not significantly influenced by aggregation, structural porosity, and soil organic matter content. Therefore, the PWP is primarily influenced by the amount and nature of clay content (Fig. 10.5).

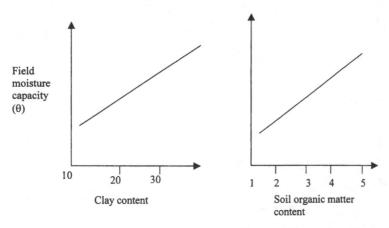

FIGURE 10.3 A schematic showing the effects of clay and soil's organic matter content on field moisture capacity.

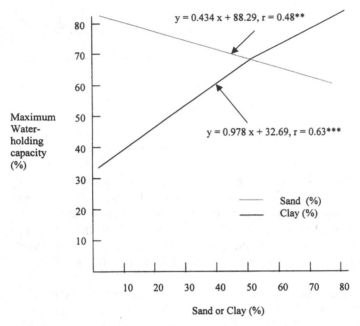

FIGURE 10.4 The effect of sand and clay content on the maximum water holding capacity of some Nigerian soils. (Redrawn from Lal, 1979.)

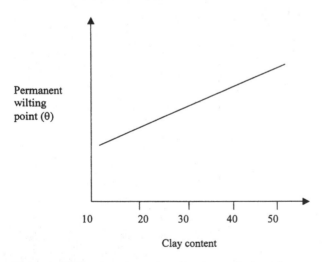

FIGURE 10.5 A schematic showing relation between clay content and the volumetric moisture content at the permanent wilting point.

10.1.3 Plant Available Water Capacity (AWC)

The available water capacity (AWC) is the difference in moisture content between FC and PWP [Eq. (10.1)].

$$AWC = FC - PWP \qquad (10.1)$$

The AWC is an important characteristic that determines a soil's physical qualities. Soils with high AWC have higher potential to produce plant biomass than those with low AWC. In contrast to the effect on FC, it is difficult to generalize the effect of clay content on soil's AWC because increase in clay content increases both the FC and the PWP (Salter et al., 1966; Salter and Hawroth, 1961; Tran-vinh-An, 1971; Pidgeon, 1972; Hallis et al., 1977; Lal, 1979a;c; Jenny, 1980; Hudson, 1994; Emerson, 1995). On the other hand, the effect of soil's organic matter on the AWC is well-defined. Increase in soil's organic matter increases the FC but not the PWP, and therefore, increases the AWC (Fig. 10.6).

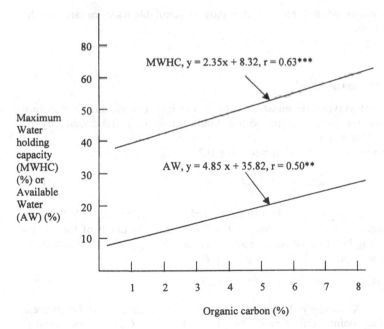

FIGURE 10.6 The relationship between organic matter content and soil water retention of some Nigerian soils. (Redrawn from Lal, 1979.)

10.1.4 Least Limiting Water Range

In addition to the moisture content of soil, AWC also depends on soil strength when moisture content is in the vicinity of the PWP and by poor aeration when close to field capacity. Letey (1985) proposed the "nonlimiting water range" (LLR) at which water uptake is neither limited by soil-resistance when too dry nor poor aeration when too wet. Keeping in view that plant growth varies in a continuous fashion with change in soil strength (see Chapter 7), matric potential (see Chapter 11), and aeration (see Chapter 18) (Dexter, 1987; Allmares and Logsdon, 1990), Da Silva et al. (1994) proposed the term "least limiting water range" (LLWR). It refers to a range of soil's moisture content at which plant growth is least limited by either soil strength or poor aeration. The LLWR is also influenced by several soil properties including particle size distribution and soil's organic matter content (Da Silva et al., 1994), bulk density, and porosity. Relative bulk density ($\rho_{b-rel} = \rho_b/\rho_{b-proctor\,max}$) may also affect LLWR (Hakansson, 1988; Carter, 1990).

Example 10.3

From the data presented in Table 10.1, calculate the available water capacity of the profile to 1-m depth.

Solution

Follow the steps shown below:

1. Convert gravimetric moisture content (w) into the volumetric moisture content (Θ) by multiplying with soil bulk density (ρ_b), and dividing by the density of water.
2. Compute actual AWC as per Eq. (10.2).

$$\text{AWC}_{\text{actual}} = (\Theta_a - \text{PWP}_\Theta)d \text{ cm} \tag{10.2}$$

where Θ_a is the antecedent or actual field moisture content, PWP_Θ is the volumetric moisture content at the PWP, and d is depth of the corresponding horizon. Obtain the sum total of $\text{AWC}_{\text{actual}}$ for all horizons.
3. Compute potential AWC as per Eq. (10.3).

$$\text{AWP}_{\text{potential}} = (\text{FC}_\Theta - \text{PWP}_\Theta)d \text{ cm} \tag{10.3}$$

where FC_Θ and PWP_Θ represent volumetric field capacity and permanent wilting point, and d is depth of the horizon. Obtain sum total of $\text{AWC}_{\text{potential}}$ for all horizons.

TABLE 10.1 Computations of Plant Available Water Capacity

Depth (cm)	ρ_b (g/cm³)	Field moisture content (w, g/g)	FC (w, g/g)	PWP (w, g/g)	Volumetric moisture content			AWC (cm)	
					Θ_a	FC_Θ	PWP_Θ	Actual	potential
0–10	1.2	0.10	0.167	0.083	0.12	0.20	0.10	1.0	0.20
10–20	1.3	0.15	0.153	0.092	0.195	0.20	0.12	0.8	0.75
20–50	1.4	0.20	0.25	0.107	0.280	0.35	0.15	6.0	3.90
50–100	1.5	0.25	0.30	0.133	0.375	0.45	0.20	12.5	8.75
							Total	20.3	13.6

Example 10.4

How deep will 5 cm of rain penetrate in the soil profile for the data shown in Table 10.1?

Solution

Compute water deficit for each horizon.

1. Water deficit for horizon $1 = (0.20 - 0.12) \times 10\,\text{cm} = 0.8\,\text{cm}$
2. Water deficit for horizon $2 = (0.20 - 0.195) \times 10\,\text{cm} = 0.05\,\text{cm}$
3. Water deficit for horizon $3 = (0.35 - 0.280) \times 20 = 1.4\,\text{cm}$
4. Water deficit for horizon $4 = (0.45 - 0.375) \times 50 = 3.75\,\text{cm}$

$\therefore$ Amount of rain needed to saturate the first 3 horizons $= 2.25\,\text{cm}$
The balance of rain water $= 5\,\text{cm} - 2.25\,\text{cm} = 2.75\,\text{cm}$
The remainder of the rain is sufficient to penetrate into the fourth horizon
to $= (2.75\,\text{cm})/(0.45 - 0.375) = 36.7\,\text{cm}$
$\therefore$ Total depth of penetration $= 10\,\text{cm} + 10\,\text{cm} + 30\,\text{cm} + 36.7\,\text{cm} = 86.7\,\text{cm}$

Example 10.5

Calculate potential and actual available water capacity from the data shown in Table 10.2.

Potential $AWC = (\Theta_{fc} - \Theta_{pwp}) \times$ depth of soil layer
Actual $AWC = (\Theta_a - \Theta_{pwp}) \times$ depth of soil layer

1. How deep will 7 cm of rain penetrate?	Balance of rain (cm)
Total deficit of the first layer $= 0.08 \times 5\,\text{cm} = 0.40\,\text{cm}$	$7 - 0.4 = 6.60$
Total deficit of the second layer $= 0.07 \times 25\,\text{cm} = 1.75\,\text{cm}$	$6.60 - 1.75 = 4.85$
Total deficit of the third layer $= 0.09 \times 50\,\text{cm} = 4.50\,\text{cm}$	$4.85 - 4.50 = 0.35$
Fractional deficit of the fourth layer $= 0.07$	
Depth of rain penetration in the fourth layer $= 0.35\,\text{cm}/0.07 = 5\,\text{cm}$	
Total depth of rain penetration $= 80 + 5\,\text{cm} = 85\,\text{cm}$	

2. How much irrigation is needed to bring the soil profile of 100 ha farm to Θ_{fc}?

$= (\text{Potential AWC} - \text{Actual AWC}) \times \text{area}$
$= (19.85 - 11.80)\,\text{cm} \times 100\,\text{ha}$
$= 805\ \text{ha} - \text{cm}$
$= 805\,\text{ha} \times 10^4 \dfrac{\text{m}^2}{\text{ha}} \times 10^{-2}\,\text{m} = 805 \times 10^2\,\text{m}^3 = 80.5 \times 10^3\,\text{m}^3$

TABLE 10.2 Computation of Plant Available Water Capacity

Soil depth (cm)	Θ_{fc}	Θ_{pwp}	Θ_a	AWC Potential	Actual
0–5	0.30	0.08	0.22	1.10	0.70
5–30	0.35	0.14	0.28	5.75	4.00
30–80	0.40	0.22	0.31	9.00	4.50
80–100	0.45	0.25	0.38	4.00	2.60

10.2 METHODS OF MEASUREMENT OF SOIL'S MOISTURE CONTENT

A quantitative measure of soil's moisture content is important to understanding soil behavior, plant growth, and soil's numerous other physical processes. Information on soil's moisture content is useful for assessing plant water requirements and scheduling irrigation, plant water uptake and consumptive use, depth of water infiltration into soil, water storage capacity of soil, rate and quantity of water movement, deep drainage and leaching of chemicals, soil-strength, soil's plastic properties, soil-compactability, soil cloddiness and consistency, and numerous other properties and processes.

Despite its numerous uses, an accurate assessment of soil's moisture content in the field has been a challenge to soil physicists and hydrologists for a long time. There are several difficulties encountered in an accurate assessment including the following:

1. Soils are highly variable even over short distances, especially in their water retention capacity as determined by differences in other soil properties, e.g., texture, soil organic matter content, and infiltration rate.
2. Actively growing roots and soil evaporation (or evapotranspiration demand) continuously alter the soil moisture status, which is a highly dynamic entity, and a constantly changing function.
3. Plant water uptake is highly variable because of differences in their growth caused by variable amounts of nutrients and water availability in the soil, and possible effects of pests and pathogens.

There is a wide range of methods used for measurement of soil moisture (Fig. 10.7). For details on these methods, readers are referred to reviews by Gardner (1986), Catriona et al. (1991), Topp (1993), Romano and Santini (2002) and Top and Ferré (2002). Most methods can be grouped under two categories: direct and indirect.

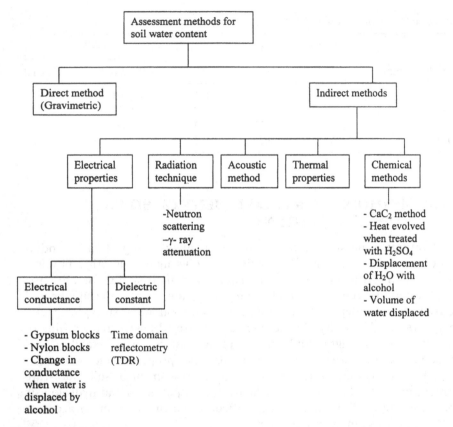

Figure 10.7 Principles underlying different methods of assessment of soil's moisture content.

10.2.1 Direct Methods

Direct methods are based on a physical or chemical technique of removing water from soil followed by its measurement. Gardner (1986) reviewed pros and cons of each direct method. Direct methods are based on three techniques: (i) removal of water by distillation or absorption by a desiccant, (ii) displacement of the water by another liquid and measuring water-induced changes in properties of the liquid, and (iii) measurement of the chemical reaction or reaction products when reactive chemicals are added to the soil. Some of these methods are also discussed under the section dealing with chemical properties related to soil moisture content.

Evaporation Method

The physical technique of removing water from soil involves its evaporation at 105°C. The chemical process of removing water involves leaching by alcohol, or other volatile compounds that can then be easily evaporated. The thermogravimetric method is simple, routine, reliable, inexpensive, and easy to use. The major limitation of this method is that it is destructive, laborious, and time consuming. Because it measures the gravimetric moisture content, it is important to know soil bulk density. Furthermore, evaporating water at 105°C does not remove all water, especially the bond water which may form a substantial amount in heavy-textured soils containing 2:1 clay minerals. There may be changes in the organic fraction of the soil due to oxidation at high temperature and in the water of hydration of the cations in soils containing high concentration of soluble salts.

Water may be present in the soil in all three states (solid, liquid, and gaseous) under cold environments, and in two states (liquid and gaseous) under normal conditions suitable for plant growth. In addition, the liquid water exists in two separate forms: (i) free water and (ii) adsorbed water. The adsorbed water, bonded by the electrostatic forces forming 1 to several molecular layers on the colloidal surfaces, is different than the free water. Most bonded water is released at a temperature of 110 to 160°C. In the conventional definition of soil moisture, therefore, water in the "bonded" state and vapor state is not considered in the definition used in this chapter and in the standard thermogravimetric evaluation. Because of the soil heterogeneity and spatial variability, large number of samples are required to obtain a representative value of soil moisture content. Soil's moisture content is expressed as a fraction and as a percentage on a gravimetric (w) or volumetric basis (θ). The gravimetric soil moisture content is determined using Eq. (10.4) and can be expressed

$$w = \frac{\text{mass of wet soil} - \text{mass of dry soil}}{\text{mass of dry soil}} \tag{10.4}$$

either as a fraction or as a percentage. In addition to soil heterogeneity, another source of error is the temperature control in the oven. Temperature in the oven may not be uniform for different shelves, and/or the temperature control may not be accurate.

Leaching Method

The soil sample is saturated with an alcohol, and then burnt (Bouyoucos, 1931; 1937). Burning evaporates the soil moisture. Repeated leaching and burning can remove the entire soil moisture to a constant weight of soil in a short period of 15 to 20 minutes. In comparison with the thermogravimetric method, this method is rapid but less accurate.

10.2.2 Indirect Methods

The following methods are based on water-induced changes in soil properties that can be measured.

Electrical Conductivity and Capacitance

Soil's moisture content influences electrical conductivity and capacitance, and these properties can be measured routinely and accurately and correlated with soil-moisture content. Attempts have been made to measure soil's electrical resistance in relation to soil moisture content (Kirkham and Taylor, 1950). However, soil heterogeneity and presence of soluble salts pose major problems. Some of these interactive problems can be overcome by using porous blocks containing suitable electrodes, and equilibrated in soil at a given depth. Electrical conductivity is measured when these blocks reach equilibrium. Commonly used material to construct porous blocks is the gypsum or plaster of Paris (Bouyoucos, 1953). Gypsum blocks, however, are progressively dissolved in soils of low pH and have to be frequently calibrated. Therefore, a wide range of porous materials has been tested ranging from nylon cloth (Bouyoucos, 1949) to fiberglass (Cummings and Chandler, 1940; Coleman and Hendrix, 1949). The method is simple, inexpensive, and nondestructive. However, each block has to be calibrated separately. While gypsum blocks are progressively dissolved in acidic soils, the method has serious limitations in soils with high salt or electrolyte concentration. The calibration curve is also affected by soil-moisture hysteresis. Further, porous blocks equilibrate with soil-moisture suction rather than with soil-moisture content. Porous blocks must be calibrated for each soil, and the calibration must be periodically checked because it changes over time. Some units are insensitive to slight changes in soil moisture, and sensitivity also depends on soil temperature.

Porous blocks can also be calibrated to relate soil's moisture content to electrical capacitance (Anderson and Edlefsen, 1942). However, electrical capacitance is more difficult to measure than electrical conductivity. The capacitance method will be discussed in relation to the electromagnetic properties and the dielectric constant.

Radiation Technique

There are two methods that use radiation techniques: one involves neutrons and the other γ-rays.

Neutron Thermalization. A neutron is an uncharged particle and almost has the same mass as that of a proton or of a hydrogen nucleus. When neutrons collide with larger nuclei, the collision is highly elastic and the loss of energy per collision is minimal. When neutrons collide with smaller nuclei, the collision is less elastic and the loss of energy is greater. Slowing down of a fast moving neutron to its thermal velocity may require 18 collisions with H, 114 with C, and 150 with O. Hydrogen in soil, in water and in organic substances (e.g., humus), has the capacity to thermalize neutrons because of elastic collisions. This characteristic is exploited in the neutron moderation technique. High-energy neutrons (5.05 MeV) emitted from a radioactive substance are slowed and changed in direction by elastic collision with the hydrogen. The process by which neutrons lose their kinetic energy through elastic collision is called *thermalization.* The loss of kinetic energy is the maximum when a neutron collides with a particle of a mass nearly equal to its own (e.g., H). The neutrons are reduced in energy to about the thermal energy of atoms in a substance at room temperature. Thermalized neutrons are counted and related to soil's moisture content. Principles and limitations of these techniques are discussed in reviews by IAEA (1970), Bell (1976), Greacen (1981), and others.

Neutron moisture meters comprise two parts: (i) probe and (ii) scalar or rate meter (Fig. 10.8). The probe contains two components: a source of fast neutrons and a detector of slow or thermalized neutrons. The scalar or

FIGURE 10.8 A neutron moisture meter with scaler/rate meter device. Some models have a rate meter built within one assembly (Ibadan, Nigeria, 1972).

rate meter is usually powered by a rechargeable battery, and is designed to monitor the flux of slow neutrons. The common source of fast neutrons used in probe is either 2–5 millicurie mixture of radium-beryllium, which in addition to neutrons also emits γ-rays. These sources have an extremely long half-life of 1620 years. The slow neutrons are monitored by a detector filled with BF_3 gas, which cause the following reaction:

$$B + \text{neutron} = \alpha \,(\text{particle with helium nucleus}) \tag{10.5}$$

The emission of α particle creates an electrical pulse on a charged wire. The number of pulses generated over a measured time interval is counted by a scalar or indicated by a rate meter.

The technique has numerous merits. It is nondestructive, facilitates monitoring soil moisture content for the same site overtime, covers a large soil volume, and monitors volume of soil moisture (Fig. 10.9). However, there are numerous limitations of the technique. It is expensive, poses health hazards, requires specialized maintenance and repair, and there are specific problems with calibration (Lal, 1974; 1979b). The equipment calibration is influenced by texture, gravel content, stoniness, clay mineralogy, and soil's chemical constituents (Fig. 10.10). Some elements present in the soil can capture neutrons. These include gadolinium, cadmium, boron, chlorine,

FIGURE 10.9 A plastic covered plot is used to assess field water capacity using a neutron moisture meter. After saturing the plot with sufficient water, the plastic cover was used to prevent evaporation. (Ibadan, Nigeria, 1971)

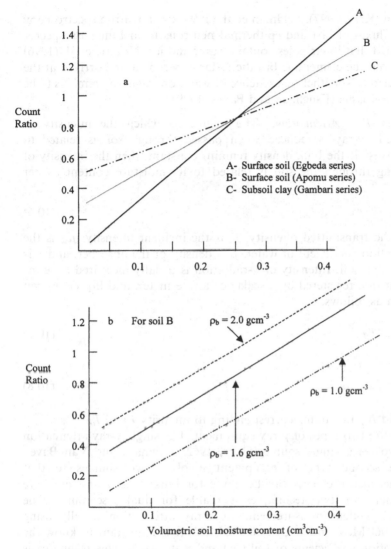

FIGURE 10.10 The effect of (a) soil texture and (b) bulk density on neutron probe calibration. (Redrawn from Lal, 1974.)

manganese, and iron. The measurements are also not very accurate for surface horizons, and in soils with high organic matter content (e.g., Mollisols, organic soils). There are, however, surface neutron meters available to measure soil's moisture content for the plow layer. Lunar Prospector using the neutron spectroscope, reported the existence of water on the moon (Kerr, 1997). Feldman et al. (1998) used neutron spectroscopy to measure fluxes of fast and epithermal neutrons from Lunar Prospector and concluded that lunar poles contain water and ice. Nozette et al. (1996) used data from the clementine bistatic radar experiment and arrived at the same conclusion. Nonetheless, existence of water on the moon remains to be a controversial issue (Eshelman and Parks, 1999).

  Gamma Ray Attenuation. The degree to which the intensity of monoenergetic γ-ray is reduced when passed through soil is related to wet soil density. If the bulk density remains constant, then the intensity of γ-ray passing through the soil is related to its moisture content as per Eq. (10.6).

$$I = I_o e^{-\mu_w} \rho x \qquad\qquad (10.6)$$

where I is the transmitted intensity, I_o is the incident intensity, μ_w is the mass absorption coefficient of water, ρ is density of the absorber, and x is thickness of the soil. Intensity of γ-radiation is usually measured in terms of the count rate registered by a scalar or a rate meter, and Eq. (10.6) can be rewritten as follows:

$$N = N_o e^{-\mu} \rho x \qquad\qquad (10.7)$$

or

$$\rho_n N / N_o = -\mu \rho x \qquad\qquad (10.8)$$

where N and N_o are counts corresponding to intensity I and I_o.

  There are two types of γ-ray equipment. The single γ-ray attenuation method involves a single source (Gurr, 1962; Reginato and Van Bavel, 1964). The second type of equipment involves two sources so that simultaneous measurements can be made for bulk density and moisture content. There are two techniques available for dual γ-scanning. One involves independent measurements of γ-ray attenuation usually using ^{241}Am at 0.060 MeV and ^{137}C$_s$ at 0.662 MeV. It is important to know the mass absorption coefficients of soil (μ_s) and water (μ_w). This technique is generally used under laboratory conditions. The second technique involves simultaneous measurement of two γ-rays at different energy levels using a multichannel analyzer. In this set up the ^{137}C$_s$ is placed behind the ^{241}Am source (Nofziger and Swartzendruber, 1974; Nofziger, 1978).

  Equation (10.8) can be solved for both moisture content and soil bulk density. Let N_p, N_s, and N_{sw} be the count rates through an empty column, through a column packed with oven dry soil, and through a column containing soil and through the column containing soil and water or wet soil, respectively. Then Eq. (10.8) can be written for dry and wet soils as Eqs. (10.9) and (10.10), respectively.

$$\ln \frac{N_p}{N_s} = \mu_s \rho_s x \tag{10.9}$$

$$\ln \frac{N_p}{N_{sw}} = -(\mu_s \rho_s + \mu_w \theta)x \tag{10.10}$$

Dividing Eq. (10.10) by Eq. (10.9) yields Eq. (10.11) and Eq. (10.12).

$$\ln \frac{N_s}{N_{sw}} = \mu_w \Theta x \tag{10.11}$$

or

$$\theta = \frac{2.3}{\mu_w x} \log_{10} \frac{N_s}{N_{sw}} \tag{10.12}$$

The γ-scanning equipment has been designed for both laboratory and field use and details of such devices are available in Gardner (1986) and Catriona et al. (1991).

Merits and limitations of the γ-scanning technique are similar to those of the neutron scattering method. Perhaps the health hazards are more with γ-scanning than with neutron scattering method.

Dielectric Properties of Soil

The dielectric constant of a material is the ratio of the value of the capacitor with the material between the plates, compared with the value with air between the plates. In comparison with a metal, a dielectric material is an insulator. When subjected to an electric field, the positive and negative charges in a dielectric material are displaced with respect to each other and tiny electric dipoles are produced. The dipoles are aligned by the electric field and the dielectric medium as a whole becomes polarized. Therefore, the dielectric constant is a measure of the polarization of a substance. Some materials (e.g., water) whose molecules have a permanent dipole moment have a large dielectric constant. The dielectric constant of water is about 80 and that of the soil about 5 to 7 (Table 10.3).

Principal properties of a dielectric material are: (i) dielectric constant, (ii) dielectric loss, and (iii) dielectric strength. The dielectric constant is

TABLE **10.3**   Dielectric Constant (E) of Some Materials at 20°C

Material	Dielectric constant K
Vacuum	1.0000
Air (1 atm)	1.0006
Paraffin	2.2
Rubber, hard	2.8
Vinyl (plastic)	2.8–4.5
Paper	3–7
Quartz	4.3
Glass	4–7
Porcelain	6–8
Mica	7
Ethyl alcohol	24
Water	80

Source: Adapted from Weast, 1987.

the factor by which the electric field strength in a vacuum exceeds that in the dielectric for the same distribution of charge. The dielectric loss is the amount of energy it dissipates as heat when placed in a varying electric field, and dielectric strength is the maximum potential gradient it can stand without breaking down.

Dielectric constant (E) is the ratio of the capacity of a condensor with that substance as dielectric to the capacity of the same condensor with a vacuum for dielectric. It is a measure, therefore, of the amount of electric charge a given substance can withstand at a given electric field strength. The dielectric constant is measured in units of hertz, which is a unit of frequency; 1 Hz equals 1 cycle/second. Two methods of soil moisture measurements are based on the dielectric properties of the soil. These methods are as follows.

The Capacitance Method.   A capacitor is a device that can store electric charge. It consists of two conducting objects placed near each other but not touching. A typical capacitor consists of parallel plates of area A separated by small distance. When voltage is applied, the capacitor becomes charged. The amount of charge acquired by each plate is proportional to the potential difference V ($Q = CV$). The constant of proportionality C is called capacitance. The capacitance method involves using the moist soil as a part of the dielectric of a capacitor. Measurement of the capacitance gives the dielectric constant, which changes with the soil's moisture content.

There is a wide range of capacitance electrodes (Schmugge et al., 1980). Rather than using probes or push-in electrodes inserted directly into the soil, electrodes or probe can be inserted into an access tube similar to that of the neutron moisture meter. However, there should be no or minimal air gaps between the access tube and the soil. Push-in electrodes are useful for measurement of soil moisture at shallow depths, where soil is highly heterogenous and measurements are extremely variable and unrepeatable. Using access tube is the best method of measurement (Thomas, 1966; Bell et al., 1987; Dean et al., 1987). The capacitance is usually measured by a bridge method at a frequency range of 30–3000 MHz.

The capacitance method has numerous advantages. It is economic, safe, without legal constraint, stable, and rapid by manual operations. Because it involves the use of an access tube, the operation is similar to that of the neutron probe but is much safer and free from legal/policy constraints. However, the techniques require calibration which may be influenced by the composition and density of soils. This method is also not sensitive to the water held by surface adsorption forces or in chemical association with humus, sesquioxides.

Time Domain Reflectometry (TR). This method is also based on the measurement of the dielectric constant of the soil (Topp et al., 1980; 1982; 1988; Topp, 1993; Dalton et al., 1984; 1986). High-energy electromagnetic pulse is fed into the soil between two metal rods. A part of the pulse is reflected back up through the soil from the bottom of the rods and the time interval for the pulse to traverse back, or the time interval between the incident and reflected pulse, is measured. This time interval is related to the soil's moisture content. Major differences between the TDR and the capacitance methods are that the TDR method

Measures an average dielectric constant over the length of the rod
Uses a pair of parallel rods inserted in the ground
Measures dielectric constant over a broad band of frequencies usually ranging from 100 to 1000 MHz
Measures electrical conductivity and dielectric constant simultaneously.

The velocity (v) of an electromagnetic wave through a transmission line in a nonmagnetic medium is given by Eq. (10.13).

$$v = C/K^{1/2} \tag{10.13}$$

where C is the velocity of light (3×10^8 m/s) and K is dielectric constant of the nonmagnetic medium, such as soil. For H_2O with a dielectric constant of 80, the v is 3.3×10^7 m/s. For applicaion to soil-moisture determinations, TDR is essentially a cable radar in which the velocity is computed to measure the time interval (t) for the wave to traverse back and forth in the rod of length L ($v = 2L/t$). Substituting $2L/t$ for v in Eq. (10.13), we can solve for dielectric K_a of the soil [Eq. (10.14)].

$$K_a = \left(\frac{Ct}{2L}\right)^2 \tag{10.14}$$

where K_a is the apparent dielectric constant of the soil which varies with soil wetness. Topp et al. (1980) observed that the dielectric constant does not vary with texture, porosity, and proposed a polynomial equation relating K_a to Θ [Eq. (10.15a)].

$$\theta = -5.3 \times 10^{-2} + 2.92 \times 10^{-2} K_a - 5.5 \times 10^{-4} K_a^2 + 4.3 \times 10^{-6} K_a^3 \tag{10.15a}$$

However, θ vs. K_a relationship is affected by soil's organic matter content especially for organic soils (Herkelrath et al., 1991), and the calibration may also be influenced by salinity (Baumhardt et al., 2000; Nadler et al., 1999). The technique can also be used for simultaneous measurement of soil's moisture content and soil-moisture potential (Noborio et al., 1999) (see Chapter 11). Details of the theoretical principles are outlined by Topp et al. (1980; 1982), Dalton et al. (1984), Catriona et al. (1991), Zegelin et al. (1992), Topp (1993); Topp et al. (2000), and Nadler et al. (2003). The technique is presently being used to assess water and solute transport, and penetrometer resistance in sols (Vaz and Hopmans, 2003; Vaz et al., 2002; Caron et al., 2002). This method has numerous advantages of the neutron scattering and γ-ray attenuation methods, yet is free from health hazard and nuclear regulation. However, calibration of the method and its reliability and reproducibility are still to be worked out.

The TDR technique is still in its evolutionary stage, and rapid progress is being made in alleviating methodological constraints (Malicki and Shierucha, 1989; Zegelin et al., 1989) and in automating the procedure (Baker and Allmaras, 1990).

Thermal Conductivity

Soil's thermal conductivity increases with an increase in soil's moisture content (see also Chapter 17), and this relationship can be used to measure soil wetness (Shaw and Baver, 1939). The temperature rise depends on the ability of the soil to conduct heat away from the source, which depends on soil's moisture content. A principal advantage of this method is that the measurement is not affected by soluble salts that are present in the soil, and the method also measures soil temperature, and the effect of soil temperature on moisture measurement can be accounted for. The technique involves placing a heating element and a temperature sensor in the soil, and the time required to increase soil temperature by a predetermined value is measured. There are two types of equipment based on: (i) encasement of the sensor and element in a porous medium (Sophocecus, 1979) and (ii) placement directly in the soil (Fritton, 1969). The first technique is more suited to measure soil-moisture's potential than moisture content because it reflects the equilibrium moisture content of the porous block. In contrast, the direct placement technique may have a limitation of the poor soil–probe contact, especially in soils with high swell–shrink capacity.

Remote Sensing

Methods of measuring soil moisture described in the previous sections are applicable at the pedon level for different depths or at plot level by simultaneous measurements at several locations. The in situ measurement of the distribution of soil moisture at a watershed scale is difficult because it requires the instruments that can remotely sense it with reasonable accuracy. Ulaby et al. (1996) described a technique of surface soil wetness. Reflectance properties (albedo) can be correlated to the degree of soil wetness. Remote sensing techniques involve use of airborne and satellite imagery procedures. Such can be used for estimating soil's moisture content of the surface layer to a maximum depth of only 0.3 m. These measurements are considerably influenced by ground cover, cloud cover, and other objects between soil and the sensing devices in the space (e.g., crop residue mulch). Remote sensing techniques estimate soil's moisture content over relatively large areas.

Potentials and limitations of remote sensing techniques have been discussed in detail by Myers (1983). These procedures are based on the following five techniques:

Digital Elevation Models (DEMs). Space borne differential interferometric synthetic aperture radar data (InSAR, C band) have the potential for measuring soil moisture at watershed scale (Nolan and Fatland, 2003). The differential InSAR is a powerful tool for making DEMs and is capable of separating surface deformations from static topography. The recent, more accurate DEMs can detect topographic noise to submillimeter range. The spatial variations of SAR are correlated in many locations where changes in soil moisture are expected such as in stream channels, farm boundary, and watershed divide. The underlying theory is that the changes in soil moisture affect soil permitivity (dielectric constant) and the penetration depth. However, penetration depth varies inversely to the soil wetness and the relationship is nonlinear. The rapid advances in the global positioning system (GPS) and inertial motion compensation technology have the potential of increasing accuracy with the added benefit of acquiring the data at any temporal resolution (Nolan and Fatland, 2003).

γ-*Radiation.* Soils natural emission of γ-rays is related to soil moisture content changes overtime. This method may be accurate within 10% for the top 30 cm layer (Grasty, 1976; Zotimer, 1971; Carroll, 1981). The γ-ray flux can be measured by a sensor placed on a low-flying aircraft at 100–200 m altitude (Salomonsen, 1983). The spatial resolution for this technique is at least 200 m. Therefore, variations in moisture content due to differences in soil at small distances cannot be detected. This technique may be useful for large tracts of extremely homogenous soils (e.g., recent alluvial or loess deposits, Andisols, etc.).

Visible and Near Infrared Spectrum. Soil's color changes with its moisture content; moist soil is darker in color. This implies that the spectral reference of soil for the visible and near infrared wavelengths decreases with increase in soil's moisture content (Condit, 1970). However, soil color and its spectral characteristics also differ due to differences in soil's organic matter content, texture, cloud cover, ground cover, and lighting conditions (Evans, 1979; Moore et al., 1975). Soil's moisture content and soil type also affect polarization characteristics of visible light. The degree of polarization of light can also be related to soil's moisture content (Stockhoff and Frost, 1972).

Thermal Infrared Radiation. Changes in surface soil temperature due to differences in soil's moisture content can be monitored and related to soil wetness. Surface soil moisture content has been related to soil temperature using an airborne thermal scanner (Cihlar et al., 1979; Elkington and Hogg, 1981).

Microwave Techniques. Changes in dielectric properties of soil at different soil moisture contents are measured in terms of the microwave energy emitted (Schmugge et al., 1974; Njoku and Kong, 1977).

Acoustic Properties

The propagation of low-energy ultrasonic waves has been used as a non-destructive method for determining moisture content of soils. Such waves propagate at certain sinusoidal frequencies (megacycles), at which the propagated energy varies with soil moisture content. Energy propagated at frequencies of 16 to 20 megacycle/s is sensitive to changes in soil's moisture content in the low range of w from 0 to 10% by weight. Energy propagated at frequencies of 114 to 142 megacycle/s is sensitive to soil moisture content in the high range of w up to 50%. The energy propagated, however, is also influenced by the presence of soluble salts in the soil (Ghildyal, 1987).

Chemical Properties

Several direct and indirect methods of soil-moisture determinations are based on soil's chemical properties. Some of these methods include the following:

1. Changes in the concentration or specific gravity of alcohol (ethyl, methyl, or propyl) when placed in contact with wet soil are related to soil's moisture content.
2. The pressure of the acetylane gas generated in a closed system when calcium carbide is mixed with a moist soil depends on soil wetness [Eq. (10.15b)].

$$CaC_2 + 2H_2 \rightleftharpoons OC_2H_2 + Ca(OH_2) \tag{10.15b}$$

    The equipment called Speedy Moisture Tester or Gas Moisture Tester is based on this principle. Known amount of soil, usually 10–25 g, is mixed with about 25 g of CaC_2 and the pressure of the gas generated is measured and related to soil's moisture content.
3. The heat evolved when the wet soil is placed in a concentrated H_2SO_4 solution is also measured and related to soil's moisture content.
4. Changes produced in the electrical conductivity of the system when water in soil is displaced with alcohol, acetone, and other organic liquids can be related to soil wetness.

Volume Displacement Method

This method is based on assessing the increase in volume of water when a known amount of wet soil is immersed in a known volume of water, and all entrapped air is removed (Prihar and Sandhu, 1968).

$$\Delta V = V_s + V_w$$

$$\Delta V = \frac{M_s}{\rho_p} + V_w$$

$$\therefore V_w = M_w \quad \text{for } \rho_w = 1.0\,\text{g/cm}^3 \text{ and } M_{ws} \text{ is mass of wet soil}$$

$$\Delta V = \left(\frac{M_{ws} - M_w}{\rho_p}\right) + M_w$$

$$\Delta V \cdot \rho_p - M_w \cdot \rho_b = M_{ws} - M_w$$

$$\therefore M_w = \frac{(\Delta V \cdot \rho_p - M_{ws})}{(\rho_p - 1)}$$

$$w = \frac{M_w}{M_{ws} - M_w} \tag{10.16}$$

10.3 COMPARATIVE ADVANTAGES AND LIMITATIONS OF DIFFERENT METHODS

Among the wide range of methods available, the choice of an appropriate method of determination of soil's moisture content depends on numerous factors including the objectives, soil properties, site accessibility, resources available, and technical expertise. Further, different methods are suitable for specific soil characteristics. Merits and limitations of different methods are outlined in Table 10.4. Special precautions should be taken for soils with gravel content. Most techniques are not suitable for soils with high gravel content. Furthermore, computations of volumetric moisture content (Θ) from gravimetric moisture content (w) require knowledge of ρ_b of the gravel-free fraction.

10.4 EXPRESSION OF RESULTS

There are numerous ways to express results of soil moisture content measurement. Among 14 methods listed in Table 10.5, the most useful and commonly used indices are those identified with an asterisk (*). Volumetric moisture content (expressed either as a fraction or a percentage) depth of soil moisture, and saturation percentage are the most useful and commonly used indices.

TABLE 10.4 Merits and Limitations of Different Methods of Determining Soil's Moisture Content

Method	Advantage	Disadvantages
Thermogravimetric	Simple, inexpensive, routine, and the most direct method	Time consuming, laborious, destructive sampling, high variability, measurement of ρ_b is necessary, same site cannot be measured.
Neutron moisture meter	Large soil volume, directly measures Θ, technically sound method, easily computerized	Expensive, health hazards, subject to nuclear regulations, not accurate for soil layers, neutron meter not suitable for organic soils.
Electrical conductance	Simple, low cost, easy to install, nondestructive	Not suitable for soils with high salt content, and soils of low pH, calibration changes with time.
TDR	Nondestructive, simple equipment (metal rods), no health hazards and nuclear regulation	Expensive, still evolving, limited depth range highly variable results.
Gamma scanner	Nondestructive, also measures soil bulk density	Very high health risks, cumbersome equipment especially with double source.
Thermal conductivity	Useful for saline soils, simultaneous measurement of soil temperature	Highly variable results due to poor contact, not applicable for soils with high swell shrink capacity due to contact problems on cracking.
Remote sensing	Large resolution, nondestructive rapid	The measurements cover a large area comprising several soils, results valid only for the surface layer, interference with cloud cover, vegetation and other land features.

TABLE 10.5 Methods to Express Soil's Moisture Content

1 and 2	Mass water fraction $(w) = M_w/(M_w + M_s)$ (fraction or %)
3 and 4*	Gravimetric moisture content $(w) = M_w/M_s$ (fraction or %)
5 and 6*	Volumetric moisture content $(\Theta) = V_w/V_t$ (fraction or %)
7*	Depth of water $(d) = (\Theta$ as fraction) × (depth of soil column/ profile/layer in units of length)
8	Soil moisture density $(\rho_m) = M_w/V_t$ (g/cm^3, Mg/m^3)
9 and 10	Saturated water holding capacity on gravimetric bases $(W_c) = M_w$ at $\Theta = s/M_s$ (fraction or %)
11 and 12	Saturated water holding capacity on volumetric bases $(\Theta) = V_w$ at $\Theta = s/V_t$ (fraction or %)
13*	Liquid ratio $(\Theta_\rho) = V_w/V_s$
14	Saturation percent $= (V_w/f_t) \times 100$

*Important and very useful.

PROBLEMS

1. Compute soil moisture content of a 20 g of wet sample that registers an increase in volume by 5 cm^3. Assume ρ_s of 2.7 g/cm^3.
2. The following soil data were obtained for an irrigation experiment with corn. Irrigation of 10 cm was applied on 6/10/88 after monitoring the soil moisture.

Depth (cm)	Bulk density (g/cm³)	Wilting point (w, g/g)	Field capacity (w, g/g)	Soil moisture content (g/g) 6/10/88	Soil moisture content (g/g) 6/20/88
0–30	1.2	0.10	0.30	0.10	0.20
31–50	1.3	0.12	0.32	0.15	0.25
51–80	1.4	0.14	0.28	0.25	0.20
81–150	1.6	0.15	0.25	0.20	0.15

(a) Calculate depth of penetration of irrigation water.
(b) Evaluate evapotranspiration of corn in mm/day.
(c) Determine drainable porosity at field capacity assuming $\rho_s = 2.65$ g/cm^3.
(d) If irrigation is withheld as from 6/20/88, how long will it take for corn crop to exhaust the entire water reserves if the ET continues at the rate computed in 'b' above?

3. Plot a calibration curve for the neutron moisture meter from the following data:

	Volumetric moisture content (Θ_v)						
	0.5	0.4	0.3	0.2	0.1	0	Standard count
Soil	CPM (10^3)						
A	34	28	24	15	8	1	20
B	60	50	48	37	30	20	

   (a) Develop an empirical relation for predictive purposes.
   (b) Estimate Θ_v for a count of 32×10^3 CPM.
   (c) Suggest possible reasons for differences in calibration curves among two soils.

4. Describe theoretical principals and practical limitations of a neutron moisture meter.

5. Prepare a matrix of the merits and demerits of different methods of moisture measurement for soil profiles with the following characteristics:

   (a) Gravelly soil
   (b) Soil with low pH
   (c) Saline/sodic soil
   (d) Peat soil
   (e) Soil with high contents of Fe and Mn
   (f) A layered profile

6. Describe the TDR method giving its principles, equipment, and merits in relation to the neutron moisture meter.

7. Why is expressing soil moisture content on volumetric basis more useful than mass or gravimetric basis?

8. How do soil structure, aeration, and soil strength influence available water holding capacity?

9. How do soil organic matter and clay contents influence plant-available water capacity?

10. What technologies do you suggest to improve waterholding capacity of coarse-textured soils?

11. Net weight of a wet soil core 7.5 cm in diameter and 7.5 cm deep is 600 g. Calculate wet and dry density and equivalent depth of water if the oven dry weight of the core is 500 g.

12. A soil clod has a volume of 100 cm^3, gravimetric moisture content of 0.20, and bulk density of 1.5 mg/m^3. Calculate the degree of saturation (s) and air-filled porosity (f_a).

REFERENCES

Allmaras, R.R. and S.D. Logsdon. 1990. Soil structural influences on the root zone and rhizosphere. In "Rhizosphere Dynamics," AAAS, Washington, D.C.: 8–54.

Baker, J.M. and R.R. Allmares. 1990. System for automating and multiplexing soil moisture measurement by TDR. Soil Sci. Soc. Am. J. 54: 1–6.

Baumhardt, R.L., R.J. Lascano and S.R. Evett. 2000. Soil material, temperature and salinity effects on calibration of multisensor capacitance probes. Soil Sci. Soc. Am. J. 64: 1940–1946.

Bell, J.P. 1976. Neutron Probe Practice. Inst. Hydrol. Report No. 19, Wallingford, U.K.

Bell, J.P., T.J. Dean, and M.G. Hodnett. 1987. Soil moisture measurement by an improved capacitance technique. II. Field techniques, evaluation and calibration. J. Hydrol. 93: 79–90.

Bouyoucos, G.J. 1931. The alcohol method of determining water content of soil. Soil Sci. 32: 173–179.

Bouyoucos, G.J. 1937. Evaporating the water with burning alcohol as a rapid means of determining moisture content of soils. Soil Sci. 44: 377–383.

Bouyoucos, G.J. 1949. Nylon electrical resistance unit for continuous measurement of soil moisture in the field. Soil Sci. 67: 319–330.

Bouyoucos, G.J. 1953. More durable plaster of Paris moisture blocks. Soil Sci. 76: 447–451.

Caron, J., L.M. Rivière, S. Charpentier, P. Renault, and J.C. Michel. 2002. Using TDR to estimate hydraulic conductivity and air entry in growing media and sand. Soil Sci. Soc. Am. J. 66: 373–383.

Carroll, T.A. 1981. Airborne soil moisture measurement using natural terrestrial gamma radiation. Soil Sci. 132: 358–366.

Carter, M.R. 1990. Relative measures of soil bulk density to characterize compaction in tillage studies on fine loamy sands. Can. J. Soil Sci. 70: 425–433.

Childs, E.C. 1969. The Physical Basis of Soil Water Phenomena. J. Wiley & Sons, London, 493 pp.

Cihlar, R., T. Sommerfeldt, and B. Patterson. 1979. Soil water content estimation in fallow fields from airborne thermal scanner measurements. Can. J. Remote Sensing 5: 18–32.

Coleman, E.A. and T.M. Hendrix. 1949. Fiberglass electrical soil-moisture instrument. Soil Sci. 67: 425–438.

Condit, H. 1970. The spectral reflectance of American soils, Photogramm. Eng., 36: 955–966.

Cummings, R.W. and R.F. Chandler, Jr. 1940. A field comparison of the electro-thermal and gypsum block electrical resistance methods with the tensiometer method for estimating soil moisture in situ. Soil Sci. Soc. Am. Proc. 5: 80–85.

Dalton, F.N., W.N. Herkelrath, D.S. Rawlins, and J.D. Rhoades. 1984. Time-domain reflectometry: Simultaneous measurement of soil–water content and electrical conductivity with a single probe. Science 224: 989–990.

Dalton, F.N. and M.T. Van Genuchten. 1986. The time-domain reflectometry method for measuring soil water content and salinity. Geoderma 38: 237–250.

Da Silva, A.P., B.D. Kay, and E. Perfect. 1994. Characterization of the least limiting water range of soils. Soil Sci. Soc. Am. J. 58: 1775–1781.

Dean, T.J., J.P. Bell, and J.B. Baty. 1987. Soil moisture measurement by an improved capacitance technique. I. Sensor design and performance. J. Hydrol. 93: 67–78.

Dexter, A.R. 1987. Advances in characterization of soil structure. Soil & Tillage Res. 11: 199–238.

Elkington, M.D., and J. Hogg. 1981. The characterization of soil moisture content and actual evapotranspiration from crop canopies using thermal infrared remote sensing. In "Geological and Terrain Analysis Studies by Remote Sensing," J.A. Allen and M. Bradshaw (eds), Remote Sensing Soc., Reading, U.K., pp. 69–90.

Emerson, W.W. 1995. Water retention, organic carbon and soil texture. Aust. J. Soil Res. 33: 241–251.

Eshleman, V.R. and G.A. Parks. 1999. No ice on the Moon. Science 285: 531.

Evans, R. 1979. Air photos for soil survey in lowland England: Factors affecting the photographic images of bare soils and their relevance to assessing soil moisture content and discrimination of soils by remote sensing. Remote Sensing Environ., 8: 39–63.

Feldman, W.C., S. Maurice, A.B. Binder, B.L. Barraclough, R.C. Elphic, and D.J. Lawrence. 1998. Fluxes of fast epithermal neutrons from Lunar Prospector: evidence for water ice at the Lunar poles. Science 281: 1496–1500.

Fritton, D.D. 1969. Resolving time, mass absorption coefficient and water content with gamma ray attenuation. Soil Sci. Soc. Am. Proc. 33: 651–655.

Gardner, W.H. 1986. Water content. In: A. Klute (ed) "Methods of Soil Analysis, Part I. Physical and Mineralogical Methods," Second Edition, ASA Monograph #9, Madison, WI, pp. 493–544.

Ghildyal, B.P. and R.P. Tripathi. 1987. Soil Physics. Wileys, Eastern Ltd., New Delhi, India, 656 pp.

Grasty, R.L. 1976. Applications of gamma radiation in remote sensing. In: "Remote Sensing for Environmental Sciences," E. Schanda (ed), Springer-Verlag, Berlin, pp. 257–276.

Greacen, E.L. (ed) 1981. Soil water assessment by the neutron method. CSIRO, East Melbourne, Victoria, Australia.

Gurr, C.G. 1962. Use of γ-rays in measuring water content and permeability in unsaturated columns of soil. Soil Sci. 94: 224–229.

Hakansson, I. 1988. A method for characterizing the state of compactness of an arable soil. In: J. Drescher et al. (ed) "Impact of water and external forces on soil structure." Catena Suppl. 2: 101–105.

Heathman, G.C., P.J. Starks, and M.A. Brown. 2003. Time-domain reflectometry field calibration in the Little Washita River Experimental Watershed. Soil Sci. Soc. Am. J. 67: 52–61.

Herkelrath, W.N., S.P. Hamburg, and F. Murphy. 1991. Automatic real-time monitoring of soil moisture in a remote field area with time domain reflectometry. Water Resources Res. 27: 857–864.

Hollis J.M., R.J.A. Jones, and R.C. Palmer, 1977. The effects of organic matter and particle size on the water retention properties of some soils in West Midlands of England. Geoderma. 17: 225–231.

Hudson, B.D., 1994. Soil organic matter and available water capacity. J. Soil Water Conserv. 49: 189–193.

IAEA. 1970. Neutron moisture gauges, IAEA, Tech. Report Series No. 112.

Jenny H., 1980. The Soil Resources, Springer-Verlag. New-York, 377 pp.

Kirkham, D. and G.S. Taylor. 1950. Some tests of a four electrode probe for soil moisture measurement. Soil Sci. Soc. Am. Proc. 14: 42–46.

Lal, R. 1974. The effect of soil texture and density on neutron and density probe calibration for some tropical soils. Soil Sci. 117: 183–190.

Lal, R. 1979a. Physical characteristics of soils of the tropics: determination and management. In: R. Lal and D.J. Greenland (eds) "Soil Physical Properties and Crop Production in the Tropics." J. Wiley & Sons, Chichester, U.K., pp. 7–44.

Lal, R. 1979b. Concentration and size of gravel in relation to neutron moisture and density probe calibration. Soil Sci. 127: 41–50.

Lal, R. 1979c. Physical properties and moisture retention characteristics of some Nigerian Soils. Geoderma. 21: 209–223.

Letey, J. 1985. Relationship between soil physical properties and crop production. Adv. Soil Sci. 1: 277–294.

Malicki, M.A. and W.M. Skierucha. 1989. A manually controlled TDR soil moisture meter operating with 300 ps rise-time needle pulse. Irrig. Sci. 10: 153–163.

Moore, F.G., M.L. Horton, J.J. Russell, and V.I. Myers. 1975. Evaluation of thermal X/5 detector Skylab S-192 data for estimating evapotranspiration and thermal properties of soils for irrigation management. In Proc. NASA Earth Resources Survey Symp., NASA Report TM-X58168, Houston, TX, pp. 2561–2583.

Myers, V.I. 1983. Remote sensing applications in agriculture. In: R.N. Colwell, D.S. Simonett, and J.E. Estes (eds) "Manual of Remote Sensing," Vol. 2, Am. Soc. Photogrammetry, Falls Church, VA, pp. 2111–2228.

Nadler, A., A. Gamliel, and I. Peretz. 1999. Practical aspects of salinity effect on TDR-measured water content: a field study. Soil Sci. Soc. Am. J. 63: 1070–1076.

Nadler, A., E. Raveh, U. Yermiyahu, and S.R. Green. 2003. Evaluation of TDR use to monitor water content in stem of lemon trees and soil and their response to water stress. Soil Sci. Soc. Am. J. 67: 437–448.

Njoku, E.G., and J.A. Kong. 1977. Theory for passive microwave sensing of near-surface soil moisture. J. Geophys. Res. 82: 3108–3114.

Noborio, K., R. Horton and C.S. Tan. 1999. Time domain reflectometry probe for simultaneous measurement of soil matric potential and water content. Soil Sci. Soc. Am. J. 63: 1500–1505.

Nofziger, D.L. 1978. Errors in γ-ray measurement of water content and bulk density in nonuniform soils. Soil Sci. Soc. Am. J. 42: 845–850.

Nofziger, D.L. and D. Swartzendruber. 1974. Material content of binary physical moistures as measured with a dual-energy beam of γ-rays. J. Appl. Phys. 45: 5443–5449.

Nolan M. and D.R. Fatland. 2003. New DEMS may stimulate significant advancements in the remote sensing of soil moisture. EOS, Trans. Am. Geophys. Un. 84(25): 233, 236–237.

Ulaby F.T., P.C. Dubois and J. van Zyl. 1996. Radar mapping of surface soil moisture. J. Hydrol. 184: 57–84.

Nozette, S., C.L. Lichtenberg, P. Spudis, R. Bonner, W. Ort, E. Malaret, M. Robinson and E.M. Shoemaker. 1996. The Clementine Bistatic radar experiment. Science. 274: 1495–1498.

Pidgeon, J.D., 1972. The measurement and prediction of available water capacity of ferralitic soils in Uganda. J. Soil Sci. 23: 431–444.

Prihar, S.S. and B.S. Sandhu. 1968. A rapid method for soil moisture determination. Soil Sci. 105: 142–144.

Reginato, R.J. and C.H.M. Van Bavel. 1964. Soil water measurement with gamma attenuation. Soil Sci. Soc. Am. Proc. 28: 721–724.

Rode, A.A. 1969. Theory of Soil Moisture Vol. 1. Moisture Properties of Soils and Movement of Soil Moisture. Translated from Russian. Israel Program for Scientific Translation. U.S. Dept. of Commerce, Springfield, VA, 560 pp.

Romano, N. and A. Santini. 2002. Water retention and storage: field. In: J.H. Dane and G.C. Topp (eds) "Methods of Soil Analysis, Part 4, Physical Methods," Soil Sci. Soc. Am., Madison, WI, pp. 721–737.

Rose, C.W. 1966. Agricultural Physics. Pergmon Press, Oxford, U.K., 226 pp.

Salter, P.J., G. Berry, and J.B. Williams. 1966. The influence of texture on the moisture characteristics of soils III. Quantitative relationship between particle size, composition, and available water capacity. J. Soil Sci. 17:93–98.

Salter, P.J. and F. Haworth, 1961. The available water capacity of a sandy loam soil. II. The effects of farm yield manure and different primary cultivations. J. Soil Sci. 12: 335–342.

Salomonsen, V.V. 1983. Water resources assessment. In: R.N. Colwell, D.S. Simonett, and J.E. Estes (eds) "Manual of Remote Sensing," Vol. 2, Am. Soc. Photogrammetry, Falls Church, VA, pp. 1497–1570.

Schmugge, T.J., P. Gloersen, T. Whilheit, and F. Geiger. 1974. Remote sensing of soil moisture with microwave radiometers. J. Geophys. Res. 79: 317–323.

Schmugge, T.J., T.J. Jackson, and H.L. McKim. 1980. Survey of methods for soil moisture determination. Water Resources Res. 16: 961–979.

Shaw, B. and L.D. Baver. 1939. An electrothermal method for following moisture changes of soil in situ. Soil Sci. Soc. Am. Proc. 4: 78–83.

Sophocecus, M. 1979. A thermal conductivity probe designed for easy installation and recovery from shallow depths. Soil Sci. Soc. Am. J. 43: 1056–1058.

Stockhoff, E.H. and R.T. Frost. 1972. Polarisation of light reflected by moist soils. In Proc. 7th Symp. Remote Sensing of Environment. Environ. Res. Inst. Michigan, Ann Arbor, MI.

Taylor, S.A. and G.L. Ashcroft. 1972. Physical Edaphology: The Physics of Irrigated and Non-Irrigated Soils. W.H. Freeman & Co., San Francisco, 533 pp.

Thomas, A.M. 1966. In situ measurement of moisture in soil and similar substances by fringe capacitance. J. Sci. Instrum. 43: 21–27.

Topp, G.C. 1993. Soil water content. In: M.R. Carter (ed) "Soil Sampling and Methods of Analysis," Lewis Publishers, Boca Raton, FL, pp. 541–557.

Topp, G.C., J.L. Davis, and A.P. Annan. 1980. Electromagnetic determination of soil water content: measurements in coaxial transmission lines. Water Resources Res. 16: 574–582.

Topp, G.C., J.L. Davis, and A.P. Annan. 1982. Electromagnetic determination of soil water content using TDR: II Evaluation of installation and configuration of parallel transmission lines. Soil Sci. Soc. Am. J. 46: 678–684.

Topp, G.C., M. Yanuka, W.D. Zebchuk, and S. Zegelin. 1988. The determination of electrical conductivity using TDR: soil and water experiments in coaxial lines. Water Resources Res. 24: 345–352.

Topp, G.C., S. Zegelin, and I. White. 2000. Impacts of the real and imaginary components of relative permitivity on time domain reflectometry measurements in soils. Soil Sci. Soc. Am. J. 64: 1244–1252.

Topp, G.C. and P.A. Ferré. 2002. Scope of methods (water content) and brief description. In: J.H. Dane and G.C. Topp (eds) "Methods of Soil Analysis, Part 4, Physical Methods," Soil Sci. Soc. Am., Madison, WI, pp. 417–533.

Tran-Vinh-An, N.H., 1971. Contribution a l'etude utile de qelques sols du Zaine. Sols Africana. 16: 91–103.

Vaz, C.M.P. and J.W. Hopmans. 2001. Simultaneous measurement of soil penetration resistance and water content with a combined penetrometer-TDR moisture probe. Soil Sci. Soc. Am. J. 65: 4–12.

Vaz, C.M.P., J.W. Hopmans, A. Macedo, L.H. Bassoi, and D. Wildenschild. 2002. Soil water retention measurements using a combined tensiometer-coiled time domain reflectometry probe. Soil Sci. Soc. Am. J. 66: 1752–1759.

Weast, R.C., (ed). 1987. Handbook of Chemistry and Physics. CRC Press, Boca Raton, FL.

Zegelin, S.J., I. White, and D.R. Jenkins. 1989. Improved field probes for soil water content and electrical conductivity measurement using time-domain reflectometry. Water Resources Res. 25: 2367–2376.

Zegelin, S.J., I. White, and G.F. Russell. 1992. A critique of the time domain reflectometry technique for determining field soil-water content. In: G.C. Topp et al. (eds) "Advances in Measurement of Soil Physical Properties: bringing theory into practice," Spec. Publ. No. 30, Soil Science Society of America, Madison, WI, pp. 187–208.

Zotimor, N.V. 1971. Use of the gamma field of the earth to determine the water content of the soil. Sov. Hydrol, 4: 313–320.

11

Soil's Moisture Potential

Soil's moisture content by itself, regardless of its method of expression in any of the 14 different ways, is not sufficient to describe the status of water in soil. There are several hydrological processes that cannot be fully explained on the basis of soil's moisture content alone. These processes include: (i) water absorption by plant roots, which differs among soils with different textures that have similar moisture content, (ii) water movement that may occur from one soil to another although their moisture contents are similar, and (iii) different soil moisture contents may occur in soils with similar management or environmental conditions. In addition to the moisture content, another property that is essential to a complete description of the soil water regime is the energy status of water in the soil. Soil's moisture content is similar to the heat content of a body. It is the index of a system's capacity in contrast to temperature, which is a measure of its intensity. Similarly, soil's moisture content is a measure of the capacity factor while the energy status of the water is an index of its intensity.

11.1 ENERGY STATUS OF SOIL MOISTURE

Soil water, similar to other natural bodies, possesses two forms of energy: (i) potential energy due to its position or configuration relative to a reference point and (ii) kinetic energy by virtue of its motion (equal to $1/2\,mV^2$ where m is mass and V is velocity). In addition, change of state of water (e.g., solid, liquid, vapor) due to differences in temperature can also affect its kinetic energy. The gravitational potential energy of soil moisture is the product of its weight (mg) and height (h) above a reference point or mgh. The gravitational potential energy is the work done by gravity in moving the mass m of water from point A to point B, h distance apart. The potential energy depends on the vertical height of soil moisture above some reference level.

In practical terms, water in soil moves at a very slow velocity, and possesses an extremely low level of kinetic energy. Further, most processes involving soil-water and plant-water systems are primarily governed by changes in potential energy of soil water and can be addressed without considering the kinetic energy. This is especially true in systems, which are isothermal. In addition, the potential energy of soil water can be substantial and an important factor governing the status of soil water. Water movement under isothermal conditions in soil, both in terms of its direction and velocity, is to a large extent governed by its potential energy. It is primarily because of the differences in this potential energy that water moves from one place to another in the direction of decreasing potential energy until it reaches an equilibrium state determined by equal potential energy at all points within a soil system connected via transmission pores. The driving force is the rate of change of potential energy with distance. It is not the absolute quantity of potential energy but the relative level of energy for one region vis-à-vis another that governs the rate, magnitude, and direction of water movement.

11.2 SOIL-MOISTURE POTENTIAL

Soil-moisture potential refers to this relative level of the potential energy contained in the soil water. It is a measure of the relative potential energy of water in the soil in comparison with pure water. In other words, soil-moisture potential is an expression or indicator of the potential energy contained in soil water relative to that of water in a standard reference state. The latter is a reservoir of pure water (no salts) at atmospheric pressure (not confined) and at the same temperature and level as the soil moisture.

Soil water is subject to the work–energy principle, which states that the work done by an object is equal to change in its energy status. If positive work is done on soil water, soil water's potential (energy) status increases equal to the work w done on it. If negative work w is done on soil water, the soil-moisture potential (energy) decreases by an amount w. In contrast with free water, soil water is held by the soil matrix because of the forces of adsorption involving cohesion, adhesion, and solution. Therefore, soil water is usually not capable of doing work W as can a reservoir of pure water. Consequently, soil water potential is usually negative.

Thus, soil water potential has the following characteristics:

Relative: It is a relative quantity.

Negative: It is usually negative.

Continuity: It is a continuous entity without any abrupt disconti- nuities.

Driving force: It is the driving force that moves soil water from one region within the soil to another.

Variability: It is highly variable even over short distances within the soil.

Dynamic: It is a highly dynamic entity.

In view of these characteristics, soil-moisture potential, hereafter desig- nated by the symbol Φ, is defined as "the amount of work that a unit quantity of water in an equilibrium soil-moisture system is capable of doing when it moves to a pool of water in the reference state at the same temperature."

Total soil-moisture potential (Φ_t) is the amount of useful work per unit quantity of pure water that must be done by means of externally applied forces to transfer irreversibly and isothermally an infinitesimal amount of water from the standard state to the soil liquid phase at the point under consideration (Bolt, 1976). Total soil water potential is measured in units of energy, which can be expressed per unit mass, volume or weight basis as follows (see also Sec. 11.6):

1. Energy per unit volume is expressed as ergs/cm^2, dynes/cm^2, N/m^2,
2. Energy per unit mass is expressed as ergs/g or J/kg, and
3. Energy per unit weight is expressed in terms of height of water as cm or m.

11.3 COMPONENTS OF TOTAL SOIL-MOISTURE POTENTIAL

Total soil-moisture potential (Φ_t) consists of several components [Eq. (11.1)]:

$$\Phi_t = \Phi_p + \Phi_m + \Phi_z + \Phi_\pi + \Phi_o \tag{11.1}$$

where t, p, m, z, π, and o refer to total, pressure, matric, height or position, osmotic, and overburden potential, respectively.

11.3.1 Pressure Potential (Φ_p)

Pressure potential (Φ_p) is defined as the water pressure exerted by the overlying saturated column of water on a specific position within a soil. It is equal to the water pressure exerted by the height of water above a specific point. If a volume Θ is transferred from a body of water where the gauge pressure is zero to one where it is p, the work done against p is [Eq. (11.2)]

$$\Phi_p = \text{work} = pv \tag{11.2}$$

The work per unit volume is $pv/v = p$. The work done by water can also be computed by assuming this water to be displaced from a tube of length l and cross-sectional area A into water at pressure p. The work done in this hypothetical case against pressure p is $W = p \cdot A \cdot l = pv$.

Therefore,

$$\text{Work } (\Phi_p) \text{ per unit volume} = pv/v = \rho g h \text{ dynes/cm}^2 \tag{11.2a}$$

$$\text{Work } (\Phi_p) \text{ per unit mass} = \frac{pv}{\rho v} = (\rho g h)\, v/\rho v = g h \text{ ergs/g} \tag{11.2b}$$

$$\text{Work } (\Phi_p) \text{ per unit weight} = \frac{pv}{\rho g v} = (\rho g h) = v/\rho g v = h \text{ cm} \tag{11.2c}$$

where p is density (g/cm^3) and g is acceleration due to gravity (cm/s^2). The pressure head is usually measured in units of length (cm, m), and exists and only under saturated soil conditions ($\Theta = s = 1$). The positive pressure potential usually occurs below the groundwater level and is called the piezometric head or the submergence potential. Under field conditions, the pressure potential is measured by a piezometric tube. A piezometer tube is a solid tube open at both ends, and a water table tube is a perforated tube open at both ends (Fig. 11.1). The pressure potential is the vertical distance from a specific point in the soil to the water surface of a piezometer

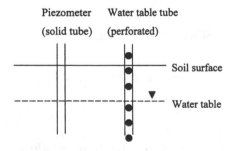

FIGURE 11.1   Piezometric pressure at different points.

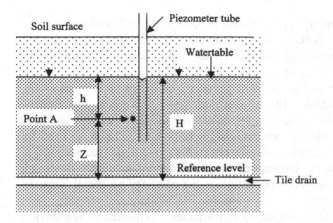

FIGURE 11.2 A piezometer tube showing the soil water pressure below the water table. At the reference point A, the pressure potential equals h ($\phi_p = h$ cm), gravitational potential z, and the total potential $H = (h + Z)$.

connected to the point in the question. The schematic in Fig. 11.2 shows the magnitude of Φ_p.

In the field situation, Φ_p is zero above and at the level of the water in the piezometer. It is positive and equal to the depth of the water column above, when the point is below the water table.

11.3.2 Matric Potential (Φ_m)

Matric potential exists only in unsaturated soils, and therefore, matric potential and pressure potential are mutually exclusive. Under specific soil-water conditions, a soil either has pressure potential (Φ_p) or matric potential (Φ_m), but not both. Soil matric potential is due to the effects of soil solids, interfacial curvature due to surface tension and forces of cohesion and adhesion of the soil matrix (Fig. 10.2). This negative pressure potential is also called *capillary potential*. Similar to the potential, the matric potential may be expressed in three units.

$$\Phi_m \text{ per unit volume} = \rho g h \text{ dynes/cm}^2 \tag{11.3a}$$

$$\Phi_m \text{ per unit mass} = g h \text{ ergs/g} \tag{11.3b}$$

$$\Phi_m \text{ per unit weight} = h \text{ cm} \tag{11.3c}$$

Some soil physicists (Jury et al., 1991) argue that Φ_m comprises tensiometric potential (capillary potential) and air potential (pneumatic potential). The tensiometric potential is the work required to transfer reversibly and isothermally an infinitesimal amount of soil solution from a reservoir in soil to the point of interest in the soil. In comparison, the air pressure potential is the gauge pressure of the soil air relative to the standard state air pressure ($P_{soil}-P_{atmosphere} = \Delta P_a$). The gauge pressure of the soil air with reference to the ambient pressure, called *pneumatic potential* (Φ_a), is usually negligible. In unsaturated soils, therefore, the matric potential is the sum of capillary potential and the pneumatic potential. Under laboratory conditions, however, Φ_a is important. The Φ_a is used to measure soil moisture retention at different matric potentials (see Sec. 11.7). In that condition, $\Phi_a = \Phi_m$. In practical terms, however, the matric potential is the same as the tensiometric potential or the capillary potential because Φ_a is practically zero.

Matric potential is measured by tensiometers. Tensiometer is a device that measures potential energy of soil water relative to free water in a porous ceramic cup in equilibrium with soil water. A graphical representation of different types of tensiometers is shown in Fig. 11.3a–c. In Fig. 11.3a, Φ_m is the vertical distance between the point in the soil and the water surface of a manometer filled with water and connected to the soil point in question via a ceramic cup. This device is called a tensiometer or a ceramic cup tensiometer (Fig. 11.3).

A tensiometer consists of a porous cup and a monometer or a pressure gauge. The ceramic (or any other suitable porous material) cup and part of the manometer are filled with deaired water and buried in soil at the desired depth. Depending on the soil wetness, the water moves from the cup into the soil and develops a negative pressure or suction, which is measured by the depression in the height of the water in the manometer tube or in the gauge pressure attached to the cup.

Depending on the system used to measure the suction created by the movement of water from the ceramic cup to the soil, there are numerous types of tensiometers.

Mercury Manometer Tensiometer

These tensiometers use a combination of H_2O and Hg to measure the Φ_m as shown in Fig. 11.3b. The use of Hg is a health hazard. Therefore, this following description is merely to explain the underlying principles.

Z = distance from top of the mercury column to the center of the ceramic cup.
Z_{Hg} = distance from top of the mercury column to the surface of the mercury in the reservoir.

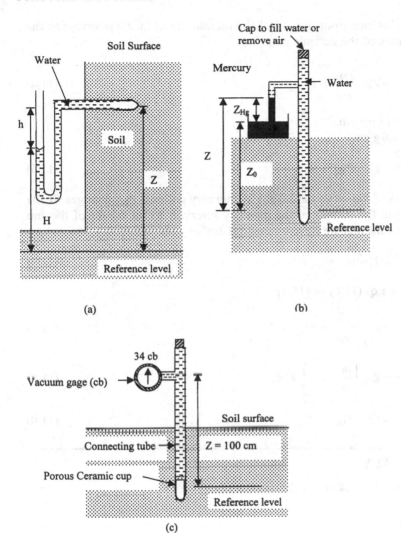

FIGURE 11.3 Different types of tensiometers: (a) a water manometer connected to a ceramic cup installed in soil at the designated depth, Φ_m equals $-h$; (b) a mercury manometer connected to a ceramic cup installed in soil at the desired depth, $\Phi_m = -Z_{Hg} \times 13.6 + Z$; and (c) a vaccum gauge tensiometer, $\Phi_m = -34 \times 10\,cm + 100\,cm = -240\,cm$.

Z_o = distance from the top of the mercury level in the reservoir to the center of the ceramic cup.

$$\Phi_m = -Z_{Hg}\frac{\rho_{Hg}}{\rho_w} + Z \tag{11.4}$$

$$\rho_{Hg} = 13.6\,g/cm^3$$
$$\rho_w = 1.0\,g/cm^3$$

$$\Phi_m = -13.6\,Z_{Hg} + Z \tag{11.5}$$

The distance Z varies as the height at mercury column Z_{Hg} changes. If the distance from the surface of the mercury reservoir to the center of the cup is kept constant (h_o) we have a constant for any tensiometer:

$$Z = Z_o + Z_{Hg} \tag{11.6}$$

Substituting Eq. (11.6) in (11.4)

$$\Phi_m = -Z_{Hg}\frac{\rho_{Hg}}{\rho_w} + Z_{Hg} + Z_o \tag{11.7}$$

$$\Phi_m = -Z_{Hg}\left[\frac{\rho_{Hg}}{\rho_w} - 1\right] + Z_o \tag{11.8}$$

$$\Phi_m = -12.6\,Z_{Hg} + Z_o \tag{11.9}$$

Example 11.1

If $Z_o = 20\,cm$, $Z_{Hg} = 14.2\,cm$, calculate Φ_m

Solution

$$\Phi_m = -12.6 \times 14.2\,cm + 20\,cm$$
$$= -17.9\,cm + 20\,cm = -159\,cm$$

Vacuum Gauge Tensiometer

In this tensiometer, the Hg is replaced by a vacuum gauge, and the reading on the dial can be converted to Φ_m (Fig. 11.3c). The units of measurement must be carefully considered.

The dial is usually calibrated from 0 to 100, which on a weight basis corresponds to a range of 0 to $-1000\,\mathrm{cm}$ (0 to -100 centibars).

Example 11.2

Calculate Φ_m in Fig. 11.3c. 1 gauge reading $= 10\,\mathrm{cm}$ of Φ_m.

Solution

$$\Phi_m = -10\,\mathrm{cm} \times (\text{gauge reading}) + Z$$
$$\Phi_m = -10 \times 34\,\mathrm{cm} + 100\,\mathrm{cm}$$
$$\Phi_m = -240\,\mathrm{cm}$$

Vacuum gauge may be also calibrated in inches of Hg rather than in cm or centibars.

Example 11.3

A tensiometer dial is calibrated from 0 to 30. If the gauge is 25 inches above the tensiometer cup and it reads 20 inches of Hg, calculate Φ_m.

Solution

$$\Phi_m = -13.6 \times 20 + 25\,\mathrm{inches}$$
$$\Phi_m = -247\,\mathrm{inches\ of\ water}$$
$$\Phi_m = -627\,\mathrm{cm\ of\ water}$$

Most commercially available tensiometers may already be calibrated for the length of the tensiometer stem. There are two principal limitations of tensiometers. First concerns with the range of suction, or Φ_m, that can be measured with a tensiometer. The useful range is about 0 to 80 kPa, or 0 to 800 cm of water suction. As soil gets drier than this range, air enters the cup and water column in the tensiometer breaks. Soil moisture content corresponding to this suction varies widely among soils, depending on the texture and organic matter content. The second limitation is due to the response time of the tensiometer. In soils with rapidly changing Φ_m, tensiometers are usually slow to respond. The response time depends on hydraulic conductivity of the porous cup and sensitivity of the gauge or the suction-measuring devices.

11.3.3 Gravitational Potential (Φ_z)

The Φ_z is due to the position of soil water. It is the energy required to move an infinitesimal amount of pure, free water from the reference elevation to the soil water elevation. Therefore, the Φ_z of soil moisture is determined by the elevation of the point relative to the reference level. Three forms of expressing Φ_z are shown by Eq. (11.10).

$$\Phi_z \text{ per unit volume} = \rho g Z \text{ dynes/cm}^2 \tag{11.10a}$$

$$\Phi_z \text{ per unit mass} = g Z \text{ ergs/g} \tag{11.10b}$$

$$\Phi_z \text{ per unit weight} = Z \text{ cm} \tag{11.10c}$$

The gravitational potential is usually measured by the height above or below an arbitrarily chosen reference point. The gravitational potential is positive if the specific point is above the reference level, and negative if the specific point is below the reference level. The Φ_z is strictly due to the position of a specific point in the soil, and is independent of the soil properties or atmospheric (ambient) conditions. Its magnitude depends on the vertical distance between the reference and the point in question.

11.3.4 Osmotic Potential (Φ_o)

Osmotic potential is due to the presence of solutes in soil moisture that affect its thermodynamic properties (e.g., entropy, enthalphy, free energy). Presence of solutes in soil lowers the vapor pressure of soil moisture and affects its Φ_o. The Φ_o refers to the change in energy per unit volume of water when solutes identical in composition to the soil solution at the point of interest in the soil are added to pure, free water at the elevation of the soil. Presence of solutes in soil moisture creates a suction that can suck water from a reservoir of pure water brought into contact with the solution through a semipermeable membrane. The ability of soil moisture to suck water from a reservoir of pure water depends on the concentration of solutes, which also determines decrease in its vapor pressure, increase in boiling point, and depression in its freezing point. The Φ_π can be expressed in three ways as per Eq. (11.11).

$$\Phi_\pi \text{ per unit volume} = \rho g h \pi \text{ dynes/cm}^2 \tag{11.11a}$$

$$\Phi_\pi \text{ per unit mass} = g h \pi \text{ ergs/g} \tag{11.11b}$$

$$\Phi_\pi \text{ per unit weight} = h \pi \text{ cm} \tag{11.11c}$$

TABLE 11.1 Components of Total Soil-Water Potential

Soil	Components of Φ_t	Remarks
Saturated soil		
Nonswelling soil	$\Phi_t = \Phi_z + \Phi_p + \Phi_\pi$	Φ_π is zero for soils in the humid region.
Swelling	$\Phi_t = \Phi_z + \Phi_p + \Phi_o$	
Unsaturated soil		
Nonswelling soil	$\Phi_t = \Phi_z + \Phi_m + \Phi_\pi$	
Swelling soil	$\Phi_t = \Phi_z + \Phi_m + \Phi_o + \Phi_a + \Phi_\pi$	Φ_a is usually 0.

The osmotic potential is also discussed in Chapter 20 in section dealing with soil salinity.

11.3.5 The Overburden Potential (Φ_o)

The Φ_o is due to the mechanical pressure exerted by the unsupported solid material on the soil water. It is the change in energy per unit volume of soil water due to the weight of the unsupported soil above the soil water. The overburden pressure is usually significant only in swelling soils (see Chapter 20).

Components of Φ_t under different situations are shown in Table 11.1. For most saturated soil situations, Φ_t comprises only two components, the gravitational potential (Φ_z) and the pressure potential Φ_p [Eq. (11.12)]. Under this case Φ_t is called the *hydraulic head*.

$$\text{Hydraulic head } (\Phi_t) = \Phi_z + \Phi_p \tag{11.12}$$

11.4 TOTAL SOIL-MOISTURE POTENTIAL UNDER FIELD CONDITIONS

Components of soil-moisture potential can be measured under field conditions for assessing the direction and magnitude of flow. A line joining all points with equal soil-moisture potential is called an *isobar*. Soil water flows perpendicular to the isobars. There is no water movement in the soil if Φ_t is equal at all points. Soil water moves in the direction of decreasing soil-moisture potential.

Example 11.4

With 10 cm of water ponding and maintained constant on the soil surface and a tile drain at 100 cm depth flowing full, plot the soil moisture potential profile.

Solution

Components of soil moisture potential in this case are pressure potential (Φ_p) and gravitational potential (Φ_z). Because it is saturated, flow Φ_m is zero. Taking soil surface as a reference point, components of Φ_t are as follows:

Φ_p (cm)	Φ_z (cm)	Φ_t (cm)
10	0	10
30	-20	10
50	-40	10
70	-60	10
90	-80	10
0	-100	-100

Example 11.5

Consider the situation in Example 11.4 when tile is plugged and not flowing. What is the Φ_t profile?

Solution

This will be a situation of steady state condition and Φ_p at 100 cm depth will be 110 giving a total water potential of 10 cm at all depths above the drain line.

Example 11.6

Consider Example 11.1 when there is no water ponded on the surface, and the drain is not flowing but a free water table exists at 100 cm depth. Calculate the Φ_t at all depths above the drain line.

Solution

Because drain is not flowing, therefore, Φ_t must be constant (same) at all points. This is based on the assumption that there is no soil evaporation. Under these conditions, components of Φ_t are as follows (all units are in cm).

Φ_z	Φ_m	Φ_p	Φ_t
0	−100	0	−100
−20	−80	0	−100
−40	−60	0	−100
−60	−40	0	−100
−80	−20	0	−100
−100	0	0	−100
−120	0	+20	−100
−140	0	+40	−100

Example 11.7

Assume a soil with water table at 100 cm depth and soil surface evaporating at a constant rate. Tensiometers are installed in the soil to measure Φ_m as shown in the Table below. Components of Φ_t are shown in the Table.

Φ_z	Φ_m	Φ_p	Φ_t
0	+800	0	+800
−20	−600	0	−620
−40	−400	0	−440
−60	0	0	−60
−80	0	20	−60
−100	0	40	−60

11.5 MEASUREMENT OF SOIL'S MATRIC POTENTIAL (Φ_m)

Techniques for measurement of Φ_m are outlined in Table 11.2, and described at length by Mullins (2001), Livingston (1993), Young and Sisson (2002), Andraski and Scanlon (2002), and Scanlon et al. (2002). *Tensiometers* are the most widely used for a low range of Φ_m from 0 to −80 KPa, and are relatively simple, inexpensive, easy to install, and have a sensitivity of about 0.1 KPa. Major limitations of tensiometers include the following: (i) insensitivity to soil solution osmotic potential rendering them unsuitable for measuring Φ_m in salt-affected soils, (ii) restricted measurement range of 0 to −80 KPa, (iii) long response time, (iv) poor soil contact in gravelly soils, (v) increase in Φ_m due to movement of water from cup into the adjacent soil as influenced by the soil's and cup's hydraulic conductivities, and (vi) the maximum limit of 4 m depth to which a

TABLE 11.2 Techniques for Measurement of Soil Matric Potential

Technique	Principle	Range (kPa)	Limitations	Reference
Tensiometers	Measurement of vacuum created in the tensiometer tube due to absorption of water by the dry soil from porous cup.	0 to −85	Low range Long response time Air entry due to poor contact	Klute and Gardner (1962)
Psychrometer	Monitoring relative humidity of vapor in equilibrium with the liquid phase in soil.	−80 to −1500	Extremely sensitive to temperature	Rawlins and Campbell (1986)
Porous material sensors (filter paper, gypsum blocks)	Evaluating changes in matric potential with change in water content of a porous material.	-1 to -10^5	Hysteresis of the material Calibration of all material	Fawcette and Collis-George (1967); Hamblin (1993); Scholl (1978); Pereira (1951)
Heat dissipation in porous blocks	Assessing the rate of heat dissipation in a porous material 0 to −100 sensor.	0 to −100		Phene et al. (1971)

tensiometer can be inserted. The absolute pressure (P) inside the tensiometer is given by Eq. (11.13):

$$P = A - \Phi_m - h \qquad (11.13)$$

where A is the atmospheric pressure and h is height above the tensiometer cup (Livingston, 1993). If a tensiometer cup is installed 3 m below the ground and the vacuum gauge is about 0.5 m above the soil surface (assuming that A is 10 m), the lowest pressure in the system will be about -0.0065 MPa. This limit is reduced for deeper installation.

Psychrometers compliment tensiometers with an upper limit of Φ_m of about -100 Kpa (Campbell and Gardner, 1971; Andraski and Scanlon, 2002). The total water potential is determined by measurement of relative vapor pressure of air in equilibrium with soil pores [Eq. (11.14)].

$$\Phi_m = \frac{\rho_w RT}{V_w \ln(p/p_o)} \qquad (11.14)$$

where Φ_m is matric potential in MPa, R is the universal gas constant $(8.314 \times 10^{-6}$ MJ/mol/K$)$, T is the absolute temperature (K), V_w is molar volume of water $(1.8 \times 10^{-5}$ m^3/mole$)$, and p/p_o is relative humidity expressed as a fraction.

There are two types of psychrometers: (i) those that can be used for in situ measurements and are placed in the soil, and (ii) those in which soil samples are placed in the sample chamber and the Φ_m is determined after about 15 minutes of equilibrium time. The former, a soil psychrometer, consists of a small ceramic cup (1 cm in diameter and 1 cm long) that contains a single thermocouple (50–100 μm in diameter) constructed of chromal and constantan wires (Fig. 11.4). The reference junction usually comprises a Cu wire. The porous ceramic cup facilitates diffusion of water vapors from soil air to the thermocouple. Accurate measurements of air and soil temperatures are critical to psychrometric evaluations. A psychrometer measures the thermal electromotive force from the cooling of the junction in an enclosed space. The force is measured in microvolts (μv) and related to Φ_m. There are two principal limitations of the psychrometric technique. One, the relative humidity of the soil air changes only slightly from 94 to 100%. Two, differences in soil temperature can lead to large errors. A difference in temperature of 1°C can lead to differences in Φ_m by 10 MPa (Campbell, 1979).

There are several miscellaneous methods of measuring Φ_m. Scanlon et al. (2002) describe seven different techniques based on heat dissipation sensors, electrical resistance sensors, frequency domain and time domain sensors, electrooptical methods, filter paper method, dew

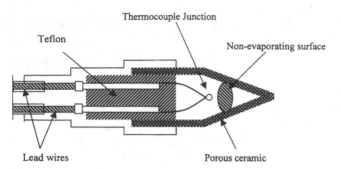

FIGURE 11.4 Soil psychrometer for measuring soil water potential in situ. (Redrawn from Campbell and Gardner, 1971.)

point potentiometer, and vapor equilibration method. A commonly used method is that of measuring electrical resistance. *Resistance blocks* for measuring Φ_m are similar to those described for soil-water measurements and are comprised of porous material such as gypsum, nylon, or fiberglass. Blocks can be used to measure Φ_m in soils drier than $-50\,\mathrm{KPa}$. These devices are simple, inexpensive, and provide nondestructive and continuous measurement of Φ_m. The electrical conductivity of porous blocks is zero for dry soil and increases with increase in Φ_m. Porous blocks have several limitations including: (i) unusable in salt-affected soils or those irrigated with saline water, (ii) change in calibration for each block over time, (iii) hysteresis of the porous material, (iv) long response time, (v) degradation of blocks over time, (vi) impact of variations in temperature, (vii) non-suitability of blocks in soils that develop large cracks, and (viii) the error may be large of the magnitude of ± 100 to $500\,\mathrm{KPa}$. Cracks are often formed in the vicinity of blocks rendering soils to dry out rapidly after the crack develops or wet quickly following rain or irrigation due to water flowing into the cracks.

The *filter paper method* uses a special type of porous material. This technique is described at length by Al-Khafaf and Hanks (1974), Hamblin (1981), and Greacen et al. (1987). The filter paper, of known porosity and soil moisture characteristic curve, is wrapped around a wedge and pushed into the soil at a desired depth. The filter paper takes 4 to 6 days to equilibrate with the soil following which it is removed and weighed to determine its moisture content. Soil matric potential is determined from the precalibrated soil moisture characteristic curve or the potential vs. θ relationship. It is a simple and inexpensive method of measuring Φ_m within the range of -50 to $-100\,\mathrm{KPa}$. Because of the long equilibration time, however, it is useful only for soils with slow changes in Φ_m.

Similar to the filter paper method, the *heat dissipation technique* also involves a porous medium. The technique is based on measuring the heat dissipation within the porous material in which is located a heat sensor. The dissipation of short heat pulse applied to the sensor depends on thermal diffusivity or its moisture content. This technique is not sensitive to salt content, and therefore, can be used for salt affected soils. Theory, design, and construction of heat dissipation devices are given by Phene et al. (1971). These devices have a measuring range of 0 to -600 KPa with an accuracy of ± 10 KPa in the low range (0 to -300 KPa) and of ± 100 KPa in the high range (-300 to -600 KPa). However, the accuracy is influenced by hysteresis and contamination of the porous material. These devices are useful for scheduling irrigation (Phene and Beale 1976).

11.6 UNITS OF MEASUREMENT OF SOIL-MOISTURE POTENTIAL

All units of soil-moisture potential are defined with regards to the unit quantity of water. The specific unit depends on the way the unit quantity of water is defined; volume basis, mass basis, or weight basis.

Relationships among different ways to express soil-moisture potential are shown in Table 11.3. A common unit to express soil-moisture potential on volume basis is a "bar." Numerous ways to express one bar are listed in Table 11.4. Similarly, a common unit to express soil-moisture potential on weight basis is pF. The latter is computed as a logarithm to the base 10 of

TABLE 11.3 Units for Expressing Soil-Water Potential

Basis	Units
Volume $(P_v = P = \rho g h)^a$	dynes/cm^2, Pa, ergs/m^2, bar, J/m^3, N/m^2
Mass $(P_m = P/\rho = gh)$	ergs/g; J/Kg
Weight $(P_w = P/\rho g = h)$	cm, m

[a]Soil water potential on volume basis (P_v) is the work done against pressure P to transfer volume V is PV/V or $P = \rho g h$.
1 Dyne $= 1\, g\, cm/s^2 = 10^{-5}\, N$
1 N $= 1\, Kg\, m/s^2$
1 Pa $= 1\, N/m^2$ (1 kPa $= 1/J\, kg^1$)
1 J $= 1\, N\, m = 10^7\, ergs = watts$
1 Bar $= 105\, Pa = 0.987$ atmosphere $= 29.53''$ Hg $= 10^6\, dynes/cm^2$
1 Atmosphere $= 1,013,250$ dynes cm^{-2} 101,325 N/m^2
1 Torr $= 1$ mm Hg $= 1/760$ atmosphere $= 1013,250/760$ dynes/cm$^2 = 133.22$ microbars
1 Watt $= J/s = 107$ erg/s
1 erg $= 1$ dyne cm $= g/cm\, s^2 = 10^{-4}\, J/kg$

TABLE 11.4   The Value of 1 Bar of Expressing Soil-Water Potential
on a Volume Basis

Unit	Equivalent quantity
One bar	100 centibars (cb, $1\,cb = 1\,J/kg$)
	1000 millibars (mb)
	1020 cm of H_2O
	401.57 inches of H_2O
	75.01 cm of Hg
	0.9869 atmosphere
	100 J/kg
	14.5 lbs inch^2 (PSI)
	10^6 ergs/g
	10^6 dynes/cm^2
	10^2 kPa

soil-moisture potential expressed in cm of water. The pF value corresponding to soil matric potential of -1, -10, -100, and -1000 cm of water is 0, 1, 2, and 3 respectively (refer to Appendix 11.1).

Example 11.8

Study the column setup shown below. Compute components of soil-moisture potential for three possible reference levels.

Solution

There are three solutions based on the choice of reference level.

1. Reference level at the top of soil column-AA.

Depth (cm)	Φ_z (cm)	Φ_m (cm)	Φ_p (cm)	Φ_t (cm)
10	−10	−200	0	−210
20	−20	−160	0	−180
30	−30	−110	0	−140
40	−40	−80	0	−120
50	−50	0	0	−50
60	−60	0	10	−50
70	−70	0	20	−50
80	−80	0	30	−50

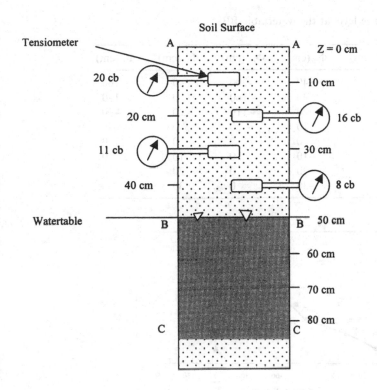

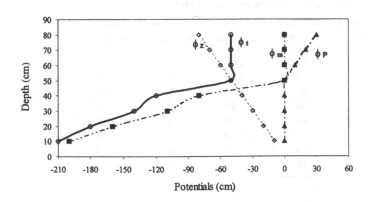

2. Reference level at the watertable-BB.

Depth (cm)	Φ_z (cm)	Φ_m (cm)	Φ_p (cm)	Φ_t (cm)
10	40	−200	0	−160
20	30	−160	0	−130
30	20	−110	0	−90
40	10	−80	0	−70
50	0	0	0	0
60	−10	0	10	0
70	−20	0	20	0
80	−30	0	30	0

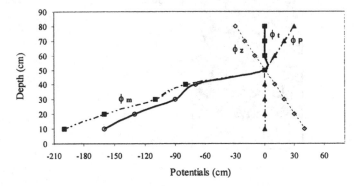

3. Reference level at the bottom of soil column-CC.

Depth (cm)	Φ_z (cm)	Φ_m (cm)	Φ_p (cm)	Φ_t (cm)
10	80	−200	0	−120
20	70	−160	0	−90
30	60	−110	0	−50
40	50	−80	0	−30
50	40	0	0	40
60	30	0	10	40
70	20	0	20	40
80	10	0	30	40

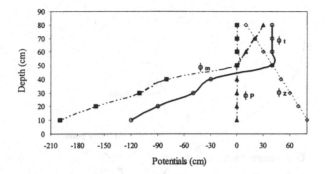

11.7   SOIL MOISTURE CHARACTERISTICS

The fundamental relationship between soil's moisture content (θ) and soil-matric potential (Φ_m) is called "soil moisture characteristics," "soil moisture characteristic curve," or "pF curve." This unique relationship depends on soil structure as determined by total porosity and the pore size distribution. Thus, change in structure and pore size distribution leads to changes in soil moisture characteristics. The unique relationship, at the time of obtaining the undisturbed core sample, may be mathematically expressed as in Eqs. (11.15) and (11.16), and graphically depicted as in Fig. 11.5.

$$\theta = f(\Phi_m) \tag{11.15}$$

$$\Phi_m = f(\theta) \tag{11.16}$$

An example of hypothetical data on soil moisture characteristic for two soils of contrasting texture is shown in Table 11.5. As expected, soil wetness increases with decrease in soil matric potential from a very high negative value for an extremely dry condition to a near zero suction for a saturated soil when all pores are full of water. As the suction increases from saturation to a low value of 10 or 20 cm of water, soil wetness remains the same in a heavy-textured soil. The suction at which soil wetness begins to decrease is called the "air entry point."

There are numerous factors that affect soil moisture characteristics. In addition to particle size distribution, soil organic matter content plays an important role, especially at low suctions (or field capacity). Soil wetness at field soil moisture capacity (pF 2.5) increases linearly with an increase in soil organic matter content (Fig. 11.6). All other factors remaining the same, soil moisture retention at a specific suction also depends on the ambient temperature. Soil moisture retention decreases

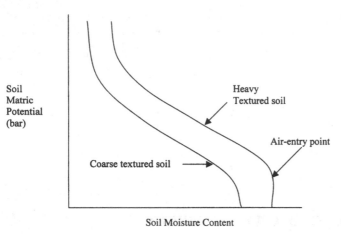

FIGURE 11.5 A schematic showing the soil moisture characteristic curve.

TABLE 11.5 Soil Moisture Characteristic for Two Soils of Contrasting Texture

Soil wetness	Soil matric potential			Volumetric wetness (θ)	
	cm of H_2O	pF		Heavy-textured	Light-textured
Saturated	1	0		0.60	0.40
	10	1	Free Water	0.60	0.38
Wet	50	1.7	↑	0.55	0.35
	100	2	↓	0.50	0.25
	330	2.5	Field Moisture Capacity	0.45	0.18
	1000	3	↑	0.40	0.15
Moist	10,000	4	↓	0.35	0.12
	15,000	4.2	Permanent Wilting Point	0.20	0.07
	30,000	4.47	↕	0.15	0.02
Dry	100,000	5	Residual water	0.10	0.005
	1,000,000	6		0.06	0
	10,000,000	7	Bonded water	0.05	0

with increase in ambient temperature (Fig. 11.7). It is important, therefore, that soil moisture characteristics are determined in a laboratory with constant temperature. Soil structure is the most important factor affecting pF curve. The soil moisture characteristic curve of a well-structured soil has a strong inflection point as indicative of change in pore size distribution. Such inflection points are not well defined in a weekly

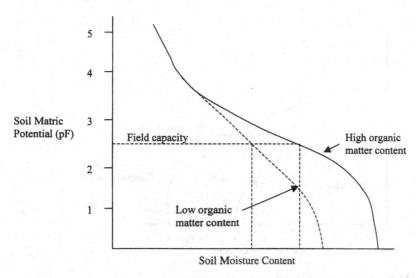

FIGURE 11.6   The pF curve of soils of similar texture but with high and low organic matter content.

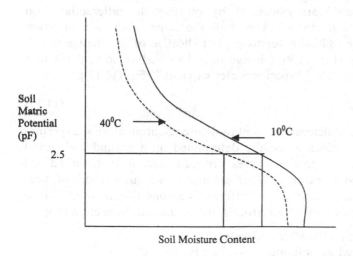

FIGURE 11.7   Soil moisture retention decreases with an increase in room temperature.

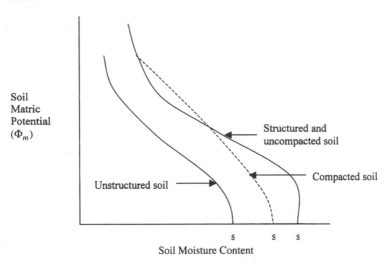

FIGURE 11.8   Effect of soil structure on pF curves.

structured or a structureless soil (single-grained sand) (Fig. 11.8). Change in soil structure due to change in land use or soil management (plow till to no till or vice versa) are evidenced by plotting the differential "soil moisture characteristic curve." A plot of the slope of the soil moisture characteristic curve $(d\Phi_m/d\theta)$ versus ϕ_m is indicative of the change in soil structure over time (Fig. 11.9). Change in soil's moisture content per unit change in ϕ_m is also called "specific water capacity" [Eq. (11.17)].

$$C_\theta = d\theta/d\Phi_m \tag{11.17}$$

With the exception of determining soil moisture retention at pF 4.2 (permanent wilting point) which is usually determined on a ground and sieved sample, soil moisture retention curves are measured on undisturbed soil either by using a core or a clod. There are also numerous methods of measuring moisture characteristics at different suctions (Table 11.6). These methods can be divided into four groups, depending on the suction range.

1. Low suction (0–60 cm of water):
    Sintered glass funnel or Haines funnel technique
    Tension table
    Sand box technique
    Plastic (porous) membranes
2. Medium suction (100–1000 cm of water):
    Pressure plate extractors
    Pressure membrane (cellophane)

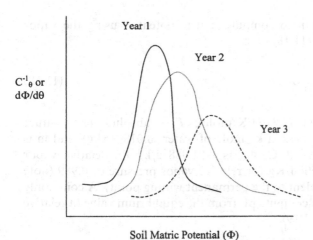

FIGURE 11.9 A plot of differential soil moisture characteristic over time indicates progressive decline in soil structure or degradation of soil physical quality.

TABLE 11.6 Methods of Determining Soil Moisture Characteristics

Matric potential (kPa)	Method use
0–10	Sand box
0–20	Haines Funnel containing porous/sintered glass plate
0–70	Suction plate
10–50	Sand/Kaolin combination
0–60	Tension table (glass or plexi glass)
1–1000	Consolidation (by applying direct load on porous disks)
10–1500	Porous plate extractors
10–3000	Centrifuge
30–1500	Osmosis using glycol and other solutions
100–2000	Psychrometers
10–10,000	Pressure membrane
3000–1,000,000	Vapor pressure equilibrium using sorption balance
1000–10,000,000	Filter paper

3. High suction (1000–20,000 cm of water):
    Pressure plate extractors
    Pressure membrane extractors

4. Very high suction (20,000–10,000 cm of water):
    Vapor pressure equilibrium using vacuum desiccators

The mathematical function to compute matric potential using the vapor pressure is shown in Eq. (11.18).

$$\Phi_m = [RT \ln(p/p_0)]/M \qquad (11.18)$$

where R is gas constant (8.3143/K/mole), T is absolute temperature ($°C + 273.16$) in K, M is mass in kg/mole of water (0.018015 kg), and ln is the natural logarithm. At 20°C, Φ_m is $-21,988$ J/kg for relative vapor pressure of 0.85 and -1500 J/Kg for relative vapor pressure of 0.989 (note the p/p_0 of 0.989 is equivalent to the permanent wilting point). A commonly used empirical formula to compute pF from the equilibrium value of relative humidity (in %) is Eq. (11.19).

$$pF = 6.5 + \log_{10}(2\text{-}\log R.H.) \qquad (11.19)$$

Using Eq. (11.19), pF for relative humidity of 98.9% is 4.2, or the permanent wilting point. High sensitivity of pF to changes in relative humidity is shown by the calculations in Tables 11.7. Some relevant models of estimating soil moisture characteristic curves are discussed in Chapter 13. Humidity values for different chemicals are shown in Appendix O.

TABLE 11.7 High Sensitivity of pF to Even Minute Changes in Relative Humidity of Soil Air

Relative humidity (%)	$PF = 6.5 + \log_{10}(2-\log_{10} R.H.)$
0.001	Undefined
1	6.8
10	6.5
20	6.35
30	6.22
40	6.10
60	5.84
80	5.50
90	5.20
95	4.80
98	4.50
98.9	4.20 (Permanent wilting point)
99	4.10
100	Undefined

11.8   SOIL-MOISTURE HYSTERESIS

Soil matric potential (Φ_m) and volumetric moisture content (Θ) as depicted by pF curve is not a unique function but depends on the prior soil wetness history. More specifically the function ($\Phi_m = f(\theta)$) depends on whether the soil is wetting (intake or absorption) or draining (withdrawal or desorption). The phenomenon of dependence of pF curve on soil-moisture history is called "soil-moisture hysteresis" (Fig. 11.10), and soils that exhibit it are called "hysteretic." The draining curve (A) represents an initially saturated soil that is drained slowly to a matric potential of 1.5 MPa or the permanent wilting point. The wetting curve (B) represents moisture content of the dry soil on wetting in increments eventually to saturation ($\Theta = s = 1$). All other factors remaining the same (e.g., temperature, solute concentration, etc.), soil moisture content (Θ) at any matric potential (Φ_m) is lower when the soil is wetting (absorbing) than draining (desorbing) (Fig. 11.10). The principal loop from saturation ($\Theta = s = 1$) to dryness is known as the main branches of pF curves. If the desorption process ceases sooner (as shown by i or ii), the wetting curve follows a different path. The relationship between Φ_m and θ over a limited range of θ (i and ii) are called scanning pF curves. Such a phenomenon has been widely observed under field conditions, and is accentuated by differences in temperature, textural and structural properties.

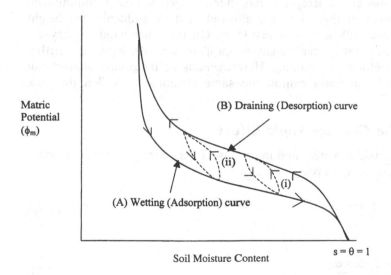

Matric Potential (ϕ_m)

(B) Draining (Desorption) curve

(ii)

(i)

(A) Wetting (Adsorption) curve

Soil Moisture Content

$s = \theta = 1$

FIGURE 11.10   A schematic of the soil moisture and scanning curves.

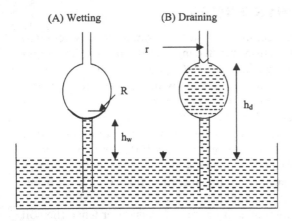

FIGURE 11.11 A schematic showing the effect of pore radii R on wetting and r on draining of a heterogeneous pore.

There are several explanations of the phenomenon of soil-moisture hysteresis. Important among these are those based on capillary theory of liquids (Cohen, 1938; 1944).

11.8.1 The Bottleneck Effect

Most soil pores are of irregular diameter and geometrically nonuniform. If the capillaries in Fig. 11.11 are allowed to drain suddenly, the height of capillary rise will be $h_d = (h_d = 0.15/r)$. On the other hand, if they are allowed to wet slowly, the height of capillary rise depends on whether capillary is wetting or draining. Heterogenous or irregularly shaped but interconnected soil pores exhibit the same phenomenon called the "ink bottle effect."

11.8.2 The Contact Angle Effect

The contact angle of water/solid interphase differs during wetting compared to the draining/drying cycle.

$$\Phi_m = -\frac{2\gamma \cos \alpha_w}{r_w} \tag{11.20}$$

$$\Phi_m = \frac{-2\gamma \cos \alpha_d}{r_d} \tag{11.21}$$

where α_w and r_w refer to contact angle and pore radius during wetting and α_d and r_d during drainage.

In fact, the contact angle may approach zero (perfect wetting) during the drainage cycle. Consequently, for specific soil moisture content, the matric potential is more negative (greater suction) during draining than wetting. Differences in contact angle may occur because of chemicals lining the pore walls (e.g., root and earthworm exudates, ash and farm chemicals transported into the pores). Differences in contact angle may depend on the degree of surface roughness, soil salinity, biomass burning, etc.

11.8.3   The Entrapped Air Effect

The entrapped air in micropores reduces soil's moisture content, which exacerbates the hysteresis effect. During the wetting cycle, the entrapped air may subsequently be dissolved but it slows the process and decreases soil wetness.

11.8.4   Swelling and Shrinking

The phenomenon of hysteresis is pronounced in soils with pronounced swell–shrink capacity, or in clayey soils containing predominantly 2:1 clay minerals. Alternate wetting/drying and freezing/thawing cause profound changes in soil structure and pore size distribution. Such soils exhibit pronounced hysteresis effect.

11.8.5   Delayed Meniscus Formation Effect

Refer to Sec. 3.1.8 and Figs. 3.19 and 3.20 with regards to packing arrangements of spheres. During the drainage (desorption) cycle, water enters the pore corresponding to the matric potential $\Phi_m = 0.15/2.83\, r_o$ at 20°C. This differential Φ_m leads to a delay in the formation of meniscus during wetting compared to the drying cycle. The phenomenon is accentuated in natural soils that comprise diverse geometric forms of interconnected pores.

11.8.6   The Entropy Effect

The magnitude of area under the scanning loop (Fig. 11.10) depends on the entropy associated with the process. Barrer et al. (1953) observed that hysteresis of soil moisture is accounted for by the irreversible work associated with swelling and with the differences in interfacial free energy related to the change in soil wetness. The area under the loop computed from a pF curve (a plot of Φ_m in units of joules/kg and gravimetric soil moisture content in kg/kg) equals the irreversible work done in the wetting/draining cycle. Entropy equals the irreversible work divided by the absolute temperature (Taylor and Ashcroft, 1972).

11.8.7 Importance of Soil-Moisture Hysteresis

The important phenomenon of soil moisture hysteresis is described by the Kelvin equation which relates free energy of the water (Φ_m expressed in joules/kg) to the radius of the pore [Eq. (11.22)].

$$RT \ln \frac{P}{P_o} = \frac{2\gamma \cos \alpha}{\rho_w r} \tag{11.22}$$

in which R is the specific gas constant (universal gas constant divided by the molecular weight of water), T is the absolute temperature, p is the vapor pressure of the water in the soil-water system, P_o is the vapor pressure of the pure free water at the same temperature, γ is the air–water surface tension, α is the contact angle, ρ_w is the density of water at that temperature, and r is the pore radius.

This phenomenon dependent upon the Kelvin equation has numerous applications in soil moisture retention and movement, especially during soil water infiltration (refer to Chapter 14) and evaporation (refer to Chapter 15). The process is particularly relevant to soil moisture redistribution. Differences in soil moisture regime in two soil profiles of similar texture and structure may occur because of differences in their wetting and drying history. Whereas the hysteretic effects are pronounced in clayey soils at all range of Φ_m, those in coarse-textured soils are prominent in the low suction range between field capacity and saturation.

11.9 APPLICATIONS OF SOIL-MOISTURE POTENTIAL

There are numerous applications of the concept of soil-moisture potential. Some important among these are listed below.

   Plant water uptake: agronomy, forestry
   Soil structure and porosity: civil engineering, agricultural engineering, physical chemistry

Solute transport: leaching of nitrates, water quality, and vulnerability of aquifers to contamination, pesticide and contaminant transport, restoration of degraded soils

Seepage: drainage, stability of dams, septic tank, urban waste disposal

Compaction: trafficability, road construction, urban/civil structure

PROBLEMS

1. Define work by a constant and varying force and explain SI units to measure work.
2. Define kinetic and potential energy and give appropriate units. What is the law of conservation of energy?
3. Explain the concept of pressure in fluids. Differentiate between atmospheric pressure and gauge pressure.
4. Describe and appropriately illustrate common devices to measure pressure.
5. Define electric "potential" and plot equipotential lines and flow lines between two points.
6. Calculate the amount of work done on a weight, volume, and mass basis to raise a unit quantity of water from point A to B 1 m apart in a vertical direction.
7. Using the data in Example 11.4 and 11.6, graphically plot distribution of total Φ and its component with depth.
8. What is the amount of work done when a 10 gram mass is moved through a distance of 2 m? Express results on a mass, volume, and weight basis.
9. Prepare a matrix showing conversion factors between $J\,m^{-3}$, $N\,m^{-2}$, and $dynes\,cm^{-2}$, $ergs\,m^{-2}$, Pa, and bar.
10. Calculate the potential gradient between two points 20 cm apart in vertical distance and reading 20 cb and 20 cm of Hg, respectively.
11. Graphically plot the soil water potential profile using the data in the following table:

Depth below the soil surface (cm)	Depression in the Hg manometer (cm)	Φ_z	Φ_m	Φ_t
10	15			
30	20			
50	15			
100	10			
150	9			
200	6			

APPENDIX 11.1   CONVERSION UNITS FOR SOIL-WATER POTENTIAL

cm of H_2O	pF	MPa	Bars	Water potential units	
				Relative humidity (p/p_o)	
10	1	0.001	0.01	0.999993	
100	2	0.01	0.1	0.999926	
330	2.52	0.033	0.33	0.999756	Field moisture capacity
1000	3	0.1	−1	0.999261	
10,000	4	1	10	0.992638	
15,000	4.18	1.5	15	0.988977	Permanent wilting point
100,000	5	10	100	0.928772	
1,000,000	6	100	1000	0.477632	
10,000,000	7	1000	10,000	0.000618	Equilibrated over P_2O_5 oven dry

There is a minus sign in front of columns 1, 3, and 4, because suction in negative.

REFERENCES

Al-Khafaf, S. and R.J. Hanks. 1974. Evaluation of the filter paper method for estimating soil water potential. Soil Sci. 117: 194–199.

Andraski, B.J. and B.R. Scanlon. 2002. Thermocouple Psychrometry. In: J.H. Dane and G.C. Topp (eds) "Methods of Soil Analysis, Part 4, Physical Methods," Soil Sci. Soc. Am., Madison, WI, pp. 609–641.

Barrer, R.M., J. Drake, and T.V. Whittam. 1953. Sorption of gases and vapors by potassium benzene sulphonate. Proc. Roy. Soc. Ser. A 219: 32–53.

Bolt, G.H. 1976. Soil Physics Terminology. Intl. Soc. Soil Sci. Bull. 23: 7–10.

Campbell, G.S. 1979. Improved thermocouple psychrometers for measurement of soil water potential in a temperature gradient. J. Phys. E. Sci. Instrum. 12: 739–743.

Campbell, G.S. and W.H. Gardner. 1971. Psychrometric measurements of soil water potential: temperature and bulk density effects. Soil Sci. Soc. Am Proc. 35: 8–12.

Cohen, L.H. 1938. Sorption hysteresis and the vapor pressure of concave surfaces. J. Am. Chem. Soc. 60: 433–435.

Cohen, L.H. 1944. Hysteresis and capillary theory of adsorption of vapors. J. Am. Chem. Soc. 66: 98–105.

Fawcett, R.G. and N. Collis-George. 1967. A filter paper method for determining the moisture characteristics of soil. Aust. J. Expl. Agric. Animal Husb. 7: 162–167.

Greacen, E.L., G.R. Walker, and P.G. Cook. 1987. Evaluation of the filter paper method for measuring soil water suction. Proc. Int'l Conference on "Measurement of Soil and Plant Water Status," Vol. 1, Logan, Utah, July, 1987: 137–144.

Hamblin, A.P. 1981. Filter paper method of routine measurement of field water potential. J. Hydrol. 53: 355–360.

Klute, A. and W.R. Gardner. 1962. Tensiometer response time. Soil Sci. 93: 204–207.

Livingston, N.J. 1993. Soil water potential. In: M.R. Carter (ed) "Soil Sampling and Methods of Analysis," Lewis Publishers, Boca Raton, FL, pp. 559–567.

Mullins, C.E. 2001. Matric potential. In: K.A. Smith and C.E. Mullins (eds) "Soil and Environmental Analysis: Physical Methods," Marcel Dekker, Inc., New York: 65–93.

Pereira, H.C. 1951. A cylindrical gypsum block for moisture studies in deep soils. J. Soil Sci. 2: 212–223.

Phene, C.J., S.L. Rawlins, and G.J. Hoffman. 1971. Measuring soil matric potential in situ by sensing heat dissipation within a porous body. I. Theory and sensor construction. Soil Sci. Soc. Am. Proc. 35: 27–33.

Phene, C.J. and D.W. Beale. 1976. High-frequency irrigation for water nutrient management in humid regions. Soil Sci. Soc. Am. J. 40: 430–436.

Rawlins, S.L. and G.S. Campbell. 1986. Thermocouple psychrometry. In: A. Klute (ed) "Methods of Soil Analysis," Part 1, Second Edition, ASA, Madison, WI, pp. 597–618.

Scanlon, B.R., B.J. Andraski and J. Bilskie. 2002. Miscellaneous methods of measuring matric or water potential. In: J.H. Dane and G.C. Topp (eds) "Methods of Soil Analysis, Part 4, Physical Methods," Soil Sci. Soc. Am., Madison, WI, pp. 643–671.

Scholl, D.G. 1978. A two-element ceramic sensor for matric potential and salinity measurements. Soil Sci. Soc. Am. J. 42: 429–432.

Taylor, S.A. and G.L. Ashcroft. 1972. Physical Edaphology: The Physics of Irrigated and Non-Irrigated Soils. Chapter 7. W.H. Freeman & Co., San Francisco, 533 pp.

Young, M.H. and J.B. Sisson. 2002. Tensiometry. In: J.H. Dane and G.C. Topp (eds) "Methods of Soil Analysis, Part 4, Physical Methods," Soil Sci. Soc. Am., Madison, WI, pp. 575–607.

12

Water Flow in Saturated Soils

Knowledge of water flow under saturated conditions is important to engineers, soil scientists, and agronomists. The data on volume and rate of flow of water in soil are needed for managing soils and plant growth. The water movement through a soil system influences aeration, nutrient availability to the plants, and soil temperature. Important applications of saturated flow in farmlands involve design of a surface and subsurface or tile, drainage system in a watershed. Most drainage designs are based on steady flow under saturated conditions. The tile spacing can be calculated from the known values of saturated hydraulic conductivity (Sec. 12.2), soil texture, and drainage design parameters. Other applications of saturated hydraulic conductivities are in the scaling of soil hydrological parameters and relationships. However, hydraulic conductivity varies strongly in space and time across the field. The magnitude of variability must be considered in all hydrological designs. The information of saturated flow is also useful for engineers constructing earthen dams, canals, waterways, etc. Failure of a dam, often caused by excessive flow through it, can have drastic consequences downstream. In agricultural fields, saturated flow occurs under rice paddy conditions. Excessive seepage losses decrease water efficiency and necessitate repeated and frequent irrigation, which lead to a rise in the water table with attendant risks of secondary salinization.

12.1   PRINCIPLES OF WATER MOVEMENT

The movement of water through a porous system occurs whenever there is a difference in potential energy of water within the porous matrix. The water content in a saturated soil system does not change during flow and only positive potentials are the driving force during the water transport. Water movement is always described in terms of potentials. The total potential of soil water is the sum of the gravitational, pressure, and osmotic potentials. The difference in potential energy of water builds a force in the system, which forces the water to move from a position of greater potential (energy) to smaller potential (energy). The osmotic potential (Φ_π) during the flow becomes important only when salt sieving phenomenon exists. Salt sieving refers to a phenomenon when water is forced through a soil and the size of the pores restricts the passage of solute (salts), but not of the water molecules. Therefore, for most water flow applications the osmotic potential is neglected, and the hydraulic potential (H) considered involves only the sum of pressure (Φ_P) and gravitational (Φ_Z) potential.

$$\Phi_t = H = \Phi_P + \Phi_Z \tag{12.1}$$

The simplest description of water movement through porous system is given by the capillary bundle concept. This concept assumes that the soil matrix is made up of bundles of small, straight capillary tubes of uniform size and shape (Fig. 12.1).

Figure 12.1a presents a single capillary tube of a small but constant radius, which is synonymous with a single pore or channel in a soil matrix. The steady water flow through this capillary varies depending upon the magnitude of hydraulic potential applied at the inlet of the capillary. Figure 12.1b shows a network of straight parallel capillaries resembling a bunch of pore channels in a soil system. Total flow across the soil system is the sum of flow through each of these capillary tubes. Since all these

(a) A single Capillary   (b) Bundle of Capillaries   (c) Complex Network

FIGURE 12.1   Flow through porous media by way of capillary tubes: (a) a single capillary; (b) a bundle of capillaries; (c) a complex network.

capillaries are essentially the replicates, the flow pattern and volumes of flow are also exactly the same. The flow across these capillaries can be calculated using Poiseuille's equation [refer to Eqs. (12.19) and (13.28)] provided that the capillary dimensions and hydraulic potential across the system are known. Soil pores are rarely straight or of equal dimensions and often form a complex network. The capillary bundle concept can to a certain extent include some of these non-uniformities in natural soils, by considering a system of capillaries, which are neither parallel to each other nor equal in size and shape (Fig. 12.1c). In reality a soil matrix may consist of pores, which have variable diameter and/or dead end pores and the micro-scopic description of flow in a single pore inside a soil matrix is difficult to describe by the capillary bundle concept. However, for a macroscopic description of flow through soil matrix, capillary bundle concept is useful and relevant.

Newton's law of viscosity (discussed in Chapter 9) can be used to calculate the flow through soil system due to a potential gradient provided the exact geometry of the soil matrix is known. Since soil consists of a complex network of pores, the pore scale description of water flow through, soil system is not practically feasible. Instead, a more macroscopic descrip-tion, which is the average flow through a cross-sectional area, is preferred.

12.1.1 Darcy's Law

Consider a salt-free soil system as given in Fig. 12.2. The system is simple and does not include the osmotic potential. The soil matrix is subjected to a hydraulic potential head (H) difference as shown in the Fig. 12.2, with head at both ends of the soil column maintained constant. For a condition when a steady flow occurs through the soil matrix from left to right, the hydraulic gradient (ΔH) across the soil matrix is given as follows [Eq. (12.2)].

$$\Delta H = \frac{H_i - H_o}{L} \tag{12.2}$$

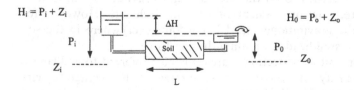

FIGURE 12.2 Flow across the cylindrical system.

where H_i and H_o are the hydraulic head maintained at inlet and outlet of the soil matrix, respectively, and L is the length of flow or soil matrix. If volume of water flowing through the soil matrix is V (L^3) in time t (T), then the volumetric flow rate $(Q/t, L^3T^{-1})$ across the column is V/t.

If the cross-sectional area of flow is A; and the soil system is homogeneous (no layering with depth, or no variation in soil properties spatially or omni direction) and isotropic (soil properties uniform in all direction) then the volumetric flow rate through soil matrix is given by the following relationships [Eq. (12.3) to (12.6)].

$$Q = \frac{V}{t} \quad \text{or} \quad Q \propto \frac{A(H_i - H_o)}{L} \tag{12.3}$$

$$Q = K_s \frac{A(H_i - H_o)}{L} \tag{12.4}$$

$$\frac{Q}{A} = K_s \frac{\Delta H}{L} \tag{12.5}$$

$$q = K_s \frac{\Delta H}{L} \quad \left[LT^{-1} = K_s \frac{L}{L} = K_s \right] \tag{12.6}$$

where q is the flow per unit cross sectional area per unit time (LT^{-1}), and is called *flux density*. The proportionality constant (K_s) in Eqs. (12.4) to (12.6) is known as "saturated hydraulic conductivity" of the soil matrix, which has the dimensions of velocity (LT^{-1}). The K_s becomes equal to the q when hydraulic gradient is unity. The constant K_s in Eqs. (12.4)–(12.6) is for a homogeneous and isotropic porous medium and is uniform and independent of the direction of flow inside the soil system. Henri Philibert Gaspard Darcy, a French hydrologist, described the relationship between the flux density and hydraulic gradient in 1856. The classical Eq. (12.6) which is the backbone of many steady saturated flow descriptions to date, is known as Darcy's law.

When flow is vertical both the pressure and gravitational head may vary and therefore, flow or gradient of flow may change. However, for horizontal flow, the gravitational head is constant everywhere in the soil system and the pressure head is the only driving force.

The flow per unit cross-sectional area q is also referred to as Darcy's velocity, or flux density. In a physical sense, q refers to the average velocity through the soil matrix. The flow of water through the soil pores is referred to as pore water velocity, which is the actual velocity of water moving through the pores. The mean pore water velocity through the soil matrix ($\bar{v}$) is given as follows:

$$\bar{v} = \frac{q}{\theta} \tag{12.7}$$

where θ is the volumetric water content of the soil matrix ($L^3 L^{-3}$). Slichter (1899) proposed a more exact and generalized differential form of Darcy's law for saturated porous media. The Slichter (1899) equation is in a vector form as follows

$$q = -K\nabla H \tag{12.8}$$

where ∇ or del, is the gradient in x, y, and z directions, and ∇H is the three-dimensional hydraulic gradient:

$$\nabla H = \frac{\partial H}{\partial x} + \frac{\partial H}{\partial y} + \frac{\partial H}{\partial z} \tag{12.9}$$

The negative sign in Eq. (12.8) indicates that water flows in the direction of decreasing potential. The second and third terms on right hand side of Eq. (12.9) are eliminated when Slichter's equation is applied to a one-dimensional flow system.

12.1.2 Validity of Darcy's Law

There are two distinct regimes of fluid flow: laminar and turbulent. Laminar flow is a state of flow when water flows like a sheet with uniform velocity throughout. Each parcel of flow is nearly parallel to adjacent ones. The forces, which can cause acceleration, are nonexistent or insignificant. In turbulent flow portions of fluid move radially and axially. The streamlines in laminar transition and turbulent flow region are given in the Fig. 12.3. Darcy's law is valid only when the flow is laminar. The validity of Darcy's

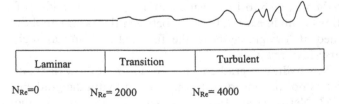

| Laminar | Transition | Turbulent |

$N_{Re}=0$ $\qquad$ $N_{Re}= 2000$ $\qquad$ $N_{Re}= 4000$

Figure 12.3 Flow regimes in pipe flow.

law is often expressed on the basis of Reynolds number (N_{Re}), which is the ratio of inertia forces to viscous forces and is a nondimensional quantity:

$$N_{Re} = \frac{\text{Forces of inertia}}{\text{Forces of viscosity}} = \frac{\rho_w r v}{\eta_w} \quad \left[\frac{ML^{-3}L\,LT^{-1}}{ML^{-1}T^{-1}} \right] \tag{12.10}$$

where ρ_w is the density (ML^{-3}) of water at a given temperature, η_w is the dynamic viscosity of water ($ML^{-1}T^{-1}$), v is the water velocity (LT^{-1}), and r is the radius of pore channel (L), and N_{Re} is dimensionless. As long as viscous forces are high enough, N_{Re} remains low and flow remains laminar. Once viscous forces become smaller, the N_{Re} becomes larger and the flow becomes turbulent. For laminar flow through straight pipes, Schneideggar (1957) and Childs (1969) reported the values of N_{Re} to be in the order of 1000–2000. However, the N_{Re} value for the laminar flow in curved and variable diameter pipes is less than 1000. Since soil pores are curved and of variable diameters, N_{Re} values of less than 1 correspond to laminar flow. Darcy's law remains valid for flow through soil for $N_{Re} < 1$. Although velocity distribution across pores of different sizes in a soil matrix is a certainty, it can be safely assumed that the shearing resistance of water balances any dissipated energy, and no part of this energy is utilized for changing the inertia or creating turbulence in the flow regime. Therefore, Darcy's law remains always valid in soils for $N_{Re} < 1$.

12.1.3 Limitations of Darcy's Law

Darcy's equation is valid when the inertial forces on the fluid are negligible compared to the viscous forces [See Eq. (12.10)] (Hubbert, 1956). For most hydraulic gradients observed in nature, such a condition generally prevails in silts, clayey, and fine-textured or structured soils. Thus, Darcy's law is valid for such soils. In coarse-textured soils (e.g., coarse sands and gravels), hydraulic gradients above unity may cause turbulence or nonlaminar flow conditions. At higher velocities the linear relationship between hydraulic gradient and flux ceases to exist and Darcy's law is no longer valid (Hubbert, 1956). In sands, especially coarse sands, it might be necessary to restrict hydraulic gradients to 0.5 to 1 to ensure laminar flow and validity of Darcy's law. Deviations from linear relationship between fluxes and applied gradients are obtained at low gradients in the fine-textured and at high gradients in the coarse-textured soils.

In clayey soils, low hydraulic gradients may result in no flow or small change in flow, which is not proportional to the applied hydraulic gradient (Miller and Low, 1963; Nerpin et al., 1966). A possible explanation for the failure of the linear relationship between flux and gradient for low flow may

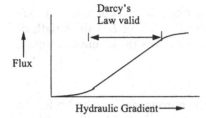

FIGURE 12.4 Deviations from Darcy's law for low as well as high velocities.

be due to the predominant adsorptive forces on water in close proximity to the soil particles compared to the remaining water and non-Newtonian conditions (refer to Chapter 9). Some soils may exhibit a threshold gradient below, which no flow conditions prevail. However, Olsen (1965) disputed some of these findings.

The validity of Darcy's law can be demonstrated by measuring flux density for a series of hydraulic gradients (Fig. 12.4). These measurements must have a linear relationship. Some of the possible explanations for the deviations from linear relationship are: non-Newtonian behavior of fluid phase changes in soil matrix under flow, electroosmotic effects, and experimental problems. Swartzendruber (1962) also presents various reasons for the failure of linear relationships. In general, Darcy's law does not apply to extreme hydraulic gradients.

12.2 SATURATED HYDRAULIC CONDUCTIVITY

The saturated hydraulic conductivity (K_s) of a porous medium, such as soil, refers to its ability to conduct water when all pores are full of water ($\theta = s = 1$). It is a compound parameter, which comprises properties of the medium and water at the specified temperature and pressure. Methods of measurements of K_s of soils are based on the direct application of Darcy's law. A saturated soil column of uniform cross-sectional area and a diameter large enough for the validity of the assumption of one-dimensional flow is subjected to a hydraulic gradient. The resulting flux of water is measured and the proportionality constant in Darcy's law gives the value of K_s of the soil column. Darcy's law expresses this procedure mathematically. After rearranging Eqs. (12.3) and (12.5), the K_s for a constant hydraulic head difference across the soil column can be calculated by the following equation [Eq. (12.11)]

$$K_s = \frac{VL}{tA\Delta H} \left[K_s = \frac{L^3 L}{TL^2 L} = \frac{L}{T} \right] \tag{12.11}$$

where A is the cross-sectional area of flow through the soil column (L^2); ΔH is the hydraulic head difference as defined in Eq. (12.6); L is the length of column; and V is volume of water (L^3) flowing across the column in time t.

12.2.1 Intrinsic Permeability

The intrinsic permeability is a property of the porous medium, and refers to the ability of the medium (e.g., soil) to transmit fluid. The intrinsic permeability of a porous medium is obtained by making the proportionality constant K_s in Darcy's law [Eq. (12.6)] more general and independent of viscosity. The latter depends on temperature and type of the fluid. Inclusion of dynamic viscosity (η, dyn sec/cm^2) results in the following form of Darcy's equation [Eq. (12.12)]

$$q = \frac{k\Delta H}{\eta_w L} \tag{12.12}$$

where k is the intrinsic permeability of the soil matrix. The intrinsic permeability (k) can be related to saturated hydraulic conductivity (K_s) as follows

$$k = \frac{K_s \eta_w}{\rho_w g} \quad \left[k = \frac{(LT^{-1})(ML^{-1}T^{-1})}{(ML^{-3})(LT^{-2})} = L^2 \right] \tag{12.13}$$

where ρ_w is the density (ML^{-3}) of water at a given temperature; g is the acceleration due to gravity (LT^{-2}), and η_w is the dynamic viscosity of water.

It is important to note here that the dimensions of intrinsic permeability are that of area (L^2), or the area of porous medium which conducts fluids.

The intrinsic permeability is the property of porous medium, whereas hydraulic conductivity is the property of both porous medium and the water. Truly speaking k is not independent of the fluid. Therefore, for most water transport applications, K_s rather than k is used.

12.2.2 Constant Head Method

The K_s can be measured in the laboratory by using a constant head or a falling head method. In the constant head method, a constant hydraulic head difference is maintained across the soil sample for the entire duration of measurement, whereas a falling head method uses hydraulic head, which varies over time.

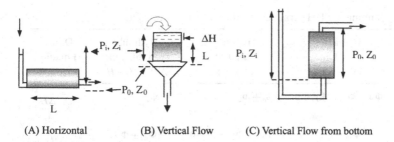

| (A) Horizontal | (B) Vertical Flow | (C) Vertical Flow from bottom |

FIGURE 12.5 Schematic of apparatus for saturated hydraulic conductivity: (A) horizontal, (B) vertical flow, (C) vertical flow from bottom.

A simple core system for the measurement of K_s is given in Fig. 12.5. The apparatus consists of a rack and clamp to hold soil cores vertically in a row, water supply tubes, a reservoir for water supply and overflow collection, and a centrifugal pump for water recirculation.

Before starting the experiment the lower end of the core is covered with a permeable material, such as cheesecloth or filter paper to retain soil. The conductance of filter paper or cheesecloth is always high so that the head loss across it is negligible compared to that across the soil core. The soil is allowed to soak water slowly through capillary rise and the saturated core sample is used for the K_s measurement. A constant head is maintained across the core and the volume of water coming out of the core is measured for specific time intervals. The flow rate along with the hydraulic head difference, length and cross-section of core are recorded and transferred into Eq. (12.11) to compute the K_s.

The three possible scenarios of conducting the experiment are horizontal (Fig. 12.5a), vertical downward (Fig. 12.5b), and upward (Fig. 12.5c) flows. Table 12.1 gives different components of head acting on the inlet and outlet of the column and hydraulic head difference, which is used in Eq. (12.11) to calculate K_s.

12.2.3 Saturating Soil Core

It is important to ensure that soil is completely saturated, and that there is no entrapped air inside the core. Therefore, use of deaerated water is recommended for saturating a soil with a small positive pressure head at the inlet. The degree of saturation can be obtained by comparing the volumetric water content (θ) of core and porosity (f_a) of soil in the core. The degree of saturation obtained by this process is also referred to as the natural saturation, which corresponds to the in situ saturation when soil is flooded with water. This state of wetting is also known as satiated. For obtaining the

TABLE 12.1 Summary of Hydraulic Head at the Inlet and Outlet

Flow	Hydraulic head at inflow, H_i	Hydraulic head at inflow, H_o	Hydraulic head difference $\Delta H = H_i - H_o$	Darcy's Equation $q =$
Horizontal	$\Phi_{Pi} + \Phi_{Zi} = \Phi_{Pi} + 0$	$\Phi_{Po} + \Phi_{Zo} = \Phi_{Po} + 0$	$\Phi_{Pi} - \Phi_{Po}$	$K_s \dfrac{\Delta H}{L}$
Vertical	$\Phi_{Pi} + \Phi_{Zi}$	$\Phi_{Po} + \Phi_{Zo} = 0 + 0$	$\Phi_{Pi} + \Phi_{Zi}$	$K_s\left(\dfrac{\Delta H}{L} + 1\right)$
Vertical from bottom	$\Phi_{Pi} + \Phi_{Zi} = \Phi_{Pi} + 0$	$\Phi_{Po} + \Phi_{Zo} = 0 + \Phi_{Zo}$	$\Phi_{Pi} + \Phi_{Zo}$	$K_s\left(\dfrac{\Delta H}{L} - 1\right)$

K_s at total saturation, a vacuum wetting procedure can also be employed. The other method, which works well with coarse-textured soils, is to flush the soil core with carbon dioxide (CO_2) followed by wetting with deaerated water. In this process the CO_2 in the soil core slowly dissolves in water and complete saturation is obtained. The CO_2 deaerated water procedure has a drawback of the acidic solution (dilute carbonic acid), which is formed when deaerated water is introduced into the soil core.

12.2.4 General Comments

Errors in volumetric flow rate measurement using the simple constant head method can be appreciable at $5\,\mathrm{ml\,h^{-1}}$ or less flow. Assuming a sample diameter of 7.5 cm and hydraulic gradient of $1.5\,\mathrm{cm\,cm^{-1}}$, the K_s is about $2 \times 10^{-5}\,\mathrm{cm\,s^{-1}}$, which is low. Therefore, a more sensitive method of measuring flow rate is required for soils of low K_s. For very large K_s, the method may also not be suitable because the siphon tubes cannot deliver water fast enough to maintain a constant head of water on the sample.

12.2.5 Alternative System

The procedures and apparatus used for measuring unsaturated hydraulic conductivity $[K(\theta)]$ can be used for K_s determination (see Chapter 13). In the absence of a high conductance porous plate relative to soil sample, hydraulic gradient is calculated by installing two piezometers at two positions along the axis of flow. A water manometer or a more sophisticated pressure transducer can be used for the hydraulic gradient measurement. The transducers are better because time to attain steady state is shorter than for a manometer system.

The head loss across the soil-porous plate can be measured to correct for the conductance of the plate (k_b). The k_b is defined as follows

$$k_b = \frac{V}{At\Delta H} \tag{12.14}$$

where ΔH is the head difference across the plate and V is the volume of water flowing through the cross-sectional area A in time t. The K_s of the soil can now be calculated as follows:

$$K_s = \frac{L}{[(L_t/K') - (1/k_b)]} \tag{12.15}$$

where K' is the conductivity of the combined soil–plate system, L is thickness of the soil sample, and L_t is the thickness of the soil–plate system. This procedure assumes that conductance of plate does not change with time and the contact resistance between the soil and the porous plate is not significant. The factor $1/k_b$, also known as resistance of the plate, should be less than 10% of L_t/K'.

12.2.6 Falling Head Method

The falling head method, as the name suggests, employs a head across the soil sample, which varies over time. The principle of the falling head method can be given by the following mathematical relationship:

$$\frac{\partial V}{\partial t} = \frac{K_s \Delta H}{L} \tag{12.16}$$

where V is the volume of water displaced in time t; Δ is the change in the magnitude of a quantity, ΔH is the total head difference, and L is the length of soil sample. Integrating the above equation between the limits t_1, H_1 to t_2, H_2 and solving for K_s leads to the relationship:

$$K_s = \frac{aL}{At} \log_e\left(\frac{H_1}{H_2}\right) \tag{12.17}$$

where A is cross-sectional area of the sample and $\log_e$ is the natural logarithm, which is equal to $2.3 \log_{10}$. A typical apparatus or arrangement for the measurement of K_s using the falling head method is given in Fig. 12.6.

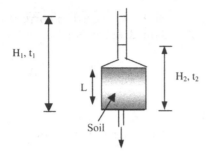

FIGURE 12.6 Schematic of falling head method for K_s measurement.

A saturated soil sample is placed on the porous plate assembly (Fig. 12.5) and water is filled in the standpipe sufficiently above the outlet end. Water from the standpoint is allowed to flow through the soil and the head difference for a given time increment is measured. The K_s is calculated by Eq. (12.17).

The K_s of the relatively impervious materials can be measured by replacing the standpipe by pressure transducers. The diaphragm type pressure transducers have the advantage of smaller volume change per unit change in pressure, hence, range of measurement can be extended up to much lower values. However, change in consolidation with change in water pressure might cause difficulty interpreting the transducer response in some cases.

12.2.7 Estimating k and K_s from Pore Geometry

Numerous studies have been carried out on the development of relationship between pore geometry and k or K_s of porous media. The soil pore geometry includes porosity, pore size distribution, and internal surface area. Darcy's law using permeability (k) in place of conductivity is given as

$$v = \frac{\rho_w g k}{\eta_w} \frac{\Delta H}{L} \tag{12.18}$$

The relationship between pore radius (r) and laminar flow through the capillaries (Q) is described by the Poiseuille equation [see also Eq. (13.28)]

$$Q = \frac{\rho_w g \pi r^4}{8\eta_w} \frac{\Delta H}{L} \tag{12.19}$$

where

$$Q = Av = \pi r^2 v = f_t \pi r^2 v; \tag{12.20}$$

and f_t is the porosity of the medium. Substituting θ in Eq. (12.20) and equating the right hand sides of the Eqs. (12.19) and (12.20) provides the relationship between intrinsic permeability, porosity, and pore radius.

$$k = \frac{f_t r^2}{8} \tag{12.21}$$

Equation (12.21) has a major drawback in terms of the true value of r, pore radius. Soil matrix consists of pores of numerous different sizes, which makes the estimation of r very difficult. Slichter (1899) attempted the determination of r by examining the geometry of pore space between spherical particles. Instead of pore radius, Kozeny (1927) derived a relationship between permeability, mean hydraulic radius (area of the flow section divided by the wetted perimeter), and surface area of the particles per unit volume of porous matrix (A_v) for a uniform pore size and isotropic material.

$$k = \frac{f_t^3}{a A_v^2} \tag{12.22}$$

where a is an empirical term and depends on tortuosity, porosity, and shape factors. Considering that the water is held in a glass tube of radius r, at suction, Φ_m, the effective radius of the largest pores to remain full of water can be calculated by equating the upward and downward forces in the tube. If γ is the surface tension of water and α is the contact angle of water with the tube, the effective radius can be obtained by the following relationship [see also Eq. (9.9c)]

$$r = \frac{2\gamma \cos \alpha}{\rho_w g \Phi_m} \tag{12.23}$$

If the contact angle α is zero then $\cos \alpha = 1$ and Eq. (12.23) reduces to:

$$r = \frac{2\gamma}{\rho_w g \Phi_m} \tag{12.24}$$

The permeability can also be estimated from the size distribution (Childs and Collis-George, 1950).

$$k = M \sum_{\rho=0}^{\rho=R} \sum_{\sigma=0}^{\sigma=R} \sigma'^2 f(\rho'') \delta r f(\sigma') \delta r \tag{12.25}$$

where M is a matching parameter obtained by experiment, σ' and σ'' are the radii of two pores forming a sequence, $f(\sigma') \delta r$ and $f(\sigma'') \delta r$ are the fractions of the cross-sectional area occupied by pores of radius r to $r + \delta r$, and R is the radius of the largest pore that is full with water. Marshall (1957; 1958) also related permeability to porosity (f_t) and average cross-sectional area of necks for pores of radius $r_1, r_2, \ldots$ and r_n in a sequence.

$$k = \frac{f_t^2 n^{-2}}{8} \sum_{i=1}^{n} (2i - 1) r_i^2 \tag{12.26}$$

where n^{-2} is the unit area between the two matching surfaces. Millington and Quirk (1959; 1961) developed relationships between gas tortuosity (ξ_g), water tortuosity (ξ_θ), air content (a) or water content (θ), and porosity (f_t)

$$\xi_g = \frac{f_a^{10/3}}{f_t^2} \tag{12.27}$$

$$\xi_\theta = \frac{\theta^{10/3}}{f_t^2} \tag{12.28}$$

The pore sizes can be measured from pressure or suction, which can be substituted in Eq. (12.26) and the k vs. Φ_m relationship similar to Eq. (12.26) can be obtained. Other efforts on relating the k to the pore radius, porosity, and water content of porous media were made by Purcell (1949), Day and Luthin (1956), Wyllie and Gardner (1958), and Elrick and Bowman (1964). Marshall and Holmes (1988) and Marshall et al. (1996) present more elaborate descriptions on these.

The modified Kozeny-Carman equation (Carman, 1939) estimates K_s from saturated water content and residual water content as follows:

$$K_s = \frac{(\theta_s - \theta_r)^3 f_m}{(100 - \theta_s)^2 \theta_r^2} \tag{12.29}$$

where f_m is the matching factor, which depends on clay content and clay activity. The θ_r can be estimated from soil organic matter content and texture (Baumer, 1989).

Brutsaert (1967) estimated the K_s from porosity (f_t), residual water content, pore size distribution index (λ), and bubbling pressure (ϕ_{mb})

$$K_s = a \frac{(f_t - \theta_r)^2}{\phi_{mb}} \frac{\lambda^2}{(\lambda + 1)(\lambda + 2)} \tag{12.30}$$

where a is constant equal to 270 and is a function of fluid parameters and gravity.

Empirical relationships are also developed for estimating K_s from easily measurable soil properties. El-Kadi (1985) related the K_s to the drainable porosity ($f_a = \theta_s - \theta_r$) as follows

$$K_s = K_o(\theta_s - \theta_r)^2 \Phi_m^\alpha \tag{12.31}$$

where θ_s, θ_r, are saturated and residual water content, respectively, K_o is a parameter related to soil, and Φ_m is the capillary suction. Campbell (1985) proposed an empirical relationship between soil texture and bulk density for the estimation of K_s.

$$K_s = C \left(\frac{1.3}{\rho_b} \right)^{1.3b} \exp(-0.025 - 3.63\text{Si} - 6.88\text{Cl}) \tag{12.32}$$

$$b = \text{GMD}_p^{-0.5} + 0.2\sigma_g \tag{12.33}$$

where C is a constant to be evaluated from data, Si is the silt fraction, Cl is the clay fraction, ρ_b is the bulk density, GMD_p is the geometric mean particle diameter, and σ_g is the geometric standard deviation. The GMD_p and σ_g can be calculated from the Si and Cl fractions of the soil (Shirazi and Boersma, 1984)

$$\text{GMD}_p = \exp(-0.025 - 3.63\text{Si} - 6.88\text{Cl}) \tag{12.34}$$

$$\sigma_g = \exp(13.32\text{Si} + 47.7\text{Cl} - \ln^2 \text{GMD}_p)^{0.5} \tag{12.35}$$

Ahuja et al. (1985) proposed a two-parameter model for estimating K_s

$$K_s = B f_a^n \tag{12.36}$$

where B and n are constants and can be set at 4 and 1015 respectively; and f_a is the air porosity at Φ_m $-33\,\text{kPa}$. Using regression techniques Rawls and Brakensiek (1985) have also proposed relationships between soil texture and porosity to predict K_s of soil.

12.3 LAPLACE EQUATION

For flow through soil systems where potential and gradient at every point in the flow domain remain constant, Darcy's law is sufficient for such a flow description. Most processes in nature do not conform to such simple steady and stationary flow processes. The potential gradient and flux may vary both in the direction and magnitude, such a process can be described by the law of conservation of matter or the continuity equation. It states that if outflow is not equal to inflow then the difference between inflow and outflow is the change in storage, which may be positive or negative depending upon the magnitude of inflow and outflow.

Consider a three-dimensional (3-D) space as shown in Fig. 12.7. The flow is q_x, q_y, and q_z in x-, y-, and z-directions, respectively.

If water content of this 3-D space is θ, then the rate of increase in q is equal to the rate of change in θ. For a one-dimensional system it can be represented as follows:

$$\frac{\partial \theta}{\partial t} = \frac{\partial q_x}{\partial x} \tag{12.37}$$

The one-dimensional flow according to Darcy's law can be given as

$$q_x = K_s \frac{\partial H}{\partial x} \tag{12.38}$$

where H is the hydraulic head and K_s is the hydraulic conductivity.

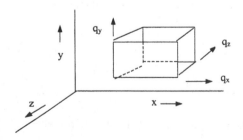

FIGURE 12.7 Flow through a 3-dimensional space.

. Combining Eqs. (12.37) and (12.38)

$$\frac{\partial \theta}{\partial t} = \frac{\partial}{\partial x}\left(K_s \frac{\partial H}{\partial x}\right) \tag{12.39}$$

The hydraulic head consists of pressure head and gravitational or positional head above a reference or datum i.e., $H = \Phi_P + \Phi_Z$ [Eq. (12.1)]

$$\frac{\partial \theta}{\partial t} = \frac{\partial}{\partial x}\left[K_s \frac{\partial(\Phi_P + \Phi_Z)}{\partial x}\right] \tag{12.40}$$

For horizontal flow if $Z = 0$ and the Eq. (12.40) reduces to

$$\frac{\partial \theta}{\partial t} = \frac{\partial}{\partial x}\left(K_s \frac{\partial \Phi_P}{\partial x}\right) \tag{12.41}$$

For vertical flow $\dfrac{\partial \Phi_Z}{\partial x} = 1$ and Eq. (12.40) changes to $\tag{12.42}$

$$\frac{\partial \theta}{\partial t} = \frac{\partial}{\partial x}\left(K_s \frac{\partial \Phi_P}{\partial x} + 1\right) \tag{12.43}$$

In saturated uncompressible soils, the rate of change of moisture content is zero. Therefore, the left hand term of Eqs. (12.39)–(12.43) is zero. If the soil is homogeneous then Eq. (12.43) reduces to

$$\frac{\partial}{\partial x}\left(K_s \frac{\partial \Phi_P}{\partial x}\right) = 0 \tag{12.44}$$

For a three dimensional system where conductivity is K_x, K_y, and K_z in X-, Y-, and Z-directions, the Eq. (12.44) can be written as

$$\frac{\partial}{\partial x}\left(K_x \frac{\partial \Phi_P}{\partial x}\right) + \frac{\partial}{\partial y}\left(K_y \frac{\partial \Phi_P}{\partial y}\right) + \frac{\partial}{\partial z}\left(K_z \frac{\partial \Phi_P}{\partial z}\right) = 0 \tag{12.45}$$

If $K_x = K_y = K_z$, i.e., for isotropic soil, the above equation reduces to a second order partial differential equation known as the Laplace equation.

$$\frac{\partial^2 \Phi_P}{\partial x^2} + \frac{\partial^2 \Phi_P}{\partial y^2} + \frac{\partial^2 \Phi_P}{\partial z^2} = 0 \tag{12.46}$$

The solution of a one-dimensional Laplace equation (boundary value problem) requires two boundary conditions and is exactly similar to Darcy's law. The solution of the Laplace equation in a two-dimensional flow domain requires four or more boundary conditions to be satisfied. One of boundary conditions for P is at the boundary while the other is for the derivative of P with respect to the normal. The remaining two boundary conditions are required in the region of application of the Laplace equation. For a more elaborate description on the Laplace equation and its solution, readers are referred to Kirkham and Powers (1972).

12.4 HOMOGENEITY AND HETEROGENEITY

Under field conditions, the K_s may vary from point to point, and in different directions. When K_s remains unchanged in one particular direction and at different locations in the same hydrologic group, the soil in that area is called homogeneous. When K_s is the same in all directions, the soil is termed as isotropic. The soil is termed anisotropic when K_s varies with change in direction. Anisotropy in the soils results from the structural variability of soils, which may result in macro or micropore flows with a directional bias. When K_s varies according to the direction of flow, it is termed as asymmetrical (Maasland and Kirkham, 1955). Consider an anisotropic soil with variable K_s as K_h and K_v in horizontal and vertical directions, respectively. The average conductivity (K_e) for this type of a soil is the geometric mean of the conductivities.

$$K_e = \sqrt{K_h \cdot K_v} \tag{12.47}$$

Non-homogeneity in soils results from layering or horizonation. The physical properties of soils in different layers vary, resulting in nonhomogeneity. For further explanation of anisotropy and inhomogeneity, readers can refer to Bear et al. (1968).

12.5 FLOW THROUGH A LAYERED SOIL

Consider a soil profile as given in Fig. 12.8. The soil profile consists of four isotropic layers. The K_s of these four layers is K_1, K_2, K_3, and K_4, respectively, from soil surface downward. The thickness of soil layers is z_1, z_2, z_3, and z_4 and the potential heads at the interface of two layers are starting from the top layer are H_0, H_1, H_2, and H_3, respectively. The total thickness of the soil profile (z) is: $z = z_1 + z_2 + z_3 + z_4$.

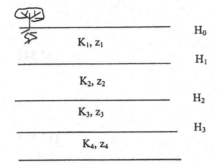

FIGURE 12.8 Hydraulic conductivity, thickness and head in a 4-layered soil.

The effective conductivity (K_e) of the soil profile can be calculated as the weighted conductivity as follows

$$K^*z = K_1^*z_1 + K_2^*z_2 + K_3^*z_3 + K_4^*z_4 \qquad (12.48)$$

$$K_e = \frac{\sum K_i z_i}{\sum z_i} \qquad (12.49)$$

Alternatively, Darcy's law can be used to calculate the overall K_s of the soil profile. The steady flow through layer 1 can be given as

$$v = K_1 \frac{(H_1 - H_0)}{z_1} \qquad (12.50)$$

Accordingly, the steady flow (v) through the entire soil profile can be given as

$$v = K_e \frac{(H_4 - H_0)}{(z_1 + z_2 + z_3 + z_4)} \qquad (12.51)$$

$$H_4 - H_0 = \frac{v(z_1 + z_2 + z_3 + z_4)}{K_e} \qquad (12.52)$$

where K_e is the effective conductivity of the soil profile. If we apply Darcy's law on each layer, following sets of equations result:

Layer 1:

$$v = K_1 \frac{(H_1 - H_0)}{z_1} \qquad (12.53)$$

Layer 2:

$$v = K_2 \frac{(H_2 - H_1)}{z_2} \tag{12.54}$$

Layer 3:

$$v = K_3 \frac{(H_3 - H_2)}{z_3} \tag{12.55}$$

Layer 4:

$$v = K_4 \frac{(H_4 - H_3)}{z_4} \tag{12.56}$$

Note as a consequence of steady flow, the flux is the same through each layer. Rearranging Eqs. (12.53)–(12.56) in terms of pressure heads and then adding results in

$$H_4 - H_0 = \frac{v(z_1 + z_2 + z_3 + z_4)}{(K_1 + K_2 + K_3 + K_4)} \tag{12.57}$$

Equating the right-hand side of Eqs. (12.52) and (12.57) gives the weighted conductivity of the soil profile, similar to Eq. (12.49).

___

Example 12.1

Consider two soil columns given in the following figure. Horizontal flow is taking place from one and the vertical flow is taking place in the other. The saturated hydraulic conductivity of sand inside both columns is 5×10^{-3} cm/sec. Estimate the hydraulic head, pressure head, and gravitation head at both ends of the columns. Calculate the Darcy's flux through both soil columns. Use the data shown in the figures. Make necessary assumptions.

Solution

Assuming a reference at the middle of soil column, the pressure head at $A = 20$ cm and gravitational head $= 0$. The pressure head and gravitational head at B are both zero. The Darcy's flux through the soil column

$$q = K \frac{\Delta H}{L} = 5 \times 10^{-3} \frac{(20 - 0)}{50} = 2 \times 10^{-3} \text{ cm/sec}$$

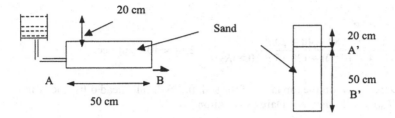

Assuming a reference point at the bottom of the vertical soil column, the pressure and gravitational head for the vertical flow are 20 cm and 50 cm, respectively, at point A'. At point B' both are zeros. The Darcy's flux through this soil column

$$q = 5 \times 10^{-3} \frac{(70 - 0)}{50} = 7 \times 10^{-3} \text{ cm/sec}$$

Example 12.2

Consider a two-layer soil profile as given in the figure below. The thickness of these two layers is 40 and 30 cm, and hydraulic conductivities are 4×10^{-4} cm/sec and 5×10^{-5} cm/sec, respectively. If the head at the top of layer one is 80 cm, calculate: (a) flux through soil profile and (b) hydraulic and pressure heads at the interface between layers.

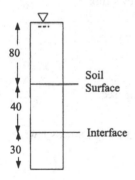

Solution

Remember the reference point is at the bottom of soil profile. For a steady flow, the flux q can be given by $q = [\Delta H/(R_1 + R_2)]$, where R_1 and R_2 are hydraulic resistances and are equal to z_1/K_{s1} and z_2/K_{s2}, respectively; where z_1 and z_2 are the thickness of layer 1 and 2, respectively. Transforming these values into an equation for q

$$q = \frac{80 + 40 + 30}{((40/(4 \times 10^{-4})) + (30/(5 \times 10^{-5})R_2))} = 2.14 \times 10^{-4} \, \text{cm/sec}$$

If we apply Darcy's law on the top layer of the soil, the hydraulic head difference will be $H_{surface} - H_{interface}$. Remember Darcy's equation

$$q = K_s^1 \frac{(H_{surface} - H_{interface})}{L_1}$$

$$2.14 \times 10^{-4} = 4 \times 10^{-4} \frac{(80 + 40 + 30) - H_{interface}}{50}$$

$$H_{interface} = 123 \, \text{cm}$$

$H_{interface}$ denotes the hydraulic head at the interface, since gravitational head at the interface is 40 cm. The pressure head at the interface will be equal to

Pressure head + Hydraulic head − gravitational head $= 123 - 40 = 83 \, \text{cm}$

Example 12.3

If the saturated hydraulic conductivities of layers in Example 12.2 are 5×10^{-2} and $6 \times 10^{-3} \, \text{cm/sec}$. Calculate the effective conductivity and Darcy's flux.

Solution

We can calculate the effective conductivity as

$$K_{eff} = \frac{40 + 30}{[40/(5 \times 10^{-2})] + [30/(6 \times 10^{-3})]} = 1.2 \times 10^{-2} \, \text{cm/sec}$$

and the flux through the entire soil column can be calculated using Darcy's law:

$$q = K \frac{\Delta H}{L} = 2.59 \times 10^{-2} \, \text{cm/sec}$$

PROBLEMS

1. Assume a 0.1 m long cylindrical capillary tube of uniform diameter of 0.01 m, which is filled with saturated sand. The tube is placed horizontally and flow is taking place from right to left. If the flow rate across the capillary is $0.2 \, \text{m}^3/\text{h}$ and

the pressure difference across the capillary is 0.02 m, calculate the hydraulic conductivity of soil in the capillary.

2. Define the conductivity obtained in the above problem. If the head difference is raised to 0.04 m. What changes do you expect in saturated hydraulic conductivity and flow? Explain.

3. What is a unit hydraulic gradient? What will be the pressure head difference in problem 1 for the condition of unit gradient flow? What will be the flow rate under unit hydraulic gradient? If the soil column in problem 1 is rotated so that the direction of flow in the column becomes vertical, what will be the head at the inlet for a unit hydraulic gradient flow?

4. Consider a soil core of equal diameter and length of 0.08 m. Water is allowed to flow vertically downward under a constant head of 0.04 m. The constant rate of effluent collected from bottom of core was 0.04 m/s. Calculate the saturated hydraulic conductivity of soil.

5. In the above problem if head at inlet varies from 0.04 to 0.02 m in time in a time interval of 2 min, calculate the saturated hydraulic conductivity of soil core. Making appropriate assumptions calculate the intrinsic permeability of soil.

6. Consider a layered soil system as shown in Fig. 12.8. The thickness of layers from top to bottom is 10, 20, 20, and 40 cm and the saturated hydraulic conductivities are 0.4, 0.8, 0.6, and 1 cm/h respectively. Calculate the overall conductivity of the soil.

7. If a constant head of 10 cm is applied on the top of the soil layer. Calculate the flux from the bottom of soil profile. (Hint: Use K value obtained in problem 6). What will be the Darcy flux at a depth of 30 cm from soil surface if steady flow takes place in the soil column?

REFERENCES

Ahuja L.R., J.W. Naney and R.D. Williams (1985). Estimating soil water characteristics from simpler properties or limited data. Soil Sci. Soc. Am. J. 49:1100–1105.

Baumer O.W. (1989). Predicting unsaturated hydraulic parameters. pp. 341–345. In: M.Th. van Genuchten, F.J. Leij and L.J. Lund (eds.). Proceeding of International workshop on Indirect Methods for estimating the hydraulic properties of unsaturated soils. Riverside California, October 11–13, 1989.

Bear J., D. Zaslavsky, S. Irmay (1968). Physical principles of water percolation and seepage. UNESCO, Paris.

Brutsaert W. (1967). Some methods of calculating hydraulic permeability. Trans. ASAE, 10:400–404.

Carman P.C. (1939). Permeability of saturated sands, soils and clays. J. Agr. Sci. 29:262.

Campbell G.S. (1985). Soil physics with BASIC: transport models for soil-plant systems. p. 150, Elsevier, New York.

Childs E.C. (1969). An introduction to the physical basis of soil water phenomenon. Wiley (Interscience), New York.

Childs E.C. and N. Collis-George (1950). The permeability of porous materials. Proc. Roy. Soc. Lond. 2(III), 134–141.

Darcy H. (1856). Les Fontaines Publiques de la Ville de Dijon. Dalmont, Paris.

Day P.R. and J.N. Luthin (1956). A numerical solution of differential equation of flow for a vertical drainage problem. Soil Sci. Soc. Am. Proc., 20, 443–447.

Elrick D.E. and D.H. Bowman (1964). Note on an improved apparatus for soil moisture flow measurements. Soil Sci. Soc. Am. Proc. 28, 450–452.

Hubbert M.K. (1956). Darcy's law and the field equations of the flow of underground fluids. Am. Inst. Min. Met. Petl. Eng. Trans. 207, 222–239.

Kirkham D. and W.L. Powers (1972). Advanced soil physics.

Kozeny J. (1927). Uber kapillare leitung des wassers im Boden. Sb. Akad. Wiss. Wien, math-naturw. Kl. Abt. Iia, 136, 271–306.

Marshall T.J. (1957). Permeability and size distribution of pores. Nature, Lond. 180, 664–665.

Marshall T.J. (1958). A relation between permeability and size distribution of pores. J. Soil Sci., 9, 1–8.

Marshall T.J. and J.W. Holmes (1988). Soil Physics, Cambridge University Press, Cambridge.

Marshall T.J., J.W. Holmes and C.W. Rose (1996). Soil Physics, 3rd Edition, Cambridge University Press, Cambridge.

Massland M. and D. Kirkham (1955). Theory and measurement of anisotropic permeability air permeability in soil. Soil Sci. Soc. Am. Proc. 19, 395–400.

Miller R.J. and P.F. Low (1963). Threshold gradients for water flow in clay systems. Soil Sci. Soc. Am. Proc. 27, 605–609.

Millington R.J. and J.P. Quirk (1959). Permeability of porous media. Nature, Lond. 183, 387–388.

Nerpin S., S. Pashkina and N. Bondarenko (1966). The evaporation from baron soil and the way of its reduction. Symp. Water Unsaturated Zone, Wageningen.

Olsen H.W. (1965). Deviations from Darcy's law in saturated clays. Soil Sci. Soc. Am. Proc. 29, 135–140.

Purcell W. R. (1949). Capillary pressure their measurement using mercury and the calculation of permeability therefrom. Trans. Am. Soc. Min. Engnrs. 186, 39–46.

Schneideggar A.E. (1957). The physics of flow through porous media. McMillan, New York.

Shirazi M.A. and L. Boersma (1984). A unifying quantitative analysis of soil texture. Solid Sci. Soc. Am. J. 48:892–898

Slichter C.S. (1899). U.S. Geol. Sur. Ann. Rep. 19-II, pp. 295–384.

Swartendruber D. (1962). Non-Darcy behavior in liquid saturated porous media. J. Geophys. Res. 67, 5205–5213.

Wyllie M.R.J. and G.H.F. Gardner (1958). The generalized Kozeny-Carman constants. Ind. Eng. Chem. 47, 1379–1388.

13

Water Flow in Unsaturated Soils

Unsaturated flow of water is a more commonly prevailing condition in the field than saturated flow. An unsaturated soil zone, or vadose zone, provides a continuum of water-unsaturated subsurface porous media connecting the soil/atmospheric interface and underlying saturated ground-water zone. It has several functions including (i) storage of water and nutrients, and (ii) transmission of water and other substances. The storage of water and nutrients is vital to the biosphere, and the water transmission is important for replenishing the aquifers. Unsaturated flow conditions are more complex and very often do not have direct solutions. Instead indirect methods, approximations, and numerical methods are more commonly used in the solution of unsaturated flow problems. This chapter describes the various laws governing the unsaturated flow through porous soil system.

13.1 UNSATURATED ZONE OR VADOSE ZONE

A soil matrix is considered unsaturated when some of the pores are filled with water and the remaining pores with air. The unsaturated zone of soil refers to that portion of the subsurface above the water table, which contains both air and water in the pores. Its thickness can vary from zero to several meters ranging from a marshy situation to an arid environment. An unsaturated zone stores the water, nutrients, and other substances and is of importance to the biosphere. It stores only a tiny fraction of fresh water and, therefore, plays a minor role in the hydrologic cycle. It is the transmission zone, which redistributes the water. Therefore, this zone controls the ground water replenishment as well as evaporation from soil surface. The unsaturated zone experiences transport processes of various kinds, chemical reactions, biological activity of roots, rodents, worms, microbiota, and other organisms. It is also a zone of human activity and is used for the cultivation and disposal of waste. This zone is also drastically disturbed by surface mining and construction of civil structures (e.g., buildings, roads, etc). The vadose zone is in direct contact with the atmosphere through gaseous fluxes of water vapor and greenhouse gases (CO_2, CH_4, and N_2O).

13.2 MECHANISMS OF UNSATURATED FLOW

The fundamental driving forces in both saturated and unsaturated flow are the potential gradient and hydraulic conductivity. As a stream of water is passed through the unsaturated soil matrix, the incoming water replaces the air present in the soil pores; it increases the total volume of water inside the soil, thus increasing the moisture content (θ) of soil. This agrees with the fundamentals of continuity equation, which states that the difference in the inflow and outflow rate is equal to the change of water storage in soil. The gradient causing flow in unsaturated soils is of negative pressure potential. The flow paths in unsaturated flow are more tortuous as several pores are filled with air. The hydraulic conductivity in an unsaturated zone can vary by as much as four to five orders of magnitude. Some of the differences in saturated and unsaturated flow are summarized in Table 13.1. A typical situation of water in an unsaturated soil media forming separate and discontinuous pockets of water given in Fig. 13.1 presents magnified pores of varying diameter. For simplicity the soil particles are sketched as either spheres or ellipses. The schematic in Fig. 13.1 shows that the empty pores must be circumvented for the water to flow through the soil matrix. This increases the length of flow path or tortuosity. Since bigger pores drain quickly most of the air-filled pores are more conductive, forcing the water to move through only smaller less conductive pores. The same is also true in aggregated soils, where large interaggregate spaces empty early leaving the small pores for water flow.

Water has an affinity for soil particle surfaces and capillary pores, which results in matric suction (Φ_m). When Φ_m is uniform throughout the soil matrix, the soil system remains in equilibrium or at a steady state

TABLE 13.1 Summary of Differences Between Saturated and Unsaturated Flow

Parameter	Saturated flow	Unsaturated flow
Water content	Constant	Variable over space and time
Air content	Zero (close to zero)	Variable over space and time
Potential gradient	Positive and constant	Negative and variable
Hydraulic conductivity	Maximum, constant	Low and variable
Vapor flow	None	Possible provided temperature gradients also exist
Water flow	Steady	Steady as well as unsteady
Flow paths	Continuous	Tortuous
Continuity equation	Inflow = outflow	Inflow = outflow + source or sink, of water
Flow descriptions	Darcy's law	Darcy–Buckingham equation Richards equation
Flow parameter	K_s	$K(\theta)$

FIGURE 13.1 Schematic of an aggregated soil.

condition. When Φ_m is not uniform within soil profile, water moves from a site of higher Φ_m till the system attains equilibrium. Another mechanism operating in the unsaturated soil system is vapor flow. In situations where there are strong temperature gradients vapor transport becomes a very important phenomenon (see Chapter 17). Depending upon the vapor pressure inside the soil matrix, soil water can move from the pockets of higher to lower moisture content, and vice versa. This is one of the fundamental reasons why pressure potentials (Φ_p) and matrix potentials (Φ_m) are used for assessing/predicting soil moisture movement rather than soil moisture content.

13.3 DARCY–BUCKINGHAM EQUATION

Unsaturated flow through a soil system is illustrated in Fig. 13.2. A saturated soil core is placed on a porous plate and the reservoir used to saturate the soil core is lowered to a position below the bottom of soil core. Since the soil core is open to the atmosphere and the water level in the reservoir is below the bottom of soil core, a Φ_m gradient is created across the soil core. As a consequence, air starts entering the soil core from the top pushing the water down and out through the bottom into the water reservoir. If the flux density measured from the bottom of soil core is q, the hydraulic head difference (ΔH) across the core equals ($H_i - H_0$) and the hydraulic gradient is $\Delta H/L$.

Flux density, as in the case of saturated flow, is proportional to the driving force (i.e., the hydraulic gradient, $\Delta H/L$)

$$q \propto \frac{\Delta H}{L} \tag{13.1}$$

or

$$q = K(\theta)\frac{\Delta H}{L}\left[\frac{L}{T} = \frac{L}{T}\cdot\frac{L}{L}\right] \tag{13.2}$$

where $K(\theta)$ is the unsaturated hydraulic conductivity of the porous medium. This equation is equivalent to Darcy's law discussed in Chapter 12. Since Buckingham (1907) was first to describe hydraulic gradient dependent flow through unsaturated media, Eq. (13.2) is known as the Darcy–Buckingham equation.

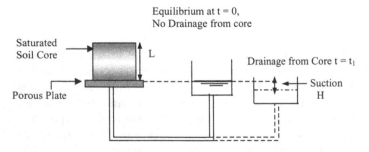

FIGURE 13.2 Schematic of unsaturated flow across a soil core.

The unsaturated hydraulic conductivity $[K(\theta)]$ is dependent on both moisture content and matric potential (ϕ_m). Equation (13.2) can be written in terms of suction (Φ_m, or its negative suction head) and gravitational component (Φ_z) as $H = \Phi_m + \Phi_z$. Contrary to saturated flow, where Φ_p is a function of z only, in unsaturated flow Φ_m is a function of both z and time (t). Therefore, the derivative ΔH is a partial derivative and Eq. (13.2) can be written as

$$q = K(\theta) \frac{\partial(\Phi_m + \Phi_z)}{\partial z} \qquad (13.3)$$

where ∂z is change in length (L). The partial derivative of Φ_m and Φ_z implies that temperature is constant during the experiment.

13.4 UNSATURATED HYDRAULIC CONDUCTIVITY

The unsaturated hydraulic conductivity $[K(\theta)]$ is a nonlinear function of both moisture content and Φ_m. Figure 13.3 presents typical curves for sand and a clay soil, and shows that at higher matric potential (i.e., near saturation, or $\Phi_m \to 0$) the sand or coarse-textured soil has higher $K(\theta)$ compared to clay soils. However, as these soils are desaturated, the hydraulic conductivity in the coarse-textured soil decreases faster than in fine textured soil and these two curves cross each other. After that for a given Φ_m, the $K(\theta)$ of coarse-textured soil is always lower than fine-textured soil. This seems logical, because coarse-textured soils have larger pores, which drain faster compared to fine-textured soils, which have relatively smaller pores. Since a greater number of pores is filled with water in

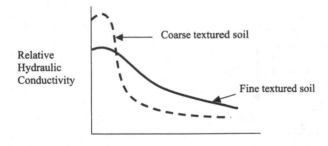

FIGURE 13.3 Schematic of relative hydraulic conductivity versus matrix potential for a fine and coarse textured soil.

fine-textured soil, the tortuosity is less and $K(\theta)$ is higher than in coarse-textured soil.

The Darcy–Buckingham equation is applicable to unsaturated flow conditions as long as moisture content (θ) remains constant over time. However, in most unsaturated flow situations in nature, it is seldom the case. Under these circumstances, Eq. (13.2) is combined with the continuity equation, which relates the time dependent rate of change of θ to the space dependent rate of change of flux density (q) in a small elemental volume of the soil. When rate of change in moisture content of a soil matrix with respect to time remains unchanged the flow is called steady flow, and when θ changes with time the flow is called unsteady flow. While steady state flow can be described by just one equation [e.g., Eq. (13.2)], two equations are necessary for describing unsteady flow through porous medium. One is the Darcy–Buckingham equation [Eq. (13.2)] and the other is the continuity equation.

13.5 CONTINUITY EQUATION

The equation of continuity combines the rate of change of moisture content of soil matrix to the changes in incoming and outgoing fluxes through the soil matrix. The continuity equation states that rate of change of incoming and outgoing flux is equal to the rate of change of storage in the soil matrix. To derive the equation of continuity for a three-dimensional and one-dimensional flow domain as given in Fig. 13.4, let us consider a three-dimensional element in an incompressible flow domain of sides dx, dy, and dz. The components of velocities in the flow element are v_x, v_y, and v_z. If the volumetric flow rate per unit cross-sectional area entering from left hand side of the flow domain as shown in Fig. 13.4, is q.

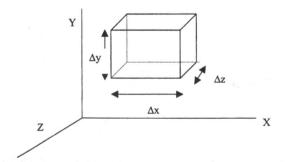

FIGURE 13.4 Water flow through a three-dimensional section of soil matrix.

Then the volumetric inflow rate from left to right is:

$$\text{Inflow} = q\partial y\partial z \tag{13.4}$$

Let us assume that volumetric flow rate at the other end of the element is less than the volumetric inflow rate. So the total outflow rate through the element is:

$$\left[q - \frac{\partial q}{\partial x}\,\partial x\right]\partial y\partial z \tag{13.5}$$

and the inflow–outflow will be equal to:

$$q\partial y\partial z - \left[q - \frac{\partial q}{\partial x}\,\partial x\right]\partial y\partial z = \frac{\partial q}{\partial x}\,\partial x\partial y\partial z \tag{13.6}$$

Since the outflow is less than inflow, it implies that there is an increase in the moisture content in the soil element. If the initial moisture content of the soil is θ the rate of change of moisture content in the flow domain in time t can be expressed as follows

$$\left[\theta - \left(\theta + \frac{\partial \theta}{\partial t}\right)\right]\partial x\partial y\partial z = -\left(\frac{\partial \theta}{\partial t}\right)\partial x\partial y\partial z \tag{13.7}$$

Since the net change in flow is equal to net change in moisture content, the Eqs. (13.6) and (13.7) are equal

$$\frac{\partial q}{\partial x}\,\partial x\partial y\partial z = -\left(\frac{\partial \theta}{\partial t}\right)\partial x\partial y\partial z \tag{13.8}$$

$$\frac{\partial \theta}{\partial t} = -\frac{\partial q}{\partial x}\left[\frac{L^3/L^3}{T} = \frac{LT^{-1}}{L} = \frac{1}{T}\right] \tag{13.9}$$

Equation (13.9) is the one-dimensional continuity equation. If we consider the volumetric fluxes in all the three directions (q_x, q_y, and q_z), Eq. (13.9) changes into a three-dimensional form of continuity equation [Eq. (13.10)].

$$\frac{\partial \theta}{\partial t} = -\left(\frac{\partial q_x}{\partial x} + \frac{\partial q_y}{\partial y} + \frac{\partial q_z}{\partial z}\right) \tag{13.10}$$

or

$$\frac{\partial \theta}{\partial t} = -\nabla q \tag{13.11}$$

where ∇ is the vector differential operator for three-dimensional gradient. It can also be called the spatial gradient of flux. It is also represented as *div*:

$$\frac{\partial \theta}{\partial t} = -div\, q = -\nabla q \tag{13.12}$$

   Richards (1931) combined the Darcy–Buckingham Eq. (13.2) and the continuity Eq. (13.11) to derive the equation of unsaturated flow.

13.6   RICHARDS EQUATION

Under most natural situations, unsteady flow occurs through the soil matrix. The $K(\theta)$ is a function of both moisture content and Φ_m, while hydraulic head includes both suction (Φ_m) and gravitational (Φ_z) components. Since pressure potentials in unsaturated soils are negative, the Φ_m is used to denote it as a positive quantity. Equation (13.2) can be reproduced in terms of suction and moisture content as follows:

$$q = K(\Phi_m)\frac{\partial H}{\partial x} \tag{13.13}$$

$$q = K(\theta)\frac{\partial H}{\partial x} \tag{13.14}$$

where K is a function of suction $[K(\Phi_m)]$ and moisture content $[K(\theta)]$. If the soil suction is fluctuating, i.e., sometimes increasing and sometimes decreasing, Miller and Miller (1956) pointed out that the Eq. (13.12) does not hold true because it does not take into account the hysteresis of soil moisture. However, if suction is omnidirectional (i.e., either increasing or decreasing continuously) then hysteresis (see Chapter 11) can be ignored. Transferring the Eq. (13.12) to the one-dimensional continuity Eq. (13.10)

$$\frac{\partial \theta}{\partial t} = \nabla[K(\Phi_m)\nabla H] \tag{13.15}$$

where

$$\nabla H = \Phi_m + \Phi_z \qquad (13.16)$$

Reproducing the continuity Eq. (13.10) here

$$\frac{\partial \theta}{\partial t} = -\left(\frac{\partial q_x}{\partial x} + \frac{\partial q_y}{\partial y} + \frac{\partial q_z}{\partial z}\right)$$

Transferring Eq. (13.19) into Eq. (13.10), while assuming soil is isotropic

$$\frac{\partial \theta}{\partial t} = \frac{\partial}{\partial x}\left(K(\Phi_m)\frac{\partial H}{\partial x}\right) + \frac{\partial}{\partial y}\left(K(\Phi_m)\frac{\partial H}{\partial y}\right) + \frac{\partial}{\partial z}\left(K(\Phi_m)\frac{\partial H}{\partial z}\right) \qquad (13.17)$$

The one-dimensional form of the Richards equation for $H = \Phi_m + \Phi_z$ becomes

$$\frac{\partial \theta}{\partial t} = \frac{\partial}{\partial z}\left(K(\Phi_m)\frac{\partial(\Phi_m + \Phi_z)}{\partial z}\right) \qquad (13.18)$$

$$\frac{\partial \theta}{\partial t} = \frac{\partial}{\partial z}\left(K(\Phi_m)\frac{\partial(\Phi_m)}{\partial z}\right) + \frac{\partial K(\Phi_m)}{\partial z} \qquad (13.19)$$

Eq. (13.19) is known as the Richards equation and was derived in 1931 (Richards, 1931). If the soil is either wetting or drying then the moisture content (θ) will be uniquely related to the suction (Φ_m). Applying the chain rule (Kaplan, 1984), rate of change of moisture content can be expressed as

$$\frac{\partial \theta}{\partial t} = \frac{d\theta}{d\Phi_m}\frac{\partial \Phi_m}{\partial t} = C_w(\Phi_m)\frac{\partial \Phi_m}{\partial t} \qquad (13.20)$$

where C_w, is known as soil-moisture capacity function (L^{-1}), which is equal to the inverse slope of the soil-moisture characteristic curve or $\Phi_m(\theta)$. It refers to the increase in Φ_m per unit increase in θ. Transferring the soil-moisture capacity function Eq. (13.20) into (13.19) gives the capacitance form of Richards equation

$$C_w(\Phi_m)\frac{\partial \Phi_m}{\partial t} = \frac{\partial}{\partial z}\left(K(\Phi_m)\frac{\partial \Phi_m}{\partial z}\right) + \frac{\partial K(\Phi_m)}{\partial z} \qquad (13.21)$$

Alternately suction component can be represented as

$$\frac{\partial \Phi_m}{\partial z} = \frac{\partial \Phi_m}{\partial \theta} \frac{\partial \theta}{\partial z} \tag{13.22}$$

Transferring Eq. (13.22) into Eq. (13.19) leads to the diffusivity form of Richards equation

$$\frac{\partial \theta}{\partial t} = \frac{\partial}{\partial z} \left(D(\theta) \frac{\partial \theta}{\partial z} \right) + \frac{\partial K}{\partial \theta} \frac{\partial \theta}{\partial z} \tag{13.23}$$

where soil-water diffusivity $D(\theta)$

$$D(\theta) = K(\theta) \frac{\partial \Phi_m}{\partial \theta} \left[L^2 T^{-1} = L T^{-1} \left(\frac{L}{L^3 L^{-3}} \right) = L^2 T^{-1} \right] \tag{13.24}$$

The dimensions for soil-water diffusivity ($D(\theta)$) from Eq. (13.24) are $L^2 T^{-1}$, and it is defined only in the absence of hysteresis. The first term on the right-hand side in Richards equation is for flow of water owing to the gradient of Φ_m. The second term is known as gravitational term. Neglecting the gravitational term in Eq. (13.23)

$$\frac{\partial \theta}{\partial t} = \frac{\partial}{\partial z} \left(D(\theta) \frac{\partial \theta}{\partial z} \right) \tag{13.25}$$

Equations (13.21) and (13.23) are highly nonlinear because of the functions $C_w(\Phi_m)$, $D(\theta)$, and $K(\theta)$ because of their dependence on both Φ_m and θ. They are also known as Fokkar–Plank equations. The equations can be solved numerically. Approximate analytical solutions are also available (Parlange et al., 1997). The soil-water diffusivity can be assumed constant for a very small range of wetness. This assumption transforms Eq. (13.25) to a form similar to Fick's second law of diffusion (See Chapter 18).

$$\frac{\partial \theta}{\partial t} = D(\theta) \frac{\partial^2 \theta}{\partial x^2} \tag{13.26}$$

The soil-water diffusivity is a complex parameter, and can be misleading. The water flow through soils is described as mass flow or convection but not diffusion. The hydraulic diffusivity expression becomes inconsistent whenever hysteresis is predominant or soil is layered. However, the range of variation of diffusivity is much smaller as compared to the hydraulic conductivity. The relationship between soil-water diffusivity

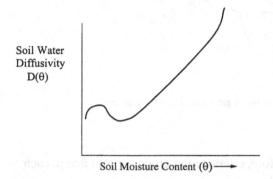

FIGURE 13.5 Schematic of the relationship between soil moisture content and soil water diffusivity.

and soil-water content is shown in Fig 13.5. It has been observed that initially when soil is very dry, diffusivity decreases with increasing water content. This is apparently due to vapor movement (Philips, 1955). Once soil attains certain moisture content, the relationship becomes positive and linear. However, near saturation, soil-water diffusivity tends to approach infinity.

13.7 ESTIMATION OF SOIL'S HYDRAULIC FUNCTIONS

The hydraulic conductivity of unsaturated soil can be calculated from the known Φ_m and θ relationship. The schematic of random distribution of pores as proposed by several authors (Childs and Collis-George, 1950; Marshall, 1958; Millington and Quirk, 1959) assumes that soil is made up of distinct randomly distributed pores of various radii. The overall conductance of such a soil system depends upon the geometric configuration and the number of pairs of interconnected pores. The random distribution of pores is shown in the Fig. 13.6.

Generally in a soil matrix, smaller pores are more numerous compared to larger pores. Therefore, smaller pores are more interconnected than larger pores. For a given volume of soil matrix, the number of pore connections N across any plane is inversely proportional to the cross-sectional area of pore. If r is the radius of a pore then N is $\propto 1/r^2$. Poiseuille's law relates the volumetric flow rate of the pore (Q) to the fourth power of

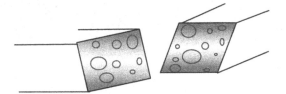

FIGURE 13.6   The random distribution of pores in a section of soil.

radius of a pore as $Q \propto r^4$. Therefore, overall conductance of soil due to each class of pores (K) can be given as

$$K(\theta) \propto N^* Q \propto r^2 \qquad (13.27)$$

By dividing and measuring the porosity of soil into several distinct classes, we can obtain the probability that a pore of various radii connects a pore of larger or smaller radius. Thus, we can obtain the conductivity values for different degrees of saturation. In practice, a soil-moisture characteristic curve (SMCC) divides porosity into pore-size classes and corresponding partial volumes, thus suction is related to moisture content. The suction can also be related to pore radius by capillary equation ($r = 2\sigma/\Phi_m$, where Φ_m is the capillary rise). The SMCC can be divided into pore radii increments and corresponding moisture volume increments. Using this technique, the $K(\theta)$ can be calculated by the Childs and Collis-George (1950) method:

$$K(\theta) = F \frac{\rho_w g}{\eta_w} \sum_{\Gamma=0}^{\Gamma=r} \sum_{\delta=0}^{\delta=R} \delta^2 f(\Gamma) \mathrm{d}r \, f(\delta) \, \mathrm{d}r \qquad (13.28)$$

where ρ_w is density of water, η_w is viscosity of water, $f(\Gamma)\,\mathrm{d}r$ the partial area occupied by pores of radii Γ to $\Gamma\,\mathrm{d}r$ and $f(\delta)\,\mathrm{d}r$ the partial area of pores with $\delta\,\mathrm{d}r$. The F in Eq. (13.28) is a matching factor, which is required to fix the hydraulic conductivity value to an experimentally predetermined $K(\theta)$ value for a known θ. In the Childs and Collis-George procedure the calculations are made successively for different values of wetness. For each case the summation terminates at the largest pore radius R, which is water filled at a specific value of soil wetness. The calculations for Childs and Collis-George procedure are cumbersome, but the $K(\theta)$ function is reportedly close to measured $K(\theta)$ function for sand and slate dust. Marshall (1958) and Millington and Quirck (1959), Kunze et al. (1968), and Jackson (1972) further simplified the calculation procedure. Changing the matching factor to the ratio of measured saturated hydraulic conductivity (K_s) and calculating saturated hydraulic conductivity Reference, Jackson (1972) gave the following formulation.

$$K(\theta)_i = K_s \left(\frac{\theta_i}{\theta_s}\right)^c \frac{\sum_{j=1}^{m} [(2j + 1 - 2i)\Phi_{mj}^{-2}]}{\sum_{j=1}^{m} [(2j - 1)\Phi_{mj}^{-2}]} \tag{13.29}$$

where $K(\theta)_i$ is the hydraulic conductivity at a moisture content of θ_i, m is the number of increments of θ, Φ_{mj} is the suction head at the midpoint of each of θ increment, and C is an arbitrary factor which is reported to be 0–4/3.

13.7.1 Measurement of Soil's Hydraulic Functions

Soil's hydraulic functions can be estimated both in the lab and field by various methods, which can be classified as: (i) steady flow methods and (ii) unsteady flow methods. In the steady state methods for the determination of $K(\theta)$ and $D(\theta)$, flux, gradient, and moisture content remain unchanged. However, in transient state methods, all three vary by parameter. Klute and Dirksen (1986) described the laboratory methods for the measurement of $K(\theta)$ and $D(\theta)$.

13.7.2 Laboratory Method

The laboratory methods are (i) steady state and (ii) transient methods. Steady state methods are those where a mean Φ_m is related to the mean θ of the soil in a core or column. Proper adjustments are made to account for the hydraulic resistance of porous media at the inlet end of the core. Transient methods include the infiltration method and pressure outflow method. The infiltration method is based on the principle that at $t \to \infty$, the steady rate of inflow into a soil (q) tends to be equal to $K(\theta)$ ($q \to K(\theta)$) for homogeneous moisture content of soil profile (θ_0). A series of measurements for q with respect to time, under different Φ_m provides the relationship between Φ_m, θ and $K(\theta)$ (Davidson et al., 1963; Youngs, 1964). A schematic of setup for the measurement of unsaturated hydraulic functions is given in Fig. 13.7.

Laboratory determination of $K(\theta)$ and $D(\theta)$ can also be made in long soil columns by inducing evaporation (Moore, 1939) or infiltration (Youngs, 1964). A series of measurements of suction gradient and moisture content can be made on a long soil column using tensiometers and gamma-ray scanning or any other nondestructive method of moisture content measurement.

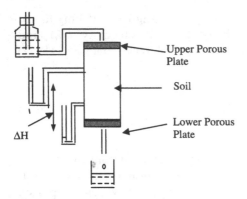

FIGURE 13.7 Schematic of a system for the measurement of unsaturated hydraulic conductivity under steady state condition.

The pressure outflow method determines the relationship between Φ_m and θ by subjecting a saturated soil core to successive increments of Φ_m applied or pneumatic pressure P. The increments in pressure heads are kept small so that D can be assumed constant over the change in θ due to change in Φ_m. Gardner (1956) solved the diffusion form of the Richards equation and gave the following solution for θ vs D relationship and unsaturated hydraulic conductivity determination.

$$\ln(V_0 - V_t) = \ln\left(\frac{8V_0}{\pi^2}\right) - \frac{\pi^2 D(\theta)t}{4L^2} \qquad (13.30)$$

$$K(\theta) = \frac{D(\theta)V_0\rho_w g}{V_s \Delta P}\left[\frac{L}{T} = \frac{L^2 T^{-1} L^3 ML^{-3} LT^{-1}}{L^3 ML^{-1} T^{-2}} = \frac{L}{T}\right] \qquad (13.31)$$

where V_0 is the total outflow volume of water, V_s is the total volume of soil in the core, V_t the outflow volume at time t, L is the length of sample, and ΔP is air pressure difference. The $D(\theta)$ obtained from Eq. (13.24) is related to $\Delta\theta$ for applied ΔP (or Φ_m equal to pneumatic pressure, P). The measurements for a tension plate device must be made for both sorption and desorption and perhaps intermediate scanning, also (Hillel, 1980). The pressure outflow method has a major drawback in that D must be constant across ΔP. Doering (1965) proposed the one step outflow method, which eliminated this limitation. The method employs

Gardner's (1962) solution and considers relatively larger ΔP, therefore relatively more accurate.

$$D(\theta) = -\frac{4L^2}{\pi^2(\theta_t - \theta_f)} \frac{d\theta_t}{dt}$$ (13.32)

where θ_t is the average value of θ across soil column and θ_f is the final moisture content. Transforming the Eq. (13.32) in terms of measured set of effluent volumes will result in the following relationship

$$D(\theta) = 0.27L^2 \left[\frac{d^2V}{dt^2} \left(\frac{dV}{dt} \right)^{-1} \right]$$ (13.33)

where V is the cumulative outflow, which can be calculated by

$$\bar{\theta} = \frac{1}{L} \int_0^L \theta \, dx$$ (13.34)

Another equation which relates $D(\theta)$ to soil water sorptivity or $S(\theta)$ was proposed by Dirksen (1975).

$$D(\theta_0) = \frac{\pi S^2}{4(\Delta\theta)^2} \left[\frac{\Delta\theta}{0.725} \frac{d(\log S^2)}{d\theta_0} - 0.198 \right]$$ (13.35)

where $\Delta\theta = \theta_o - \theta_i$ and θ_i is soil's initial moisture content. The $d(\log S^2)/d\theta$ in the above equation can be evaluated by fitting a polynomial to the measured sets of S and θ values. Measurement of unsaturated hydraulic function ($K(\theta)$ and $D(\theta)$) is easier in the lab; however, the lab measurements are made on a discrete and small sample, away from the natural continuum. Since the variability of soil and, therefore, that of soil's hydraulic functions is well established and well known, field methods for the measurement of soil's hydraulic functions appear as useful techniques if a preassessment of variability is made.

13.7.3 Unsteady State Boltzmann Transform Method

The Boltzmann transform method assumes a semi-infinite uniform soil. The flow through the soil is one-dimensional and horizontal or vertical with the gravity component assumed insignificant. The initial and final moisture contents are constant. If initial moisture content (θ_0) is higher than final (θ_1), outflow or drainage occurs, and if $\theta_1 > \theta_0$, infiltration happens. Substituting the Boltzmann variable $\lambda = xt^{-1/2}$ in Eq. (13.24), and upon integration, following equation for soil–water diffusivity is obtained.

$$D(\theta') = -0.5 \frac{d\lambda}{(d\theta)_{\theta=\theta'}} \int_{\theta_1}^{\theta'} \lambda(\theta)\, d\theta \tag{13.36}$$

where $D(\theta)$ and $d\lambda/d\theta$ are evaluated at the moisture content θ'. Methods proposed by Bruce and Klute (1956) or Whisler et al. (1968) can be used to estimate the diffusivity function.

13.7.4 In Situ Methods

The in situ methods address several of the disadvantages of the laboratory methods. However, the spatial and temporal variations make it difficult to determine exact boundary conditions. For example in case of a steady infiltration through soil profile, the essential and limited condition is that flow should be steady both at the upper and lower end of soil profile, which is extremely difficult to achieve in a heterogeneous field. However, keeping the fluxes large, so that errors due to tortuosity of flow paths are small, can minimize the problems due to heterogeneity.

Sprinkler Method

The sprinkler method, or sprinkler infiltrometer method, makes a uniform application of water on the soil surface (Peterson and Bubenzer, 1986). Water is applied at a rate slightly lower than the effective hydraulic conductivity (K_e) of the soil. Both θ and Φ_m increase gradually and suction gradients become zero. The flow through soil profile is only due to gravity and hydraulic gradient of flow becomes unity. Under this situation flux through soil profile is essentially equal to the hydraulic conductivity of soil. This provides one set of value of Φ_m and θ. The experiment can be repeated for different steady rates of water application and corresponding values of Φ_m and θ can be obtained. This method, although simple, requires large number of closely spaced sprinklers supplying water at a constant rate. Since measurement of Φ_m and θ need to be made for a wide range, the method poses a problem for sprinkler application rate of less than 1 mm/h. The exposed soil surface tends to disperse and seal the pores due to the impact of rain droplets, thus reducing the infiltrability of soil. The steady flux is very difficult to achieve for a layered soil.

Crust-Topped Method

The crust-topped method employs a less permeable crust of topsoil, which reduces the flux density, soil wetness, and corresponding $K(\theta)$ and $D(\theta)$ values of the infiltrating profile due to the steep hydraulic gradient across the less permeable crust of topsoil. The impeding layer induces the suction in the subsoil, which increases with the increasing hydraulic resistance of the crust. Once steady infiltration is established flux and conductivity of subsoil becomes equal. The Φ_m and θ can also be measured simultaneously using tensiometer and nondestructive moisture content measurement device. This method can work for a wide range of Φ_m and θ measurements, thus eliminating the range limitation of the sprinkler method described above. For very high suctions, the measurements may take long time and accurate measurement of flux may become difficult to achieve. Evaporation may also become significant if proper care is not taken. The crust-topped method can be applied while using double ring infiltrometers, tension infiltrometers, or disc permeameters (Perroux and White, 1988).

Internal Drainage Method

The internal drainage method for the measurement of $K(\theta)$ and $D(\theta)$ where water movement is purely by drainage and not by evaporation was developed by Watson (1966) and known as the instantaneous profile method. In this method, plots 5 m × 5 m or 10 m × 10 m in cross section are selected and instrumented with a tensiometer and nondestructive moisture content measurement devices at various depths. Water is applied to the field till the tensiometer readings at each depth become constant. The top of the plot is covered with plastic or any other material so that loss of water due to evaporation is not significant. The readings of Φ_m and θ are recorded at various depths simultaneously at specified time intervals. It has been generally found that during internal drainage in a deep, wetted soil, the hydraulic gradient deep in the soil layer is unity. Therefore, $K(\theta)$ is equal to the time rate of change of moisture content as given below:

$$K(\theta) = \left(\frac{dW}{dt}\right)_z \tag{13.37}$$

where, W is the total moisture content of profile to depth z

$$W = \int_0^z \theta \, \mathrm{d}z \tag{13.38}$$

and the total moisture content change per unit time can be obtained by integrating between successive soil layers up to the depth z.

$$\left(\frac{\mathrm{d}W}{\mathrm{d}t}\right)_z = K\left(\frac{\mathrm{d}H}{\mathrm{d}z}\right)_z \tag{13.39}$$

The $D(\theta)$ can be determined by the time rate of change of matric suction, and hydraulic gradient (Gardner, 1970)

$$D(\theta) = L\,\frac{\mathrm{d}\Phi_m}{\mathrm{d}t}\left(\frac{\mathrm{d}H}{\mathrm{d}z}\right)^{-1} \tag{13.40}$$

If the hydraulic gradient is near unity, then the inverse term in Eq. (13.40) becomes one and $D(\theta)$ can be calculated by the time rate of change of suction only.

13.7.5 Functional Relationships

Some common functional relationships are described below.

Soil Water Retention Models

One of the most simple and popular functions for describing $\theta(\Phi_m)$ has been the equation of Brooks and Corey (1964), herein referred to as the BC equation

$$s = \frac{\theta - \theta_r}{\theta_s - \theta_r} = \left(\frac{\Phi_m}{\Phi_{ma}}\right)^{\lambda} \rightarrow \Phi_m < \Phi_{ma} \quad \text{or} \quad s = \begin{cases} (\alpha\Phi_m)^{-\lambda} \rightarrow \alpha\Phi_m > 1 \\ 1 \rightarrow \alpha\Phi_m \leq 1 \end{cases}$$
$$= 1 \rightarrow \qquad \Phi_m \geq \Phi_{ma} \tag{13.41}$$

where s is the effective degree of saturation, also called the reduced moisture content ($0 \leq s \leq 1$), θ_r and θ_s are the residual and saturated water contents, respectively; Φ_{ma} is the air entry or bubbling pressure, α is an empirical parameter (L^{-1}) whose inverse is often referred to as the air entry value or (Φ_{ma}), and λ is a pore-size distribution parameter affecting the slope of the retention function. Remember for notational convenience, α and Φ_m are taken as positive for unsaturated soils.

Several continuously differentiable (smooth) equations have been proposed to improve the description of soil-moisture retention near saturation. A related smooth function with attractive properties is the equation of van Genuchten (1980), herein referred to as the VG equation:

$$s = [1 + (\alpha \Phi_m)^n]^{-m} \qquad (13.42)$$

where α, n, and m are empirical constants affecting the shape of the SMCC. The limiting curve follows from Eq. (13.42) by removing the factor 1 from the denominator. This shows that the VG- and BC-functions become equivalent at low s when $\lambda = mn$.

Hydraulic Conductivity Models

The model of Mualem (1976) for predicting the relative hydraulic conductivity, K, is

$$K(s) = K_s s^l \left(\frac{f(s)}{f(1)} \right)^2 \qquad (13.43)$$

with

$$f(s) = \int_0^s \frac{1}{\Phi_m(x)} dx \qquad (13.44)$$

where K_s is the saturated hydraulic conductivity, s is degree of saturation, and l is a pore-connectivity parameter equal to 0.5 as an average for many soils estimated by Mualem (1976). Substituting the inverse of the VG equation into Eq. (13.43) then integrating and then substituting the $K = 0$ leads to the restriction $m = 1 - 1/n$, and the Eq. (13.43) reduces to the following expression for K:

$$K(s) = K_s s^l [1 - (1 - s^{1/m})^m]^2 \qquad (13.45)$$

In terms of pressure head

$$K(\Phi_m) = \frac{K_s \{(1 + \alpha \Phi_m)^{mn} [1 + (\alpha \Phi_m)^n]^{-m}\}^2}{[1 + (\alpha \Phi_m)^n]^{ml}} \qquad (13.46)$$

and for soil-water diffusivity, following equation can be derived from Eqs. (13.23) and (13.42)

$$D(s) = \frac{(1-m)K_s s^{l-1/m}}{\alpha m(\theta_s - \theta_r)}[(1 - s^{1/m})^{-m} + (1 - s^{1/m})^m - 2]$$ (13.47)

When the BC retention function is substituted into (13.44) the following hydraulic conductivity function with respect to moisture content, pressure head and soil-water diffusivity equations are obtained

$$K(s) = K_s s^{l+2+2/\lambda}$$ (13.48)

$$K(\Phi_m) = \frac{K_s}{(\alpha \Phi_m)^{\lambda(l+2)+2}}$$ (13.49)

$$D(s) = \frac{K_s s^{l+1+1/\lambda}}{\alpha \lambda(\theta_s - \theta_r)}$$ (13.50)

The predictive equations for $K(\theta)$ used thus far assume that K_s is a well-defined and easily measured soil hydraulic parameter. This assumption is probably correct for many repacked, coarse-textured, and other soils characterized by relatively narrow pore-size distributions. However, direct field measurement of K_s is generally very difficult for undisturbed and especially structured field soils. Also $K(\theta)$ near saturation is determined primarily by soil's structural properties, which are subject to considerable spatial variability in the field (van Genuchten et al., 1991). However, soil's textural properties are less variable and have a more dominant effect on $K(\theta)$ in the dry range. The rapid decrease of the predicted $K(\theta)$ near saturation when n is relatively small is intuitively realistic. It suggests that $K(\theta)$ near saturation is determined by only a very few large macropores or cracks which may have little relation to the overall pore-size distribution that determines the general shape of the predicted conductivity curve at intermediate moisture contents. Thus, it seems more accurate to match the predicted and observed unsaturated hydraulic conductivity functions at moisture content somewhat less than saturation. The same holds for the θ_s, which is best regarded as an empirical parameter to be used in the context of a specific water retention model, and hence must be fitted to observed unsaturated soil water retention data points.

The model of Burdine (1953) can be written in a general form as follows

$$K(s) = K_s s^l \frac{g(s)}{g(l)}$$ (13.51)

where

$$f(s) = \int_0^s \frac{1}{[\Phi_m(x)]^2} \, dx \tag{13.52}$$

as in Eq. (13.50) the pore-connectivity parameter l accounts for the presence of a torturous flow path. A variety of values have been suggested for l; Burdine (1953) assumed a value of 2. Results analogous to those for Mualem's model can also be derived for Burdine's model and can be referred in the user's manual for RETC code (van Genuchten et al., 1991).

Empirical Approaches to Estimating Unsaturated Hydraulic Conductivity

The hydraulic properties can also be expressed in terms of relatively simple mathematical expressions (Mualem, 1989). A list of empirical models for unsaturated hydraulic conductivity is given in Table 13.2.

TABLE 13.2 Empirical Expressions for Unsaturated Hydraulic Conductivity

Model	Boundary condition	Reference
$K(\theta)_r = s^n \ n = 3.5$ $s = (\theta - \theta_r)/(\theta_s - \theta_r)$		Averjanov (1950)
$K(\theta) = \alpha \lvert \Phi_m \rvert^{-n}$		Wind (1955)
$K(\theta)_r = \exp(\alpha \Phi_m)$		
$K(\theta) = \dfrac{a}{(\lvert \Phi_m \rvert^n + b)}$		Gardner (1958)
$K(\theta) = K_s$	For $\Phi_m \geq \Phi_m$	Brooks and Corey (1964)
$K(\theta)_r = (\Phi_m / \Phi_{mCT})^{-n}$	For $\Phi_m < \Phi_m$	
$K(\theta)_r = \alpha \dfrac{\cosh[(\Phi_m / \Phi_{m_1})^n]}{\cosh[(\Phi_m / \Phi_{m_1})^n]}$		King (1964)
$K(\theta) = K_s$	For $\Phi_m \geq \Phi_{mCT}$	
$K(\theta)_r = \exp[\alpha(\Phi_m - \Phi_{mCT})]$	For $\Phi_{m1} \leq \Phi_m \leq \Phi_{mCT}$	Rijtema (1965)
$K(\theta) = K(\theta)_1 (\Phi_m / \Phi_{m1})^{-n}$	For $\Phi_m < \Phi_{m1}$	

Note: $\sinh(x)$, $\cosh(x)$, $\tanh(x)$, $\coth(x)$, $\mathrm{sech}(x)$, and $\mathrm{csch}(x)$ are hyperbolic trigonometric functions and are defined in terms of the natural exponential function e^x (e.g., hyperbolic sine of x or $\sinh(x) = [e^x - e^{-x}]/2$ and $\cosh(x) = [e^x + e^{-x}]/2$).

Example 13.1

Calculate the unsaturated hydraulic conductivity parameters from $\Phi_m-\theta$ relationship for a sandy loam soil given in the following table. The saturated hydraulic conductivity of the soil was $4 \times 10^{-3}\,\mathrm{cm\,sec^{-1}}$.

Solutions

For water content $= 0.44$

The pore class increment (i) and denominator index (j) both are equal to 1. Also $\theta_i = \theta_s$, therefore, the hydraulic conductivity is $K = K_s = 4 \times 10^{-3}\,\mathrm{cm\,sec^{-1}}$.

For water content $= 0.43$

The pore class $i = 2$, therefore in the numerator $2j + 1 - 2i = 1,\ 3,\ 5,\ 7 \ldots 25$ [Eq. (13.29]

$$K_2 = 4 \times 10^{-3} \left(\frac{0.43}{0.44}\right) \frac{\frac{1}{15^2} + \frac{3}{25^2} + \frac{5}{35^2} + \cdots + \frac{25}{5000^2}}{\frac{1}{10^2} + \frac{3}{15^2} + \frac{5}{25^2} + \cdots + \frac{27}{5000^2}} = 0.002$$

For water content $= 0.40$

The pore class $i = 3$, therefore in the numerator $2j + 1 - 2i = 1,\ 3,\ 5,\ 7 \ldots 23$

$$K_3 = 4 \times 10^{-3} \left(\frac{0.40}{0.44}\right) \frac{\frac{1}{25^2} + \frac{3}{35^2} + \frac{5}{40^2} + \cdots + \frac{23}{5000^2}}{\frac{1}{10^2} + \frac{3}{15^2} + \frac{5}{25^2} + \cdots + \frac{27}{5000^2}} = 0.001$$

These calculations can progress till we get to the last moisture content value of 0.02. The conductivity for this class will be

$$K_{14} = 4 \times 10^{-3} \left(\frac{0.02}{0.44}\right) \frac{\frac{1}{5000^2}}{\frac{1}{10^2} + \frac{3}{15^2} + \frac{5}{25^2} + \cdots + \frac{27}{5000^2}} = 1.27 \times 10^{-10}$$

These calculations can be easily made on a spreadsheet or by writing a simple computer program. Remember the denominator for all the calculations remains the same. The calculated values of conductivity are listed in the table below. At least twenty sets of moisture content vs. suction readings at a regular interval are required. However, if more sets are available, the better.

Water content $(cm^3 cm^{-3})$	Suction head (cm)	Pore class increment (nr.)	Denominator index $2j-1$	Hydraulic conductivity $(cm\ sec^{-1})$
0.44	10	1	1	0.004
0.43	15	2	3	0.002
0.40	25	3	5	0.001
0.34	35	4	7	0.0006
0.30	40	5	9	0.0003
0.26	50	6	11	0.0002
0.22	60	7	13	7.05E–5
0.18	70	8	15	2.64E–6
0.14	80	9	17	6.92E–6
0.08	150	10	19	7.47E–7
0.05	600	11	21	5.24E–8
0.04	1000	12	23	9.73E–9
0.03	3000	13	25	1.1E–9
0.02	5000	14	27	1.27E–10

Example 13.2

Plot the soil water characteristic curve from the Φ_m, θ data given in Example 13.1. If volumetric moisture content is 0.44 at saturation, calculate (a) bulk density of soil, (b) moisture content at 1/3 bar, (c) moisture content at 15 bar, (d) available water, (e) water release from 0.5 m soil profile, and (f) volumetric water capacity.

Solution

The soil water characteristic plotted on a semilog paper is given in figure (a) below.

(a) Assuming moisture content at saturation corresponds to porosity of soil. Bulk density can be estimated from volumetric water content.

$$\theta = 1 - \rho_b/\rho_s; \quad -\rho_b = -\rho_s(1 - \theta) = 2.65^*(1 - 0.44) = 1.48\ g\,cm^{-3}$$

(b) Water content at 1/3 bar or 333 cm (read from SWC) = 7%
(c) Water content at 15 bar or 15,000 cm (read from SWC) = 1%
(d) Available water = $b-c$ = 6%
(e) Water release from 0.5 m soil profile = $0.06^*0.5 = 0.03\ m = 30\ mm$
(f) Volumetric water capacity $(d\theta/dh)$ curve is plotted and shown in following Fig. (b).

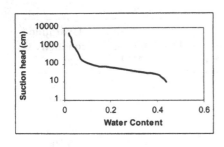

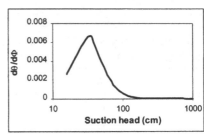

(a)   (b)

Example 13.3

Calculate the unsaturated hydraulic conductivity and diffusivity function for the SWC data given in Example 13.2. Use the VG equation for the above calculation. Assume $\theta_r = 0$ for the above calculations.

Solution

Step1: Optimize the van Genuchten parameters α, n, m, and θ_s for the given soil, using RETC code (van Genuchten et al., 1991)

(a) Assume initial values of α, n, and θ_s and assume $m = 1 - 1/n$.
(b) Use the above data and Eqs. (13.41) and (13.42).
(c) Calculate the value of Φ_m for each θ.
(d) Calculate the sum of squares between predicted and given h values in Example 13.2.
(e) Minimize the sum of squares by using any nonlinear optimization program. (solver subroutine in excel, RETC program of van Genuchten et al., 1991, etc.)
(f) Obtain the final estimates of α, n, and θ_s.

Step 2: Calculate K function using Eq. (13.47) and D function using Eq. (13.48).

The typical K and D functions obtained for the above data are presented in the following figure.

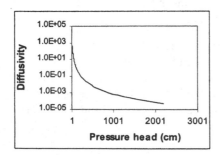

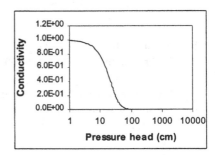

PROBLEMS

1. Use the Φ_m–θ data in example 2 and plot Φ_m–θ curve. Read the data from curve for a small interval so that at least 40 sets of h–θ readings are obtained. If the saturated hydraulic conductivity of the soil is 5.4×10^{-4} cm sec^{-1}, calculate the K functions and plot it with respect to pressure head.

2. Use the same data and optimize parameters of van Genuchten equation [(13.42)–(13.43)]. Assume θ_s as 0.46. Obtain and Plot the K and D functions.

3. Using Eqs. (13.31) and (13.32), calculate the dimensions of K and D.

4. Consider a 50 m long horizontal sandy loam column. Atmospheric pressure is maintained at one end of this column, while –150 cm is maintained at the other. Calculate the direction and magnitude of flux density. If $K(\Phi_m)$ function is given as (a) $K=2$, (b) $K=2+0.01$ (Φ_m), and (c) $K=3\exp(0.05\Phi_m)$ cm h^{-1}. Plot a graph between $K(\Phi_m)$ and Φ_m (x).

REFERENCES

Averjanov S.F. (1950). About permeability of subsurface soils in case of incomplete saturation. pp. 10–21. In English Collection, Vol. 7, (1950), as quoted by P. Ya. Palubarinova, 1962. The theory of Ground Water Movement (English translation by I.M. Roger DeWiest, Princeton University Press).

Brooks R.H. and A.T. Corey (1964). Hydraulic properties of porous media. Hydrology paper 3, Colorado St. Univ., Fort Collins.

Bruce R.R. and A. Klute (1956). The measurement of soil water diffusivity. Soil Sci. Soc. Am. Proc. 20:458–462.

Buckingham E. (1907). Studies on the movement of soil moisture. Bulletin 25, U.S. Department of Agriculture Bureau of soils. Washington, DC.

Burdine N.T. (1953). Relative permeability calculation size distribution data. Pet. Trans. Am. Inst. Min. Metal. Pet. Eng. 198-71-78.

Childs E.C. and G.N. Collis-George (1950). The permeability of porous materials. Proc. Roy. Soc. London, Ser. A. 201:392–405.

Davidson J.M., D.R. Nielsen and J.W. Biggar (1963). The measurement and description of water flow through Columbian silt loam and Hesperia sandy loam. Hilgardia 34:601–617.

Dirksen C. (1975). Determination of soil water diffusivity by sorptivity measurements. Soil. Sci. Soc. Am. Proc. 39:22–27.

Doering E.J. (1965). Soil water diffusivity by the one-step method. Soil Sci. 99:322–326.

Gardner W.R. (1956). Calculation of capillary conductivity from pressure plate outflow data. Soil Sci. Soc. Am. Proc. 20:317–320.

Gardner W.R. (1962). Note on the separation and solution of diffusion type equations. Soil. Sci. Soc. Am. Proc. 33:404

Hillel D. (1980). Applications of soil physics. Academic Press, New York.

Jackson R.D. (1972). On the calculation of hydraulic conductivity. Soil Sci. Soc. Am. Proc. 36:380–382.

Kaplan W. (1984). Advanced Calculus. Addison-Wesley, Reading, MA.

King L.G. (1964). Imbibition of fluids by porous solids. Ph.D. thesis. Colorado State University, Fort Collins.

Klute A. and C. Dirksen (1986). Hydraulic conductivity and diffusivity: laboratory methods. p. 687–734. In: A. Klute (Ed): Methods of soil analysis, Part I Agronomy Monograph Series No. 9 (2nd edition), ASA and SSSA, Madison, Wisconsin.

Kunze R.J., G. Uehara and K. Graham (1968). Factors important in calculation of hydraulic conductivity. Siol Sci. Soc. Am. Proc. 32:760–765.

Marshall T.J. (1958). A relation between hydraulic permeability and size distribution of pores. J. Soil Sci. 9:1–8.

Millington R.J. and J.P. Quirk (1961). Permeability of porous solids. Trans. Faraday Soc. 57:1200–1207.

Moore R.E. (1939). Water conduction from shallow water tables. Hilgardia 12:383–426.

Mualem, Y. (1976). A new model for predicting the hydraulic conductivity of unsaturated porous media. Water Resour. Res. 12:513–522.

Parlange J.Y., D.A. Barry, M.B. Parlange, W.L. Hogarth, R. Haverkamp, P.J. Ross, L. Ling, and T.S. Steenhuis (1997). New approximate analytical technique to solve Richards equation for arbitrary surface boundary condition. Water Resour. Res. 33(4):903–906.

Perroux K.M. and White I. (1988). Designs for Disc Permeameter. Soil Science Society of America Journal. 52, 1205–1215.

Peterson, A.E. and G.D. Bubenzer (1986). Intake rate: sprinkler infiltrometer. In: A. Klute (Ed): Methods of soil analysis, Part I Agronomy Monograph Series No. 9 (2nd edition), ASA and SSSA, Madison, Wisconsin, pp. 845–870.

Philip J.R. (1955). Numerical solution of equations of the diffusive type with diffusivity concentration-dependent. Trans. Faraday Soc. 51:885–892.

Richards L.A. (1931). Capillary conduction of liquids through porous media. Physics 1:318–333.

Rijtema P. E. (1965). An analysis of actual evapotranspiration. Agric. Res. Rep. 659. Center for Agricultural Publications and Documentation, Wageningen, The Netherlands.

van Genuchten M.Th., F.J. Leij, and S.R. Yates, (1991). The RETC code for quantifying the hydraulic functions unsaturated soils. Soil Salinity Lab, USDA, Riverside California.

van Genuchten M.Th. (1980). A closed form equation for predicting the hydraulic conductivity of unsaturated soils. Soil Sci. Soc. Am. J. 44:892–898.

Watson K.K. (1966). An instantaneous profile method for determining the hydraulic conductivity of unsaturated porous materials. Water Resour. Res. 1:557–586.

Whisler F.D., A. Klute and R.J. Millington (1968). Analysis of steady state evapotranspiration from soil column. Soil Sci. Soc. Am. Proc. 32:167–174.

Wind G.P. (1955). Field experiment concerning capillary rise of moisture in heavy clay soil. Neth. J. Agric. Sci. 3:60–69.

Youngs E.G. (1964). An infiltration method of measuring the hydraulic conductivity of unsaturated porous materials. Soil Sci. 97:307–311.

14

Water Infiltration in Soil

Water is one of the most precious natural resources for the survival and advancement of civilization. Depending upon the intensity, a part of the rainfall enters the soil surface and the remaining runs off as an overland flow. An important soil/ecosystem function is to enhance infiltration of precipitation water into the soil with less runoff and erosion. This process is controlled by biological and physical interactions, which create a stable soil structure with enough macropores to rapidly transmit water. Soil that is continually disturbed with tillage and other anthropogenic activities often develops poor structure, leading to surface sealing of pores and crusting, and consequently less infiltration and high runoff. This chapter describes the concept of steady and unsteady state infiltration and different methods of measurement or prediction of infiltration rate and parameters affecting it.

14.1 DEFINITIONS AND BASIC CONCEPT

When water is applied on the soil surface either as irrigation or as rainfall, depending upon the hydraulic conductivity, soil's antecedent moisture content, and soil aggregation, a part or whole of the applied water enters the soil matrix. At the same time the remaining water flows over the soil surface as overland flow. If the soil surface has depressions then some of that water is initially ponded in those depressions and subsequently enters the soil profile (Fig. 14.1). The entry of water into the soil matrix through air–soil interface is called infiltration. The water entry is generally referred to as vertical downward infiltration. The rate of infiltration of water into soil matrix governs the amount of water storage in soil, which is available for plants. It also influences the amount of runoff and erosion. Therefore, knowledge of water infiltration into soil is essential for soil and water conservation, and minimizing the risk of nonpoint source pollution.

How much water can infiltrate the soil? An answer to this question can be obtained by applying water using a sprinkler system. When a sprinkler is run initially at a low discharge, all the water enters or infiltrates into the soil profile. With an increase in discharge of a sprinkler, a stage comes when water either begins to pond on the soil surface or begins to move down a slope as overland flow, also called the surface runoff. The start of runoff or ponding means the rate of water application has exceeded the rate of water infiltration into the soil. This limiting rate of water entry into the profile is known as the soil infiltration capacity, and is defined as maximum rate of infiltration into soil. The infiltration rate is the volume flux of water entering through a unit soil surface area. According to Richards (1952), the term *infiltration rate* is to be preferred over infiltration capacity because capacity refers to the volume rather than to the flow rate.

The infiltration rate of a soil depends on texture, structure, antecedent soil moisture content (i.e., the moisture content of soil profile before rainfall or irrigation begins), continuity and stability of pores, and soil matric potential (Lal and Vandoren, 1990). Soil management including tillage, rotations, residue management, etc., influences these factors. With low

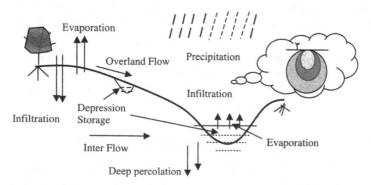

FIGURE 14.1 A schematic of processes during a rainfall or irrigation event along with the components of hydrological cycle.

antecedent moisture content, (e.g., initially dry soil) applied water rapidly enters into the soil matrix. With a continuous supply of water by rainfall or irrigation, the rate of entry of water or infiltration rate decreases over time until it reaches a steady state or a constant rate. The constant rate is also termed as steady state or equilibrium infiltration rate. Figure 14.2 presents a typical time dependent infiltration rate and cumulative infiltration curve.

The high initial infiltration rate in Fig. 14.2a can be explained by large suction gradients. When water is supplied to an initially dry soil, the suction gradients across the soil surface become very high, which results in a high infiltration rate. As the wetting front moves downward, the suction gradient across the soil profile decreases, which limits the rate of water infiltration into the soil surface. Eventually, after a long time, the infiltration rate approaches zero. However, in actual practice, if ponding on soil surface continues for a long time, the infiltration rate gradually becomes steady gravity driven flow and is equal to the saturated hydraulic conductivity (K_s) of the homogeneous rigid soil. The decrease in infiltration rate may also be caused by dispersion of aggregates or slaking, soil compaction and surface sealing, or clogging of soil pores (Shukla et al., 2003a). Slaking is a term used to describe the initial fragmentation of soil aggregates several millimetres in diameter, which may disintegrate further to become micro-aggregates. For details on soil aggregates, readers are referred to Chapter 4.

14.2 MECHANISM OF WATER MOVEMENT AT THE WETTING FRONT

As long as the rate of water application to the soil surface is less than the instantaneous infiltration rate of soil, all the water infiltrates into the soil

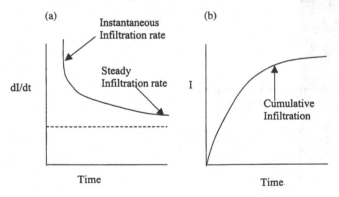

Figure 14.2 Typical (a) time dependent (dI/dt) or rate and (b) cumulative infiltration (I) curves.

profile. Under this circumstance, supply rate determines the infiltration rate and the process of infiltration is called "flux controlled." On the other hand, if water is applied at a rate higher than the instantaneous infiltration rate of soil, the soil water transmission properties determine the rate of actual infiltration and cumulative infiltration. Under this circumstance, the infiltration is called "profile-controlled." According to Childs (1969), the infiltration process is dependent upon both hydraulic conductivity (K_s) and the hydraulic gradient ($\Delta H/L$) of the soil profile. Therefore, the entire soil profile rather than just the surface layer governs the infiltration rate in a profile-controlled process. During infiltration, a clear water divide is often seen during the wetted region overlying the drier region. This distinct sharp boundary between wet and dry regions is known as the "wetting front."

A soil-water profile of a homogeneous soil column under ponded infiltration can be divided into three distinct zones: saturated zone ($\theta = s = 1$), transmission zone ($\theta < \theta s$; $\Delta H/L = 1$), and wetting front (Fig. 14.3). The wetting front is a visible wet/dry soil boundary and may be smooth in a clayey soil and a diffused/fingered in a coarse-textured or non-homogeneous soil. The water movement at the wetting front occurs through a condensation–evaporation process. With the advance of the

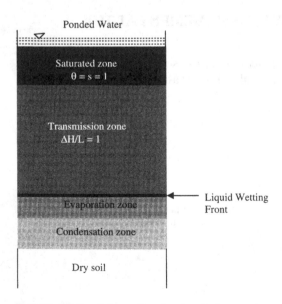

FIGURE 14.3 Schematic of saturated zone, transmission zone, and wetting front advance through a soil profile. (Adapted from Collis-George and Lal, 1971, 1972.)

liquid wetting front, the water vapor moves ahead of the liquid front and condenses. The condensation releases heat of wetting and increases soil temperature in the zone immediately ahead of the liquid wetting front. According to Collis-George and Lal (1972), the average thickness of the zone just ahead of wetting front where condensation occurs varies with soil type. This zone can be identified by the color change due to vapor condensation and the exponential temperature rise from the initial ambient to a maximum (Fig. 14.4). The magnitude of temperature rise depends on soil texture, clay minerals, and the degree of wetness. The rise in temperature is due to the heat of wetting and latent heat of condensation. After attaining a maximum, the temperature declines prior to the arrival of the liquid wetting front. The decline in temperature is due to evaporation of water and the attendent loss of heat representing the latent heat of evaporation (Fig 14.4). Thus, the vapor-wetting front comprises evaporation zone and a condensation zone. There is a critical soil matric potential (Φ_{mc}) at which soil begins to lose water vapors through evaporation. The water movement at the liquid wetting front is not regular in that it remains at one point for a few seconds before it colonizes the granules and pores below (Collis-George and Lal, 1970). The wetting front actually behaves in a quantum-like fashion (Fig. 14.5). The length of each step is time dependent, the life of each step being smaller in the initial phases of infiltration. The discontinuous movement of wetting front is more pronounced in soils with macro- than microaggregates.

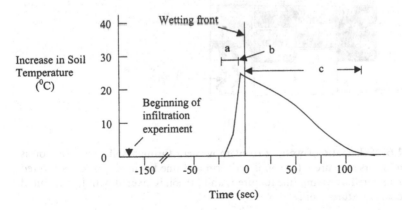

FIGURE 14.4 Zone of vapor condensation and temperature rise (a), zone of temperature decline between temperature maximum and wetting front (b), and zone of decreasing temperature of high to near saturated moisture content (c) for a chernozem soil at 176 cm. The wetting front reached 40 cm at 176 s with velocity of 0.17 cm s^{-1}. (Redrawn from Collis-George and Lal, 1972.)

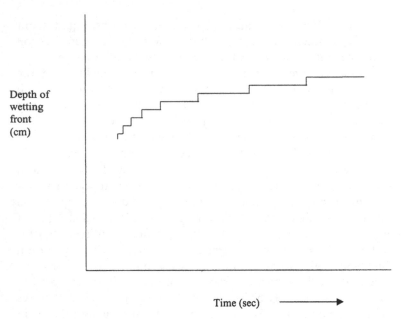

FIGURE 14.5 Schematic of depth of wetting front varying with time for 1.0–2.0 mm size fraction. These measurements were taken from still prints with a magnification of 13 years. (Adapted from Collis-George and Lal, 1970.)

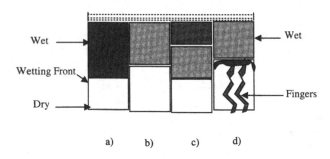

FIGURE 14.6 Schematic of wetting front movement through soils when (a) soil is uniform and coarse textured; (b) soil is uniform and fine textured; (c) soil is layered with coarse textured overlying fine textured; and (d) soil is layered with fine textured overlying coarse textured soil.

The water flow through a homogeneous soil profile, which consists of either coarse-textured soils or fine-textured soils (Fig. 14.6) is very uniform and closely resembles a piston flow. If the coarse-textured soil overlies a fine-textured soil the flow through this layered soil system is again relatively uniform. However, if fine-textured soil overlies a coarse-textured soil, the flow through coarse-textured soil occurs as pipe flow or fingers within the layers. Under these circumstances a large portion of the soil is bypassed. According to Hill and Parlange (1972), the finger formation occurs due to instability at the gravity driven air–water interface. The size of the fingers depends on pore size and therefore, increases with an increase in coarseness of soil texture.

The theory of moisture movement in soil under temperature and potential gradients, which incorporates the interaction of vapor flow with liquid and solid phases, was developed by Philips and De Vries (1957) and DeVries (1958). In chemical reactions the change in enthalpy (a thermodynamic state function) is related to the changes in the free energy and entropy by the Gibbs equation [see Eq. (17.4)]. Entropy is a thermodynamic quantity and is a measure of the degree of disorder within any system. The greater the degree of disorder, the higher is the entropy.

The movement of vapor ahead of the liquid wetting front is demonstrated by observing the temperature fluctuations at the wetting front (Anderson and Linville, 1962; Anderson et al., 1963; Collis-George and Lal, 1970; 1971). The vaporization curves of most liquids have similar shape, and the vapor pressure steadily increases as the temperature increases. The process can be understood by developing a mathematical model for the pressure increase as a function of temperature. Laboratory experiments under controlled conditions have showed that the pressure P, enthalpy of vaporization, ΔH_{vap}, and temperature T are related, and can be represented by the Clausius–Clapeyron equation as follows:

$$P = A \exp\left(-\frac{\Delta H_{vap}}{RT}\right) \tag{14.1}$$

where R ($=8.3145$ J/mol/K) is the gas constant and A is an unknown constant. The Clausius–Clapeyron equation can be used to estimate the vapor pressure at another temperature, if the vapor pressure is known at some temperature, and if the ΔH_{vap} is known. Collis-George and Lal (1970; 1971; 1972) suggested that for unstable slaking soils the heat of wetting and its rate of production controls the infiltration rate. The water potential at known moisture content, in appropriate units relative to a free water surface, is the partial specific free energy ($\Delta \bar{G}_w$) of the soil water at that moisture content. Therefore, using the Clausius–Clapeyron equation a plot of $\Delta \bar{G}_w/T$ versus $1/T$, where T is the absolute temperature, produces a straight line of slope, ΔH_w, the partial specific enthalpy at that moisture content, and with intercept equal to the partial specific entropy (ΔS_w).

TABLE 14.1 Thermodynamic Description of an Oven-Dried (o.d.) Chernozemic Soil

	$\Delta \bar{G}_w$						
	$(cal\,g^{-1}\,(H_2O)\,g^{-1}\,(o.d.\,soil))$						
$w\,(g\,g^{-1})$	20°C	25°C	30°C	35°C	ΔH_w	ΔS_w	$\Delta H\,(cal\,g^{-1}\,soil)$
0.02	−95.2	−91.3	−87.0	−82.4	−324	−0.8	−25.3
0.10	−13.7	−12.3	−11.1	−11.1	−92	−0.27	−3.32
0.20	−0.52	−0.45	−0.34	−0.34	−6	−0.019	−0.12

Source: Adapted from Collis-George and Lal, 1972.

The thermodynamic description of a chernozemic soil used by Collis-George and Lal (1972) in laboratory infiltration experiments is given in Table 14.1. It shows that as the moisture content of soil increases the attendant $\Delta \bar{G}_w$ (at 20°, 25°, 30°, and 35° C), ΔH_w, ΔS_w and integral enthalpy or heat of wetting (ΔH) also increase.

14.3 CONCEPTUAL INFILTRATION MODELS

There are several models to predict the infiltration rate into soil. Some of these models are conceptual and based on basic processes, while others are empirical or statistical (Shukla et al., 2003b).

14.3.1 Conceptual Models

There are two principal conceptual models:

Green and Ampt Model

Green and Ampt (1911) made several simplifying physical assumptions to develop a mathematical relationship for predicting the infiltration rate (Kutilek and Nielsen, 1994). The soil-water profile during infiltration was assumed to be a steplike function and infiltration into the soil was assumed to be piston flow going progressively deeper with time. The wetted soil profile at time t was replaced with a uniformly wetted region of thickness L. The soil profile is considered homogeneous and isotropic; therefore, saturated hydraulic conductivity (K_s) is a unique parameter. Darcy's law can be used to predict the infiltration rate as follows:

$$q \quad or \quad i = K_s \frac{\Delta H}{L} \tag{14.2}$$

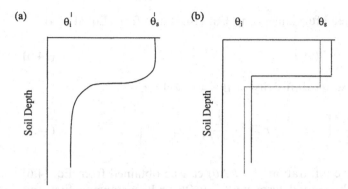

(a) θ_i θ_s (b) θ_i θ_s

Soil Depth   Soil Depth

FIGURE 14.7 Infiltration through (a) a natural homogeneous soil (b) Green–Ampt approximation.

where K_s is the saturated hydraulic conductivity (LT^{-1}) of wetted region $0 < x < L$, ΔH is the hydraulic head difference between wetting front (W_f) and soil surface (W_0). Figure 14.7 explains the natural soil water profile wetting and the corresponding Green–Ampt profile.

The infiltration will increase the moisture content of the soil profile. Assuming a uniformly wetted zone up to the wetting front with cumulative infiltration (I) equal to the product of wetting front depth (L) and increase in moisture content ($\Delta\theta$). Similarly, infiltration rate equals the rate of change in water storage per unit time:

$$i = \frac{dI}{dt} = \frac{d}{dt}[(\theta_s - \theta_i)L] \tag{14.3}$$

where I is the total or cumulative infiltration within a selected time interval, θ_s and θ_i are the saturated moisture content and initial moisture content, respectively ($\Delta\theta = \theta_s - \theta_i$). Equating the right-hand side of Eqs. (14.2) and (14.3):

$$\Delta\theta \frac{dL}{dt} = K_s \frac{\Delta H}{L} \tag{14.4}$$

After rearranging the equation so that factors dependent on L and time are on one side, Eq. (14.4) can be integrated to the limits of 0 to L and 0 to t as follows

$$\int_0^L L\,dL = K_s \frac{\Delta H}{\Delta\theta} \int_0^t dt \tag{14.5}$$

Eq. (14.6) is obtained after integrating Eq. (14.5) and from Eq. (13.24)

$$\frac{L^2}{2} = K_s \frac{\Delta H}{\Delta \theta} t = D(\theta)_0 t \tag{14.6}$$

where $D(\theta)_0$ is soil-water diffusivity of the wet soil region

$$D(\theta)_0 = K_s \frac{\Delta H}{\Delta \theta} \left[L^2 T^{-1} = LT^{-1} \frac{L}{L^3/L^3} = L^2 T^{-1} \right] \tag{14.7}$$

The cumulative infiltration ($I = L\Delta\theta$) can be obtained from Eq. (4.6) after multiplying the second term with $\Delta\theta/D\theta$ and rearranging first and second terms.

$$I = \Delta\theta(2D(\theta)_0 t)^{0.5} \tag{14.8}$$

and the infiltration rate

$$i = \frac{dI}{dt} = \Delta\theta \left(\frac{D(\theta)_0}{2t}\right)^{0.5} \left[\frac{L}{T} = \frac{L^3}{L^3}\left(\frac{L^2 T^{-1}}{T}\right) = \frac{L}{T}\right] \tag{14.9}$$

The infiltration rate of soil is proportional to $t^{-0.5}$. Once the gravity component of total hydraulic head is taken into consideration, or in other words, for vertical infiltration, Eqs. (14.2) and (14.3) can be written as

$$q \quad \text{or} \quad i = K_s \left(\frac{\Delta H + L}{L}\right) \tag{14.10}$$

$$i = \Delta\theta \frac{dL}{dt} \tag{14.11}$$

equating the right-hand side of Eqs. (14.10) and (14.11) and after rearrangement of terms and applying integration limits

$$\int_0^L \frac{L\,dL}{\Delta H + L} = \frac{K_s}{\Delta\theta} \int_0^t dt \tag{14.12}$$

$$\frac{K_s t}{\Delta\theta} = L - \Delta H \ln\left(1 + \frac{L}{\Delta H}\right) \tag{14.13}$$

As t increases, the second term in Eq. (14.13) increases much more slowly as compared to the first term. So at a very large time from the start of the infiltration experiment, the Eq. (14.13) can be approximated by

$$L \cong \frac{K_s t}{\Delta \theta} + b \tag{14.14}$$

$$I \cong K_s t + b \tag{14.15}$$

and in terms of infiltration rate

$$i \approx K_s \tag{14.16}$$

where b is a constant. The Green and Ampt model uses an approximate description of actual flow regimes. It also requires the value of effective wetting from suction (Φ_{mf}). According to Green and Ampt (1911) and Hillel and Gardner (1970), the value of Φ_{mf} for initially dry soil may be of the order of 50–100 cm. However, in actual field conditions, where initial moisture content of soil profile is not uniform, Φ_{mf} may be difficult to define. One of the alternatives is the indirect evaluation of Φ_{mf} (Chong et al., 1982). According to Philip (1966), if diffusivity is assumed to be concentrated at the wet end of the moisture range, the Green–Ampt equation corresponds to the nonlinear diffusion description of infiltration. However, real soils do not manifest a delta function $D(\theta)$. A common form of Green-Ampt equation with $i = dI/dt$ and $I = \int_0^t i \, dt$ is given below:

$$i = K_s + \frac{b}{I} \Leftrightarrow i = i_c + \frac{b}{I} \left(LT^{-1} = LT^{-1} + \frac{b}{L} \Rightarrow b = L^2 T^{-1} \right) \tag{14.17}$$

Philip's Model

The Philip algebraic infiltration equation uses mathematical approximation for the infiltration process in a soil matrix. There are two forms of the Philip model: horizontal and vertical infiltration.

For horizontal infiltration, Philip (1957) used the Richards equation with $K(\theta)$ (second term on right-hand side) removed and showed that infiltration rate can be expressed as

$$i(t) = \frac{1}{2} S t^{-0.5} \tag{14.18}$$

where S is soil-water sorptivity and $S=f(\theta_i, \theta_o)$. θ_o and θ_i are boundary and initial moisture content. Since S is constant over time, the cumulative infiltration

$$I(t) = S\, t^{0.5} \tag{14.19}$$

The sorptivity, which will be more important during the initial period, can be defined as

$$S = \frac{I}{t^{0.5}} \left[S = \frac{L}{T^{0.5}} = LT^{-0.5} \right] \tag{14.20}$$

For vertical infiltration, the Philip equation consists of two separate parts: one for a short duration and the other part for long times after infiltration commenced. The solution describes the time dependence of cumulative infiltration by means of an infinite power series.

$$I(t) = \sum_{n=1}^{\infty} j_n(\theta) t^{n/2} \tag{14.21}$$

where coefficient $j_n(\theta)$ are calculated from $K(\theta)$ and $D(\theta)$, the coefficients are termed as sorptivity (S). Transferring S into Eq. (14.21) and expanding

$$I(t) = S\, t^{0.5} + (A_2 + K(\theta)_0)t + A_3 t^{3/2} + A_4 t^2 + \cdots + A_n t^{n/2} \tag{14.22}$$

Differentiating equation (14.22) with respect to time gives the solution for infiltration rate.

$$i(t) = \frac{1}{2} S\, t^{-0.5} + (A_2 + K(\theta)_0) + \frac{3}{2} A_3 t^{1/2} + 2A_4 t + \cdots + \frac{n}{2} A_n t^{n/2-1} \tag{14.23}$$

In practice the Eqs. (14.22) and (14.23) can be approximated by two parameter models as follows

$$I(t) = S\, t^{0.5} + A\, t \tag{14.24}$$

$$i(t) = \frac{1}{2} S\, t^{-0.5} + A \tag{14.25}$$

where A is a constant known as soil-water transmissivity and it approaches K_s of the soil profile as $t \to \infty$. It has the dimensions of LT^{-1}. According to Philip (1969), as t approaches infinity, the infiltration rate decreases to a finally asymptotic value and the coefficient A can be replaced with saturated hydraulic conductivity of upper layer (K_s)

$$I(t) = St^{0.5} + K_s t \tag{14.26}$$

$$i(t) = \frac{1}{2}St^{-0.5} + K_s \tag{14.27}$$

For very large time

$$i(t) = K_s \tag{14.28}$$

The magnitude of A in equation (14.25) is $(A_1 + K_o(\theta) + \varepsilon)$, where $K_o(\theta)$ is the hydraulic conductivity and ε is the truncation error (Kutilek and Nielson, 1994). The value of S estimated from Eq. (14.27) is quite reliable, however, truncation errors influence the estimated value of A. To overcome this problem Kutilek and Crejka (1987) proposed to use first three terms of the Philip's series solution [Eq. (14.23)] as follows

$$I = St^{1/2} + C_1 t + C_2 t^{1\frac{1}{2}} \tag{14.29}$$

$$i = \frac{1}{2}St^{-1/2} + C_1 + C_2 t^{1/2} \tag{14.30}$$

where C_1 is the estimate of $(A_2 + K_i)$ and C_2 the value of $(A_3 + \varepsilon_1)$ where ε_1 is truncation error for having used three terms. Another model proposed by Swartzendruber (1987) uses the adjusted Philip time series solution to derive the following infiltration models

$$I = \frac{S}{A_0}[1 - \exp(-A_0 t^{1/2})] + K_s t \tag{14.31}$$

$$i = \frac{S}{2}t^{-0.5}\exp(-A_0 t^{1/2}) + K_s \tag{14.32}$$

where A_0 is a constant. Equations (14.26), (14.27), (14.29), (14.30), (14.31), and (14.32) also provide an initial infinite infiltration at $t=0$ and a constant equilibrium infiltration rate at large t. Substituting $4K_s/3S$ for A_0 into Eqs. (14.31) and (14.32) results in two-parameter Stroosnijder (1976) infiltration model. Using the horizontal solution of Philip (1957), Brutsaert (1977) added a correction for force of gravity to arrive at following infiltration equations

$$I = K_s t + \frac{S^2}{BK_s}\left[1 - \frac{1}{\{1 + (BK_s t^{1/2})/S\}} \right] \tag{14.33}$$

$$i = K_s + \frac{1}{2}St^{-1/2}\left[\frac{1}{\{1 + (BK_s t^{1/2})/S\}^2} \right] \tag{14.34}$$

Brutsaert (1977) proposed B values to be 1/3, 2/3, or 1, but for most practical purposes recommended $B=1$ (Kutilek and Nielsen, 1991). All these models are time dependent and provide an infinite initial infiltration at $t=\emptyset$ and a finite steady state infiltration at large t.

14.3.2 Empirical Models

There are three principal empirical models.

Kostiakov Model

The Kostiakov (1932) equation for cumulative infiltration (I) and infiltration rate (i) can be expressed as follows

$$I = Bt^{-n} \tag{14.35}$$

$$i = B't^{-n-1} \tag{14.36}$$

where parameters B, B', and n are constants. These parameters do not have a physical meaning and can be obtained by fitting the equation to the experimental data. It can be inferred from Eq. (14.35) that at $t=0$, I approaches ∞. However, as t increases further, i tend to become zero. Therefore, Eqs. (14.35) and (14.36) explain the horizontal infiltration. However, for vertical infiltration, the Kostiakov (1932) equation is inadequate. To overcome the problem, Kostiakov proposed a maximum time range of application with $t_{max} = (B/K_s)^{1/n}$ and Mezencev (1948) included another coefficient, i_c, which essentially shifts the axis for infiltration rate equations and for large times infiltration approaches a finite steady state infiltration rate.

$$I = i_c t + Bt^{-n} \tag{14.37}$$

$$i = i_c + B't^{-n-1} \tag{14.38}$$

Horton Model

Horton (1940) equations for cumulative infiltration (I) and infiltration rate (i) are:

$$I = i_c t + \frac{i_0 - i_c}{k}\left[1 - e^{-kt}\right] \tag{14.39}$$

$$i = i_c + (i_0 - i_c)e^{-kt} \tag{14.40}$$

where i_0, is initial infiltration rate at $t = 0$, and i_c is final constant infiltration rate after a long time from the start of infiltration. The constant "k" determines the rate at which i_c approaches i_0. Unlike other equations, the Horton equation has a finite infiltration rate, i_0 at $t = 0$. The equation is somewhat cumbersome as it has three constants, which need be evaluated experimentally. According to Horton (1940), the decrease in infiltration rate with time can be described by a number of factors, such as, the closure of soil pores by a swelling soil or erosional deposit, compaction due to raindrop impact, etc.

Holtan Model

The Holtan equation (1961) for infiltration rate in soil matrix is again a two-form mathematical equation as given below

$$i = i_c + af_a' \tag{14.41}$$

where i_c is the final constant rate of infiltration and f_a' is the available porosity as depleted by infiltration volumes, which can be expressed as

$$f_a' = M - I \tag{14.42}$$

where M is the moisture storage capacity of the soil above the first impeding stratum or control layer. It can also be expressed as total porosity—antecedent moisture content of soil, in depth units. I is the cumulative infiltration at that time.

$$i = i_c \quad \text{for} \quad I > M \tag{14.43}$$

As long as $0 \leq I \leq M$, the Eq. (14.40) is consistent, however, for $I > M$, the $(M-I)^n$ becomes positive or negative depending upon the value of exponent n. Also in absence of an impeding layer, Holtan (1961) did not discuss the meaning of M. Huggings and Monk (1967) reported that effective depth is a function of land use and soil management. According to Holtan and Creitz (1967), the control depth could be the depth down to B-horizon and the parameter n could be assumed constant equal to 1.4 for all soils. The parameters of empirical infiltration equations are time dependent. The Horton (1940) and Holtan (1961) provide a finite infiltration rate both at $t = 0$ and $t = \infty$. A comparison of all the infiltration models discussed above is given in Table 14.2 and can also be referred in Davidson and Selim (1986); Haverkamp et al. (1988) and Shukla et al. (2003b) among others.

14.4 INFILTRATION INTO LAYERED PROFILE

The infiltration process occurs in natural soils, which are neither uniform in texture nor homogeneous. They comprise several layers or horizons of different texture, bulk density, and moisture content extending up to variable depths. This nonhomogeneity of soil profile has a pronounced effect on the water infiltration process. Infiltration primarily depends on the relationship between hydraulic conductivity $[K(\theta)]$ and gradients of suction (Φ_m) in each layer. According to Miller and Gardner (1962), metric suction (Φ_m) and hydraulic head (H) remain continuous throughout the soil profile regardless of the sequence of layers. However, soil's moisture content (θ) and hydraulic conductivities $[K(\theta)]$ at the soil layer interface exhibit sharp discontinuity (Fig. 14.8). For a two-layer soil system, Takagi (1960) demonstrated that if an upper soil layer is less pervious than a lower, a continuous negative pressure develops in the lower profile, which extends deep inside the soil matrix (Fig. 14.8a).

The layering, as shown in Fig. 14.8, can also cause the instability of wetting front. At the interface, pressure head is too small in the upper layer to force the entry of water into coarser pores of the impeding layer. As θ increases in the upper layer, the pressure head also increases and water enters the impeding soil via finer pores. Since the supply of water from

TABLE 14.2 Comparison of Various Infiltration Equations

Parameter	Green–Ampt (1911)	Philip (1957)	Kostiakov (1932)	Horton (1940)	Holtan (1961)
Number	2	2	2	3	4
Theory	Physically based	Physically based	Empirical	Empirical	Empirical
Surface ponding	Required	Required	Not required	Not required	Not required
Initial infiltration	Infinite	Infinite	Infinite	Finite	Finite
At large time	Steady infiltration = hydraulic conductivity (K_s)	Steady infiltration = hydraulic conductivity (K_s)	Zero	Steady infiltration = hydraulic conductivity (K_s)	Steady infiltration = hydraulic conductivity (K_s)

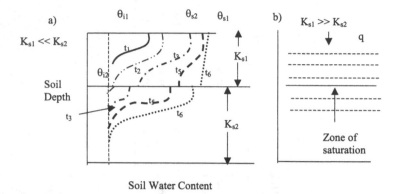

Soil Water Content

FIGURE 14.8 (a) The discontinuity of water content across a layer when conductivity of top layer was smaller than the lower layer (based on Takagi, 1960; Kutilek and Nielsen, 1994); (b) zone of saturation when conductivity of top is higher than the impeding layer.

topsoil is limited by the K_s for that soil, water conduction in the impeding layer is mainly through preferential domains. This type of flow is also termed as "finger flow" and points out towards the hydrodynamic instabilities or inabilities to describe infiltration through layered soil profile (also refer Fig. 14.6d).

If we reverse the layers in Fig. 14.8, topsoil is more pervious than the impeding soil (Fig. 14.8b). A zone of $\Phi_m > 0$, is formed on top of the impeding low conductivity layer. After a large time since infiltration began, a zone of saturation as shown in Fig. 14.6b is formed on either side of the layer. This temporary water table development is also termed the perched water table. In dealing with nonhomogeneous soil profiles it is usually convenient to divide the profile into homogeneous layers. The simplest example is a crust-topped profile, which we will discuss now.

14.5 INFILTRATION INTO CRUSTED SOILS

A surface crust is created on the soil surface mainly by the impact of raindrops. This can also take place due to the spontaneous slaking or breakdown of soil aggregates during rapid wetting. The impact of raindrops may be partly reduced if the soil surface has mulch or other vegetative cover on it. These processes are discussed in Chapter 6. A surface crust is normally characterized by a relatively less porosity and higher bulk density. Consequently, crust has a lower hydraulic conductivity than the soil strata beneath. The crust can also be formed if the infiltrating solution has a high sodium adsorption ratio (SAR) by inducing dispersion and swelling of clay particles. The dispersion of aggregates separates the finer particles from the macroaggregate, which then occupy the interaggregate space in the soil domain, thus reducing the hydraulic conductivity of the crust layer. At steady infiltration rate, flux through the crust equals flux through the impeding soil. These two fluxes can be equated using Darcy's law:

$$K_C \left(\frac{dH}{dz}\right)_C = K_I \left(\frac{dH}{dz}\right)_I \tag{14.44}$$

where K_C and K_I are the conductivity of crust and impeding layer respectively, and $K_c = L_c / R_c$, L_c is the thickness and R_c the resistance of crust. The second term on either side defines the gradient for both layers, respectively, with H as total hydraulic head. As steady infiltration is approached, the gradient in the impeding layer or subcrust zone tends to become unity and the gravitational gradient is the only effective driving gradient. So flux through subcrust soil can be written as

$$q = K(\theta)_I (\Phi m_I) \tag{14.45}$$

where $K(\theta)$ is the unsaturated hydraulic conductivity of subcrust zone, which is also a function of matric potential head in the subcrust zone. If the crust remains saturated the flux through the crust can be expressed as

$$q = K_c \left(\frac{H_0 + \Phi_{mc} + \Phi_{zc}}{z_c}\right) \tag{14.46}$$

where H_0 is the depth of ponding on crust surface, Φ_{mc} is suction head, and z_c is the depth or thickness of crust. If depth of ponding and crust thickness are very small and can be neglected then

$$q_I = q_c = K_c \frac{\Phi_{mc}}{z_c} \tag{14.47}$$

In terms of hydraulic resistance, the Eq. (14.46) can be written as follows provided the hydraulic gradient of the impeding layer is unity.

$$\frac{K_I}{\Phi_{mI}} = \frac{K_C}{z_C} = \frac{1}{R_C} \tag{14.48}$$

14.6 PRECIPITATION INFILTRATION

The infiltration models assume that soil surface is held at a fixed potential (Φ_m) rather than at a fixed rate of water flux. When rainfall is the source of water, the fixed potential condition occurs if rainfall intensity exceeds the soil infiltration rate. As a result, ponding or positive potentials occurs on the soil surface. Ponding does not occur immediately, because the initial infiltration into the soil is very rapid. However, if rainfall continues at a constant rate, the infiltration rate decreases and the rainfall rate exceeds infiltration rate, thus creating ponding on soil surface.

Many researchers have studied the infiltration process under rainfall or sprinkler irrigation [Youngs (1964), Rubin and Steinhardt (1964), Rubin (1966, 1968), Parlange and Smith (1976), Boulier et al. (1987)]. Rubin (1968) carried out a series of rainfall-infiltration tests under different surface boundary conditions. Under ponding conditions the wetted profile consists of two parts, a water-saturated part and the water-unsaturated part. The depth of the saturated zone and the steepness of infiltration curve increase continuously downward. However, after a long time the steepness gradually decreases and tends to be asymptotic to the time axis (Fig. 14.9). The infiltration curves when surface ponding does not occur remain parallel to time axis (Fig. 14.9) till the rainfall or sprinkler rate exceeds the infiltration rate. The infiltration curve becomes steep and takes the form of a ponded infiltration curve. From Fig. 14.9, two inferences can be made: (i) the

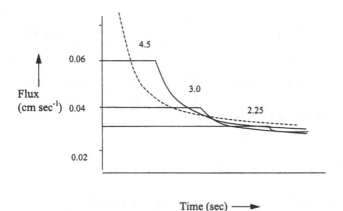

FIGURE 14.9 Relation between surface flux and time during infiltration into Rehovot sand due to rainfall (solid line) and flooding (dashed line). The numbers labeling the curve indicate the ratio of the rainfall rate to the saturated hydraulic conductivity. (Redrawn from Rubin, 1968.)

constant rate of infiltration limited by rainfall or infiltration occurs as long as the ratio of rainfall rate to hydraulic conductivity is less than one; (ii) the shapes of ponded infiltration curve and the rain limited curves are similar and attain same limiting infiltration rate, however, they do not coincide.

14.7 UNSTEADY INFILTRATION

There are numerous situations under field conditions when the infiltration rate is unsteady, transient, or instantaneous. It never attains a steady state condition.

14.7.1 Dirichlet's Boundary Condition

The soil surface remains saturated ($\theta = s = 1$) as long as water is ponded on it. If the depth of ponding is small ($d \to 0$), only the surface soil remains saturated. Such a situation is defined as Dirichlet's boundary condition (DBC) for infiltration into semi-infinite homogeneous soil (Kutilek and Nielsen, 1994). Mathematically DBC can be represented by Eqs. (14.49) to (14.51):

$$t \geq 0 \quad z = 0 \quad \theta = \theta_s \tag{14.49}$$

or

$$t \geq 0 \quad z = 0 \quad \Phi_m = \Phi_{m0} \tag{14.50}$$

$$t \geq 0 \quad z = 0 \quad \Phi_m = 0 \tag{14.51}$$

and

$$t = 0 \quad z > 0 \quad \theta = \theta_i \tag{14.52}$$

The boundary condition specified by Eq. (14.49) is employed when the diffusivity form of Richards' equation is used [see Eq. (13.23)]. The boundary conditions (14.50) and (14.51) are used with capacitance form of Richards' equation [see Eq. (13.21)]. The condition (14.52) is the initial condition. Equations (14.50) and (14.51) have the advantage that it specifies the depth of ponding of water on soil surface. DBC is applicable on the infiltration data obtained through double ring infiltrometer. The infiltration rate by a double ring infiltrometer is measured either by measuring the fall of water in the inner ring or the fall of water in a Marriotte bottle used to maintain a constant head in the inner ring. The role of outer ring is limited to keep the divergence of flow paths in inner cylinder to a minimum.

14.7.2 Neumann's Boundary Condition

While describing the rainfall distribution, it is assumed that rainfall is evenly distributed on the entire soil surface and the rainfall flux density passes either partially or fully through the soil surface. The boundary condition for such a situation is described by Darcy–Buckingham equations [Eqs. (14.53) and (14.54)]

$$q_0 = -K(\theta)\frac{\partial H}{\partial z} \quad z = 0 \quad t \geq 0 \tag{14.53}$$

$$q_0 = K(\theta) - D(\theta)\frac{\partial \theta}{\partial z} \quad z = 0 \quad t \geq 0 \tag{14.54}$$

Equation (14.54), which is also the diffusivity form of the Darcy–Buckingham equation, is known as Neumann's boundary condition (NBC), and is used to describe the rainfall or sprinkler infiltration. For these experiments regardless of how NBC is achieved the initial condition is kept the same as that for DBC (see Sec. 14.71).

14.8 SOIL WATER REDISTRIBUTION

After the cessation of infiltration, it is commonly observed that moisture content of the topsoil gradually decreases, even when the topsoil is covered and evaporation is either nonexistent or insignificant. Downward movement of water in the soil profile accompanies the decrease in moisture content of topsoil. In the absence of a water table, this phenomenon, where wetted topsoil wets the drier subsoil, is known as "soil moisture redistribution". It is defined as the continued movement of water through a soil profile. However, in case a shallow water table exists, the phenomenon is known as "drainage to groundwater".

Water distribution is a complex process, as topsoil loses water and becomes drier while subsoil gets wetter. Therefore, soil-moisture hysteresis (refer to Chapter 11) may have important influence on the shape and dynamics of moisture content profile.

Figure 14.10 shows the schematic of a typical soil moisture redistribution curve after an irrigation or rainfall event. Once the rainfall or irrigation is stopped, the soil moisture potential from surface up to the wetting front is close to zero ($\theta = s = 1$). Immediately below the wetting front the value of soil moisture potential is very low, thus a large gradient exists between wetting front and dry soil. Therefore, the waterfront starts moving downward and the moisture content of the surface profile gradually

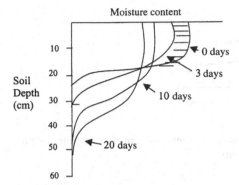

FIGURE 14.10 Schematic of water content redistribution cycles following a rainfall or irrigation event.

decreases. There is no abrupt change in moisture content of top profile. As surface drying proceeds, the waterfront gradually moves downwards. The waterfront creates a sharp continuous boundary between the wet region and the dry soil. The shape of the water profile is very close to being called a rectangle, with a height equal to the thickness of wetted profile and width equal to the difference in moisture content between wet and dry regions. As the moisture content is fairly uniform within the wetting front, the flow can be described as gravity-driven flow.

As moisture content of wetted region continuously decreases and moisture content of drier region increases, the soil profile tends to move towards equilibrium. The rate of decrease of moisture content depends upon the depth of the original wetting front, gradient, and hydraulic conductivity of soil. As soil-moisture potential decreases the gradient also decreases and hydraulic conductivity also decreases simultaneously. The schematic of the waterfront positions after 3 days, 10 days, and 24 days are shown in Fig. 14.10.

14.9   FIELD WATER CAPACITY

The moisture content at which the internal drainage completely ceases is known as the field moisture capacity of soil (refer to Chapters 10 and 11). The field moisture capacity is an arbitrary and not an intrinsic physical property of a soil. A working definition of field capacity is the moisture content two days after infiltration or rainfall event. The field capacity concept can be easily applied to coarse textured soils where an initially high infiltration and redistribution slows down considerably owing to the large decrease in hydraulic conductivity than in fine textured soils, where slow but

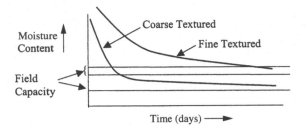

FIGURE 14.11 Schematic of decrease in water content with respect to time for an initially wetted sail profile during redistribution.

appreciable amount of water movement can persist for a much longer duration (Fig. 14.11). Field capacity of soil is a function of soil texture, type, and amount of clay content, organic matter content, antecedent soil moisture status, evapotranspiration, depth of impervious layer, and depth of wetting of soil profile. Richards et al. (1956) proposed the following equation for calculating the decrease of moisture content with respect to time [Eq. (14.55)]

$$-\frac{\mathrm{d}W}{\mathrm{d}t} = at^{-b} \tag{14.55}$$

where W is the moisture content of soil profile, t is time, and a and b are constants related to the conductance properties of soil and boundary conditions, b is also related to soil diffusivity.

14.10 MEASUREMENT OF INFILTRATION

Under laboratory conditions, infiltration is measured using soil columns with the Mariotte bottle technique (Fig. 14.12). The infiltration may be measured in either vertical or horizontal columns. The cumulative infiltration is expressed in units of cm (volume of water read from a Mariotte bottle divided by the cross section area of soil column). Infiltration rate under field conditions can be measured by a double ring infiltrometer under positive pressure head (Fig. 14.13). The infiltration rate under tension can be measured by a tension infiltrometer. Details on various infiltrometer, are presented by Perroux and White (1988), Everts and Kanwar (1992). If the soil has a low infiltration rate, the change in height of water can also be measured using a water stage recorder (Fig. 14.14). Under field conditions, infiltration rate may also be measured using a rainfall simulator (Fig. 14.15). Different types of rainfall simulators are explained by Meyer (1994).

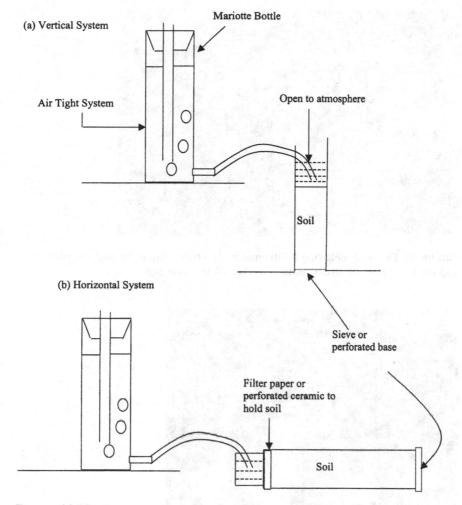

FIGURE 14.12 Measurement of infiltration in soil columns using a Marriotte bottle technique.

14.11 MANAGEMENT OF INFILTRATION

The infiltration into a soil profile is a function of structural and textural properties, and soil moisture of a soil. The land use and soil management practices have a profound influence on the infiltration of water into the soil profile (Lal and Vandoren, 1990; Shaver et al., 2002; Shukla et al., 2003a;b). Tillage practices (Fig. 14.16), which alter soil structure and increase porosity of the upper layer, enhance the initial infiltration into

FIGURE 14.13 A double ring infiltrometer. The outer ring is 30–50 cm in diameter and the inner ring 20–30 cm. Both rings are 20 to 30 cm high.

FIGURE 14.14 The change in height of water in a column used to assess infiltration can also be measured using a water stage recorder.

FIGURE 14.15 Rainfall simulators are commonly used to assess infiltration rate under nonponded condition.

FIGURE 14.16 Para plow can alleviate compaction without causing inversion.

FIGURE 14.17   Residue mulch and no-till farming enhance infiltration rate in soil.

FIGURE 14.18   Furrow irrigation increases water infiltration into the root zone.

the soil. However, subsoil compaction often results in a net decrease in total infiltration. The mulching affects both physical and chemical properties of soil. Mulching controls soil erosion by avoiding the direct impact of a raindrop on soil surface, decreasing the runoff rate and increasing infiltration of rainwater (Fig. 14.17). Furrow irrigation is practiced to enhance water infiltration in the root zone (Fig. 14.18).The no-till farming practices with mulching also increase the infiltration rate of soil. The soil remains undisturbed with respect to the farming operations and this results in enhanced worm activities in the soil (Butt et al., 1999). No-till also improves soil aggregation, which increases the total porosity of soil. All these factors not only result in more aeration but also in high infiltration rate of irrigation or rainwater through larger pores around aggregates and macropores or biopores formed by earthworms.

Example 14.1

The infiltration rate was measured in a no-till silt loam soil with corn and soybean rotation as given in the following table. The data in first three columns are given. Calculate the (a) depth of actual infiltration, (b) cumulative infiltration, (c) infiltration rate, (d) parameters of the Green and Ampt equation (14.17) and the Philip equation [(14.26) and (14.27)].

Solutions

(a)   The actual depth of infiltration is given in column (5)
(b)   The cumulative depth of infiltration is given in column (6)
(c)   Infiltration rate is given in column (9)

T (min)	Reading in infiltrometer (cm)	Water filled upto (cm)	dT (min)	Infiltration I (cm)	Cumulative infiltration CI (cm)	CI_{-1}	$t_{(-1/2)}$	dI/dt (cm/min)
1	2	3	4	5	6	7	8	9
0	1		0	0	0		0	0
2	3		2	2	2	0.500	0.707	1.000
5	5.2	1	3	2.2	4.2	0.238	0.447	0.733
10	4		5	3	7.2	0.139	0.316	0.600
15	6	1	5	2	9.2	0.109	0.258	0.400
25	5.2	1	10	4.2	13.4	0.075	0.200	0.420
35	4.8	1	10	3.8	17.2	0.058	0.169	0.380
50	5.4	1	15	4.4	21.6	0.046	0.141	0.293
65	5.5	1	15	4.5	26.1	0.038	0.124	0.300
85	5.6	1	20	4.6	30.7	0.033	0.108	0.230
115	6.8	1	30	5.8	36.5	0.027	0.093	0.193
145	6.5	1	30	5.5	42	0.024	0.083	0.183
180	7		35	6	48	0.021	0.075	0.171

(d) The parameters of the Philip equation are obtained by plotting column (8) and (9) and fitting a linear relationship:

$$S = 2.66; \text{ and } A = 0.11$$

(e) The parameters of the Green–Ampt equation are obtained by plotting column (7) and (9) and fitting a linear regression line

$$i_c = 0.22; b = 1.74$$

Example 14.2

Use the same data as given in Example 14.1 and calculate the parameters of the Kostiakov and Horton infiltration equations.

Solutions

The parameters of the Kostiakov equation: $B = 1.387$ and $n = 0.397$
The parameters of the Horton equation: $i_c = 0.239 \, \text{cm min}^{-1}$, $i_0 = 1.068 \, \text{cm min}^{-1}$, and $k = 0.08$

PROBLEMS

1. Consider the following data from the double ring infiltrometer test conducted in plow till vs. no-till experiments over a 3-hr period. Fit Green and Ampt, Philip, Kostiakov, Mezencev, Horton, and Holtan models to these data, and compute the parameters of these equations.

Time (min)	Cumulative infiltration (cm)		Time (min)	Cumulative Infiltration (cm)	
	Plow till	No-till		Plow till	No-till
0	0	0	100	64.6	137.1
10	9.2	21.6	110	68.5	148.8
20	19.6	39.2	120	74.4	158.6
30	27.4	54.8	130	78.3	166.4
40	32.2	68.5	140	82.2	178.2
50	39.2	82.2	150	86.2	186.0
60	45.0	94.0	160	90.1	193.8
70	49	105.7	170	94.0	201.7
80	54.8	115.5	180	97.9	209.5
90	58.7	127.3			

2. Horizontal infiltration test was conducted in a tube of $50\,\text{cm}^2$ cross-sectional are. After 10 minutes, the cumulative infiltration was 1.2 ml. Calculate the expected cumulative infiltration and infiltration rate after 30 min, 1 hr, 3 hrs, and 6 hrs. Use Philip equation and calculate S for the given time intervals.

3. Infiltration rate monitored as a function of cumulative rainfall was found to be $25\,\text{mm h}^{-1}$ when a total of 80 mm of water had infiltrated. If final steady state of infiltration is $4\,\text{mm h}^{-1}$, calculate infiltration rate at a cumulative infiltration of 150 and 300 mm.

4. Using the necessary data from Problem 3, calculate how much water can be delivered to the root zone of a crop without exceeding the soil's infiltrability if the sprinkler irrigation rate is 20 and $25\,\text{mm h}^{-1}$. What will be the highest steady state infiltration rate for 250 mm depth of irrigation in the shortest possible time?

5. A double ring infiltrometer study yielded the following results.

Cumulative infiltration (mm)	Infiltration rate (mm h^{-1})
50	10
100	6

Use the Green and Ampt equation and compute infiltration rate corresponding to a cumulative infiltration of 400 mm. Assume that the steady rate of infiltration is $2\,\text{mm h}^{-1}$.

6. Using the information in Problem 5, calculate the amount of water provided to the root zone of a strawberry crop being irrigated with an overhead sprinkling irrigation system at the rate of $10\,\text{mm h}^{-1}$. At what rate should the irrigation be supplied to provide 200 mm of water in the shortest time?

7. Assuming the K_s of soil used in Problem 2 to be $1\,\text{cm h}^{-1}$, estimate cumulative infiltration and infiltration rate for a vertical column at the end of time intervals given in Problem 2.

8. Refer to Problems 2 and 7 above and assume that the soil characteristics defined above are applicable under field conditions with a natural slope of 5%. How much runoff will occur in storms with constant rainfall intensities of 20 and $50\,\text{mm h}^{-1}$ for 15 minutes and 1 h each?

REFERENCES

Anderson D.M. and Linville A. (1962). Temperature fluctuations at a wetting front. I. Characteristic temperature–time curves. Proc. Soil Sci. Soc. Am. 26, 14–18.

Anderson D.M., G. Sposito and Linville A. (1963). Temperature fluctuations at a wetting front. II. The effect of initial water content of the medium on the magnitude of the temperature fluctuations. Proc. Soil Sci. Soc. Am. 27, 367–369.

Brutsaert W. (1977). Vertical infiltration in dry soil. Water Resour. Res. 13:363–368.

Boulier J.F., J.Y. Parlange, M. Vauclin, D.A. Lokington, and R. Haverkamp (1987). Upper and lower bounds of the ponding time for near constant surface flux. Soil Sci. Soc. Am. J. 51:1424–1428.

Butt K.R., M.J. Shipitalo, P.J. Bohlen, W.M. Edwards, and R.W. Parmelee (1999). Long-term trends in earthworm populations of cropped experimental watersheds in Ohio, USA. Pedobiologia 43:713–719.

Childs E.C. (1969). An introduction to the physical basis of soil water phenomenon. Wiley, New York.

Chong S.K., R.E. Green, L.R. Ahuja (1982). Infiltration prediction based on estimation of Green–Ampt wetting front pressure head from measurement of soil water redistribution. Soil Sci. Soc. Am. J. 46:235–238.

Clothier B.E., (2001). Infiltration. In K.A. Smith and C.E. Mullins (eds) "Soil and Environmental Analysis, Physical Methods." Marcel Dekker, Inc, New York: 239–280.

Collis-George N. and R. Lal (1970). Infiltration into columns of swelling soil as studied by high speed photography. Aust. J. Soil Res., 8:195–207.

Collis-George N. and R. Lal (1971). Infiltration and structural changes as influenced by initial moisture content. Aust. J. Soil Res., 9:107–116.

Collis-George N. and R. Lal (1972). The temperature profiles of soil columns during infiltration. Aust. J. Soil Res., 11:93–105.

Davidoff B. and H.M. Selim (1986). Goodness of fit for eight water infiltration models. Soil Sci. Soc. Am. J., 50:759–764.

De Vries D.A. (1957). Simultaneous transfer of heat and moisture in porous media. Trans. Am. Geophysical Union. 39(5):909–916.

Everts C.J. and R.S. Kanwar (1992). Interpreting tension-infiltrometer data for quantifying soil macropores: Some practical considerations. Trans. ASAE 36:423–428.

Green W.H. and G.A. Ampt (1911). Studies on soil physics: I. Flow of air and water through soils. J. Agr. Sci. 4, 1–024.

Haverkamp R., M. Kutilek, J.Y.Parlange, L. Rendon, and M. Krejca (1988). Infiltration under ponded conditions: 2. Infiltration equations tested for parameter time-dependence and predictive use. Soil Sci., 145(5):317–329.

Hill E.D. and J.Y. Parlange (1972). Wetting front instability in layered soils. Soil Sci. Soc. Am. Proc. 36, 697–702.

Hillel D. and W.R. Gardner (1970). Steady infiltration into crust topped profiles. Soil Sci. 108, 137–142.

Holtan H.N. (1961). A concept for infiltration estimates in watershed engineering. U.S. Dept. Agr. Agr. Res. Service Publication 41–51.

Holtan H.N. and N.R. Creitz (1961). Influence of soils, vegetation and geomorphology on elements of the flood hydrograph. Proc. Symposium on floods and their computation, Leningrad, Russia.

Horton R.E. (1940). An approach towards a physical interpretation of infiltration capacity. Soil Sci. Soc. Am. Proc., 5, 399–417.

Huggings L.F. and E.J. Monk (1967). A mathematical model for simulating the hydrologic response of a watershed. Water Resour. Res. 4:529–539.

Kostiakov A.N. (1932). On the dynamics of the coefficient of water percolation in soils and on the necessity of studying it from a dynamic point of view for purposes of amelioration. Trans. Com. Int. Soc. Soil Sci., 6th, Moscow Part A, 17–21.

Kirkham D. and W.L. Powers (1972). Advanced soil physics. Wiley Interscience, John Wiley & Sons, Inc., New York.

Kutilek M. and D.R. Nielsen (1994). Soil Hydrology. Catena Verlag, 38162 Cremlingen-Destedt, Germany.

Kutilek M. and M. Krejca (1987). A three-parameters infiltration equation of the Philips type solution (in Czech). Vodohosp. Cas. 35:52–61.

Lal R. and D.M. Jr. Vandoren (1990). Influence of 25 years of continuous corn production bt three tillage methods on water infiltration for two soils in Ohio. Soil Till. Res. 16:71–84.

Meyer, L.D. (1994). Rainfall simulators for soil erosion research. In R. Lal (ed.) Soil Erosion Research Methods. SWCS, Ankeny, IA:83–103.

Mezencev V.J. (1948). Theory of formation of the surface runoff (in Russian). Meteorologia I gidrologia 3:33–40.

Miller D.E. and W.H. Gardner (1962). Water infiltration into stratified soils. Soil Sci. Soc. Am. Proc. 26, 115–118.

Parlange J.Y. and R.E. Smith (1976). Ponding time for variable rainfall rate. Canadian J. Soil Sci., 56:121–123.

Perroux K.M. and I. White (1988). Designs for disc permeameters. Soil Sci. Soc. Am. J. 52:1205–1215.

Philip J.R. (1957). Theory of infiltration:4. Sorptivity and algebraic infiltration equations. Soil Sci. 84, 257–264.

Philip J.R. (1966). Absorption and infiltration in two and three-dimensional systems. In 'Water in the unsaturated zone" R.R. Rijtema and H. Wassink ed., Vol. 2, pp 503–525. IASH/UNESCO Symp. Wageningen.

Philip J.R. (1969). Theory of infiltration. Adv. Hydrosci. 5, 215–290.

Philip J.R. and D.A. De Vries (1957). Moisture movement in porous materials under temperature gradients. Trans. Am. Geophysical Union. 38(2)222–232.

Richards L.A. (1952). Report of the subcommittee on permeability and infiltration, committee on terminology, Soil Science Society of America. Soil Sci. Soc. Am. Proc. 16:35–88.

Richards, L.A., W.R. Gardner and G. Ogata (1956). Physical processes determining water loss from soil. Soil Sci. Soc. Am. Proc. 20:310–314.

Rubin J. (1966). Theory of rainfall uptake by soils initially drier than their field capacity and its application. Water Resourc. Res. 2, 739–749.

Rubin J. (1968). Numerical analysis of ponded rainfall infiltration. pp. 440–450. In: Proceedings of the UNESCO symposium on Water in the unsaturated zone. Vol. I, Wageningen.

Rubin J. and R. Steinhardt (1964). Soil water relations during rain infiltration. III. Water uptake at incipient ponding. Soil Sci. Soc. Am. Proc. 28:614–619.

Santanello J.A. and T.N. Carlson (2001). Mesoscale simulation of rapid soil drying and its implications for predicting daytime temperature. J. Hydrometeor. 2:71–88.

Shaver T.M., G.A. Petrson, L.R. Ahuja, D.G. Westfall, L.A. Sherrod, and G. Dunn (2002). Surface soil physical properties after twelve years of dry land no-till management. Soil Sci. Soc. Am. J. 66:1296–1303.

Shukla M.K., R. Lal, L.B. Owens, and P. Unkefer (2003a). Land Use and Management Impacts on Structure and Infiltration Characteristics of Soils in the North Appalachian Region of Ohio. Soil Science 168 (3) 167–177.

Shukla M.K., R. Lal and P. Unkefer (2003b). Experimental Evaluation of Infiltration Models for Different Land Use. Soil Science, 168 (3) 178–191.

Stroosnijder L. (1976). Cumulative infiltration and infiltration rate in homogeneous soils. Agric. Research report 847:69–99.

Swartzendruber D. (1987). A quasi-solution of Richards equation for the downward infiltration of water into soil. Water Resour. Res. 23:809–817.

Takagi S. (1960). Analysis of vertical downward flow of water through a two layered soil. Soil Sci. 90, 98–103.

van de Griend A.A. and M. Owe (1994). Bare soil surface-resistance to evaporation by vapor diffusion under semiarid conditions. Water Resour. Res. 30:181–188.

Youngs, E.G. (1964). An infiltration method of measuring the hydraulic conductivity of unsaturated porous material. Soil Sci. 109, 307–311.

15

Soil Water Evaporation

Soil water evaporation is an important component of surface energy balance. The rate and quantity of evaporation from a soil surface is a complicated process affected by many soil characteristics, tillage, and environmental interactions. Evaporation also affects plant available water content of soil and causes salinization in irrigated lands. It is known that energy and water availability largely dominate the process of evaporation, thus on an average these broad principles can be used to estimate direct soil water evaporation.

15.1 INTRODUCTION

Evaporation is the process of change of the state of water from a liquid to gaseous phase. It is a principal process of water cycling in the hydrosphere. Evaporation from a landscape may occur from plant canopies, free water surface, or soil surface. Evaporation of water from bare soil (i.e., in the absence of vegetation) is the process by which water is lost from the soil to the atmosphere. If the evaporation process is not controlled, a considerable amount of water can be lost from an irrigated or a rainfed cropland. Evaporation takes place from plowed land, shallow farmland, from soil between tree and row crops, and agricultural lands with no vegetation.

During planting and germination period, evaporation can reduce soil water content significantly and can hamper plant growth.

There are four conditions for evaporation from soil to occur. One, the evaporation from a bare soil takes place continuously, provided there is a continuous supply of energy. The amount of energy required is the latent heat of water for evaporation, which is about 590 cals/g of water evaporated at 15°C. The soil body itself, which gets cooler after rather than before evaporation, can supply this energy. Alternately, it can come from the advected or radiated energy from the surroundings. The second and third physical conditions for evaporation from bare soils are: the vapor-pressure gradient between the soil and the atmosphere, and the transport of vapor away from soil by diffusion and/or convection. The energy for evaporation and vapor removal are generally external to the evaporating soil and are greatly influenced by meteorological factors (i.e., air temperature, humidity, wind velocity, and radiation). The evaporation rate is determined by the external evaporativity or water conductivity of the soil. The fourth condition is that there is a continuous supply of water within the soil body to the evaporating surface. This condition depends on both physical and conductance properties of soil. Some of them are: water content of soil body, soil water potential, hydraulic conductivity of soil, texture, compaction, soil horizonation or layering, and depth to the water table.

An important condition in poorly drained soils is evaporation in presence of a shallow ground water table. The evaporation occurs at a nearly steady state level and water content of profile remains almost the same. In the absence of a water table, the evaporation will dry the topsoil and the process is mostly in an unsteady state. The flow domains during the evaporation process may be one-dimensional or three-dimensional. The flow processes may be isothermal or non-isothermal and there may be the interactions between liquid flow and temperature gradients, conduction of heat, and vapor in the soil domain. The cracks inside the soil matrix form the secondary evaporation planes. The evaporation also depends upon the environmental conditions, which can be regular (i.e., diurnal, seasonal) or irregular (i.e., spells of cool and warm weather and of rewetting and drying). The presence of surface mulch, depth of soil, and degree of homogeneity of soil also alter evaporation from the soil surface.

15.2   THE EVAPORATION PROCESS

Following three processes summarize the evaporation of water from soil surface.

15.2.1 Transport of Water to the Soil Surface

As the evaporation process begins at the soil surface, a suction gradient is established between the surface soil and the layer beneath. This gradient forces the water to move upwards through capillary rise and supplies the water to the soil surface. This process continues as long as the soil underneath has enough water storage. The transfer of water from soil underneath is facilitated easily and for a much longer duration immediately after irrigation or a rainfall event, when soil water content is high.

15.2.2 Uninterrupted Supply of Heat to Change the State of Water

Solar energy is the most predominant source of heat for water evaporation. There are other minor sources of energy also, e.g., exothermic reactions, microbial activity, etc. The energy balance of the soil depends on physical and thermal properties. The latent heat of vaporization changes the water from a liquid to vapor state.

15.2.3 Transfer of Water Vapor from Soil Surface to Atmosphere

The vapor pressure immediately above the evaporating soil surface is lower than the vapor pressure inside the soil. This differential pressure creates a vapor pressure gradient, which enables water vapor to escape from soil to the atmosphere through the process of convection and diffusion.

15.3 SOIL DRYING DURING EVAPORATION

Evaporation leads to the loss of water, with attendant drying and depletion of the soil moisture reserves. This process of soil drying occurs in three distinct stages (Fig. 15.1) (Fisher, 1923; Pearse et al., 1949).

15.3.1 Initial Stage

When soil is very wet, evaporation of soil water is governed by external atmospheric conditions rather than soil properties. The soil has enough water, therefore, conductivity and supply of water to soil surface are at the potential rate. The evaporation rate during this stage is denoted as "potential evaporation." This stage is sustained over time because as the water content of soil profile decreases the hydraulic conductivity also decreases. However, hydraulic gradient increases and compensates for the reduction in hydraulic conductivity. This situation is analogous to the

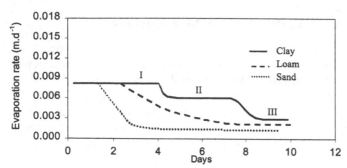

FIGURE 15.1 Three stages of evaporation during simultaneous drainage and evaporation from initially saturated profiles of sand, silt, and clay. (Modified from Hillel and van Bavel, 1976.)

flux-controlled stage of water infiltration into soil. Some soil properties, which influence the meteorological or atmospheric factors, include soil surface reflectance, mulch, ground cover, etc. The duration of the first stage of the drying process is lower for coarse-textured than fine-textured soils because fine-textured soils retain high water content and have more conductivity than coarse-textured soils. Figure 15.1 shows that the duration of the first stage of drying is in the order clay > loam > sand. The duration for first stage is also lower for a structureless than for a structured soil. Mulching increases the water content profile but shortens the duration of first stage of drying (Fig. 15.1). The duration of first stage is inversely proportional to soil water diffusivity (D_O).

15.3.2 Intermediate Stage

The evaporation rate during this stage is no longer at the potential rate but starts decreasing gradually with time. Soil starts to heat up and is not able to conduct water to the surface at the potential rate. The water content of the soil profile is decreased further as is the hydraulic conductivity. The hydraulic gradient can no longer increase significantly because the soil water pressure head is close to the partial water vapor pressure. The time at which the decrease in hydraulic conductivity is not compensated by hydraulic gradient denotes the end of first stage of drying. The depth of dry zone increases as does the hydraulic resistance of soil to water transport. The rate of evaporation during this stage is directly proportional to soil water diffusivity.

15.3.3 The Final Stage

The evaporation rate during this stage is relatively steady at a low rate and can continue up to several days. During this stage the liquid-water conductance totally ceases. This stage is also known as the vapor diffusion stage, since water transmission is primarily due to a slow process of vapor diffusion. The evaporation rate is determined by soil properties (affinity of the soil for water) rather than the evaporative demand of the atmosphere.

15.4 THEORY OF EVAPORATION

Specific processes of evaporation depend upon the presence of the water table at shallow depth, horizonation, and soil temperature region.

15.4.1 Steady Evaporation in the Presence of a Water Table

The essential and necessary condition for steady state flow is that the rate of change of flux density with depth is zero ($dq/dz = 0$). Moore (1939) first studied the vertical steady state flow of water from a water table through soil profile during evaporation. Philip (1957), Gardner (1958), Ripple et al. (1972) discussed theoretical solutions for steady state evaporation. Mathematically, the steady upward flow can be described by the Darcy–Buckingham equation, for boundary condition $z = 0$, $\Phi_m = 0$ and z is positive upward from the ground water table. The steady state upward flow or evaporation (q_e) can be expressed as follows [see also Darcy law for unsaturated flow] and see Eq. (13.3):

$$q_e = K(\Phi_m)\left(\frac{d\Phi_m}{dz} - 1\right) \tag{15.1}$$

In terms of soil-water diffusivity [see Eq. (13.24)], the steady state evaporation is given by

$$q_e = D(\theta)\frac{d\theta}{dz} - K(\theta) \tag{15.2}$$

where $D(\theta)$ is hydraulic diffusivity, θ is water content, $K(\theta)$ is hydraulic conductivity, z is height above the water table, and Φ_m is suction head. Eq. (15.1) shows that for $d\Phi_m/dz = 1$, $q = 0$. Eq. (15.1) can be rearranged as follows

$$\frac{q_e}{K(\Phi_m)} + 1 = \frac{d\Phi_m}{dz} \tag{15.3}$$

separation of variables results in

$$dz = \frac{K(\Phi_m)}{q_e + K(\Phi_m)} d\Phi_m \tag{15.4}$$

Integrating Eq. (15.4) to depths between 0 to z and suction gives the following expression in terms of depth of soil profile

$$z = \int \frac{K(\Phi_m)}{q_e + K(\Phi_m)} d\Phi_m \tag{15.5}$$

Similarly, Eq. (15.2) can be integrated for depths between 0 to z and following relationship in terms of z is obtained

$$z = \int \frac{D(\theta)}{q_e + K(\theta)} d\theta \tag{15.6}$$

Solutions of Eqs. (15.5) and (15.6) require prior knowledge of the functional relationships between $K(\Phi_m)$ and Φ_m and $K(\theta)$ and $D(\theta)$. For solving Eq. (15.5), Gardner (1958) proposed following relationship between Φ_m and K:

$$K(\Phi_m) = \frac{a}{\Phi_m^n + b} \tag{15.7}$$

where parameters a, b, and n are constants and are functions of type of soil. These parameters are soil-specific and need to be determined for each soil separately. Transferring Eq. (15.7) into Eq. (15.1) gives the following equation for evaporation rate (e) estimation

$$e = q_e = \frac{a}{\Phi_m^n + b} \left(\frac{d\Phi_m}{dz} - 1 \right) \tag{15.8}$$

Transferring Eq. (15.7) into (15.5) gives the following equation, which provides the suction distribution with depth of soil for different fluxes

$$z = \int \frac{a/(\Phi_m^n + b)}{q_e + a/(\Phi_m^n + b)} d\Phi_m = \int \frac{a}{q_e(\Phi_m^n + b) + a} d\Phi_m \tag{15.9}$$

Steady rate of upward flow and evaporation rate from water table as a function of the suction prevailing at the soil surface for a sandy loam soil with $n = 3$ is presented in Fig. 15.2 (Gardner, 1958). Fig. 15.2 shows that a

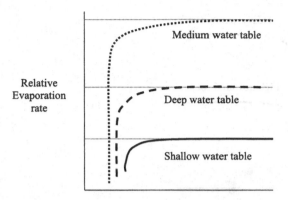

Relative
Evaporation
rate

Suction head at the soil surface

FIGURE 15.2 Steady rate of upward flow and evaporation from a water table as a function of the suction prevailing at the soil surface. The soil is sandy loam with $n = 3$ (Modified from Gardner, 1958.)

steady rate of evaporation depends on the depth of this water table. The maximum possible evaporation for a ground water table at z will be for the lowest water content at the soil surface. Under this situation suction will tend to be infinite (pressure head infinity with negative sign). However, the extraction of water from soil profile is limited by the capacity of soil profile to transmit water or in other words by the hydraulic conductivity of soil. Ignoring the constant b, in Eq. (15.7), which leads to $K(\Phi_m) = a[\Phi_m^n]$. Gardner (1958) derived the following relationship between the depth of water table (d) and the maximum or limiting rate of transmission of water (q_{max}) by soil to the surface layer

$$d = \left(\frac{Aa}{q_{e\text{-}max}} \right)^{1/n} \tag{15.10}$$

where A is a constant and is a function of n and $q_{e\text{-}max}$, a and n are constants from Eq. (15.7). Equation (15.10) shows that evaporation and depth of water table are inversely related. Gardner (1958) showed that the evaporation rate from the soil is dependent on soil texture and is greater from medium textured soils as compared to the coarse textured soils (Fig. 15.3). Value of n is greater in coarse textured soils as compared to fine textured soils and therefore maximum evaporation rate

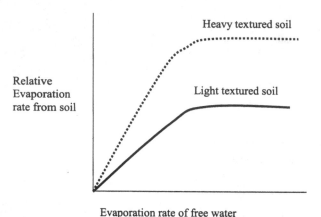

Evaporation rate of free water

FIGURE 15.3 Schematic of evaporation rate as affected by texture of soil (water table depth, 60 cm). (Modified from Gardner, 1958.)

decreases more rapidly with depth in coarse textured soils as compared to the fine ones.

15.4.2 Evaporation in the Absence of a Water Table

Evaporation from soils in the absence of a water table is a transient process. Steady evaporation from soils is not always true because water table depths do not always remain constant for very long time. Therefore, a transient process describes the evaporation from soil surface more realistically. The transient condition implies that water content of soil profile does not remain constant, instead it decreases with evaporation as soil becomes drier. Another common assumption for steady state evaporation is that the external conditions (i.e., atmospheric evaporativity) remain constant, which is seldom true. However, for the sake of simplicity, constant atmospheric evaporativity is generally assumed for describing transient evaporation from the soil surface.

The process of transient evaporation or drying is already described as having three stages: initial constant rate stage, falling rate stage, and slow rate stage (Fig. 15.4). It is clear from Fig. 15.4 that the transition from the first to second stage of drying is sharp. However, transition from second to third stage is gradual and difficult to separate. In the initial stage of drying soil moisture depletion on the soil surface is compensated by soil underneath and evaporation remains by and large constant. Gradually, suction gradient inside the soil becomes larger with corresponding decrease in soil

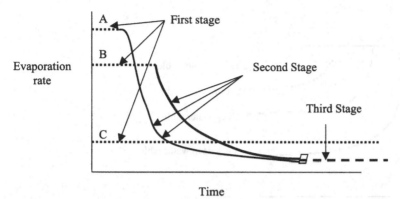

FIGURE 15.4 Stages of evaporation from a soil during steady atmospheric condition.

conductivity. This results in a decrease in evaporation rate with respect to time. The length of time the initial stage of drying can go on depends on the evaporativity. A low evaporativity will increase the duration of first stage of drying.

15.4.3 Evaporation from Layered Soils

Steady evaporation from layered soils can be determined similar to that from a homogeneous profile. Willis (1960) carried out the analysis by assuming that steady flow through layered profile depends upon the transmission property of soil. He further assigned that the suction or matric potential is continuous through the entire soil profile, although water content and conductivity are discontinuous using the relationship between $K(\Phi_m)$ and Φ_m [Eq. (15.7)] (Gardner, 1958) and assuming that each layer is internally homogeneous, he proposed the following relationship:

$$\int_0^{d_2} dz + \int_{d_2}^{d_1+d_2} dz = \int_{\Phi_{m0}}^{\Phi_{mL}} \frac{d\Phi_m}{1 + e/K_1(\Phi_m)} + \int_{\Phi_{mL}}^{\Phi_{m(L+d)}} \frac{d\Phi_m}{1 + e/K_2(\Phi_m)}$$

$$(15.11)$$

where d_1 and d_2 are the thickness of top and bottom layers respectively. Eq. (15.11) relates depth of water table to suction for a given evaporation rate. The limiting evaporation rate for a known water table depth can be calculated from above equation by assuming the suction (Φ_m) to

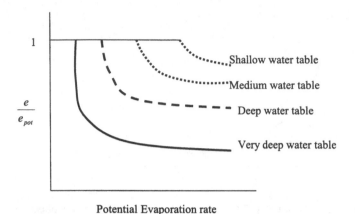

Potential Evaporation rate

FIGURE 15.5 Dependence of relative evaporation rates, (e/e_{pot}) upon potential evaporation rate (evaporativity, e_{pot}) for a clay soil. Numbers labeling the curve indicate the depth to water table (cm). (Modified from Ripple et al., 1972.)

be infinite at soil surface. Ripple et al. (1972) proposed a graphical method to measure the steady state evaporation from a multilayer soil profile. They included both the soil properties (i.e., water retention and transmission, vapor flow, depth of water table) and the meteorological factors (i.e., humidity, air temperature, and wind velocity) (Figs. 15.5 and 15.6).

15.4.4 Mathematical Modeling of Stages of Drying

The difference in suction at soil surface and a location with the soil body supplying water is much higher as compared to the depth of soil involved in the process of drying. Therefore, gravity effects are generally neglected for evaporation calculations. Most analysis is based on soil water content and the hydraulic diffusivity relationship. The first and second stages of drying depend upon the hydraulic diffusivity. In order to derive approximate description of drying in the first stage (Fig. 15.6), Gardner (1959) assumed that the evaporation rate from a soil profile of depth (L) could be expressed as

$$e = -L\frac{\partial \theta}{\partial t}$$

(15.12)

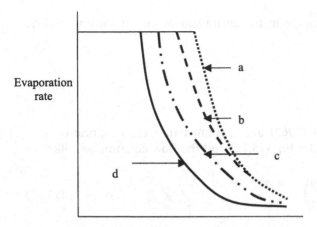

FIGURE 15.6 Effect of horizonation and water table depth on the evaporation rate: (a) limiting curve for soil water evaporation from for homogeneous soil; (b) a two-layer soil with the upper layer thickness of 3 cm; (c) thickness 10 cm; (d) a three-layer soil with thickness of intermediate and uppermost layers equal to 10 cm each. (Modified from Ripple et al., 1972.)

if the soil water diffusivity $(D(\theta))$ can be expressed by the following relationship (Gardner and Mayhugh, 1958)

$$D(\theta) = D(\theta)_0 \exp[\beta(\theta - \theta_0)] \tag{15.13}$$

where $D(\theta)_0$ correspond to θ_0 and β ranges from 1 to 30. Gardner (1959) combined Eqs. (15.12) and (15.13) and after further approximation proposed that the total water content of soil profile can be approximated by the following relationship:

$$W = \frac{\beta W}{L} \tag{15.14}$$

and

$$D(\theta) = \frac{2D(\theta)_0}{e_{\text{pot}}\beta L} \tag{15.15}$$

where W is the water storage in the entire soil profile at the end of first stage $t = t_1$.

$$W = \int_0^L \theta(z, t_1)\mathrm{d}z \tag{15.16}$$

Gardner and Hillel (1962) also assumed that the evaporation rate from soil profile is given by Eq. (15.12) and the flow equation as follows

$$-\frac{e}{L} = \frac{\partial \theta}{\partial t} = \frac{\partial}{\partial z}\left(D\frac{\partial \theta}{\partial z}\right) \tag{15.17}$$

where z is the height above the bottom of soil profile. Eq. (15.17) was integrated once. The constant of integration was assumed zero since flow through bottom of the soil profile ($z = 0$) is zero. Assuming D can be represented by Eq. (15.13), Gardner and Hillel (1962) found that actual evaporation rate ceases to be equal to potential rate when at $z = L$, $\theta = \theta_0$. They proposed the following equation for total water content (W) of profile.

$$W = \frac{L}{\beta}\ln\left(1 + \frac{e\beta L}{2D_0}\right) \tag{15.18}$$

For a long soil column dependence of e on t is only approximately valid. For a soil column of finite length (L), Gardner and Hillel (1962) proposed the following relationship to calculate evaporation during second stage (Fig. 15.5).

$$e = -\frac{\mathrm{d}W}{\mathrm{d}t} = D(\bar{\theta})\frac{W\pi^2}{4L^2} \tag{15.19}$$

where $\bar{\theta}$ is average water content, the $D(\theta)$ is known diffusivity function. Eq. (15.19) can be integrated to obtain cumulative infiltration.

Gardner (1959) presented the analytical solution for the second stage of evaporation by using the solution for diffusion by Crank (1956). According to Crank (1956), the weighted mean diffusivity for desorption ($\bar{D}(\theta)$) is

$$\bar{D}(\theta) = \frac{1.85}{(\theta_i - \theta_0)^{1.85}}\int_{\theta_0}^{\theta_i} D(\theta)(\theta_i - \theta)^{0.85}\mathrm{d}\theta \tag{15.20}$$

where $\bar{D}(\theta)$ for sorption process is higher than that for desorption process. The weighing is done differently because in infiltration maximal flux occurs at the wet end of column, where diffusivity is the highest. However, in drying the greatest flux is through the dry end, where diffusivity is the lowest. This is also the reason, why sorption processes are faster as compared to desorption. Gardner (1959) assumed that initial evaporation is infinitely high and soil surface is instantaneously brought to the final stage of drying. Therefore, $e_{pot} \to \infty$, at $t = 0$ when second stage of drying starts. Using the diffusivity form of Richards' equation and assuming that influence of gravity is negligible, the evaporation rate for a semi-infinite soil column can be given as

$$e = (\theta_i - \theta_0)\left(\frac{\bar{D}(\theta)}{\pi t}\right)^{1/2} \qquad (15.21)$$

and cumulative evaporation (E) can be given as

$$E = 2(\theta_i - \theta_0)\left(\frac{\bar{D}(\theta)t}{\pi}\right)^{1/2} \qquad (15.22)$$

Using the sorptivity concept of Philip (1975), Rose (1966) presented the following relationship for evaporation calculation in the second stage of drying:

$$E = S't^{1/2} + A't \qquad (15.23)$$

or

$$e = \frac{1}{2}S't^{-1/2} + A' \qquad (15.24)$$

where S' is the soil evaporativity (which is equivalent to soil water sorptivity in infiltration, since the process is drying it can be termed as desorptivity, $LT^{-1/2}$) and A' is a constant (LT^{-1} comparable to trasmissivity [see Eq. (14.24)]). The value of S' is positive whereas b is negative. The assumption of a zero flux at the bottom of the profile or at depth L, although simple, implies that in the absence of this condition evaporation will accompany redistribution. This will reduce both the evaporation rates and the duration of the first stage. Solutions of evaporation considering isothermal conditions differ from the nonisothermal condition. The concept of three stages of evaporation does not strictly hold in field conditions (Jackson et al., 1973). The diurnal temperature fluctuations and other atmospheric process largely affect the evaporation rate. When air temperatures are low the upward heat flow is accompanied with water flow. When temperatures are high, the downward heat flow is accompanied with water flow and/or vapor flow. All these effects make sure that the second stage of drying starts well before the moisture content of soil has reached hygroscopic coefficient or the final dry value. Another factor, which can influence evaporation by as much as 50% is the presence of cracks in the soil. The cracks or similar soil inhomogeinities have totally different thermal fields compared to homogeneous soils. Downward vapor flow due to thermal gradients is observed within the cracks of small sizes (Hatano et al., 1988). The cracks may not increase the evaporation rate during the early stage of drying, but can increase the duration of that stage. Cracks can also increase the evaporation rate of subsequent profile controlled drying period (or second stage).

15.4.5 Nonisothermal Evaporation

The isothermal flow equation is assumed to predict the constant and falling rate of evaporation reasonably well. The role of nonisothermal conditions is explained by comparing the solutions of an isothermal process to the solutions of nonisothermal process (Milley, 1984). According to Jackson et al. (1974), in wet soils the thermal and isothermal vapor fluxes are approximately equal and opposite in direction for diurnal variation of temperature. For a dry soil surface layer, the thermal vapor pressure increases the evaporation from soil profile during night. However, neglecting thermal effects over a month introduce only about 1% error.

15.5 MANAGEMENT OF EVAPORATION

Evaporation from bare soil surface needs to be reduced so that moisture status of soil can be maintained at a stage favorable for crop growth and production. The evaporation management can be done by: (i) reducing the total amount of incident radiations or sources of energy responsible for evaporation; (ii) modifying the color of soil by applying amendments and changing the albedo parameters; and (iii) reducing the upward flux of water by either lowering the water table, or decreasing the diffusivity and conductivity of the soil profile. The methods of evaporation reduction from bare soils depend on the stage of drying. The first stage requires modifications, which will alter meteorological conditions of the surroundings. The second stage requires measures, which will change water transmission properties of the soil profile. Covering or mulching the surface with vapor barriers or with reflective materials can reduce the intensities of the incoming radiations and reduce the evaporation in the first stage of drying. A deep tillage may change the variation of diffusivity with changing water content of soil profile and may change the rate at which water can be supplied to the soil surface from underneath for evaporation.

15.5.1 Mulching

Mulch is any material placed on a soil surface primarily to cover the surface for the purpose of reducing evaporation, controlling weeds, and obtaining beneficial changes in soil environment. The other benefits of mulching are: (i) reducing soil erosion; (ii) sequestering carbon; (iii) providing organic matter and plant nutrition; (iv) regulating and moderating soil temperature; (v) increasing earthworm population and improving soil structure; and (vi) reducing soilborne diseases.

Mulches can consist of many different types of materials, such as sawdust, manure, straw, leaves, crop residue, gravels, paper, and plastic sheets, etc. (Fig. 15.7) (Lal, 1991). Paper or plastic mulches, especially light colored, are effective in reducing the effects of meteorological variables, which influence the evaporative demand during the first stage of soil evaporation (Figs. 15.8 and 15.9). Black paper and plastic mulches are effective in weed control (Fig. 15.10). The temperature of the soil under plastic mulch can be 8 to 10°C higher than under straw mulch. Soil thermal regime is a function of the contact coefficient, which is a product of thermal conductivity and volumetric heat capacity of the soil (refer to Chapter 17). A mulched plot with dry crop residue is equivalent to a two-layered profile of which the upper layer has a lower contact coefficient. Therefore, temperature variations in the soil underlying the mulched layer are reduced (Figs. 15.11 and 15.12). High temperature may be beneficial to the crops on temperate regions during germination in spring. However, high temperature during summer and in the tropics may adversely affect the growth of temperature-sensitive crops. Other mulch materials may include preparations of latex, asphalt, oil, fatty acids, and alcohols. These materials can be used as mulches for reducing evaporation from soil surface. Hillel (1976) proposed that uppermost layer of soil be formed by clods or a rough seedbed, which are treated with water proofing materials (e.g., silicones). These waterproof clods act as dry mulch and reduce evaporation and erosion from soil surface.

Vegetative mulch must have sufficient thickness to be effective in reducing evaporation and risks of soil erosion. The porosity and hydraulic conductivity of the vegetative mulches are high, and therefore diffusion or

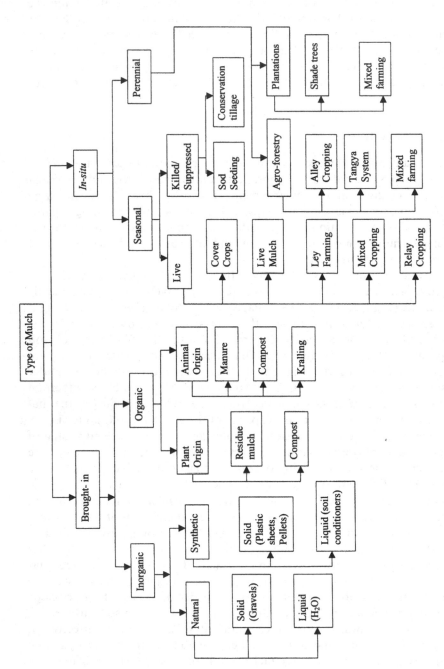

FIGURE 15.7 Type of mulches on the basis of the source of the material. (Modified from Lal, 1991.)

FIGURE 15.8 Clear plastic mulch used on cassava grown at IITA in western Nigeria to conserve soil water.

FIGURE 15.9 Clear plastic mulch used on a ridged seed bed. Note holes in the plastic for seedling emergence.

FIGURE 15.10 Black plastic mulch to conserve water and control weeds in strawberries grown in California.

airflow through the vegetative mulch is also high. A mulch of small thickness may be mostly ineffective. Vegetative mulches are light colored and reflect most of the incident radiations. Therefore, the initial evaporation rate under mulch is generally less. Gravel mulching is a common practice of water conservation, as it enhances the infiltration and simultaneously suppresses evaporation and reduces erosion of soil. Disadvantages of gravel mulch are that gravel cannot be removed from the field after application and can adversely affect future land uses.

15.5.2 Tillage

Among the various soil management practices for weed control and seedbed preparation, tillage is an important technique of soil manipulation. Tillage operations generally result in opening up of soil, changes in structure, loosening of tilled soil, and compaction of soil immediately below the tilled layer (Fig. 15.13) (Lal 1989, 1990). The opening of the topsoil enhances the evaporation from the tilled soil layer. However, the compaction of layers underneath might reduce the upward transmission of water and subsequently make the water availability limiting and reduce evaporation. The reduction of diffusivity in the soil layer also reduces the evaporation. The discontinuity of pore channels due to the

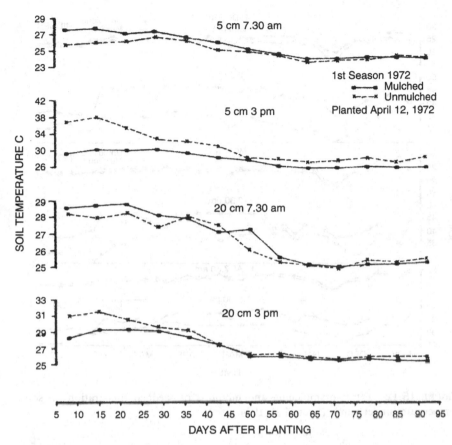

FIGURE 15.11   Effects of mulching on soil temperature under maize. (From Lal, 1974.)

tillage operations does not reduce the upward flow of water and does not reduce the total evaporation. More recent trends have indicated that management practices involving minimum tillage are better for efficient soil management. The tillage is beneficial under two situations: (i) in soils with high swell–shrink capacity and where frequent wetting and drying produces cracks. These cracks are the sources of secondary evaporation from soil. Cultivation may prevent development of or help obliterate cracks. (ii) Tillage eliminates weeds and may reduce the rate of application of herbicides. Burning crop residue and the presence of ash on the soil surface can influence soil temperature by altering albedo and soil moisture regime (Figs. 15.14 and 15.15).

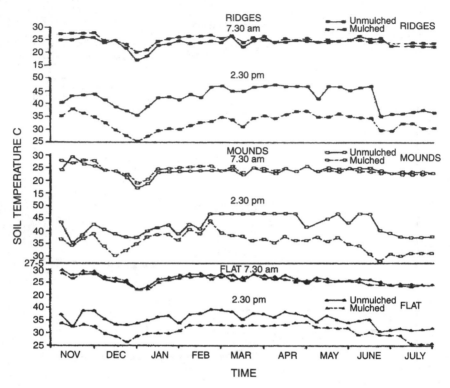

FIGURE 15.12 Effect of mulching and methods of seedbed preparation on soil temperature under yams. (From Lal, 1974.)

15.5.3 Conservation Tillage

Conservation tillage practices leave a high percentage of the residues from previous crops on the soil surface (Fig. 15.16). Plant residues left on the soil surface are effective in reducing evaporation and conserving soil moisture. A conservation tillage practice widely used in semiarid and humid regions is stubble mulching where wheat stubbles or corn stalks from previous crops are uniformly spread over the soil surface. The land is then tilled with special implements, which leave most of the residue on the soil surface. The next crop is planted through the stubble, which results in a healthy environment (temperature, water, and air) for seed germination. No tillage, or zero tillage, is another conservation tillage system that leaves residue on the soil surface and a new crop is planted directly through the residue of the previous crop with no plowing or disking (Lal, 2003).

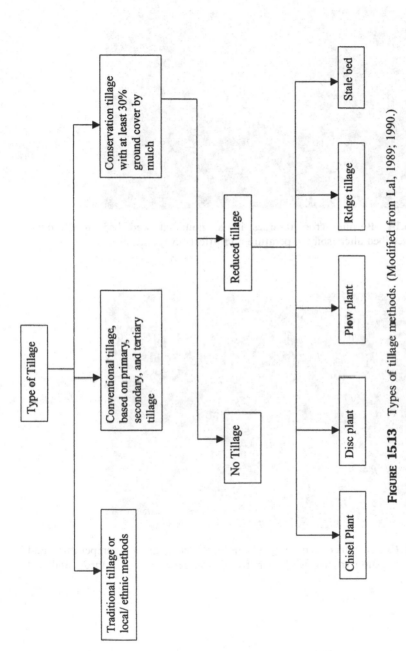

FIGURE 15.13 Types of tillage methods. (Modified from Lal, 1989; 1990.)

FIGURE 15.14 Burning crop residues in a mounded seed bed in Ethiopia. Mounded seedbed alters soil temperature and affects evaporation rate.

FIGURE 15.15 A mulch cap on yam mounds decreases soil temperature and reduces evaporation (right), while ash from crop residue alters albedo and soil temperature.

FIGURE 15.16 No-till farming with crop residue mulch reduces soil evaporation.

Example 15.1

Assume average daily steady state evaporation is 1 cm in a saturated loam soil in a high water table area. Estimate (a) threshold depth beyond which water table must be lowered, (b) water table depth at which evaporation will fall to 20% of potential value, and (c) plot daily evaporation rate with respect to water table depth. Use Eq. (15.10), assuming Aa to be equal to 4.5 cm^2.sec and $n = 3$.

Solution

According to Eq. (15.10)

$$q_{max} = \left(\frac{Aa}{d^n}\right) \Leftrightarrow d = \left(\frac{4.5}{1/86400}\right)^{1/3} = 72.99 = 73 \text{ cm}$$

where d is the maximum depth of water table below the soil surface, which can supply water to maintain a steady flux for evaporation. Hence

(a) Threshold water table depth is 73 cm.
(b) The water table depth ($d_{0.2}$) at which evaporation rate falls by 20% can be calculated from again Eq. (15.10) as follows:

$$d_{0.2} = \left(\frac{4.5}{1 * 0.2/86400}\right)^{1/3} = 124.8 = 125 \text{ cm}$$

(c)

D (cm)	q_{max} (cm)
0–73	1
80	$4.5/80^3 = 0.76$
90	$4.5/90^3 = 0.53$
100	0.39
120	0.23

Example 15.2

Consider an infinite sandy loam soil profile, which is initially saturated with water. The initial moisture content of soil is $0.52 \, cm^3 \, cm^{-3}$ and final moisture content of $0.2 \, cm^3 \, cm^{-3}$. If weighted mean diffusivity of soil is $80 \, cm^2 \, d^{-1}$, calculate evaporation and the evaporation rate for each day during the next 10 days.

Solution

From Eq. (15.21) the evaporation rate (e), and from Eq. (15.22), the cumulative evaporation (E), can be calculated for days 1, 2, 3 ... 10 as follows:

Mid-day	e (cm d^{-1})	Day	E (cm)
0.5	2.28	1	3.23
1.5	1.32	2	4.57
2.5	1.02	3	5.59
3.5	0.86	4	6.46
4.5	0.76	5	7.22
5.5	0.69	6	7.91
6.5	0.63	7	8.54
7.5	0.59	8	9.13
8.5	0.55	9	9.69
9.5	0.52	10	10.21

PROBLEMS

1. If the composite coefficient Aa is $4.5 \, cm^2/s$, n $= 3$, potential rate of evaporation is 8 mm/d to what depth must the water table be lowered for reducing evaporation? Also calculate the watertable depth at which the evaporation rate drops by 10%, 30%, and 70% of potential evaporation rate.

2. Assume an infinitely deep, saturated sandy loam soil profile under very high evaporativity. If initial volumetric water content of soil is 0.50, final volumetric water content is 0.10 and weighted mean diffusivity is $2 \times 10^4 \, \text{mm}^2 \text{d}^{-1}$. Calculate the evaporation and evaporation rate, for the next 6 days.

3. If an impermeable layer exists at the end of a uniform wetted soil of depth 1.2 m, initial volumetric water content (θ_o) 0.24, and initial diffusivity $(D(\theta_o))$ $4 \times 10^4 \, \text{mm}^2 \, \text{d}^{-1}$. If evaporativity is 10 mm/d, calculate evaporation rate during the first 10 days if diffusivity $(D(\theta))$ is given by Eq. (15.17) are assuming $B{=}15$, calculate $D(\theta_o)$ for the next 6 days.

4. Briefly outline techniques of regulating soil evaporation and explain the principle of their effectiveness in reducing evaporation.

5. What should be the irrigation strategy in arid environments and why?

REFERENCES

Hatano R., H. Nakamoto, T. Sakuma, and H. Okajima (1988). Evapotranspiration in cracked clay field soil. Soil Sci. Plant Nutr. 34: 547–555.

Gardner W.R. (1958). Some steady state solutions of the unsaturated moisture flow equation with application to evaporation from a watertable. Soil Sci. 85(4): 228–232.

Crank J. (1956). The mathematics of diffusion. Oxford University Press, London and New York.

Fisher R.A. (1923). Some factors effecting the evaporation of water from soil. J. Agr. Sci. 13: 121–143.

Gardner W.R. (1959). Solutions of the flow equation for the drying of the soils and other porous media. Soil Sci. Soc. Am. Proc. 23: 183–187.

Gardner W.R. and D. Hillel (1962). The relation of external evaporative conditions to the drying of soils. J. Geophys. Res. 67: 4319–4325.

Gardner W.R. and M.S. Mayhugh (1958). Solutions and tests of the diffusion equation for the movement of water in soil. Soil Sci. Soc. Am. Proc. 22: 197–201.

Hillel D. (1976). On the role of soil moisture hysteresis in the suppression of evaporation from bare soil. Soil Sci. 122: 309–314.

Hillel D. (1980). Fundamentals of soil physics. Academic Press, New York.

Hillel D. and C.H.M. van Bavel (1976). Dependence of profile water storage on soil hydraulic properties: a simulation model. Soil Sci. Soc. Am. J. 40: 807–815.

Jackson R.D., B.A. Kimball, R.J. Reginato, and S.F. Nakayama (1973). Diurnal soil water evaporation: comparison of measured and calculated soil water fluxes. Soil Sci. Soc. Am. Proc. 38: 861–866.

Lal R. (1974). Role of Mulching Techniques in Tropical Soil and water management. Tech. Bulletin no. 1, International Institute of Tropical Agriculture, Nigeria.

Lal R. (1989). Conservation tillage for sustainable agriculture. Adv. Agron. 42: 85–197.

Lal R. (1990). Soil erosion in the tropics: Principle and Management. McGraw Hill Book Co., 579 pp.

Lal R. (1991). Soil Structure and sustainability. Journal of sustainable Agriculture. 1(4): 67–92.

Lal R. (2003). Historical development of no-till farming. In R. Lal, P. Hobbs, N. Upofl, and D.O. Hansen (eds) Sustainable agriculture and the rice wheat system. Marcel Dekker, New York.

Milley P.C.D. (1984). A linear analysis of thermal effects on evaporation from soil. Water Resour. Res. 20: 1075–1086.

Moore R.E. (1939). Water conduction from shallow watertable. Hilgardia 12: 383–426.

Pearse J.F., T.R. Oliver, D.M. Newitt (1949). The mechanisms of the drying of solids: Part I. The forces giving rise to movement of water in granular beds during drying. Trans. Inst. Chem. Eng. (London) 27: 1–8.

Philip J.R. (1957). Evaporation, moisture and heat fields in soil. J. Meteorol. 14: 354–366.

Ripple C.D., J. Rubin, and T.E.A. van Hylckama (1972). Estimating steady state evaporation rates from bare soils under conditions of high water table. US Geol. Surv. Water-Supply Paper 2019-A.

Rose C.W. (1966). Agriculture Physics. Pergamon Oxford.

Willis O.W. (1960). Evaporation from layered soils in presence of a watertable. Soil Sci. Soc. Am. Proc. 24: 239–242.

16

Solute Transport

16.1 INTRODUCTION

Water entering the soil profile from rain or irrigation is essentially a dilute solution. Rainwater is pure when it condenses to form clouds; during descent it absorbs atmospheric gases (i.e., CO_2, N_2, products of sulfur and O_2, etc.). When water flows on soil surface as overland flow and/or through the soil matrix, it also dissolves solutes (e.g., salts, fertilizers, pesticides). These solutes not only move with soil water but also within the soil matrix mainly due to the concentration gradients. Sometimes, solutes react among themselves and/or with soil material according to a range of physical and chemical processes.

In agricultural ecosystems, solutes may be categorized on the basis of their function (e.g., nutrients, pesticides, waste compounds, salts, organic chemicals, heavy metals, viruses, and bacteria). Understanding transport of solutes in soil is important to many management problems in agriculture. It can help when developing procedures for maximizing the effective use of fertilizers or pesticides and other chemicals within the root zone while minimizing their movement into groundwater. Knowledge of these processes is important to understanding the problems of contamination of natural water through leaching or redistribution within a vadose zone to groundwater, availability of solutes for plant uptake, surface runoff, salt intrusion in coastal aquifers, seepage from storage or disposal systems, and chemical residues.

Depending upon chemical stability and reactivity, the solutes are broadly classified into two categories: (i) conservative solutes, which remain unchanged physically and chemically, and do not undergo irreversible reactions, such as chloride (Cl) and bromide (Br); and (ii) nonconservative solutes, which can undergo irreversible reactions and change their physical or chemical phase. The nonconservative solutes can be divided into labile solutes and reactive solutes. The labile solutes can undergo reversible or irreversible physiochemical, biochemical, or microbial reactions and can change their physical or chemical phase with time. The examples of labile solutes are: nitrate, sulfate, and ammonia, which are involved in mineralization, immobilization, or redox reactions. Some pesticides are also labile and their lability is quantified by their half-life (White et al., 1998). Reactive solutes undergo reversible or irreversible reactions with soil constituents by way of adsorption (adsorption of cations, e.g., Ca^+, Mg^{++}, on clay particles), precipitation or dissolution (e.g., precipitation of calcium as calcium sulfate or calcium carbonate). The anions (e.g., such as nitrate (NO_3^-), Cl^-, and Br^-), which are weakly adsorbed on positively charged sites, are known as nonreactive solutes. The transport of reactive and nonreactive solutes through soil is affected relative to the movement of water (Nielsen et al., 1986).

Some solutes are already present in the water-filled pore space of the soil. These solutes may be present in the soil owing to: (i) mineralization of organic matter, (ii) saline groundwater intrusion, (iii) fertilizer and/or pesticide application, (iv) atmospheric deposition, and (v) weathering of mineral. When solute-free water flows through the soil matrix, the concentration of these preexisting solutes is the highest in those pores experiencing the lowest water flux. Apart from the preexisting, solutes are also applied on soil surface (e.g., fertilizer, pesticides, etc.). Basically solute transport within a soil matrix occurs by two physical processes: diffusion and convective flow. Several simple and complicated mathematical models have been developed in the past, which can reproduce the experimental results very well. Most of these models are developed for the macroscopic scale (Nielsen et al., 1986), although pore scale description is available (e.g., Navier–Stokes equation). This chapter describes the transport mechanisms in more detail and discusses the transport models on a macroscopic scale.

16.2   SOLUTE TRANSPORT PROCESS

The movement of solutes inside the soil matrix is caused by "mass flow" or "convection." This type of flow is also called *Darcian flow* (see Chapter 12). The velocity at which solutes travel through soil matrix is generally known as "pore water velocity" and is the ratio of volumetric flow of solute through a unit cross-sectional area and volumetric moisture content of the soil matrix. In other words, the pore water velocity is the ratio of Darcian velocity and moisture content. In general, pore water velocity accounts for the straight-line length of path traversed in the soil in a given time. In reality, the flow paths are not always straight but are irregular or tortuous. This property is known as "tortuosity" of soil pores. Solutes do not always flow with water but sometimes go ahead of it due to the twin process of diffusion and dispersion or exclusion, lag behind due to adsorption or retardation, or get precipitated or volatilized. The movement of solute from the higher concentration to the lower concentration gradient is also known as the process of "diffusion." This process commonly occurs within gaseous and liquid phases in the soil matrix due to the random thermal motion, also called "Brownian movement." There is another simultaneous process that tries to mix and eventually even out the concentration gradients known as "hydrodynamic dispersion." Diffusion is an active process, whereas dispersion is a passive process. However, in most practical applications these two solute transport processes are considered additive.

Some chemicals, which are soluble in water and have a nonnegligible vapor phase, can exist in three different phases in a soil matrix: as a dissolved solute in soil water, as a gas in soil air, and as an ion absorbed on the soil organic matter or charged clay mineral surfaces. Therefore, all solute concentration terms are not equal in dimensions and depend on the concentration in these soil phases and the partitioning of these phases. The total solute resident concentration ($C, g\,cm^{-3}$) in a soil matrix can be mathematically expressed as

$$C = \rho_b C_a + \theta C_1 + f_a C_g \tag{16.1}$$

where ρ_b is the soil bulk density ($g\,cm^{-3}$), C_a is adsorbed concentration ($g\,g^{-1}$), θ is volumetric soil moisture content ($cm^3\,cm^{-3}$), C_1 is dissolved solute concentration ($g\,cm^{-3}$), f_a is the volumetric air content ($cm^3\,cm^{-3}$), and C_g is gaseous solute concentration ($g\,cm^{-3}$). Soil physical parameters (ρ_b, θ and f_a) weight the solute concentrations in the three phases of soil on a volume basis, and convert different reference dimensions to cm^3 of soil. The resident concentration is the volume-averaged concentration in soil, which is measured by extracting a known volume of soil in water. The resident concentration is expressed as the mass of solute per unit volume of soil water to make it comparable to flux-averaged concentration. The flux concentration is the solute concentration in water flowing through the soil.

16.3 MACROSCOPIC MIXING

Several different mechanisms operating in the porous media during transport of solute are responsible for the mixing at macroscopic level. Some of these include the following (Greenkorn, 1983):

1. *Molecular diffusion*: If the process is stationary or slow moving and the time required for the solute to move through the porous media is sufficiently long (i.e., for sufficiently long time scale) molecular diffusion is the primary source of macroscopic mixing.

2. *Tortuosity*: The tourtuous flow paths inside the soil profile causes the fluid element to remain at different distances from the same starting position even when they travel at the same pore water velocity (ratio of Darcy velocity and soil moisture content).

3. *Connectivity of pores*: If the pores are not well interconnected or if some of the pores in the porous media are not accessible to the fluid element flowing through that pore, they cause macroscopic mixing and dispersion.

4. *Hydrodynamic dispersion*: The solute element near the wall of pore travels at a different velocity than the element at the center of pore (Fig. 16.1a). This results in a velocity gradient inside the pore and solute elements move relative to each other at different velocity.

5. *Immobile zones*: The immobile water zones normally causes the fluid element to move quicker and out in the effluent solution earlier (early breakthrough), and at the same time, increases the tail of the breakthrough curve mainly due to the slow release of solute element trapped inside immobile water (see Sec. 16.12).

6. *Turbulence*: If the size of the pore abruptly changes, the flow inside a pore may become turbulent and mixing is caused by eddies.

7. *Adsorption*: When the concentration front looses some ions abruptly as they are removed from solution by the process known as adsorption, the unsteady state flow occurs and the concentration profiles becomes flat.

(a)

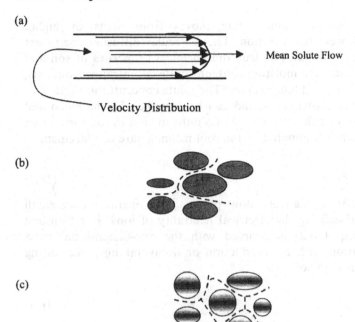

Mean Solute Flow

Velocity Distribution

(b)

(c)

FIGURE 16.1 The physical mechanisms for hydrodynamic dispersion of solutes through soil matrix: (a) influence of velocity distribution within a soil pore; (b) influence of size of pore, and (c) influence of microscopic flow direction.

16.4 FICK'S LAW

There are two Fick's laws, which describe diffusion of substances in porous media. The movement of ions from areas of higher concentration to lower concentration is proportional to the concentration gradient, the cross-sectional area available for diffusion, and the elapsed time during the solute transport. The net amount of solute crossing a plane of unit area in unit time is known as the solute flux density ($J; \mathrm{g\,cm^{-2}\,s^{-1}}$), which is given by Eq. (16.2) known as Fick's first law (1855) for steady state one-dimensional solute transport:

$$J = -D_m \frac{\partial C}{\partial x} \tag{16.2}$$

where D_m is the ionic or molecular diffusion coefficient of the porous media ($\mathrm{cm^2\,s^{-1}}$), C is the solute concentration ($\mathrm{g\,cm^{-3}}$) and x is the distance (cm). The concentration gradient ($\partial C/\partial x$) in Eq. (16.2) is the driving force and the minus sign indicates that solute moves from areas of higher concentration to lower concentration. The molecular diffusion coefficient in Eq. (16.2) varies with soil physical and chemical properties of soil and solute (i.e., soil texture, soil moisture content, solute cocentration, and pH), soil solute interactions, and temperature. The solute concentration follows a normal, or Gaussian, distribution and can be described by the mean and variance. The depth of penetration (X_p) of a diffusing ion in soil for a given time duration (t) can be estimated by the root mean-square displacement as follows:

$$X_p = (2D_m t)^{1/2} \tag{16.3}$$

Diffusion in soils is a relatively slow process and operates over small distances, thus maintaining the electrical neutrality of ions. For transient state condition, Eq. (16.2) is coupled with the one-dimensional mass conservation equation with no production or decay taking place during solute transport through soil

$$\frac{\partial C}{\partial t} = -\frac{\partial J}{\partial x} \tag{16.4}$$

Equation (16.4) implies that the net change in solute concentration is as a result of net change in rate of flow. Combining Eqs. (16.2) and (16.4) and assuming that D_m is independent of solute concentration and depth, results in Fick's second law for one-dimensional transient solute flow

$$\frac{\partial C}{\partial t} = D_m \frac{\partial C^2}{\partial x^2} \tag{16.5}$$

16.5 TRANSPORT EQUATIONS

When a solute enters a soil matrix (which can be in a soil core, repacked soil column, or agricultural soil in a field) the initial sharp boundary between the resident and displacing solute starts diminishing mainly due to the twin processes of diffusion and dispersion. The transport of a solution through soil matrix consists of three main components: convection, diffusion, and dispersion, which are briefly described below.

16.5.1 Convection or Mass Transport

Convective or advective transport of a solution inside a soil matrix is known as the passive movement with flowing soil water. If the transport process has only convective transport without any diffusion, the water and solute move at the same average flow rate. Mathematically convective transport (J_m) can be expressed as

$$J_m = q_s C \tag{16.6}$$

where J_m is the flux density for convective or mass transport ($ML^{-2}T^{-1}$), q_s is the volumetric fluid flux density with dimensions of velocity (LT^{-1}), and C is the volume averaged solute concentration (ML^{-3}). The flux density of water can be calculated by the Darcy equation for a steady state flow of water. The q_s is also analogous to θ, where v is the pore water velocity (LT^{-1}).

16.5.2 Diffusive Transport

Diffusion is a spontaneous process resulting from the random thermal motion of dissolved ions and molecules. In general, the diffusion is an active process and diffusive transport tends to decrease the existing concentration gradients and moves the process towards homogeneity rather rapidly. Fick's law defines the diffusive transport and for one-dimensional steady state transport is given as:

$$J_D = -D_m \frac{\partial C}{\partial x} \tag{16.7}$$

where J_D is solute flux density for diffusive transport of solute ($ML^{-2}T^{-1}$), θ is the volumetric moisture content (L^3L^{-3}). The diffusion coefficient in soils (D_m) is slightly less than the diffusion coefficient in pure water (D_0) mainly due to the tortuous flow paths in soils.

$$D_m = D_0 \theta \xi \tag{16.8}$$

where ξ is the dimensionless tortuosity factor ranging roughly from 0.3 to 0.7 for most soils.

16.5.3 Dispersive Transport

The soil matrix consists of pores of different shapes, sizes, and orientation. This heterogeneity of pore structure causes a large deviation of local pore water velocities inside each individual pore. Consider a one-dimensional flow through a single capillary tube of constant radius R. According to Poiseuille's law, the flow rate through each pore varies proportional to the fourth power of the radius R (Kutilek and Nielsen, 1994). However, the flow velocity (v) through the tube is a decreasing function of radial distance (r) from the center of tube. If average velocity is v' then $v = 2v'(1 - (r^2/R^2))$, when $r = R$, i.e., at the wall of pore $v = 0$, and at $r = 0$, i.e., at the center of pore $v = 2v'$. It is, therefore, clear that microscopic scale variations of pore water velocity in the soil matrix are very important and large.

Dispersive transport occurs because of the velocity variations in soil matrix with respect to average pore water velocity. The velocity variations in a soil matrix is caused by several factors such as zero velocity at the particle surface, which increases gradually and is the maximum at the center of pore or at air water interface under unsaturated conditions (Fig. 16.1a). Pore sizes also create velocity gradients with the velocity in larger pores greater than the velocity in smaller pores (Fig. 16.1b). The other possible reason is the fluctuation of flow paths of an element of water with respect to the mean direction of flow (Fig. 16.1c). Macroscopically, dispersion process is similar to the diffusion process, however, unlike diffusion, it occurs only during water movement. Field and laboratory experiments have shown that the dispersive transport can be described by an equation similar to diffusion as follows:

$$J_h = -\theta D_h \frac{\partial C}{\partial x} \tag{16.9}$$

where D_h is the mechanical dispersion coefficient (Bear, 1972) and is assumed to be a function of fluid velocity as follows:

$$D_h = \lambda v^n \tag{16.10}$$

where λ is the dispersivity and exponent "n" is an empirical constant generally assumed equal to 1.

The mixing or dispersion that occurs along the direction of flow path is called longitudinal dispersion and that in the direction normal to flow is known as transverse dispersion. Diffusion is an active process whereas dispersion is passive, in spite of this, most analysis on solute transport considers both processes to be additive because macroscopically both processes are similar.

$$D = D_m + D_h \tag{16.11}$$

where D is the longitudinal hydrodynamic dispersion coefficient (Bear, 1972) or apparent dispersion coefficient (Nielsen et al., 1972).

Combining Eqs. (16.6), (16.7), (16.9), and (16.11) leads to the following expression for solute flux, J_s

$$J_s = -\theta D \frac{\partial C}{\partial x} + qC \tag{16.12}$$

The equation of continuity states that:

$$\frac{\partial J_s}{\partial x} = -\frac{\partial}{\partial t}(\theta C + \rho_b S_s) \tag{16.13}$$

where S_s is adsorbed concentration (MM^{-1}), ρ_b is the bulk density (ML^{-3}), and t is time (T). Combining Eqs. (16.12) and (16.13) gives the following solute transport equation

$$\frac{\partial}{\partial t}(\theta C + \rho_b S_s) = \frac{\partial}{\partial x}\left(\theta D \frac{\partial C}{\partial x} - qC\right) \tag{16.14}$$

It is well known that adsorption and exchange processes are usually nonlinear and also depend on the competing species in the soil system. Still, one of the most common approaches to describe the relationship between adsorbed and solution concentrations has been to assume instantaneous adsorption and linearity between C and S of the form (forcing the constant or intercept to zero)

$$S_s = K_D C \tag{16.15}$$

where K_D is the empirical distribution coefficient. Inserting Eq. (16.15) into Eq. (16.14) and dividing both sides with θ results in Eq. (16.16):

$$\frac{\partial}{\partial t}\left(C + \frac{\rho_b K_D C}{\theta}\right) = \frac{\partial}{\partial x}\left(D \frac{\partial C}{\partial x} - \frac{q}{\theta}C\right) \tag{16.16}$$

Assuming that the soil profile is homogeneous and moisture content and flux density are constant in time and space, Eq. (16.16) reduces to

$$R \frac{\partial C}{\partial t} = D \frac{\partial^2 C}{\partial x^2} - v \frac{\partial C}{\partial x} \tag{16.17}$$

where R is the retardation factor and is given by

$$R = 1 + \frac{\rho_b K_D}{\theta} \tag{16.18}$$

K_D in Eq. (16.15) can be obtained from the slope of sorbed concentration (MM^{-1}) versus solution concentration (ML^{-3}). A zero value of K_D in Eq. (16.18) reduces R to 1, which indicates no interactions between solute and soil. A negative value of K_D makes R less than one, which indicates anion exclusion or immobile water, which does not contribute to convective transport. In case of anion exclusion, $(1 - R)$ is known as anion exclusion volume. A positive K_D results in $R > 1$, which indicates sorption.

16.6 BREAKTHROUGH CURVES

When a fluid (or solute) is passed through a soil matrix containing another liquid in its pore space, the introduced fluid, which can also be called the displacing liquid or applied liquid, gradually displaces the preexisting liquid (displaced liquid). Analysis of the collected effluent from soil matrix at a given depth (or from one end of a repacked soil column) shows a change in composition of effluent solution with respect to time. If the displacing and displaced solutions are not mutually soluble, the process is called "immiscible" displacement (e.g., oil and water). On the other hand, if both solutions are soluble, the process is called "miscible" displacement (e.g., aqueous solutions). The graphical representation of the concentration of these solutes with respect to time or cumulative effluent volume or pore volume is known as "breakthrough curves" (BTC). Pore volume is the ratio of cumulative effluent volume (cm^3) at a specified time and total volumetric moisture content of soil (cm^3). Pore volume is a nondimensional number and is zero at time zero.

16.6.1 Solute Input

As is evident in Figs. 16.2a–c, BTCs can have different shapes depending upon the solute application. Figure 16.2a shows a BTC where effluent solute concentration increases and reaches a maximum and then remains constant thereafter. The y-axis on Fig. 16.2 is the relative solute concentration (C/C_0), which is the ratio of concentration of effluent solute collected at a given time (C) and the concentration of displacing or incoming solution (C_0). The BTC in Fig. 16.2a is for a step input of displacing solute or tracer, where applied solution displaces all the preexisting solution gradually.

(a)

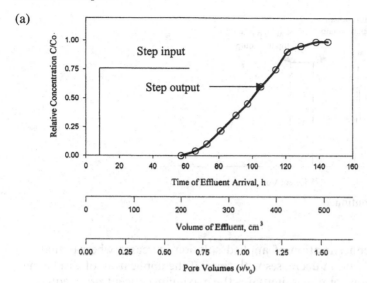

(b)

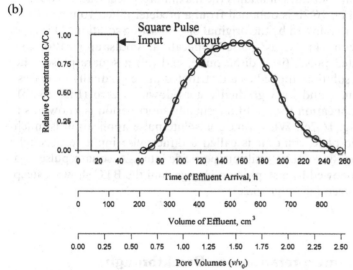

FIGURE 16.2 Breakthrough curves with respect to time of effluent arrival, volume of effluent, and pore volumes. (a) Chloride application as a step input through a 10 cm loam soil column (pore water velocity $= 0.11\ cm.h^{-1}$); (b) chloride application as a pulse input through a 10 cm loam soil column (pore water velocity $= 0.1\ cm.h^{-1}$); and (c) schematic for a Dirac and square pulse input and output. (Modified from Shukla et al., 2002.)

(c)

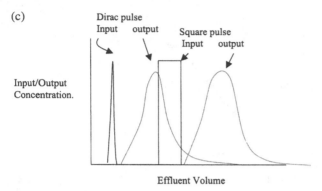

FIGURE 16.2   Continued.

Therefore, the concentration of applied solution increases whereas that of the preexisting solution decreases with time. If the application of displacing or applied solution continues, it attains the maximum concentration equal to C_0. The BTC in Fig. 16.2b is obtained from a predetermined volume of the displacing solution followed by the original or preexisting solution. This type of solute application is known as "pulse" application. A pulse application can be: (i) a distributed pulse, (ii) a dirac pulse, and (iii) a square pulse. The concentration of solution applied as a distributed pulse gradually increases, attains a maximum, and then gradually goes down to zero (Fig. 16.2b). A solute pulse application for an infinitesimally short period is known as a "dirac pulse" (Fig. 16.2c). When time for solute pulse application is much smaller than time of leaching, it is called a dirac pulse input (e.g., single application of highly soluble fertilizer, pesticide, etc.). A square pulse is a step-up change followed by a step-down change, and the BTC shows a steep rise followed by steep fall (Fig. 16.2c).

16.6.2 Some Interpretations of Breakthrough Curves

Pore volumes are defined as the ratio of the volume of displacing water (V, water entered or flowed out at a given time), and the volumetric moisture content of the soil (V/V_0). Assuming that the moisture content of soil in a repacked column is $0.5\,cm^3\,cm^{-3}$ (or 50%) and the total volume of soil column is $100\,cm^3$, therefore, volumetric moisture content of the repacked soil column is $50\,cm^3$. Once $50\,cm^3$ of displacing solution is passed through the soil column, it corresponds to a pore volume of 1.

Soil–Solute Interactions

The BTC in Fig. 16.3a depicts a condition when a solute of a given concentration displaces another solute (such as water) in such a way that all the soil pores start contributing at the same time and the solute concentration jumps from zero to the maximum (C_0) as soon as 1 pore volumes of displacing solution passes through the soil column. This type of flow is known as a "piston flow," which corresponds to pure mass flow or convection. In piston flow the entire center of solute front arrives at the end of column at the same time. Piston flow occurs in the absence of diffusion or dispersion or any type of interactions between solute and soil and solute and water move at the same velocity inside soil matrix. This type of flow is rare or near impossible under natural conditions. For known moisture content of soil and column dimensions, it is possible to calculate the number of pore volumes required before piston flow begins. The time required for a displacing solute to reach the other end of a column is known as breakthrough time, residence time, or travel time (t^*), and is equal to L/v, where L is the length of soil column. For sorbing solutes the total travel time is obtained by multiplying R and L/v.

The BTC in Fig. 16.3b shows an early arrival of displacing solute in the effluent solution (less than 1 pore volume). This process takes place because of the difference in the velocity at which water and solute travel inside the soil domain. The solute travels ahead of water because of "molecular diffusion and hydrodynamic dispersion." The BTC presented in Fig. 16.3b passes through C/C_0 of 0.5 at pore volume of 1. The area A and area B of this figure are numerically equal. This BTC represents a "convective dispersion process" with no interaction between solute and soil. The BTC in Fig. 16.3c is slightly shifted or retarded towards the x-axis. This type of shift is known as "sorption." Opposite of sorption is "repulsion" or a phenomenon of "anion exclusion" when BTC moves away from x-axis (Kutilek and Nielsen, 1994) (Fig. 16.3d).

Influence of Displacement Length

With increasing displacement length, the tortuosity and pore size distributions of the soil also increases. For a given pore water velocity, the total resident time of the solute in the soil increases with increasing displacement length. Therefore, the total mixing by convection and diffusion also increases (Nielsen and Biggar, 1962). Figure 16.4 makes it abundantly clear that if a pulse of same amount is passed through soil columns of 10, 20, and 30 cm length, the progressive attenuation of the initial concentration takes

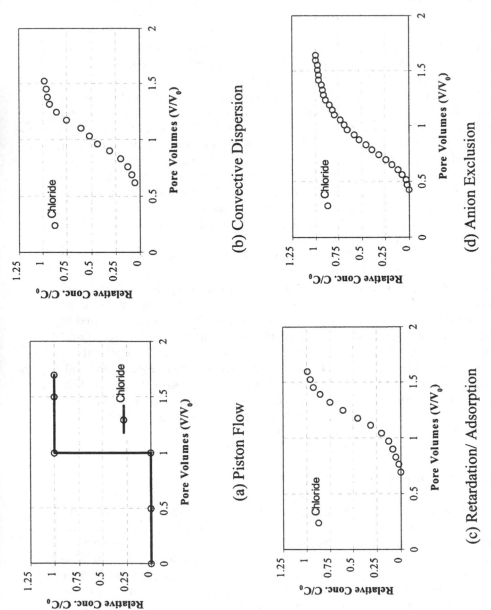

Figure 16.3 Interpretations from experimental breakthrough curves. (Modified from Kutílek and Nielsen, 1994; Shukla et al., 2002.)

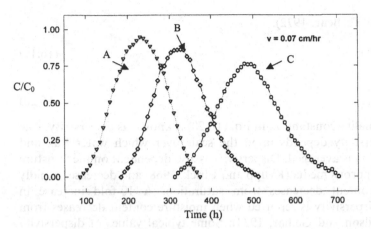

FIGURE 16.4 The progressive attenuation of BTCs for 10 cm (A), 20 cm (B), and 30 cm (C) soil columns for a pulse type chloride application through laboratory soil columns. (Redrawn from Shukla et al., 2000.)

place. This attenuation is the direct result of dilution. Therefore, solute applied as a pulse cannot carry its total mass beyond a certain depth. The total volumes of solution and total time required to completely displace the applied pulse increases with displacement length (Kutilek and Nielsen, 1994) (Fig. 16.4).

16.7 DISPERSION PROCESSES

Assuming the random capillary bundle concept (see Chapter 12), the classical dispersion theory was developed and a dispersion equation was suggested, which is similar to Fick's law and takes into account both dispersive and diffusive fluxes (Taylor, 1953; De Josselin De Jong, 1958; Bear and Bachmat, 1967; Fried and Combarnous, 1971). There are several mechanisms that cause macroscopic mixing and are generally accounted for in the dispersion coefficient. Some of them are mixing due to tortuosity, inaccessibility of pore water, recirculation due to flow restrictions, macroscopic and hydrodynamic dispersion, and turbulence in flow paths (Greenkorn, 1983). In addition, molecular diffusion, the presence of dead-end pores, sorption, exclusion, and physical nonequilibrium affect the degree of asymmetry in BTCs in different proportions (Nielsen et al., 1986).

The hydrodynamic dispersion coefficient (D) is proportional to the pore water velocity of a solute under steady state flow conditions (Biggar and Nielsen, 1967; Bear, 1972).

$$D \propto \frac{q}{\theta} \text{ or } v \tag{16.19}$$

$$D = \lambda v \tag{16.20}$$

The proportionality constant, λ, in Eq. (16.20) is known as dispersivity. The value of dispersivity depends upon the scale over which water flux and solute convection is averaged. Dispersivity is also dependent on the moisture content of the porous media (Krupp and Elrick, 1968) and decreases rapidly as moisture content decreases from saturation. A 10-fold increase in longitudinal dispersivity is reported when moisture content decreases from saturation (Wilson and Gelhar, 1974). Some typical values of dispersivity for laboratory soil columns range from 0.5 to 2 cm (Jury et al., 1991), 0.11 to 0.37 cm (for loam soil) and 0.14 to 0.22 cm (for sandy loam soil; Shukla et al., 2003). The dispersion processes are site specific and depend upon the subtler factors, which are related to the experimental conditions (Flury et al., 1998). The longitudinal dispersivity values are measured in field soils by placing a suction cup at different depths and measuring solute breakthough as a function of time. The dispersivity calculated for field soils by one-dimensional convective dispersion Eq. (16.23) or method of moments (Jury and Roth, 1990) are given in Table 16.1.

TABLE 16.1 Dispersivity Values Measured in Field Soils Using Suction Cups

Soil	Tracer	Application rate, cmd^{-1}	Dispersivity cm	Reference
Clay, silty clay	Cl, Tritium	2	9.4	Van de Pol et al. (1977)
Clay loam	Cl, NO$_3$	—	8.3	Biggar and Nielsen (1976)
Clay loam	Br	96	5.2–23	Fleming and Butters (1995)
Clay loam	Br	30,33,41,67	16–38	Jaynes (1991)
Loam	Cl	9.6–19.2	29	Roth et al. (1991)
Loamy sand	Br	1.1	3.2–15.8	Butters et al. (1989)
Loamy sand	Cl, NO$_3$, BO$_3$	1.3	1–2	Ellsworth et al. (1996)
Sand	Cl	84	0.7–1.6	Hamlen and
		132	0.8–2	Kachanowski (1992)
Sand	Cl	84, 117	17, 2.7	Van Wesenbeck and Kachanowski (1991)

Source: Modified from Flury et al., 1998.

16.8 RELATIONS BETWEEN DISPERSION COEFFICIENT AND PORE WATER VELOCITY

The effective dispersion coefficient generally varies with mean microscopic flow velocity. Based upon the magnitude of the Peclet number (P, defined as vL/D, where L is a characteristic length), within the range of average pore water velocities, molecular diffusion dominates the dispersion of the solute at smaller displacement velocities and gives way to convective dispersion at greater velocities. Hence, for relatively small average pore water velocities we expect the apparent diffusion coefficient to have values close to that of the diffusion coefficient (D_o) in the soil solution, and to be only somewhat dependent on pore water velocity. At relatively large velocities, the dispersion coefficient is strongly related to pore water velocity.

The five dispersion regimes can be identified in Fig. 16.5 as (i) pure molecular diffusion; (ii) molecular diffusion and kinematic dispersion; (iii) predominant kinematic dispersion and (iv) and (v) as pure kinematic dispersion regimes (Shukla et al., 2002). In regimes ii–v, an increase in average pore water velocity increases mixing and reduces the impact of molecular diffusion in the direction of flow. Using mixing cell approximations, it can be shown that in the region $0.01 < P < 50$, dispersion is directly proportional to pore water velocity (Perkins and Johnston, 1963). Further

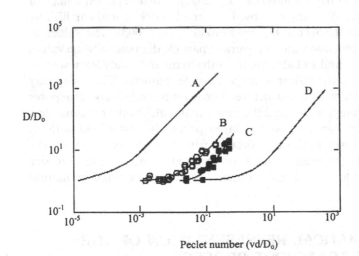

FIGURE 16.5 The relationship between DD_0^{-1} and Peclet number for: (A) field soil (A) (from Biggar and Nielsen, 1976), (B) loam, (C) sandy loam (from Shukla et al., 2002), and (D) single grain material (from Pfannkuch, 1963). (Redrawn from Shukla et al., 2002.)

increases in P results in a nonlinear relation to velocity ($P \propto v^n$, with $n > 1$). Pfannkuch (1962) and Torelli and Scheidegger (1972) reported an n value of 1.2, Taylor (1953) of 2, Biggar and Nielsen (1976) of 1.11, and Shukla et al. (2002) of 1.71 for sandy loam and 1.21 for loam.

The relations between D/D_o and Peclet number (vd/D_o) given as solid lines in Fig. 16.5 for natural undisturbed field soil by Biggar and Nielsen (1976), and for laboratory columns of loam and sandy loam soils (Shukla et al., 2002) and for graded sands and other single-grained materials (Pfannkuch, 1962) satisfy Eq. (16.21)

$$\frac{D}{D_o} = 1 + m\left(\frac{vd}{D_o}\right)^n \tag{16.21}$$

where

$$D_o = 0.66\,\theta\,D_m \tag{16.22}$$

with D_m being the diffusion coefficient for free solution. The D_0 in Eq. (16.22) can be obtained from known values of θ and D_m (see also Chapter 18). For loam and the sandy loam soils D_0 is 0.0222 and $0.0216\,\mathrm{cm^2\,h^{-1}}$, respectively (Shukla et al., 2002). Other reported values of D_0 in literature are: $0.02\,\mathrm{cm^2\,h^{-1}}$ by Jury et al. (1991), $0.01\,\mathrm{cm^2\,h^{-1}}$ by Sposito (1989), and $0.0203\,\mathrm{cm^2\,h^{-1}}$ by Shukla et al. (2003). The values of m increase with decreasing values of average particle diameter d while values of n range between 1 and 2 (Table 16.2). In the loam and sandy loam soils as well as the field soil, decreasing average particle diameter (increasing clay content) is associated with soil structure. The loam has relatively large pores as a result of microaggregates, and the sandy loam, although containing less clay than the loam, nevertheless has large pores also associated with its microaggregates as well as those associated with its high sand content. The field soil manifests the greatest value of m because of its large pore size distribution owing to its high clay content, its aggregation and its natural field structure.

16.9 MATHEMATICAL REPRESENTATION OF THE SOLUTE TRANSPORT PROCESS

The simplest form of one dimensional convective–dispersive equation (CDE), assuming macroscopic steady state water flow, constant soil–moisture content, and no interactions between the chemical and the solid

TABLE 16.2 Parameters for Eq. (21) for the Results Illustrated in Fig. 16.5

Soil	m	n	d mm	D_o cm^2 h^{-1}
Laboratory columns (mostly sand) (Pfannkuch 1962)	0.5	1.2	0.156	0.022
Sandy loam (Shukla et al. 2001)	70.5	1.71	0.0508	0.0216
Loam (Shukla et al. 2001)	141	1.21	0.0158	0.0222
Field soil (more clay) (Biggar and Nielsen 1976)	17780	1.11	0.00272	0.0250

phase was developed by Lapidus and Amundson, (1952), which is similar to Eq. (16.17), for $R = 1$

$$\frac{\partial C}{\partial t} = D \frac{\partial^2 C}{\partial x^2} - v \frac{\partial C}{\partial x} \tag{16.23}$$

One additional term is added to Eq. (16.23) when chemical adsorption is included. Following is the one-dimensional solute transport equation describing transport through a homogeneous medium during steady state flow with adsorption [same as Eq. (16.17)]

$$\frac{\partial C}{\partial t} + \frac{\rho_b}{\theta} \frac{\partial S_s}{\partial t} = D \frac{\partial^2 C}{\partial x^2} - v \frac{\partial C}{\partial x} \tag{16.24}$$

The solution of Eq. (16.24) depends upon the knowledge of the relationship between adsorbed concentrations, S_s, and the solution concentration, C. Adsorption or exchange reactions perceived as instantaneous are described by equilibrium isotherms $S_s(C)$, which can be of the mass action, linear, Freundlich, Langmuir, or any other functional form (Nielsen et al., 1986). Besides adsorption, the reactive process such as first-order degradation and zero-order production can also be taken into account during miscible displacement process. Therefore, the comprehensive CDE for one-dimensional transport of reactive solutes, subject to adsorption, first-order degradation, and zero-order production, in a homogeneous soil, is written as:

$$\frac{\partial}{\partial t}(\theta C_r + \rho_b S_s) = \frac{\partial}{\partial x}\left(\theta D \frac{\partial C_r}{\partial x} - vC\right) - \theta \mu_\ell C_r - \rho_b \mu_s S + \theta \gamma_\ell(x)$$

$$+ \rho b \gamma_s(x) \tag{16.25}$$

where C_r is the volume-averaged or resident concentration of liquid phase (ML^{-3}), S_s is the concentration of the adsorbed phase (MM^{-1}), v is the volumetric water flux density (LT^{-1}), μ_l and μ_s are first-order decay coefficients for degradation of the solute in the liquid and adsorbed phases respectively (T^{-1}), γ_1 $(ML^{-3}T^{-1})$, and γ_s $(MM^{-1}T^{-1})$ are zero-order production terms for the liquid and adsorbed phases, D, θ, ρ_b, x, and t are the same as defined above. Assuming reversible equilibrium adsorption [Eq. (16.16)] and steady state flow in a homogeneous soil, Eq. (16.25) is modified to:

$$R\frac{\partial C_r}{\partial t} = D\frac{\partial^2 C_r}{\partial x^2} - v\frac{\partial C_r}{\partial x} - \mu C_r + \gamma(x) \tag{16.26}$$

where μ and γ are combined first- and zero-order rate coefficients

$$\mu = \mu_\ell + \frac{\rho_b K_D}{\theta} \tag{16.27}$$

$$\gamma(x) = \gamma_\ell(x) + \frac{\rho_b \gamma_s(x)}{\theta} \tag{16.28}$$

16.10 SORPTION PHENOMENON

Adsorption is a process where ions or molecules are attached to the surfaces of soil solids. This results in a higher concentration of solute at the surface of solid phase than in the bulk solution. The opposite of adsorption is anion exclusion where concentration in a soil solution is higher than the solid phase. Sorption and exclusion processes are important in modifying the movement of chemicals through a soil domain. The plot between amount adsorbed and the amount in solution is known as the adsorption isotherm (Fig. 16.6). The forces active at soil–water interface and at molecular level are electrical and are the same at both levels. These forces vary as the reciprocal of the separation distance raised to a power. Equilibrium sorption (Fig. 16.6) of organic molecules is dominated by the organic fraction of soil.

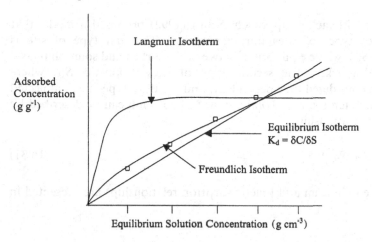

FIGURE 16.6 A schematic of adsorption isotherms. (Modified from July et al., 1994)

To account for this effect, value of K_D [Eq. (16.15)] is divided by soil organic carbon content (SOC) as below:

$$K_D = SOC * f_{OC} \qquad (16.29)$$

The Freundlich adsorption model is given as $S_s = K_D C_l^n$, where n is close to 1 (Fig. 16.6). The Freundlich model is based on the assumption that there is no limiting concentration of adsorbate as solution concentration is increased without limit. This is unrealistic because available surfaces in soil domain are limited for adsorption to occur. The Langmuir adsorption model was developed from kinetics of gas adsorption on solid surfaces and has a sound conceptual basis. The model assumes that the energy of adsorption is constant and independent of surface coverage, the adsorbed molecules do not interact with each other, and the maximum possible adsorption is that of a complete monolayer (Fig. 16.6). The equilibrium adsorption (S_s) by Langmuir model is as follows:

$$S_s = \frac{aQC_l}{1 + aC_l} \qquad (16.30)$$

where a is the ratio of adsorption rate constant, Q is the total number of available adsorption sites, and C_l is the solute concentration in solution. Several sorption models are available in literature, some are derived from the adsorption of gases by solids while others are either empirical or kinetic.

Based upon multireaction approach, Selim (1992) proposed a model that involves three types of sites during sorption. The first type of site is equilibrium (S_{se}), where equilibrium between the sorbed and solution phases is established quickly. The second type of sites is kinetic, S_{sk}, where adsorption is considered time dependent, and the third type-site is subjected to irreversible retention S_{sir}. Total amount of sorption can be described by the following relationship

$$S_s = S_{se} + S_{sk} + S_{sir} \tag{16.31}$$

Some of the equilibrium and kinetic sorption relationships are presented in Table 16.3.

16.11 EQUILIBRIUM ANION EXCLUSION MODEL

Certain anions interact with the negatively charged solid surfaces of the soil (such as clay or ionizable organic matter) and are excluded from the liquid region adjacent to the soil particle surfaces. This phenomenon is known as anion exclusion or negative adsorption. Eq. (16.17) represents the anion exclusion phenomenon for $R < 1$. In the presence of a soil solution, the negative charge extends from the surfaces of particles into the solution and forms diffuse double layer (Bolt, 1979). The existence of the negative charge causes repulsion of anions from this region. The resulting concentration gradient reduces the concentration of anion at the soil surface to zero, which increases exponentially with distance and at the limit of diffuse double layer becomes equal to the concentration of bulk solution (Bolt, 1979). Assuming that effective exclusion volume (θ_{ex}) expressed as volumetric moisture content is evenly distributed over the particle surface, the one-dimensional transport of an anionic solute exhibiting anion exclusion can be described as follows (Bresler, 1973; James and Rubin, 1986)

$$(\theta - \theta_{ex}) \frac{\partial C}{\partial t} = D \frac{\partial^2 C}{\partial x^2} - q \frac{\partial C}{\partial x} \tag{16.32}$$

The observed concentration (C) is less than the concentration of bulk solution (C_0) because of the exclusion volume, which does not contain ions. This interrelationship between C and C_0 can be expressed as follows:

$$C = C_0 \left[1 - \frac{\theta_{ex}}{\theta} \right] \tag{16.33}$$

TABLE 16.3 Equilibrium and Kinetic Models for Sorption in Soils[a]

Model	Formulation
EQUILIBRIUM TYPE	
Linear	$S_{se} = K_D C_l$
Freundlich	$S_{se} = K_D C_l^n$
General Freundlich	$S_s/S_{smax} = [\omega C/(1 + \omega C)]^\beta$
Rothmund-Kornfeld ion exchange	$S_{si}/S_{sT} = K_D(C_i/C_T)^n$
Langmuir	$S_s/S_{smax} = (\omega C)^\beta/(1 + \omega C)$
General Langmuir–Freundlich	$S_s/S_{smax} = (\omega C)^\beta/(1 + \omega C)^\beta$
Langmuir with sigmoidicity	$S_s/S_{smax} = (\omega C)/(1 + \omega C + \omega/C)$
KINETIC TYPE	
First order	$\partial S_s/\partial t = K_D(\theta/\rho_b)(C_l - K_{D1}S_s)$
nth order	$\partial S_s/\partial t = K_D(\theta/\rho_b)(C_l^n - K_{D1}S_s)$
Irreversible (sink/source)	$\partial S_s/\partial t = K_D(\theta/\rho_b)(C - C_p)$
Second-order irreversible	$\partial S_s/\partial t = K_D(\theta/\rho_b)C(S_{smax} - S_s)$
Langmuir kinetic	$\partial S_s/\partial t = K_D(\theta/\rho_b)C(S_{smax} - S_s) - K_D S_s$
Elovich	$\partial S_s/\partial t = A\exp(-BS_s)$
Power	$\partial S_s/\partial t = K_D(\theta/\rho_b)C^n S_s^m$
Mass transfer	$\partial S_s/\partial t = K_D(\theta/\rho_b)(C - C^*)$

[a]Where k, A, B, n m, S_s, S_{smax}, C^*, C_p, and ω are adjustable model parameters.
Source: Modified from Selim and Amacher, 1997.

If the sufficient volume of input solution (concentration $= C_0$) infiltrates in a soil column for a long duration, the excluded water content can be calculated by using Eq. (16.34) (Bond et al., 1982)

$$\theta_{ex} = \theta \left[1 - \frac{C}{C_0} \right] \tag{16.34}$$

The C in the soil profile is always lower than C_0 when anion exclusion is occurring. The anion exclusion also increases the average velocity of travel of anions in the soil profile. By excluding the anions from the diffuse double layer where water is either moving slowly or is immobile, the rate of transport is greater than given by q/θ. Bolt (1979) assumed anion exclusion to be evenly distributed over the soil surface of thickness d_{ex}.

$$d_{ex} = \frac{Q}{\sqrt{\beta N}} - \delta \tag{16.35}$$

where β is a constant (1.06×10^{19} mkeq^{-1} at 25°C), N is the total normality of bulk solution (keq m^{-3}), Q is a factor for ionic composition of bulk solution (m^{-1}), and δ a correction term. The specific surface area (A_r) can be calculated from exclusion volume and bulk density (ρ_b; Mg m^{-3}) as follows

$$A_r = \frac{\theta_{ex}}{d_{ex}\rho_b} * 10^{-6} \tag{16.36}$$

16.12 NONEQUILIBRIUM TRANSPORT

The application of Eq. (16.24) or (16.26) to transport through laboratory soil columns or in fields having relatively uniform soils involving nonreactive or weakly reactive solutes was found to be fairly successful (Biggar and Nielsen, 1976; Jaynes, 1991; Ellsworth et al., 1996; Shukla et al., 2003). The BTCs for these tracers are symmetrical and mass recoveries are relatively high (Fig. 16.7). However, for strongly adsorbed chemicals and aggregated soils these equations do not perform very well (van Genuchten and Wierenga, 1976; Nkedi-Kizza et al., 1984).

During solute transport in heterogeneous soils, the assumption of local equilibrium implies instantaneous interchange of mass, large residence time sufficient to make concentration gradients negligible, and high degrees of interactions between macroscopic transport properties and microscopic soil

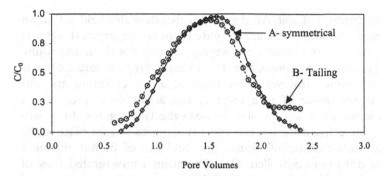

FIGURE 16.7 Schematic of equilibrium and nonequilibrium transport of a tracer through laboratory soil columns, the BTC "A" is symmetrical and mass recoveries are higher than a asymmetrical BTC "B".

physical properties. Some of the macroscopic transport properties are water flux, apparent dispersion, and moisture content, and microscopic properties are aggregate size, exchange, pore geometry. The microscopic properties impose a rate limiting effect on solute transport through heterogeneous soils and deviations from local equilibrium conditions are observed. The mass recoveries, for these asymmetrical and nonsigmoidal concentration distributions or BTCs, are less and the BTCs have a long tail (Fig. 16.7). Such a deviation is caused by a number of physical and chemical nonequilibrium processes. The physical nonequilibrium is caused by a heterogeneous flow regime and a chemical nonequilibrium by the kinetic adsorption. This paves the way for the examination of diffusion controlled or chemically controlled kinetic rate reactions or both of the form $\partial S/\partial t = f(S_s, C)$. The following sections will examine briefly the nonequilibrium processes arising out of physical or chemical nonequilibrium.

16.13 TWO-REGION NONEQUILIBRIUM TRANSPORT MODEL

There are several factors responsible for physical nonequilibrium conditions occurring in a soil system during solute transport. Some of them are: (i) heterogeneity of pore size distribution or aggregation; (ii) heterogeneous diffusion into the Neurst film of water surrounding soil particles than soil bulk solution. Physical nonequilibrium is represented by a two-region (dual porosity) type formation. In this case, the medium is assumed to contain two distinct mobile (flowing) and immobile (stagnant) liquid regions. The simplest explanation of a two region mobile and immobile formation is the water inside an aggregated soil. All the intraaggregate water held within an aggregate is immobile and the interaggregate (between aggregates) water is mobile. The water flowing around dry aggregates imbibes them and solute entry inside aggregate is by convection. For moist aggregate, solute entry is governed by diffusion. However, there must be a concentration gradient from outside to the inside of an aggregate, and a first-order process can adequately describe the mass transfer between the two regions. In a two-region model, convective diffusion transport is assumed to take place in the mobile region while transfer of solutes into and out of mobile region is assumed to be diffusion controlled. One-dimensional unsaturated flow of conservative nonsorbing solute in a soil is given as follows (Coats and Smith, 1964):

$$\theta_m \frac{\partial C_m}{\partial t} + \theta_{im} \frac{\partial C_{im}}{\partial t} = \theta_m D_m \frac{\partial^2 C}{\partial x^2} - \theta_m v_m \frac{\partial C_m}{\partial x} \qquad (16.37)$$

$$\theta_{im} \frac{\partial C_{im}}{\partial t} = \alpha(C_m - C_{im}) \qquad (16.38)$$

where t is time (T); C_m and C_{im} are the solute concentrations in the mobile and immobile liquid phases (ML^{-3}) with corresponding volumetric moisture contents θ_m and θ_{im} (L^3/L^3) respectively; D_m is apparent diffusion coefficient of mobile liquid phase (L^2T^{-1}); x is the distance from the inflow boundary in the direction of flow (L); v_m is the average mobile pore water velocity in (LT^{-1}); and α is the first order rate coefficient (T^{-1}).

In Eqs. (16.37) and (16.38) as the ratio of mobile water fraction (θ_m) to total moisture content (θ) increases (i.e., θ_m increases), more and more of the wetted pore space is included in the transport, which causes greater and more complete mixing, and the BTC shifts further to the right. At the extreme end, the $\theta_m = \theta$, where the above equation reduces to one-dimensional CDE [Eq. (16.23)]. The parameter α, which has the dimensions of T^{-1}, can vary from 0 to ∞. A zero value of α indicates no mixing between mobile and immobile water fractions. Therefore, the term on left-hand side of Eq. (16.38) equals zero and Eq. (16.37) reduces to one-dimensional CDE, similar to Eq. (16.23) but with total moisture content of θ_m. When $\alpha = \infty$, the two concentrations mix instantaneously and $C_m = C_{im}$. In this case Eq. (16.37) reduces to Eq. (16.23).

One-dimensional solute transport for an exchanging solute during steady-state flow through a homogeneous porous medium, where the liquid phase is presumed to consist of a mobile and immobile region and includes a

Freundlich-type equilibrium adsorption–desorption processes (van Genuchten and Wierenga, 1976) can be described by a two-region model as follows:

$$\theta_m \frac{\partial C_m}{\partial t} + f\rho_b \frac{\partial S_{sm}}{\partial t} + \theta_{im} \frac{\partial C_{im}}{\partial t} + (1-f)\rho_b \frac{\partial S_{sim}}{\partial t}$$

$$= \theta_m D_m \frac{\partial^2 C}{\partial x^2} - \theta_m v_m \frac{\partial C_m}{\partial x} \qquad (16.39)$$

and

$$\theta_{im} \frac{\partial C_{im}}{\partial t} + (1-f)\rho_b \frac{\partial S_{sim}}{\partial t} = \alpha(C_m - C_{im}) \qquad (16.40)$$

where S_{sm} and S_{sim} are concentration of adsorbed phase in mobile and immobile phase respectively (MM^{-1}); R_m and R_{im} are retardation factors accounting for equilibrium type adsorption processes in mobile and immobile regions, respectively; and parameter f represents the mass fraction of solid phase that is in direct contact with the mobile liquid phase. If the exchange process in both the dynamic (S_{sm}) and stagnant (S_{sim}) region is assumed to be instantaneous, linear and reversible process (van Genuchten, 1981) then,

$$S_{sm} = K_D C_m \quad \text{and} \quad S_{sim} = K_D C_{im} \qquad (16.41)$$

and the total adsorption can be represented by

$$S_s = f S_{sm} + (1-f) S_{sim} \qquad (16.42)$$

For equilibrium adsorption, transferring these into Eqs. (16.39) and (16.40) results in following set of equations

$$(\theta_m + \rho_b f k) \frac{\partial C_m}{\partial t} + [\theta_{im} + (1-f)\rho_b k] \frac{\partial C_{im}}{\partial t} = \theta_m D_m \frac{\partial^2 C_m}{\partial x^2} - \theta_m v_m \frac{\partial C_m}{\partial x}$$

$$(16.43)$$

$$[\theta_{im} + (1-f)\rho_b K_D] \frac{\partial C_{im}}{\partial t} = \alpha(C_m - C_{im}) \qquad (16.44)$$

16.14 TWO-REGION ANION EXCLUSION MODEL

The two-region anion exclusion model divides the total soil–water phase into two compartments, (i) mobile water and (ii) immobile water, and anion exclusion is assumed to take place in the immobile region (van Genuchten, 1981). This assumption is analogous to the assumptions made by Krupp et al. (1972), as anion exclusion takes place in the smaller pores inside the dense aggregate or in the immobile water along the pore wall. An equivalent exclusion distance (d_{ex}) exists near the pore wall where concentration remains zero. Therefore, specific exclusion volume (V_{ex}; cm^3 water g^{-1} of soil) is related to specific surface area (A_m; cm^2 g^{-1}) and d_{ex} as follows:

$$V_{ex} = d_{ex} A_m \tag{16.45}$$

The θ_{ex} is obtained by multiplying Eq. (45) by soil bulk density (ρ_b)

$$\theta_{ex} = V_{ex} \rho_b \tag{16.46}$$

The part of liquid phase unaffected by anion exclusion (θ_a) can be calculated as follows:

$$\theta_a = \theta_{im} - \theta_{ex} \tag{16.47}$$

Using Eq. (16.47), the following physical nonequilibrium equation representing the anion exclusion process is obtained (van Genuchten, 1981)

$$\theta_m \frac{\partial C_m}{\partial t} + \theta_a \frac{\partial C_a}{\partial t} = \theta_m D \frac{\partial^2 C_m}{\partial x^2} - \theta_m v_m \frac{\partial C_m}{\partial x} \tag{16.48}$$

$$\theta_a \frac{\partial C_a}{\partial t} = \alpha(C_m - C_a) \tag{16.49}$$

where C_a is concentration in the part of immobile zone unaffected by exclusion. The model described above assumes anion exclusion taking place inside the immobile water zone. Therefore, convective transport in mobile zone remains unaffected by the exclusion process and C_m never exceeds input concentration C_0 (van Genuchten, 1981).

16.15 TWO-SITE NONEQUILIBRIUM TRANSPORT MODEL

Considering that the solid phase of soil is made up of various constituents (i.e., soil minerals, organic matter, aluminum, and iron oxides), and chemical react with these different constituents at different rates and intensities. Selim et al. (1976) and Cameron and Klute (1997) proposed a two-site chemical nonequilibrium model where adsorption term consists of two components, equilibrium adsorption, and first-order kinetics. The sorption or exchange sites in this model are assumed to have instantaneous adsorption (type-1 sites) and time-dependent kinetic adsorption (type-2 sites). At equilibrium, adsorption on both types of sorption sites is described by the following linear equations:

$$S_{s1} = K_{De}C = FK_DC \tag{16.50}$$

$$S_{s2} = K_{Dk}C = (1 - F)K_DC \tag{16.51}$$

where subscript "e" refers to type 1 or equilibrium site and subscript "k" refers to type 2 or kinetic sites, respectively, and F is the fraction of all sites occupied by type 1 sorption sites. Total adsorption at equilibrium is

$$S_s = S_{se} + S_{sk} \tag{16.52}$$

Because type 1 sites are always at equilibrium therefore,

$$\frac{\partial S_{se}}{\partial t} = FK_D \frac{\partial C}{\partial t} \tag{16.53}$$

The adsorption rate for type 2 kinetic nonequilibrium sites can be given by a linear and reversible first order equation of following form

$$\frac{\partial S_{sk}}{\partial t} = \alpha[(1 - F)K_DC - S_{sk}] \tag{16.54}$$

where α is the first order rate coefficient. Combining above equations with Eq. (16.14) lead to following formulation (van Genuchten, 1981; Nkedi-Kizza et al., 1984):

$$\left(1 + \frac{F\rho_b K_D}{\theta}\right) \frac{\partial C}{\partial t} + \frac{\rho_b}{\theta} \frac{\partial S_{sk}}{\partial t} = D \frac{\partial^2 C}{\partial x^2} - v \frac{\partial C}{\partial x} \tag{16.55}$$

$$\frac{\partial S_{sk}}{\partial t} = \alpha[(1 - F)K_DC - S_{sk}] \tag{16.56}$$

16.16   INITIAL AND BOUNDARY CONDITIONS FOR STEP INPUT EXPERIMENTS

The analytical solutions of Eqs. (16.23), (16.24), (16.25), (16.43), (16.44), (16.48), (16.49), (16.55), and (16.56) are available for a large number of initial and boundary conditions for both finite and semi-infinite systems for both step and pulse type solute application (van Genuchten 1981, van Genuchten and Alves, 1982). This section briefly describes some of the initial and boundary conditions required for solving solute transport equations. The most common initial condition for any soil is:

$$C(x, 0) = C_i \qquad (16.57)$$

At the upper boundary of the soil surface or (or inflow into the soil column; i.e. at $x = 0$), two different boundary conditions can be considered. The first type or constant concentration boundary condition is of the form as follows:

$$C(0, t) = C_0 \qquad (16.58)$$

For column displacement experiments, where chemical is applied at a constant rate, the boundary condition (16.58) leads to mass balance errors, which become quite significant for large values of (D/v) (van Genuchten, 1981, Parker and van Genuchten, 1984). The other boundary condition is a third type, or constant flux type, that leads to the conservation of mass inside the soil column provided dispersion outside the soil can be ignored is given as follows:

$$\left[-D\frac{\partial C}{\partial x} + vC \right]\Big|_{x=0} = vC_0 \qquad (16.59)$$

A third type inlet condition is usually preferred over first type inlet condition (van Genuchten and Parker, 1984, Toride et al., 1993). In order to describe the outlet conditions, it is assumed that the concentration is macroscopically continuous at the outlet and no dispersion occurs outside the soil. Parker and van Genuchten, (1984) suggested that by assuming that the upstream solute concentrations are not affected by the outlet boundary, solutions for an infinite outlet condition can be applied to the finite region. The outlet condition for a semi infinite profile $(0 \leq x < \infty)$ and a finite system of length L can be specified in terms of zero concentration gradient as below

$$\left(\frac{\partial C}{\partial x}\right)(\infty, t) = 0 \qquad (16.60)$$

$$\left(\frac{\partial C}{\partial x}\right)(L, t) = 0 \qquad (16.61)$$

The boundary condition [Eq. (16.60)] assumes a semi-infinite soil column and is commonly used. When effluent curves from finite columns are calculated using analytical solutions based on boundary condition [Eq. (16.60)], some errors may be introduced. Therefore, zero concentration gradient at the upper end of the column as specified by Eq. (16.61) is frequently used for column displacement studies. However, there is no evidence available to prove that the boundary condition Eq. (16.61) leads to a better description of physical processes at and around $x = L$. On the other hand, the boundary condition [Eq. (16.59)] gives a discontinuous distribution at the inlet, which is against the requirement of a continuous distribution at $x = L$ (van Genuchten, 1981).

16.17 DIMENSIONAL INITIAL AND BOUNDARY CONDITIONS FOR PULSE APPLICATION

Assuming that the concentrations are continuous across the inlet boundary and that input solution is well mixed, a first type boundary condition across the inlet boundary for a pulse type injection can be specified as (van Genuchten, 1981):

$$C(0, t) = C_0 \qquad 0 < t \le t_0$$
$$C(0, t) = 0 \qquad t > t_0 \tag{16.62}$$

A third type boundary condition for the pulse input for a well mixed input solution can be specified as

$$[-D(\partial C/\partial x) + vC]|_{x=0} = vC_0 \qquad 0 < t \le t_0$$
$$[-D(\partial C/\partial x) + vC]|_{x=0} = 0 \qquad t > t_0 \tag{16.63}$$

The two-site model [Eqs. (16.55) and (16.56)] can be solved for the boundary and initial conditions given by Eqs. 16.57 to 16.61. One additional initial condition for the solution is

$$S_{sk}(x, 0) = (1 - F)K_D Ci \tag{16.64}$$

The initial condition and the boundary conditions at exit remain the same as described by Eqs. (16.52), (16.55), and (16.56). The boundary condition at inlet, Eq. (16.57), becomes inappropriate when the input solution is not well mixed. Other arguments against the applicability of Eq. (16.57) can be that the plane considered as a macroscopic boundary has no physical relevance at the microscopic level, as irregularity in pore structure and morphology become manifest at this level. Also the medium properties vary continuously over a finite transition zone of $l/2$, where l is the representative elementary volume (REV) of the porous medium (Parker and van Genuchten, 1984).

16.18 THE COMBINED NONDIMENSIONAL TRANSPORT EQUATIONS

Nonequilibrium transport Eqs. (16.31), (16.32), (16.36), (16.37), (16.43), and (16.44) are mathematically equivalent and transferring nondimensional quantities listed in Table 16.4 reduces them to the following combined nondimensional equations (van Genuchten, 1981; Nkedi-Kizza, 1984)

$$\beta R \frac{\partial C_1}{\partial T} + (1 - \beta) R \frac{\partial C_2}{\partial T} = \frac{1}{P} \frac{\partial^2 C_1}{\partial Z^2} - \frac{\partial C_1}{\partial Z} \tag{16.65}$$

$$(1 - \beta) R \frac{\partial C_2}{\partial T} = \omega(C_1 - C_2) \tag{16.66}$$

where β is partition coefficient, ω is nondimensional mass transfer parameter and P is peclet number. Initial and boundary conditions for a step type input are

$$C_1(x, 0) = C_2(x, 0) = 0 \tag{16.67}$$

$$-\frac{1}{P} \frac{\partial C_1}{\partial x} + C_1|_{x=0} = 1 \tag{16.68}$$

$$\frac{\partial C_1}{\partial x}(\infty, T) = \frac{\partial C_2}{\partial x}(\infty, T) = 0 \tag{16.69}$$

For $\beta = 1$, Eqs. (16.65) and (16.66) reduce to the nondimensional CDE. Some of the analytical solutions of Eqs. (16.17), (16.65), and (16.66) are given in Table 16.5.

16.19 ESTIMATION OF SOLUTE TRANSPORT PARAMETERS

The equilibrium solute transport equation [refer to Eq. (16.17)] has two parameters: (i) the apparent diffusion coefficient (D) or P (vL/D) and (ii) the retardation factor (R).

TABLE 16.4 Nondimensional Variables Introduced in the Solute Transport Equations

The nondimensional variables

FOR ALL THE EQUATIONS

$$T = \frac{vt}{L} \qquad Z = \frac{x}{L} \qquad P = \frac{vL}{D} \qquad R = 1 + \frac{\rho K_D}{\theta} \qquad C_1 = \frac{C - C_i}{C_0 - C_i}$$

FOR TWO SITE EQUATION

$$R_m = 1 + \frac{F\rho K_D}{\theta_m} \qquad \beta = \frac{\theta + F\rho K_D}{\theta + \rho K_D} = \frac{R_m}{R} \qquad \omega = \frac{\alpha_2(1-\beta)RL}{v} \qquad C_2 = \frac{S_{s2} - (1-F)K_D C_i}{(1-F)K_D(C_0 - C_i)}$$

FOR TWO REGION EQUATION

$$C_1 = C_m/C_0 \qquad P = \frac{v_m L}{D_m} \qquad \phi_m = \frac{\theta_m}{\theta} \qquad \omega = \frac{\alpha L}{q}$$

$$R_m = 1 + \frac{f\rho K_D}{\theta_m} \qquad R_{im} = 1 + \frac{(1-f)\rho K_D}{\theta_{im}} \qquad \beta = \frac{\theta_m + f\rho_b K_D}{\theta + \rho_b K_D} = \frac{\phi_m R_m}{R}$$

$$C_1 = \frac{C_m - C_i}{C_0 - C_i} \qquad C_2 = \frac{C_{im} - C_i}{C_0 - C_i} \qquad T = \frac{v_m \phi_m t}{L} \qquad q = \theta_m v_m$$

Source: Modified from van Genuchten, 1981.

TABLE 16.5 Analytical Solutions of Equilibrium CDE and Nonequilibrium (NE) Transport Equations

Dimensionless exit concentration	Concentration-type boundary conditions	Flux-type boundary conditions
	$C_1(0,T) = 1$ $\dfrac{\partial C_1(\infty, T)}{\partial Z} = 0$	$\left(-\dfrac{1}{P}\dfrac{\partial C_1(Z,T)}{\partial Z} + C_1(Z,T)\right)\Big|_{z=0} = 1$ $\dfrac{\partial C_1(\infty, T)}{\partial Z} = 0$

CDE

C_e

$$C_e = \frac{1}{2}\,\text{erfc}\left[\left(\frac{P}{4RT}\right)^{1/2}\cdot(R-T)\right] + \frac{1}{2}\exp(P)\cdot\text{erfc}\left[\left(\frac{P}{4RT}\right)^{1/2}\cdot(R+T)\right]$$

$$C_e = \frac{1}{2}\,\text{erfc}\left[\left(\frac{P}{4RT}\right)^{1/2}\cdot(R-T)\right] + \left(\frac{PT}{\pi R}\right)^{1/2}\cdot\exp\left[-\frac{P}{4RT}\cdot(R-T)^2\right] - \frac{1}{2}\cdot\left(1+P+\frac{PT}{R}\right)\cdot\exp(P)\cdot\text{erfc}\left[\left(\frac{P}{4RT}\right)^{1/2}\cdot(R+T)\right]$$

NE-models

$G(\tau)$

$$G(\tau) = \frac{1}{2}\,\text{erfc}\left[\left(\frac{P}{4\beta R\tau}\right)^{1/2}\cdot(\beta R-\tau)\right]$$
$$+\frac{1}{2}\exp(P)\cdot\text{erfc}\left[\left(\frac{P}{4\beta R\tau}\right)^{1/2}\cdot(\beta R-\tau)\right]$$

$$G(\tau) = \frac{1}{2}\,\text{erfc}\left[\left(\frac{P}{4\beta R\tau}\right)^{1/2}\cdot(\beta R-\tau)\right]$$
$$+\left(\frac{P\tau}{\pi\beta R}\right)^{1/2}\cdot\exp\left[-\frac{P}{4\beta R\tau}\cdot(\beta R+\tau)^2\right]$$
$$-\frac{1}{2}\left(1+P+\frac{P\tau}{\beta R}\right)\cdot\exp(P)\cdot\text{erfc}\left[\left(\frac{P}{4\beta R\tau}\right)^{1/2}\cdot(\beta R+\tau)\right]$$

$F(\tau)$

$$F(\tau) = \frac{\beta}{\tau}\left(\frac{PR}{4\pi\beta\tau}\right)^{1/2}\cdot\exp\left[-\frac{P}{4\beta R\tau}\cdot(\beta R-\tau)^2\right]$$

$$F(\tau) = \left(\frac{P}{\pi\beta R\tau}\right)^{1/2}\cdot\exp\left[-\frac{P}{4\beta R\tau}\cdot(\beta R-\tau)^2\right]$$
$$-\frac{P}{2\beta R}\cdot\exp(P)\cdot\text{erfc}\left[\left(\frac{P}{4\beta R\tau}\right)^{1/2}\cdot(\beta R+\tau)\right]$$

Source: Modified from van Genuchten, 1981.

16.19.1 Retardation Factor (*R*)

From Measured Breakthrough Curve

The retardation factor (*R*) can be estimated by locating the number of pore volumes ($T = R$) at which the relative concentration of the measured BTC is 0.5. For the measured chloride BTC in Fig. 16.8, the value of *T* at C/C_0 of 0.5 is 1.2. Therefore, the value of *R* is also 1.2. Both pore volumes (*T*) and retardation factor (*R*) are dimensionless.

From Batch Experiment

The batch experiments for solute adsorption are performed by mixing air-dried soil and solution (1:1). At least six different initial solution concentrations, which are within the experimental range, are usually selected. Generally three to four replications for each concentration are made. The mixture is stirred, and after equilibrating for 24 hours, is centrifuged and the concentration of the extracted solution is measured. The difference between the initial solution concentration and that in the supernatant (centrifuge) is assumed to be the result of adsorption. A graph is plotted between the solution concentration and the adsorbed concentration (Fig. 16.6) and the slope of the line gives the value of distribution coefficient (K_D). The R can be calculated from Eq. (16.18) for known values of bulk density and water content of soil in the experiment.

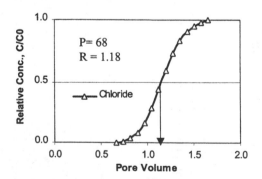

FIGURE 16.8 The estimation of retardation coefficient from a measured BTC ($R = T = 1.2$) where *T* is pore volumes. (Redrawn from Shukla and Kammerer, 1998.)

By Fitting Flow Velocity

The pore water velocity can be used as a fitting parameter in the trial and error method while keeping R a constant and equal to 1. Therefore, fitted velocity will effectively be a v/R value. The slope of the plot between fitted and measured pore water velocity gives an effective R-value.

From Travel Time Analysis

Time moment analysis provides a model independent tool for characterizing the solute BTCs. The first temporal moment provides the mean break-through time, the second central temporal moment (i.e., the variance) describes the solute spreading, and the third (skewness) describes the degree of asymmetry of the BTCs (Valocchi, 1985). These numerical estimates can be compared to the CDE theoretical travel time moments to provide estimates of the CDE model parameters, in contrast to least-squares fitting of the analytical solution to Eqs. (16.23) and (16.24). For a finite pulse, the expected or theoretical mean travel time to depth L is:

$$\frac{RL}{v} + \frac{t_0}{2} \tag{16.70}$$

and the theoretical travel time variance is

$$\frac{2DR^2L}{v^3} + \frac{(t_0)^2}{12} \tag{16.71}$$

where R is retardation factor, D is apparent dispersion coefficient $(cm^2 h^{-1})$, v is pore water velocity $(cm h^{-1})$, t_0 is the duration of pulse (h), and L is the displacement length (cm). For the step input experiments, a smooth cubic spline to each BTC can be fitted, and then the derivatives can be computed with respect to time. The center of mass of an inert solute pulse under steady flow at a given average measured pore water velocity (v) is model independent and moves at the same rate as the average v. However, different process models often result in quite different rates of spreading or dispersion but these do not affect the mean travel time (Valocchi, 1985). The slope of the best-fit curve between observed and theoretical travel times provides the effective R-value with intercept equal to zero. For details on travel moment analysis readers are advised to refer Jury and Roth (1990).

16.19.2   Apparent Dispersion Coefficient

The apparent dispersion coefficient (D) can be estimated by the following methods.

Trial and Error Method

The parameters D (or P) can be estimated by comparing the experimentally measured BTC with a series of calculated distributions. The distributions can be calculated for a known value of R ($=T$) by selecting several values of P (1, 2, 4, 5, 10, 20, 50, 100, 300, etc.). The value of P, which provides the best fit between the experimental and calculated BTC is chosen, and D is calculated from the known values of displacement length and pore water velocity ($D = vL/P$).

From Slope of an Effluent Curve

The apparent diffusion coefficient can be approximated by an experimental BTC from the following equation (Kirkham and Powers, 1972)

$$D = \frac{vL}{4\pi m^2} \tag{16.72}$$

where m is the slope of BTC at one pore volume, i.e.,

$$m = \frac{\partial(C/C_0)}{\partial p}\Big|_{p=1}. \tag{16.73}$$

Log Normal Plot of Effluent Curve

In this method the inverse complimentary error function of relative concentration (see Table 16.5) from the experimentally determined BTC is plotted against log of pore volumes (T). The value of P is estimated from the slope (m) of above straight line ($P = 4*m^2 - b$, where b is a correction factor) (van Genuchten and Wierenga, 1986).

Least Square Analysis

The trail and error method is expanded into a more rigorous approach by continuously adjusting the values of P and R until the sum of the

TABLE 16.6 Merits and Demerits of Approximate Methods of Solute Transport Parameter Estimation

Method	Merits	Demerits
Trial and error	Provide first estimates of P and R quickly	Method is not necessarily reproducible
From slope of BTC	Method is simple and based upon analytical solution. For conservative solutes works reasonably well.	Method is not suitable for small values of P and for nonconservative solutes
Log normal plot	Results are more accurate than the above two methods	Straight line is not generally obtained. Method is not suitable for aggregated or structured soils
Least square analysis	Results are the most accurate among all the methods described above. Computer programs are available and easy to use. Number of fitting parameters can be varied according to the need	User judgment is necessary for reporting fitted values of parameters

squared deviations between measured and fitted concentrations are minimized in a least square sense (van Genuchten and Wierenga, 1986). The merits and demerits of all the methods described above are presented in Table 16.6.

16.19.3 Parameters of TRM

The physical nonequilibrium model or two-region model (TRM) requires specification of four dimensionless parameters P (v_m, L, D), $R(\rho_b, K_D, \theta)$, β $(\phi, \theta_m, \theta_{im}, f)$ and ω $(\alpha, L, \theta_m, v_m)$ [refer Eqs. (16.65) and (16.66)]. The parameters of TRM can be estimated by a number of ways:

Least Square Fitting

The first option is to use a trial-and-error method and fit all the four-nondimesional parameters to the measured breakthrough curve, also known as "inverse modeling technique," by minimizing the sum of squares between measured and fitted breakthrough curves using a nonlinear least square method. The second option is to determine R from the batch experiment and obtain the remaining three-nondimensional parameters by least square fit. It should be remembered while using the least square method that for P values > 5, the least square fitting method is appropriate, however for $P < 5$, the problems associated with conservation of mass become important and trial and error method remains no longer appropriate. The lower P values also suggest extremely broad range in pore water velocity distributions in mobile water region, which renders division of flow domain into two flow regions inadequate. A possible solution is to divide flow domain in more compartments (Morisawa et al., 1986) or consider pore water velocity to be a continuous function (White et al., 1986).

Mobile (θ_m) and Immobile (θ_{im}) Water Contents

The total moisture content (θ) of the soil is the sum of the mobile (θ_m) and immobile (θ_{im}) moisture contents. The mobile and immobile water can be estimated in a number of ways: (i) all the water held at field capacity (24 h after the infiltration test or at suction of 330 kPa) can be considered as immobile water. Therefore, mobile water (θ_m) can be obtained by subtraction the θ_{im} from total water content of soil (θ) as follows:

$$\theta_m = \theta - \theta_{im} \qquad (16.74)$$

(ii) The total concentration in soil after infiltration test is given by a mass balance equation as follows:

$$\theta C = \theta_m C_m + \theta_{im} C_{im} \qquad (16.75)$$

A conservative tracer such as bromide (Br) or chloride (Cl) of known initial concentration (C_0) used as a solute is infiltrated into the soil. After the steady state infiltration with tracer solution is achieved, the concentration of the solute extracted from soil sample (C) below the infiltration can be measured. If all the soil moisture is mobile than C equals C_0. If immobile moisture is present $C < C_0$ and θ_{im} can be obtained as follows (Clothier, et al., 1992):

$$\theta_m = \theta \frac{C}{C_m} = \theta \frac{C}{C_0} \qquad (16.76)$$

alternately

$$\theta_{im} = \theta\left(1 - \frac{C}{C_0}\right)$$ (16.77)

The above equation assumes that transfer coefficient (α) in Eq. (16.56) is small and very little solute diffuses into the immobile region.

The θ_{im} and α

The θ_{im} and α can also be estimated simultaneously by applying a sequence of nonconservative nonreactive tracers for varying periods of time (Jaynes et al., 1995).

Eq. (16.37) after separating the variables can be written as follows:

$$\ln\left(1 - \frac{C}{C_0}\right) = \frac{-\alpha}{\theta_{im}}t + \ln\left(\frac{\theta_{im}}{\theta}\right)$$ (16.78)

where t is defined as the application time and varies for different tracers, Plotting the $ln(1 - C/C_0)$ versus t, for all the tracers, gives straight lines with negative slopes (Fig. 16.9). The intercept at $t=0$ gives natural log of the

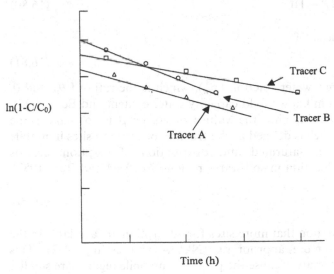

FIGURE 16.9 A schematic of normalized concentration of tracers and the time of application.

ratio of immobile water and total moisture content [the second term on right-hand side of Eq. (16.78)]. For a known θ, θ_{im} can be estimated by multiplying the intercept with θ and making appropriate ln transformations. The first term of Eq. (16.78) gives the slope and for a known θ_{im}, α can be easily calculated. The tracer front will reach a given sampling depth (d) slightly earlier than specified by t. Therefore, t in Eq. (16.78) can be replaced by "$t - d/v_m$" and Eq. (16.77) becomes (Jaynes and Horton, 1998):

$$\ln\left(1 - \frac{C}{C_0}\right) = \frac{d\alpha}{\theta_{im}v_m} + \ln\left(\frac{\theta_{im}}{\theta}\right) - \frac{\alpha}{\theta_{im}}t \qquad (16.79)$$

The θ_{im} and α can again be measured by plotting the $\ln(1 - C/C_0)$ versus t. It should clearly understood that the assumption $C_m = C_0$ associated with Eqs. (16.76) to (16.79) may not be correct for $\alpha > 0$.

By Making Approximations

The partition coefficient "β" can be obtained by using the inverse modeling technique from a measured breakthrough curve. The β, f and ϕ_m are related by the following equation, which shows that from a known value of β, the f and ϕ_m cannot be calculated directly:

$$\phi_m = [R\beta - f(R - 1)] \qquad (16.80)$$

If R is close to 1 then

$$\phi_m = \beta \qquad (16.81)$$

For $R \neq 1$, the mobile water fraction (ϕ_m), which is the ratio of θ_m and θ, can be calculated from known field capacity water content and Eq. (16.74). A better option for obtaining the values of ϕ_m or β is to make some assumptions on f, which is defined as the fraction of sorption sites in mobile region. When soil is saturated and distribution of sorption sites is independent of the location in soil-water regions (Seyfried and Rao, 1987)

$$f = \phi_m = \beta \qquad (16.82)$$

However, the assumption that more sites for adsorption are available in the immobile region is more appropriate (Nkedi-Kizza et al., 1982). This assumption is appropriate because the pores in immobile regions are smaller and have higher exposed surface area than in mobile region. Therefore, f can be assumed to vary from 0 to $\phi_m/2$ (Seyfried and Rao, 1987).

Aggregate Geometry Models

The nondimensional mass transfer parameter (ω) is not directly related to any specific soil characteristic or property and is difficult to determine. The α is a function of time of diffusion, sphere radius (particles constituting the porous medium), molecular diffusion coefficient, intraaggregate water content (θ_{im}), macroporosity (fraction of total porosity); therefore, apart from Eqs. (16.78) and (16.79), the α can be calculated for the known or assumed geometry of aggregates. For spherical aggregates α can be calculated as follows (Rao et al., 1980).

$$\alpha = \frac{D_e(1-f)\theta}{r^{20}}\alpha^* \tag{16.83}$$

where D_e is the effective diffusion coefficient, r is the radius of sphere, and α^* is time dependent variable. The α values of cubic aggregates can be obtained by replacing "a" with an equivalent spherical radius "$r = 0.6203l$", where l is the length of the side of the cube (Rao et al., 1982). Another widely used formula for the estimation of α based on soil geometry is (van Genuchten, 1985; van Genuchten and Dalton, 1986):

$$\alpha = \frac{nD_e(1-f)\theta}{a_e^2} \tag{16.84}$$

where n is a geometry factor, and a_e is an average effective diffusion length. If a soil matrix, with overall conductivity of K_e, can be divided into a two-flow domain physical nonequilibrium model. The water contents of these flow regions are θ_A and θ_B for velocities v_A and v_B, respectively. For steady flow condition the α can be estimated as follows (Skopp et al., 1981):

$$\alpha = \frac{2K_e\theta_A(v_A - v_B)^2}{gdr_p\pi} \tag{16.85}$$

where d is the aggregate size (cm), r_p is the interaggregate pore size (cm), and g is acceleration due to gravity (cm h^{-2}). It should be remembered here that α values estimated using aggregate geometry models does not necessarily fit the measured breakthrough curves very well. The α values depend on the experimental conditions (Ma and Selim, 1998). In general α values increase with flow velocity probably as a result of turbulent mixing at high velocities (van Genuchten and Wierenga, 1977). However, α values decrease if greater pore connectivity exists in the flow domain (Skopp and Gardner, 1992).

16.20   Land Use Effects on Flow and Transport

The flow and transport properties of soils often vary with time due to
the influence of land use and soil management practices. The soil remains
mostly undisturbed under no-till, which enhances the organic matter
accumulation at the soil surface and development of macropores (cracks
between aggregates and pores). The macropore channels in no-till
system increase the leaching of nutrients and pesticides by bypassing the
water-filled micropores unless the sources are located within the soil
micropores. These cracks increase the hydraulic conductivity of soil and
decrease reactivity of dissolved chemicals due to the low pore surface area
and short residence time. The increase in organic matter increases the
reactivity of chemicals in the soil matrix and the soils start behaving as a
multireaction, multiregion soil. It is important to know this shift in flow and
transport processes due to macropores as failure to take these into account
can lead to erroneous conclusions. For an example: In a macroporous soil
system, a zero-tension lysimeter was installed at a 90-cm depth, which
captured 50% of the applied pesticide leached out of root zone system via
macropore channels. The analysis of soil samples at different depth
increments showed very little traces of pesticides. Therefore, without
having the knowledge of preferential flow of pesticides, an inaccurate
conclusion that pesticides had limited mobility due to high degradation rates
can be drawn. Similarly, an increase in organic matter provides kinetic
adsorption sites for some solutes, which would lead to inaccurate results
if lumped into instantaneous equilibrium adsorption terms (Wilson et al.,
2000).

Example 16.1

Concentration of a solute is 30 mg/g of soil and bulk density is 1.35 Mg/m^3.
Assuming steady flow conditions, solute free soil profile, and solute diffusion
coefficient 3×10^{-10} m^2/s, calculate flux density at a vertical distance of 0.1 m and
amount of solute in 1 ha that diffuses across this boundary in 2 months. (Hint: Use
Fick's first law.)

Solution

Solute concentration $= \rho_b * C_a = 30 * 1.35 = 40.5$ M/gm^3 or 4.05×10^7 mg/m^3
Concentration gradient at 0.1 m below soil surface

$$\delta C / \delta z = \Delta C / \Delta z = (0 - 4.05 \times 10^7)/(0 - 0.1) = 4.05 \times 10^8 \text{mg/ m}^4$$

The flux density of solute is obtained by using equation 7

$$J = -(3 \times 10^{-10}) * (4.05 \times 10^8) = -0.122 \, \text{mg m}^{-2}\text{s}^{-1}$$

The negative sign implies that solute is moving downward. The total quantity of solute moved below 0.1 m in one month (Q) can be calculated as

$$Q = 0.122 * 10000 * 30 * 24 * 3600 = 3.16 \times 10^9 \text{mg ha}^{-1}$$

Example 16.2

Nitrate-N was applied in a field at volumetric moisture content of 0.35. If soil water flux density was 0.05 cm d^{-1} and soil solution concentration of NO$_3$-N was 4 mg L^{-1}, calculate the pore water velocity and amount of NO$_3$-N leached per unit area by convective flow below the root zone in 2 days.

Solution

Pore water velocity $= v = q/\theta = 0.05/0.35 = 0.143$ cm/d
The flux density for convective transport (J_m) can be calculated from equation 6.

$$J_m = qC = 0.05 * 4 * 1000/1000 = 2.0 \, \text{mg/m}^2 \, \text{d}$$

Therefore, amount of NO$_3$-N (Q) leached through root zone in 2 days

$$Q = J_m * A * t = 2 * 1 * 2 = 4 \, \text{mg}$$

Example 16.3

Assuming steady condition and piston flow through a soil column at moisture content of 0.35 cm^{3}cm^{-3}, calculate the total time required to transport chloride from the bottom of the root zone to groundwater at 50 m below when average daily drainage rate is 0.25 m/d.

Solution

Total depth of water in the vadose zone $= 0.35 * 50 = 17.5$ m
The breakthrough time (total time required) to transport all the chloride to groundwater $= 17.5/0.25 = 70$ d
Alternately, pore water velocity of chloride $(v = q/\theta) = 0.25/0.35 = 0.71$ m/d
Breakthrough time $(t^* = L/v) = 50/0.71 = 70$ d

Example 16.4

Using the information in Example 3, calculate the velocity and breakthrough time for chloride if bulk density of soil was $1.4\,Mg/m^3$ and the slope of equilibrium isotherm was $0.06\ m^3\ Mg^{-1}$.

Solution

The retardation factor $(R) = 1 + k^* \rho_b/\theta = 1 + 0.06 * 1.4/0.35 = 1.24$
Average chloride velocity $= 0.25/(0.35 * 1.24) = 0.58$ m/d
The breakthrough time $= 50/0.58 = 86$ days

PROBLEMS

1. In a repacked loam soil column with total porosity (ϕ of 0.5, the measured dispersivity (λ) was 1.2. Assuming that diffusion coefficient of solute in water (D_0) is $1\,cm^2\ day^{-1}$, calculate remaining parameters given in the table below.

   Note: Tortuosity factor (ξ) is given as $\xi = \theta^{10/3}/\phi^2$ (known as the Millington–Quirk formula, 1961). Effective-dispersion diffusion coefficient (D) is given by $D_e = D_h + D_m$.

q (cm.d^{-1})	θ	v (cm.d^{-1})	ξ	D_h	D_m	D
0.2	0.25					
1	0.3					
2	0.35					
5	0.4					

2. Assume that average volumetric water content (θ) of soil is 0.2; and bulk density (ρ_b) is $1.5\,g\,cm^{-1}$. The average annual drainage rate (dr) is $0.5\,m\,yr^{-1}$. If a pesticide, $Kd = 2\,cm^2g^{-1}$, is applied to this soil, calculate how long (breakthrough time) it will take to move the pesticide to the groundwater at (L) 12 m depth.

3. Chloride solution was applied as a step input to a 10 cm long soil column initially saturated with water. The flux density of chloride (q) was $0.5\,cm\ h^{-1}$, and average water content of column was $0.45\,cm^3\,cm^{-3}$. The chloride BTC can be plotted on an Excel spreadsheet with X-axis as pore volumes (p) and relative chloride concentration (C/C_0). The pore volumes are 0.2, 0.4, 0.6, 0.8, 1.0, 1.2, 1.4, and 1.6 and corresponding C/C_0 are 0.01, 0.06, 0.15, 0.3, 0.54, 0.8, 0.96, and 0.99 respectively. Calculate the apparent diffusion coefficient (D) and retardation coefficient (R).

REFERENCES

Bear J. (1972). Dynamics of fluid in porous media. Elsevier Science, New York.

Bear J. and Y. Bachmat (1967). A generalized theory on hydrodynamic dispersion in porous media. Symp. Artificial Recharge Management Aquifers, Haifa, Int. Assoc. Sci. Hydrol., 72:7–36.

Biggar J.W. and D.R. Nielsen (1967). Miscible displacement and leaching phenomenon. Agronomy 11:254–274.

Bolt G.H. (1979) (ed.). Soil chemistry. B: Physio-chemical models. Elsevier Scientific Pub. Co., New York.

Bond W.J., B.N. Gardiner, and D.E. Smiles (1982). Constant flux adsorption of a tritiated calcium chloride solution by a clay soil with anion exclusion. Soil Sci. Soc. Am. J. 46:1133–1137.

Bresler E. (1973). Anion exclusion and coupling effects in a nonsteady transport through unsaturated soils: I. Theory. Soil Sci. Soc. Am. Proc. 37:663–669.

Butters G.L., W.A. Jury, and F.F. Ernst (1989). Field scale transport of bromide in an unsaturated soil. 1. Experimental methodology and results. Water Resour. Res. 25:1575–1581.

Cameron D.R. and A. Klute (1977). Convective–Dispersive solute transport with a combined equilibrium and kinetic adsorption model. WRR, 13(1):183–188.

Clothier B.E., M.B. Kirkham, and J.E. Mclean (1992). In situ measurements of the effective transport volumes for solute moving through soil. Soil Sci. Soc. Am. J. 56:733–736.

Coats K.H. and B.D. Smith (1964). Dead end pore volume and dispersion in porous media. SPE J. 4:73–84.

De Josselin De Jong G. (1958). Longitudinal and transverse diffusion in granular deposits. Trans. Amer. Geophys. Union 59:67.

Ellsworth T.R., P.J. Shouse, T.H. Skaggs, J.A. Jobes, and J. Fargerlund (1996). Solute transport in unsaturated soil: experimental design, parameter estimation, and model discrimination. Soil Sci. Soc. Am. J. 60:397–407.

Fleming J.B. and G.L. Butters (1995). Bromide transport detection in tilled and nontilled soil:solute samplers vs. soil cores. Soil Sci. Soc. Am. J. 59: 1207–1216.

Flury M., W.A. Jury, and E.J. Kladivko (1998). Field scale solute transport in vadose zone: Experimental observations and interpretation. In: H.M. Selim and L. Ma (eds.), Physical Nonequilibrium in Soils. Ann Arbor Press, Michigan, p 349–365.

Fried J.J. and M.A. Combarnous (1971). Dispersion in porous media. Ad. Hydroci. 7:169–282.

Greenkorn R.A. (1983). Flow phenomena in porous media. Marcel Dekker Inc., New York and Basel, p 190.

Hamlen C.J. and R.G. Kachanowski (1992). Field solute transport across a soil horizon boundary. Soil Sci. Soc. Am. J. 56:1716–1720.

James R.V. and J. Rubin (1986). Transport of chloride ion in a water unsaturated soil exhibiting anion exclusion. Soil Sci. Soc. Am. J. 50:1142–1149.

Jaynes D.B. (1991). Field study of bromacil transport under continuous flood irrigation. Soil Sci. Soc. Am. J. 55:658–664.

Jaynes D.B., S.D. Logsdon, and R. Horton (1995). Field method for measuring mobile/immobile water content and solute transfer rate coefficient. Soil Sci. Soc. Am. J. 59:352–356.

Jury A.W. and K. Roth (1990). Transfer functions and solute movement through soil. Birkhaeuser Verlag, Basel, Germany.

Jury A.W., W.R. Gardner, and W.H. Gardner (1991). Soil physics, 5th Edition, John Wiley, New York. Torelli L. and A.E. Scheidegger (1972). Three-dimensional branching type models of flow through porous media. J. Hydro. 15:23.

Kirkham D. and W.L. Powers (1972). Advanced soil physics. Wiley Interscience, John Wiley & Sons, Inc., New York.

Krupp H.K. and D.E. Elrick (1968). Miscible displacement in an unsaturated glass bead medium. Water Resour. Res. 4:809–815.

Krupp H.K., J.W. Biggar, and D.R. Nielsen (1972). Relative flow rates of salt and water in soil. Soil Sci. Soc. Am. Proc. 36:412–417.

Kutilek M. and D.R. Nielsen (1994). Soil Hydrology. Catena Verlag, Cremlingen-Destedt, Germany.

Lapidus L. and N.R. Amundson (1952). Mathematics of adsorption in beds. J. Phys. Chem. 56:584.

Ma L. and H.M. Selim (1998). Physical nonequilibrium in soils: modeling and application. In: H.M. Selim and L. Ma (eds.), Physical Nonequilibrium in Soils. Ann Arbor Press, Michigan, p 83–115.

Morisawa S., M. Horiuchi, T. Yamaoka, and Inoue Y. (1986). Evaluation of solute transport in unsaturated soil column by multicompartment flow model. (In Japanese) Proc. Environ. Sanitary Engg. Res. 22:9–22.

Nielsen D.R., M.Th. van Genuchten, and J.W. Biggar (1986). Water flow and solute transport processes in the unsaturated zone. Water Resour. Res., 22(9): 89S–108S.

Nkedi-Kizza P., J.W. Biggar, H.M. Selim, M.Th. van Genuchten, P.J. Wierenga, J.M. Davidson, and D.R. Nielsen (1984). On the equivalence of two conceptual models for describing ion exchange during transport through an aggregated oxisol. Water Resour. Res., 20(8):1123–1130.

Nkedi-Kizza P., P.S.C. Rao, R.E. Jessup, and J.M. Davidson (1982). Ion exchange and diffusive mass transfer during miscible displacement through and aggregated Oxisol. Soil Sci. Soc. Am. J. 46:471–476.

Parker J.C. and M.Th. van Genuchten (1984a). Flux-averaged and volume-averaged concentrations in continuum approaches to solute transport. Water Resour. Res. 20(7):866–872.

Perkins T.K. and O.C. Johnston (1963). A review of diffusion and dispersion in porous media. Pet. Trans. AIME 228, SPEJ 70.

Pfannkuch H.O. (1962). Contribution a L'etude des deplacement de fluides miscible dans un milieu poreux. Rev. Inst. Fr. Petrol. 18(2):215.

Rao P.S.C., R.E. Jussup, D.E. Rolston, J.M. Davidson, and D.P. Kilcrease (1980). Experimental and mathematical description of nonadsorbed solute transfer by diffusion in spherical aggregates. Soil Sci. Soc. Am. J. 44(4):684–688.

Rao P.S.C., R.E. Jessup, and T.M. Addiscott (1982). Experimental and theoretical aspects of solute diffusion in spherical and nonspherical aggregates. Soil Sci. 133:342–349.

Roth K., W.A. Jury, H. Fluehler, and W. Attinger (1991). Transport of chloride through an unsaturated field soil. Water Resourc. Res. 27:2533–2541.

Selim H.M. (1992). Modeling the transport and retention of inorganics in soil. Adv. Agron. 47:331–384.

Selim H.M., J.H. Davidson, and R.S. Mansell (1976). Evaluation of a two-site adsorption desorption model for describing solute transport in soils. Proceedings Summer Computer Simulation Conference, Washington D.C., 444–448.

Seyfried M.S. and P.S.C. Rao (1987). Solute transport in undisturbed columns of an aggregated tropical soil: Preferential flow effects. Soil Sci. Soc. Am. J. 51: 1434–1444.

Shukla M.K. and G. Kammerer (1998). Comparison between two models describing solute transport in porous media with and without immobile water. Austrian Journal of Water Management 50(9/10):254–260.

Shukla M.K., F.J. Kastanek, and D.R. Nielsen (2000). Transport of chloride through water-saturated soil columns. The Bodenkulture, Austrian Journal of Agricultural Research 51(4):235–246.

Shukla M.K., F.J. Kastanek, and D.R. Nielsen (2002). Inspectional analysis of convective dispersion equation and application on measured BTCs. Soil Sci. Soc. of Am. J. 66(4):1087–1094.

Shukla M.K., T.R. Ellsworth, R.J. Hudson, and D.R. Nielsen (2003). Effect of water flux on solute velocity and dispersion. Soil Sci. Soc. Am. J. 67:449–457.

Skopp J. and W.R. Gardner (1992). Miscible displacement: an interacting flow region model. Soil Sci. Soc. Am. J. 56:1680–1686.

Sposito G. (1989). The chemistry of soils. Oxford Press, p 277.

Taylor G.I. (1953). The dispersion of matter in solvent flowing slowly through a tube. Proc. R. Soc. London, Ser. A 219:189–203.

Toride N., F.K. Leij, and M.Th. van Genuchten (1993). A comprehensive set of analytical solutions for nonequilibrium solute transport with first-order and zero-order production. Water Resour. Res. 29(7):2167–2182.

Valocchi, A.J. (1985). Validity of local equilibrium assumption for modeling sorbing solute transport through homogeneous soils. Water Resour. Res. 21: 808–820.

Van de Pol R.M., P.J. Wierenga, and D.R. Nielsen (1977). Solute movement in field soil. Soil Sci. Soc. Am. J. 41:10–13.

van Genuchten M.Th. (1985). A general approach for modeling solute transport in structured soils. Proc. 17 The Int. Congress. IAH, Hydrogeology of Rocks of Low Permeability. Jan 7–12, 1985, Tucson, AZ. Mem. Int. Assoc. Hydrogeol. 17:512–526.

van Genuchten M.Th. and F.N. Dalton (1986). Models for simulating salt movement in aggregated field soils. Geoderma. 38:165–183.

van Genuchten M.Th. and P.J. Wierenga (1976). Mass transfer studies in sorbing porous media I Analytical solutions. SSSA Proceedings 40(4):473–480.

514 Chapter 16

van Genuchten M.Th. (1981). Non-equilibrium transport parameters from miscible displacement experiments. Research report 119, USDA, US Soil salinity lab Riverside, California.

van Genuchten M.Th. and J.C. Parker (1984). Boundary conditions for displacement experiments through short laboratory soil columns. Soil Sci. Soc. Am. J. 48: 703–708.

van Genuchten M.Th. and P.J. Wierenga (1977). Mass transfer studies in sorbing porous media. II. Experimental evaluation with tritium (^3H_2O). Soil Sci. Soc. Am. J. 41:272–277.

van Genuchten M.Th. and P.J. Wierenga (1986). Solute dispersion coefficients and retardation factors. In: A. Klute, (ed.), Methods of Soil Analysis, Part 1: Physical and Mineralogical Methods, 2nd Edition, American Society of Agronomy, Madison, Wisconsin.

van Genuchten M.Th. and W.J. Alves (1982). Analytical solutions of the one-dimensional connective dispersion solute transport equation. USDA Tech. Bull.1661.

Van Wesenbeck I.J. and R.G. Kachanowski (1991). Spatial scale dependence of in situ solute transport. Soil Sci. Soc. Am. J. 55:3–7.

White R.E., L.K. Heng, and R.B. Edis (1998). Transfer function approaches to modeling solute transport in soils. In: H.M. Selim and L. Ma (eds.), Physical nonequilibrium in soils. Ann Arbor Press, Michigan, p 311–346.

White R.E., J.S. Dyson, Z. Gerstl, and B. Yaron (1986). Leaching of herbicides through undisturbed cores of a structured clay soil. Soil Sci. Soc. Am. J. 50: 277–283.

Wilson J.L. and L.W. Gelhar (1974). Dispersive mixing in a partially saturated porous medium, Persons Laboratory Report 191, Massachusetts Institute of Technology, Cambridge.

Wilson G.V., H.M. Selim, and J.H. Dane (2000). Flow and transport processes. In: H.D. Scott, (ed.), Water and Chemical Transport in Soils of the Southeastern USA. SCSB-395. Department of Plant and Soil Sciences, Oklahoma State University.

17

Soil Temperature and Heat Flow in Soil

17.1 TEMPERATURE

Temperature is a measure of the thermal state of a body with respect to its ability to transfer heat. It is also defined as the measure of intensity or potential energy or heat. Temperature is the driving force for heat flow as pressure head is for water flow. Temperature is measured in three scales: Celsius (°C), Fahrenheit (°F), and Kelvin (K). The conversion from one scale to another is given in Table 17.1.

17.2 THE DEVELOPMENT OF THERMOMETER AND TEMPERATURE SCALES

One of the first attempts to make a standard temperature scale occurred about 170 AD, when Galen proposed a standard neutral temperature made up of equal quantities of boiling water and ice with four degrees of heat and cold on either side of this temperature, respectively. The earliest device used to measure the temperature was known as a "thermoscope" and consisted of a glass bulb having a long tube, which extended downward into a container of colored water. Before filling the liquid, some of the air in the bulb was

515

TABLE 17.1 Mathematical Expressions and Relations for
Temperature Scales

Temperature scales	Mathematical formula
Celsius (°C)	$°C = (5/9)*(°F - 32)$
Fahrenheit (°F)	$°F = (9/5)*°C + 32$
Kelvin (K)	$K = °C + 273.15$

removed, causing the liquid to rise into the tube. As the remaining air in the bulb was heated or cooled, the level of the liquid in the tube would vary reflecting the change in the air temperature. An engraved scale on the tube allowed for a quantitative measure of the temperature fluctuations.

The first sealed thermometer using liquid rather than air as the thermometric medium was developed for Ferdinand II in 1641. The thermometer was a sealed alcohol-in-glass device, with 50 "degree" marks on its stem, but no "fixed point" was used to zero the scale. This device was referred to as a "spirit" thermometer. Robert Hook (1664) used a red dye in the alcohol and the scale needed only one fixed point, for which every degree represented an equal increment of volume equivalent to about 1/500 part of the volume of the thermometer liquid, which was the freezing point of water. Hook demonstrated that a standard scale could be established for thermometers of a variety of sizes. Hook's original thermometer was known as the standard of Gresham College, and was used by the Royal Society until 1709.

Ole Roemer of Copenhagen, Denmark, developed the thermometer scale in 1702 based upon two fixed points: snow (or crushed ice) and the boiling point of water, and recorded the daily temperatures at Copenhagen in 1708 and 1709. Gabriel Fahrenheit, an instrument maker in Amsterdam, The Netherlands, was the first to use mercury as the thermometric liquid in 1724. Mercury's thermal expansion is large and uniform, and does not stick to the glass, and remains a liquid over a wide range of temperatures. The silvery appearance also makes it easy to read. Fahrenheit measured the boiling and freezing points of water to be 212 and 32, respectively, and designated temperatures in degrees Fahrenheit (°F).

In 1745, Carolus Linnaeus of Uppsala, Sweden, described the freezing point of water as zero, and the boiling point as 100, making it a "centigrade" (one hundred steps) scale. Anders Celsius (1701–1744) used the reverse scale in which 100 represented the freezing point and zero the boiling point of water, still, with 100 degrees between the two defining points. In 1948 use of the centigrade scale was dropped in favor of a new scale using degrees Celsius (°C). A degree Celsius equals the same temperature change as a degree on the ideal-gas scale. An "ideal gas" is one whose physical behavior is accurately described by the ideal-gas equation*. On the Celsius scale, the boiling point of water at standard atmospheric pressure is 99.975°C in contrast to the 100 degrees defined by the centigrade scale.

In 1780, J. A. C. Charles, a French physician, showed that for the same increase in temperature, all gases exhibited the same increase in volume. Because the expansion coefficient of gases is about the same, it is possible to establish a temperature scale based on a single fixed point rather than the two fixed-point scales, such as the Fahrenheit and Celsius scales. This brings us back to a thermometer that uses a gas as the thermometric medium.

P. Chappuis in 1887 conducted extensive studies of gas thermometers with constant pressure or with constant volume using hydrogen (H_2), nitrogen (N_2), and carbon dioxide (CO_2) as the thermometric medium. Based on his results, the Comité International des Poids et Mesures adopted the constant-volume hydrogen scale based on fixed points at the ice point (0°C) and the steam point (100°C) as the practical scale for international meteorology.

17.3 MEASUREMENT OF TEMPERATURE

The temperature of a substance (such as soil) is generally measured indirectly by measuring a property that responds to changes in its heat content. Some of these instruments are the liquid-in-glass thermometer, electric resistance thermometer, bimetallic thermometer, thermocouple, and remote-sensing thermometer (www.temperatures.com; Childs et al., 2000; Scott, 2000).

17.3.1 Liquid-in-Glass Thermometer

The liquid-in-glass thermometer is placed in close contact with soil or any substance, the conduction of heat between thermometer and its surrounding soil causes the change in volume of liquid in the glass thermometer (Childs et al., 2000; Scott, 2000). The traditional liquid-in-glass thermometer consists of a reservoir and capillary tubes and is based on the design proposed by Danial Fahrenheit in 1714. This type of thermometer is used very commonly in the field and is sufficient for a reliable measurement of soil temperature provided good contact between reservoir and soil is

---

*$PV = nRT$, where P is the pressure (atm), V is the volume (m^3), n is number of moles, T is temperature (K), and R universal gas constant.

ensured. The accuracy of these devices ranges from ±0.01 to ±4°C. For the measurement of maximum and minimum temperatures, alcohol, toluene, or mercury is used as thermometric liquid. Since mercury vapors are toxic to human (ATSDR, 1999), cheaper resistance devices giving a digital readout have replaced mercury-in-glass thermometers.

17.3.2 Electric Resistance Thermometer

Electric methods are mostly based on the thermoelectric effect of temperature or change in resistance of a metal with a change in temperature. The motion of free electrons and atomic lattice vibrations are temperature dependent, which makes it possible to relate the resistance of a conductor to temperature. Resistance thermometers consist of a thin platinum or nickel wire, which is spiraled on a cylinder. The resistance measured using a bridge circuit generally increases by 0.4–0.5% per °C rise in temperature. A semiconductor known as a thermistor is a special type of resistance thermometer whose resistance decreases exponentially with an increase in temperature as follows (Scott, 2000):

$$R = B \exp\left(\frac{a}{T}\right) \qquad\qquad (17.1)$$

where a and B are constants and T is the absolute temperature. The advantage of electric thermometers is that they can be easily used for continuous and rapid temperature measurements and can be highly accurate (Childs et al., 2000). However, these thermometers need to be frequently calibrated during use (Scott, 2000).

17.3.3 Bimetallic Thermometer

Bimetallic thermometers have two metals strips, which are joined together. These strips have different thermal expansion coefficients. The strips are also connected to a pointer. When temperature changes, the metal strips get deformed, which moves the pointer on a temperature scale. These thermometers are commonly used in thermographs and their accuracy is few tenths of a °C. An advantage of these devices is that they do not require a power supply (Childs et al., 2000; Scott, 2000).

17.3.4 Thermoelectric Thermometer

Sir William Siemens, in 1871, proposed a thermometer whose thermometric medium is a metallic conductor, whose resistance changes with temperature. The element platinum does not oxidize at high temperatures and has a relatively uniform change in resistance with temperature over a large range. The platinum resistance thermometer is now widely used as a thermoelectric thermometer and covers the temperature range from about $-260°C$ to $1235°C$. It defines the international temperature scale between the triple point of hydrogen (H_2), 13.8023 K, and freezing point of silver, 1234.93 K, within an accuracy of $±0.002$ K. Errors associated with platinum resistance thermometers are self heating, oxidation, corrosion, and strain of sensing element (Childs et al., 2000). If accuracy is less critical, a cheaper form of resistance thermometer known as a "thermistor" can be used. This utilizes a semiconductor (e.g., mixtures of oxides of nickel, magnesium, iron, copper, cobalt, manganese, titanium, etc.) in place of platinum. The accuracy of these devices for commercial application is $±1°C$.

17.3.5 Thermocouple

Thermocouple, the most widely used soil temperature measurement instrument, is made up of two wires of different metals (commonly copper–constantan, iron–constantan, or chromel–constantan) welded together at two places with the welds kept at different temperatures. The temperature difference causes a roughly proportional electric potential difference between the welds and current flows through the circuit formed by two wires. This effect is known as the thermoelectric effect. For measurement of soil temperature, one of the welds is kept at reference temperature while the other is kept in contact with soil. The compensation method measures the thermoelectric potential difference, and a galvanometer, the thermoelectric current between welds. Thermocouples are less economic, robust, and capable of monitoring temperatures between -270 and $3000°C$. The sensitivity and speed of these devices is sufficient for many applications but are less accurate than resistance temperature devices (Childs et al., 2000; Scott, 2000).

17.3.6 Remote Sensing Thermometer

Temperature measurement devices based on thermal radiation monitoring can measure temperatures from 50 to 6000 K (Childs et al., 2000). Infrared thermometry is the most popular methods of estimating the temperature of the surfaces of soil, plant leaves, and crop canopies. According to the Stefan–Boltzmann equation the infrared radiations emitted by the surface are expressed as follows [see also Eqs. (17.15) and (17.16)]

$$R_l = e\sigma T^4 \tag{17.2}$$

where R_l is the long wave radiation, e is emissivity, which is close to 1 for most soil and plant surfaces, σ is Stefan–Boltzmann constant ($5.675 \times 10^{-8}\,Wm^{-2}K^{-4}$), and T is absolute temperature (Scott, 2000). An

infrared measurement system comprises a source, a medium through which
heat energy is transferred (e.g., gas), and a measurement device (e.g., optical
system, a detector, a control and analysis system).

17.4 TEMPERATURE AS A THERMODYNAMIC PROPERTY (see also Chapter 14)

Experiments with gas thermometers have shown that there is very little
difference in the temperature scale for different gases. Thus, it is possible
to set up a temperature scale that is independent of the thermometric
medium if it is a gas at low pressure. In this case, all gases behave like
an "ideal gas" and have a very simple relation between their pressure,
volume, and temperature:

$$PV = (\text{constant})T \tag{17.3}$$

where P is partial pressure of gas, V is volume of gas, T is temperature
(also known as thermodynamic temperature), which is defined as the
fundamental temperature and whose unit is the Kelvin (K), named after
Lord Kelvin. Note that there is a naturally defined zero on this scale that
is the point at which the pressure of an ideal gas is zero, making the
temperature zero. With this as one point on the scale, only one other
fixed point needs to be defined. In 1933, the International Committee of
Weights and Measures adopted this fixed point as the triple point of water,
the temperature at which water, ice, and water vapor coexist in equilibrium;
its value is set as 273.16 K.

17.4.1 Entropy

Entropy (S_e) is a thermodynamic quantity, which is a measure of the
degree of disorder within any system. The greater the degree of disorder,
the higher the S_e. For an increase in disorder, S_e is positive and has the
units of joules per degree K per mole. The entropy has a standard that is
fixed by the third law of thermodynamics (see the following section).

17.4.2 Enthalpy

Enthalpy (H) is a thermodynamic state function, generally measured in
kilojoules per mole. In chemical reactions the enthalpy change (ΔH)
is related to changes in the free energy (ΔG) and entropy (ΔS_e) by the
Gibbs equation:

$$\Delta G = \Delta H - T\Delta S_e \tag{17.4}$$

The enthalpy of an element has an internationally defined value at 298.15 K and 101.32 kPa and its entropy is zero at 0 K and 101.32 kPa. The temperature that is most often used for recording thermodynamic data is 298.15 K, and by international convention the enthalpy of a pure element at 298.15 K and standard pressure is zero.

17.5 HEAT AND THERMODYNAMICS

Heat is the kinetic energy of random thermal motion of soil particles. Prior to the nineteenth century, it was believed that the sense of how hot or cold an object felt was determined by how much "heat" it contained. Heat was envisioned as a liquid that flowed from a hotter to a colder object, this weightless fluid was called caloric, and no distinction was made between heat and temperature. Black was the first to distinguish between the quantity (caloric) and the intensity (temperature) of heat. Joule (1847) conclusively showed that heat was a form of energy.

The zeroth law of thermodynamics states that if two bodies (e.g., masses of soil; A and B) are at the same temperature, and a third body C has the same temperature as body B, then the temperature of body C is equal to the temperature of body A.

$$\text{Temperature } A = B = C \tag{17.5}$$

The first law of thermodynamics is the conservation of energy and it states, "When heat is transformed into any other form of energy, or when other forms of energy are transformed into heat, the total amount of energy (heat plus other forms) in the system (plus surrounding) remains constant." To express it another way, the law states, "It is in no way possible either by mechanical, thermal, chemical, or other means, to obtain a perpetual motion machine; i.e., one that creates its own energy." At the same time, it is not possible to construct a cyclic machine that does nothing but withdraw heat energy and converts it into mechanical energy. No cyclic machine can convert heat energy wholly into other forms of energy, because efficiency of a cyclic machine can never be 100%. In the simplest form, the first law states "energy can neither be created nor destroyed." It can change from one form to another, for example, electricity to heat, heat that can boil water and make steam, hot steam that can push a piston (mechanical energy) or turn a turbine that makes electricity, which can be changed into light (in a light bulb) or can change to sound in an audio speaker system, and so forth. If the total energy of a system is E, then between any two equilibrium states (E_1 for system 1 and E_2 for 2), the change in internal energy is equal to the difference of heat transfer (Q) into a system and work done (W) by the system

$$E_2 - E_1 = Q - W \tag{17.6}$$

A process, which does not involve heat transfer, is known as an adiabatic process. The second law of thermodynamics implies that there is an irreversibility of certain processes—that of converting all heat into mechanical energy. The law states that "there exists useful state variable called entropy (S_e) and the change in entropy is equal to the heat transfer divided by the temperature." For a given physical process, the entropy of the system and the environment will remain constant if the process can be reversed.

$$\Delta S_e = \frac{\Delta Q}{T} \tag{17.7}$$

If we denote the initial and final states of the system by "i" and "f", then for a reversible system the change in entropy is zero (i.e., $S_{ef} = S_{ei}$) and for a reversible system the entropy will increase (i.e., $S_{ef} > S_{ei}$).

An example of a reversible process is ideally (no boundary layer losses) forcing a flow through a constricted pipe. As the flow moves through the constriction, the pressure, temperature and velocity would change, but these variables would return to their original values downstream of the constriction. The state of the gas would return to its original conditions and the change of entropy of the system would be zero. The second law states that if the physical process is irreversible, the entropy of the system and the environment must increase and the final entropy must be greater than the initial entropy. An example of an irreversible process is when a warm soil is kept in contact with a cold one and after some time both achieve the same equilibrium temperature. If we then separate two soils, they do not naturally return to their original (different) temperatures. The process of bringing them to the same temperature is irreversible.

The third law of thermodynamics was formulated by Walter Nernst and is also known as the Nernst heat theorem. The law states, "at absolute zero, all bodies have the same entropy." In other words, a body at absolute zero could exist in only one possible state, which possesses a definite energy, called the zero-point energy. This state is defined as having zero entropy, which is the entropy of a pure perfect crystal at $0\,K$. At $0\,K$, the atoms in a pure perfect crystal are aligned perfectly and do not move. Moreover, there is no entropy of mixing since the crystal is pure. For a mixed crystal containing the atomic or molecular species A and B, there are many possible arrangements of A and B and there is, therefore, entropy associated with the arrangement of the atoms/molecules.

17.5.1 Heat Capacity

The amount of temperature change in a body in response to heat adsorption or release is known as heat capacity. There are two types of heat capacities. The gravimetric heat capacity (C_g) is "the amount of heat energy required to raise the temperature of 1 kg of a substance by 1 K." The volumetric heat capacity (C_v) is "the amount of heat required to raise the temperature of $1\,m^3$ of a substance by 1 K." The units of C_g and C_v in SI system are $Jkg^{-1}K^{-1}$ and $Jm^{-3}K^{-1}$, respectively. These two heat capacities are related by soil bulk density (ρ_b) as follows

$$C_g * \rho_b = C_v \tag{17.8}$$

The specific heat of a substance is the ratio of the heat capacity of substance and water, and is dimensionless (see also the section on heat capacity of soils).

17.5.2 Blackbody

A blackbody is assumed to satisfy the ideal conditions, such as, (i) absorbs all incident radiation regardless of wavelength and direction, (ii) for a prescribed temperature and wavelength, no surface can emit more energy than a blackbody, and (iii) radiation emitted by a blackbody is a function of wavelength and is independent of direction. A blackbody is also known as a diffuse emitter.

17.6 FACTORS AFFECTING INSOLATION AT THE SOIL SURFACE

Radiations received at the soil surface are affected by a number of physical factors, which include vegetation, albedo, exposure, distribution of land and water, etc.

17.6.1 Vegetation

Vegetation cover buffers the soil beneath against sudden fluctuations in temperature. Bare soil is unprotected from the direct rays of the Sun and gets warm during the day and loses heat to atmosphere during the night. However, a good vegetative cover intercepts significant amount of solar radiation and prevents soil from getting warmer in summer. During winter or cold seasons, it prevents soil from losing heat as well, thereby reducing the daily variation of soil temperature as well as frost penetration and depth of freezing. The vegetation alters the soil energy balance in a number of ways, which include (i) altering albedo, (ii) insulating soil surface to prevent heat exchange, (iii) reducing depth of penetration of solar radiation, and (iv) increasing the removal of latent heat by evapotranspiration. Application of mulches on soil surface also alters the heat exchange in bare soil. The light colored mulches transmit short wave thermal energy to soil but prevent the loss of long wave thermal radiation and keep the soil warm by producing a green house effect.

17.6.2 Albedo

The fraction of all incoming solar radiations reflected back into space at the crop or soil surface is known as albedo. Albedo depends upon the nature of soil surface, angle of sunlight, and latitude. The albedo increases significantly with the distance from the equator. Water surfaces generally reflect 10% of the incoming radiations and therefore have lower albedo as compared to crop or soil surface. The albedo for canopy surfaces ranges from 5% for forest canopies to 25% for nonequatorial crops at full ground cover (Jury et al., 1991). The color of the soil is an important factor and affects the amount of reflection, for example, a light-colored soil has higher albedo than a dark-colored soil. Similarly a dry soil has higher albedo than a wet soil. The value of albedo for some soils and crops is presented in Table 17.2.

17.6.3 Latitude

The angle at which the Sun's rays meet the earth influences the amount of radiation received per unit area because of two reasons: (i) the albedo is high because of the angle, and (ii) radiation is subject to higher scattering reflection and adsorption, since they move through more atmospheres. Albedo is the highest in polar areas, decreases slightly in the middle latitudes, and is the lowest in tropical regions.

17.7 SOIL TEMPERATURE

Soil temperature is one of the most important factors affecting plant growth. Until soil reaches a certain critical temperature neither seeds

TABLE 17.2 Albedo from Soil, Forest, and Crops

Cover/surface	Albedo (%)	Reference
Light sand (Dry)	30–60	Geiger (1965)
Serozem (Dry)	25–30	Chudnovskii (1966)
Serozem (Wet)	10–12	Chudnovskii (1966)
Chernozem (Dry)	14	Chudnovskii (1966)
Chernozem (Wet)	8	Chudnovskii (1966)
Clay (Dry)	23	Chudnovskii (1966)
Clay (Wet)	16	Chudnovskii (1966)
Forest	5–20	Geiger (1965)
Corn (New York)	23.5	Chang (1968)
Sugar cane (Hawaii)	5–18	Chang (1968)
Pineapple (Hawaii)	5–8	Chang (1968)
Potato (Russia)	15–25	Chang (1968)

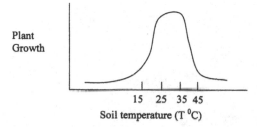

Plant Growth

15 25 35 45
Soil temperature (T ^{0}C)

FIGURE 17.1 A graph of plant growth with soil temperature.

germinate nor plants have a normal growth because it affects root and shoot growth and availability of water and nutrients (Fig. 17.1). The optimum range of soil temperature for plant growth is between 20 and 30°C. The rate of plant growth declines drastically when temperature is less than 20°C (suboptimal) and above 35°C (supraoptimal) (Figs. 17.2–17.4). Further, all soil processes are temperature dependent. Consequently, the thermal regime of soil strongly influences the edaphic environment. The release of soil nutrients for root uptake is also dependent upon soil temperature regime. The biological processes (such as respiration by plants) are temperature dependent. Respiration rate (R_T) at a temperature (T) is expressed as follows:

$$R_T = R_0 Q_{10}^{(T-T_0)/10}$$

(17.9)

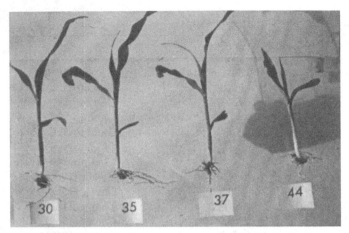

FIGURE 17.2   Corn seedling growth in relation to constant soil temperature maintained from 30 to 44°C in the root zone. (From Lal, 1972, greenhouse experiments.)

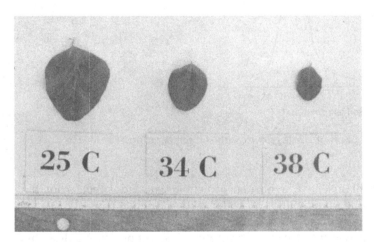

FIGURE 17.3   Soybean leaflets of seedlings grown at constant soil temperature of 25 to 38°C. (From Lal, 1972, greenhouse experiments.)

where R_0 is rate at reference temperature T_0, Q_{10} is the factor which relates respiration to each 10°C change in temperature. The reaction rate (K_{rea}) in soil is mostly described by the Arrhenius equation as follows:

$$K_{rea} = A \exp\left(-\frac{E}{RT}\right) \qquad (17.10)$$

FIGURE 17.4 Chloratic symptoms of nutrient imbalance in corn seedlings grown at constant soil temperature of 38°C. (From Lal, 1972, greenhouse experiments.)

where A is the preexpopnetial factor, E is activation energy (J), R is gas constant ($8.314\,\mathrm{Jmol^{-1}K^{-1}}$), and T is absolute temperature. A plot of $\log K_{rea}$ vs. $1/T$ provides the values of empirical constants E (as the slope) and A (as the intercept).

Soil-water movement, soil-water availability, evaporation, and aeration are also governed by soil temperature. Heat stored near the soil surface has a strong influence on evaporation from soil. A drier soil warms up relatively more quickly and cools down faster than wetter soil because heat capacity of water is several times more than that of soil. Soil temperature also influences the properties of water, such as surface tension, and to a lesser degree, viscosity and density (Table 17.3). Hence, soil-water characteristic curves and hydraulic conductivity functions are also temperature dependent. Bouyoucus (1915) was among the first to observe water movement caused by the soil temperature gradient. He imposed temperature gradients across soil columns, which were at different water contents and contained different soil materials. He found that the difference in water between two halves of column was dependent on both soil material and initial temperature.

Soil temperature varies as a result of radiant, thermal, and latent heat energy exchange processes, which take place primarily through the soil

TABLE 17.3   Density and Viscosity of Water at Various Temperatures

Temperature (°C)	Density (g cm^{-3})	Viscosity (cp)
0	0.99987	1.787
3.98	1.00	1.568
5	0.9999	1.519
10	0.9997	1.307
20	0.9982	1.002
30	0.9957	0.7975
50	0.988	0.5468
80	0.971	0.3547
100	0.9584	0.2818

Source: Adapted from Handbook of Chemistry and Physics, 1988–89.

surface. Soil characteristics, which govern temperature regime, include bulk density, degree of wetness, soil heat capacity, and thermal sources and sinks present in soil matrix.

17.8   SOIL TEMPERATURE REGIMES

Soil temperature continuously varies in response to the changing meteorological regimes acting upon the soil atmosphere interface. The meteorological regimes are characterized by periodic succession of days and nights and winters and summers. Daytime heating is by short-wave radiation from the sun and sky, whereas nighttime cooling is from long-wave radiation emitted by soil. The temperature regimes of soil surface have two cyclical periods, namely diurnal and annual cycles.

17.8.1   The Diurnal Cycle

The variations in soil temperature owing to daytime heating and night-time cooling are known as diurnal variations. In the morning before sunrise, the minimum temperature of soil is the lowest at surface and increases with increase in depth. Similarly, the temperature continues to rise in the lower layers even after the top layer starts to cool down. However, the amplitude of the diurnal wave continues to decrease with soil depth (Fig. 17.5). The amplitude of the surface temperature fluctuation is the range from maximum or minimum to the average temperature (Fig. 17.5).

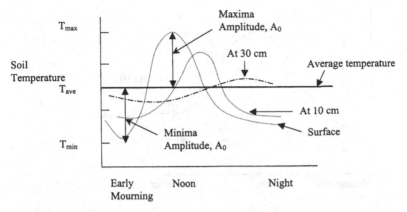

FIGURE 17.5 Schematic of diurnal variations in temperature measured at different depths.

17.8.2 The Annual Cycle

The annual variations in soil temperature result from the variations in short-wave radiation throughout the year. As one goes farther away from the equator, the annual variations in soil temperatures become significant. The summer months in June and July in the Northern Hemisphere represent the peak of global radiations and temperatures, whereas winter months have effects similar to nocturnal daily temperatures. During summer months, the soil temperature at surface is less than that of deeper layers (Fig. 17.6) (Smith, 1932).

The diurnal and seasonal variations of heat can be mathematically represented by assuming that soil temperature oscillates as a pure harmonic (sinusoidal) function of time around an average temperature. Let us also assume that average temperature of soil for all depths is the same. Assuming starting temperature as 0°C, the temperature at the soil surface and at any time t [$T(0, t)$] can be expressed as

$$T(0, t) = \overline{T} + A_0 \sin \omega t \qquad (17.11)$$

where A_0 is the amplitude of surface temperature fluctuation, ω is the angular or radial frequency (1 radian $=$ 57.3 degrees), which is 2π times the actual frequency, $\overline{T}$ is the average temparture and t is time. Assuming that at infinite depth ($z = \infty$), the temperature is constant and equal to $\overline{T}$ (Fig. 17.5). Therefore, temperature at any depth [$T(z, t)$]

$$T(z, t) = \overline{T} + A_z \sin(\omega t + \phi(t)) \qquad (17.12)$$

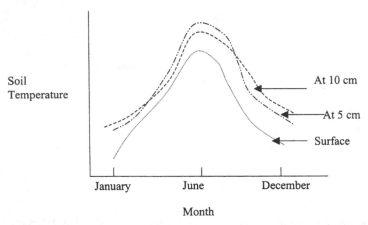

FIGURE 17.6 Schematic of annual variations in temperature measured at different depths in the Northern Hemisphere.

where A_z the amplitude at depth z and $\phi(z)$ are the functions of z, but not time. Incorporating the characteristic depth, also known as damping depth (d), which is defined as the depth at which temperature amplitude decreases to the fraction $1/e^* A_0$, or $1/2.718^* A_0$, or $0.37^* A_0$, provides the following equation

$$T(z,t) = \overline{T} + \frac{A_0 \sin(\omega t + \phi(t))}{e^{z/d}} \qquad (17.13)$$

The damping depth is also related to the thermal properties of the soil and the frequency of temperature fluctuation by the following relationship

$$d = \left(\frac{2k_t}{C_v\omega}\right)^{1/2} = \left(\frac{2D_T}{\omega}\right) \qquad (17.14)$$

where C_v is volumetric heat capacity and D_T is thermal diffusivity of soil.

17.8.3 Soil Temperature Classes

Based upon the mean annual soil temperature, soil temperature regime is expressed in six categories (SSSA, 1987) namely pergelic, cryic, frigid, mesic, thermic, and hyperthermic (Table 17.4). In pergelic soils, mean annual temperature is lower, whereas in cryic soils, it is higher than 0°C. If mean annual temperature is lower than 8°C, the soils are known as frigid,

TABLE 17.4　Classes of Soil Temperature According to Taxonomy of Soils

Class	Mean annual temperature	Remarks
Pergelic	$<0°C$	Permafrost is present
Cryic	$0°C < T < 8°C$	Mean $(T_{summer} - T_{winter}) = 5°C$ at depth 0.5 m
Frigid	$<8°C$	Mean $(T_{summer} - T_{winter}) > 5°C$ at depth 0.5 m
Mesic	$8°C < T < 15°C$	Mean $(T_{summer} - T_{winter}) > 5°C$ at depth 0.5 m
Thermic	$15°C < T < 22°C$	Mean $(T_{summer} - T_{winter}) > 5°C$ at depth 0.5 m
Hyperthermic	$>22°C$	Mean $(T_{summer} - T_{winter}) > 5°C$ at depth 0.5 m

Source: Modified from SSSA, 1987; Scott, 2000.

otherwise as mesic, provided mean annual temperature is below 15°C. For isofrigid, isomesic, isothermic, and isohyperthermic soils, the temperature differs by less than 5°C.

17.9　HEAT TRANSFER IN SOIL

There are three principle heat transport processes: radiation, conduction, and convection.

17.9.1　Radiation

Radiation is the process of heat transfer in which the emission of energy is expressed in the form of electromagnetic waves. The energy of the radiation field can also be transmitted through a vacuum since it does not require a carrier. The energy travels as discrete packets called quanta or photons, whose energy content depends on their wavelength or frequencies. According to the Stefan–Boltzmann law, the total energy emitted by a body, J_t, integrated over all wavelengths is proportional to the fourth power of the absolute temperature of the body, T, and can be expressed as below:

$$J_t = \sigma A T^4 \tag{17.15}$$

where A is the cross-sectional area of body (m²), and σ is the Stefan–Boltzmann constant and in SI units is expressed as $5.675 \times 10^{-8}\, Wm^{-2}K^4$. Eq. (17.15) gives the maximum energy flux that can leave a surface area A at any absolute temperature T. The ratio of radiant energy emitted by soil and maximum amount of radiant energy emitted ($\varepsilon_s/\varepsilon_b$) is known as emissivity coefficient, which equals one for a perfect emitter. Normally a blackbody transmits the maximum and is known as a perfect emitter. Soils emit much less radiant energy. The ε varies as a function of the wavelength of radiation and serves as a correction factor or indicator of the efficiency of natural resources. Equation (17.15) can be modified to Eq. (17.16):

$$J_t = \varepsilon \sigma A T^4 \qquad (17.16)$$

The T also determines the wavelength distribution of the emitted energy and is inversely proportional to maximum radiation intensity, λ_m, micrometers (μm), which is also known as Wien's law.

$$\lambda_m = 2900/T \qquad (17.17)$$

Assuming the temperature of the soil as 300 K, the radiations emitted by soil surface [Eq. (17.15)] have peak intensity at about 10 μm [refer to Eq. (17.17)] and its wavelength distribution is over the range of 3–50 μm, which falls in the infrared region. The Sun is a blackbody and has a temperature of 6000 K. The radiation emitted by the Sun has a peak intensity of about 500 nm [2900/6000]. The Sun's radiation includes a visible light range from 400 to 700 nm [400–425—Violet; 425–490—blue; 490–575—green; 575–585—yellow; 585–650—orange; 650–700—red; and invisible light range from 100–400—ultraviolet and 700–1400 nm—infrared (WHO, 1979), where 1 nm=10^{-9} m]. Planck's law describes the actual intensity distribution as a function of the wavelength, λ, and temperature T as follows:

$$E_\lambda = \frac{C_1}{\lambda^5}\left(e^{C_2/\lambda T} - 1\right) \qquad (17.18)$$

where E_λ is the energy emitted for a given wavelength or range and C_1 and C_2 are constants. In general, the incoming solar radiations are referred to as short-wave radiations and the spectrum emitted by Earth comprises long-wave radiation. Most of the solar radiation reaching Earth's atmosphere is dissipated before it strikes the soil surface. The dissipation occurs partially as a result of the reflection of radiation by clouds, absorption by water vapor, oxygen, carbon dioxide*, and ozone, and diffusion by molecules and particles in air. Solar radiation reaching Earth's

---

*Greenhouse gases allow incoming solar radiation to pass through Earth's atmosphere, but prevent most of the outgoing infrared radiation from the surface and lower atmosphere from escaping into outer space. The greenhouse effect is the rise in temperature that Earth experiences because certain gases in the atmosphere (water vapor, carbon dioxide, nitrous oxide, methane, halogenated fluorocarbons, ozone, perfluorinated carbons, and hydrofluorocarbons) trap energy from the Sun. Without these gases, heat would escape back into space and Earth's average temperature would be about 33°C colder. Because of how they warm Earth, these gases are referred to as greenhouse gases (http://www.epa.gov/globalwarming).

surface is partly direct and partly in the form of scattered beams. After striking the crop or canopy, a fraction of incoming radiation is reflected back to the atmosphere, which is known as albedo (α). The thermal radiations are also transmitted from soil surface into the atmosphere R_{earth}, and onto soil surface from clouds R_{sky}, therefore net radiation (R_N) is:

$$R_N = (1-a)R_s + R_{nt} \qquad (17.19)$$

where R_s is global solar radiation (sum of direct and scattered beam) and R_{nt} is net long-wave thermal radiation ($R_{sky} - R_{earth}$). The R_N varies significantly with climate, latitude, and surface cover. For a known value of emissivity, both R_{sky} and R_{earth} can be calculated by the Stefan–Bolzmann equation (17.15).

17.9.2 Conduction

Conduction is the primary heat transfer mechanism in soil and refers to the propagation of heat within a soil or another body by molecular motion. It is the transfer of translational, rotational, and vibrational energy from molecule to molecule. The process of heat conduction is analogous to diffusion and both try to equilibrate, or even out, mixer's distribution of molecular kinetic energy. The heat flow by conduction in soil takes place from warmer locations towards the cooler regions. Fourier law explains the heat flow by conduction and macroscopically one-dimensional conduction of heat energy through a soil section is described as follows

$$q_h = -k_T A \left(\frac{\partial T}{\partial z} \right) \qquad (17.20)$$

where q_h is the heat flux (Js^{-1}), k_T is proportionality constant or thermal conductivity ($Jm^{-1}s^{-1}K^{-1}$), A is the area of cross section (m^2), T is the temperature in $^\circ K$, and $\partial T/\partial z$ is the temperature gradient in degrees per unit length and the slope of the temperature–distance curve. The negative sign in Eq. (17.20) indicates that heat transfer occurs in the direction of decreasing temperature. Thermal conductivity of solids (Table 17.5) varies from $1\,Jm^{-1}s^{-1}K^{-1}$ to $100\,Jm^{-1}s^{-1}K^{-1}$. For liquids and gases, it ranges from $0.01\,Jm^{-1}s^{-1}K^{-1}$ to $1.0\,Jm^{-1}s^{-1}K^{-1}$ and $0.001\,Jm^{-1}s^{-1}K^{-1}$ to $0.1\,Jm^{-1}s^{-1}K^{-1}$, respectively. The ratio of q_h and A is also known as heat flux density ($Jm^{-2}s^{-1}$).

17.9.3 Convection

The transfer of heat energy in a convection process involves the movement of a heat-carrying mass. Infiltration of warm water into an initially cold soil

TABLE 17.5   Thermal Conductivity of Certain Metals (at 25°C)

Metal	Thermal conductivity (cal cm^{-1}s^{-1}°C^{-1})
Aluminium	9.56
Copper	16.25
Gold	2.44
Iron	2.24
Platinum	0.75
Silver	7.05
Tungstun	4.88

Source: Adapted from Handbook of Chemistry and Physics, 1988–89.

results in heat transfer by the process of convection. Newton's first law of cooling can be used to calculate energy fluxes in and out of the system

$$q_v = C_v A v (T_s - T_0) \quad \text{or} \quad L * E \tag{17.21}$$

where q_v is the heat flux of convection (J s^{-1} or W), C_v is the volumetric heat capacity (Jm^{-3}K^{-1}), v is the velocity of fluid (m s^{-1}), T_s is the temperature of the soil in contact with fluid (K), T_0 is the temperature of the fluid far away from the surface (K), L is latent heat of vaporization, and E is evaporation rate. The convection phenomenon is probably more important in the atmosphere, where there is a consistent circulation of warm and cold air and heat exchange. In soils, the heat convection phenomenon is less important in general, however, during infiltration and redistribution of water in the soil profile, which is cooler than the incoming water, convectional heat energy transport becomes important.

17.10   OTHER PROCESSES OF HEAT EVOLUTION AND TEMPERATURE IN SOIL

17.10.1 Condensation

The conversion of water vapor to a liquid state is known as condensation. Condensation is an exothermic process, and the heat energy released in condensation warms the soil surface. A similar phenomenon is observed when liquid water freezes. About 600 cal g^{-1} of heat energy is released when water vapor condenses, whereas 80 cal g^{-1} of heat (of fusion) is taken up when soil freezes. The six-phase changes that water can undergo and the heat gained and lost is given in Table 17.6.

TABLE 17.6 The Six Phase Changes That Water Can Undergo and the Heat Gained and Lost

Process	From	To	Heat gained/lost $(calg^{-1})$
Condensation	Vapor	Liquid	600
Evaporation	Liquid	Vapor	−600
Freezing	Liquid	Ice	80
Melting	Ice	Liquid	−80
Deposition	Vapor	Ice	680
Sublimation	Ice	Vapor	−680

Source: Modified from http://www.usatoday.com/weather/wlatent1.htm.

17.10.2 Microbial Processes and Heat Evolution

Diverse communities of organisms are present in soil. Variations of microbial population, distribution and activity are a function of depth and type of soil including structure, texture, and water status (Misthustin, 1956). Metabolism is defined as the sum of all chemical reactions occurring within a living organism. Catabolic reactions are exergonic or energy-releasing reactions, which break down more complex molecules, usually by hydrolysis, into simpler components (e.g., chemical processes of digestion). Anabolic reactions are endergonic or energy requiring and build more complex molecules, usually by condensation, from subunit components. The energy for anabolic reactions is provided by catabolic reactions. Micro-organisms decompose the organic matter present in the soils and heat energy is released. An example of an exergonic reaction is the fermentation of alcohol as follows:

$$C_6H_{12}O_6 \rightarrow 2C_2H_5OH + 2CO_2 + 226\,kJ \qquad (17.22)$$

17.10.3 Chemical Reactions and Heat Evolution

Oxidation and reduction reactions (REDOX reactions) are always coupled in biological systems. Oxidation reactions are exothermic (or exergonic) and release energy, whereas reduction reactions are endothermic (or endergonic) and harness energy. Chemical reactions may be viewed in terms of the amount of energy required by the reaction at various stages. A convenient way to do this is with an energy hill diagram. In these diagrams, the total amount of energy, both kinetic and potential in the chemicals involved in the reaction, is plotted as a function of time. Conversion of

sulfur (S) to sulfur dioxide (SO_2) and to sulfuric acid (H_2SO_4) are examples
of exothermic processes.

$$S + O_2 \rightarrow SO_2 \qquad (17.23)$$

$$2SO_2 + O_2 \rightarrow 2SO_3 \qquad (17.24)$$

$$SO_3 + H_2O \rightarrow H_2SO_4 \qquad (17.25)$$

The above chemical reaction releases about 178 Kcal/mole of
energy (Bohn et al., 1934). Chemical reactions are known as *exothermic*
when the chemical products of the reaction have less energy than the
starting materials. Another example of an exothermic reaction is burning
wood or a burning match. Wood is mainly cellulose and has a lot of
chemical energy. The products of burning (e.g., CO_2 and H_2O) have much
less energy because the net balance of the energy is converted into light
and heat.

Chemical reactions that lead to products having more energy at the
end than at the beginning are called *endothermic*. Endothermic reactions
typically involve the synthesis of complex molecules from simple ones.
Examples of endothermic reactions are cells making proteins from
amino acids and photosynthesis in plant cells. In photosynthesis, CO_2 and
H_2O, which are the starting materials for photosynthesis, have less energy
than the final product, i.e., carbohydrates (http://old.jccc.net/~pdecell/
metabolism/energyhill.html).

17.11 ENERGY BALANCE OF SOIL

Net radiation is the sum of all incoming minus all outgoing radiation on
Earth's surface. Steady state one-dimensional heat energy balance at the soil
surface or crop canopy can be written as

Net heat energy arriving at surface
$$- \text{net heat energy leaving surface} = 0^* \qquad (17.26)$$

Equation (17.26) disregards the transient energy changes due to heating
or cooling of soil surface and lateral heat energy inputs. Heat transfer

---

*Except in a greenhouse, where net heat energy leaving the surface is smaller, therefore, Eq.
(17.26) is not satisfied.

from the soil surface takes place as (i) convective heat flux (H_c), (ii) soil heat flux (J_H) and (iii) latent heat flux ($L*E$). H_c represents the transport of warm air from the soil surface to the atmosphere vertically above it. The J_H represents the vertical transport of heat into the soil, and $L*E$ denotes evaporation and subsequent transport of water vapor from the soil surface (L is the latent heat of vaporization and E is the evaporation rate). The net radiation received by the soil surface is transformed into heat, which warms soil and air and vaporizes water. Therefore, under steady state conditions the heat balance equation can be written as

$$R_N = H_c + J_H + L*E \tag{17.27}$$

Combining Eqs. (17.19) and (17.27) provides the total surface energy balance as follows

$$(1 - \alpha)Rs + Rnt - (H_c + J_H + L*E) = 0 \tag{17.28}$$

17.12 HEAT CAPACITY OF SOIL

The heat capacity of soil as defined in Sec. 17.5.1 is the amount energy required to change the temperature of a body by 1°C by heat adsorption or release. The relationship between volumetric and gravimetric heat capacities for a dry soil is described in Eq. (17.8), which is rewritten below:

$$C_v = \rho_b * C_g \tag{17.8}$$

and for a wet soil

$$C_v = \rho_b' * C_g = \rho_b(1 + w) * C_g = (\rho_b + \theta) * C_g \tag{17.29}$$

where ρ_b' is wet bulk density, w is gravimetric water content, and θ is volumetric water content. The C_v is dependent on composition of solid phase, which constitutes mineral and organic matter, bulk density, and water content of soil. The total C_v is calculated by summing the heat capacities of various constituents, weighted according to their volumes (de Vries, 1975)

$$C_v = \sum(f_{si}C_{si} + f_W C_W + f_a C_a) \tag{17.30}$$

TABLE 17.7 Thermal Conductivity and Heat Capacity of Gases (Oxygen at 25°C, 1 atm), Water (26.7°C), Quartz (37.8°C), and Sandstone (100°C)

Gas	Thermal conductivity (k_T) (10^{-6} cal cm^{-1} s^{-1} °C^{-1})	Heat capacity (C_g) (cal g^{-1} °C^{-1})
Air	62.2	0.25
Carbon dioxide	39.67	–
Oxygen	63.64	0.219
Water	42.57	0.998
Quartz (C-axis)	6.4	0.18
Sandstone	3.82	0.26
Iron		

Source: Adapted from Handbook of Chemistry and Physics, 1988–89.

where f is the volumetric fraction of each constituent phase, subscript s, w, a, and i stand for solid, water, air, and number of components in a given phase. The gravimetric heat capacity is the ratio of volumetric heat capacity and particle density (Table 17.7). The heat capacity of loam, sandy loam, sandy clay loam, a clay loam, and clay soil is given in Table 17.8. The heat capacity for each of the components solid, water, and air is the product of their particle density and specific heat or heat capacity per unit mass, i.e., $C_{si} = \rho_{si} C_{mi}$, $C_w = \rho_w C_{mw}$, and $C_a = \rho_a C_{ma}$. In general, the contribution of air is almost negligible because of the very small density and is ignored. The solid phase is divided into two components: mineral (m) and organic matter (o). Eq. (17.30) can be rewritten as

$$C_v = \sum (f_m C_m + f_o C_o + f_W C_w) \tag{17.31}$$

17.13 THERMAL CONDUCTIVITY

The thermal conductivity (k_T) is defined as the quantity of heat transferred through a unit cross-sectional area in unit time under a unit temperature gradient. The K_T of soil depends upon the volumetric proportions of the solid, liquid, and gaseous phase of the soil medium. The other factors, which influence K_T, are the size and arrangement of solid particles, and interfacial contact between solid and liquid phases. The K_T values of some materials is presented in Tables 17.7 and 17.9, which show that air has a much lower K_T than water and solid, therefore, high air content reduces the thermal contact between soil

TABLE 17.8 Heat Capacity of Some Nigerian Soils

Soil texture	Heat capacity ($cal g^{-1} {}^{\circ}C^{-1}$)
Sandy loam	0.322
Sandy clay loam	0.350
Loam	0.279
Clay loam	0.224
Clay	0.248

Source: Modified from Ghuman and Lal, 1985.

TABLE 17.9 Range and Averages of Thermal Diffusivity and Thermal Conductivity of Soil Particles

Soil type	Moisture state	Thermal diffusivity $\times 10^{-3}$ ($cm^2 s^{-1}$)		Thermal conductivity $\times 10^{-3}$ ($cal\,cm^{-1} s^{-1} {}^{\circ}C^{-1}$)	
		Range	Average	Range	Average
Sand	Dry	3.5–1.5	2.23	0.55–0.37	0.42
	Wet	12.6–4.4	8.0	4.35–3.7	4.02
Clay	Dry	1.8–1.2	1.5	0.37–0.17	0.26
	Wet	11–3.2	5.97	3.5–1.4	2.69

Source: From Geiger, 1965; Nakshabandi and Kohnke, 1965; and van Duin, 1963.

particles and reduces the K_T of soil. On the other hand, an increase in bulk density of soil lowers the porosity and improves the thermal contact between soil particles and increases K_T and D_T ($cm^2 s^{-1}$). The increase in water content of the soil also improves the thermal contact between soil particles and increases K_T as well as D_T.

17.14 HEAT FLOW IN SOILS

Fourier (1822) analyzed the heat conduction in solids and developed a mathematical relationship, which is analogous to the diffusion equation by Fick (see Chapters 16 and 18), conduction of fluid flow in porous media by Darcy (refer to Chapters 12 and 13), and conduction of electricity by Ohm (Table 17.10). The first law of heat conduction is known as Fourier's law and under steady state condition in one-dimension, the heat flux density (q_h, $Jm^{-2}s^{-1}$) equation and the heat energy balance equation for

TABLE 17.10   The Analogous Laws of Water, Air, Heat, and Electricity Flow

Process	Law	Equation
Water movement	Darcy's	$q = -K\dfrac{\partial H}{\partial x}$
Air movement	Fick's	$q_x = -D\dfrac{\partial C}{\partial x}$
Heat movement	Fourier's	$h = -k\dfrac{\partial T}{\partial z}$
Electric flow	Ohm's	$a = -k\dfrac{\partial \phi}{\partial x}$

homogeneous soils is expressed as follows:

$$q_h = -K_T \frac{\partial T}{\partial z} \tag{17.32}$$

$$\frac{\partial H}{\partial t} = -\frac{\partial q_h}{\partial z} \tag{17.33}$$

where H is the volumetric heat content (Jm^{-3}), and volumetric heat capacity of soil (C_v) is equal to dH/dT. The early developments of heat conduction dealt with dry media, i.e., solids only (Carslaw and Jaeger, 1959). The theory was later expanded to soils containing water. At first the heat flow was studied for homogeneous soils with constant C_v and K_T.

Combining these two equations [(17.32) and (17.33)], which ignores the existence of a sink or source term, results in general heat conduction equation. These equations can be solved numerically for nonhomogeneous soil profiles by assuming that heat transfer takes place by conduction only. The addition of water complicates the process as water may evaporate and condense. As the heat energy inside a soil matrix is transported by convection (by flowing water, air, and latent heat), conduction, and radiation. The first two are the most important heat transport process through soil. The convection can be represented as given by Eq. (17.21). The expression for net flux of heat through soil can be written as

$$q_h = -k_T \frac{dT}{dz} + L * E_v \tag{17.34}$$

where E_v is the water vapor mass flux ($\text{gcm}^{-2}\,\text{s}^{-1}$). The temperature gradient across a moist soil results in movement of water both as liquid and gas along with heat, therefore, the effective value of K_T exhibits a temporal variation and therefore cannot be measured in the soil directly (de Vries, 1958). If D_v is the thermal vapor diffusivity then water vapor flux in soil in one-dimension flow where relative humidity is above unity can be written as

$$E_v = -D_v \frac{\mathrm{d}T}{\mathrm{d}z} \tag{17.35}$$

Transferring Eq. (17.35) into Eq. (17.34) results in

$$q_h = -(K_T + D_v L)\frac{\mathrm{d}T}{\mathrm{d}z} \cong -K_{Te}\frac{\mathrm{d}T}{\mathrm{d}z} \tag{17.36}$$

where K_{Te} is the effective thermal conductivity of the porous medium, which includes the effects of conduction and convection of latent heat.

17.15 HEAT CONSERVATION EQUATION

The heat conservation equation for a small cubic soil matrix (Fig. 17.7) can be derived by accounting for the amount of heat energy entering a system, leaving a system, and change in the heat energy of a system. Mathematically a heat balanced equation can be written as follows (Jury et al., 1991)

The amount of heat energy entering soil matrix

= amount of heat leaving the soil matrix

+ increase in heat energy of the soil matrix

+ loss of heat energy from the soil matrix. (17.37)

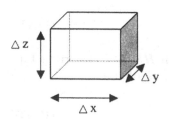

FIGURE 17.7 Schematic of a soil matrix for heat conservation equation.

Assuming the heat flow is in z-direction, the one-dimensional heat flow in vertical direction during time interval Δt through the soil matrix (Jury et al., 1991; Scott, 2000)

$$= h(x, y, z, t + \frac{1}{2}\Delta t)\Delta x \Delta y \Delta t \qquad (17.38)$$

where h is the heat flux at the average time $t + (\Delta t/2)$ and Δx and Δy are the cross-sectional area of the soil matrix. The amount of heat flowing out of the matrix for the same time interval

$$= h(x, y, z + \Delta z, t + \frac{1}{2}\Delta t)\Delta x \Delta y \Delta t \qquad (17.39)$$

The net change in H within the soil matrix

$$[H(x, y, z + \frac{1}{2}\Delta z, t + \frac{1}{2}\Delta t) - H(x, y, z + \frac{1}{2}\Delta z, t)]\Delta x \Delta y \Delta z$$
$$= \Delta H \Delta x \Delta y \Delta z \qquad (17.40)$$

where H is at the middle or $z + (\Delta z/2)$. If the amount of heat lost from the soil matrix per unit system is r_s, the total heat loss

$$= r_r \Delta x \Delta y \Delta z \Delta t \qquad (17.41)$$

Transferring Eqs. (17.38), (17.39), (17.40), and (17.41) into Eq. (17.37), and after rearranging and assuming that ΔZ and $\Delta t \rightarrow 0$, the resultant equation is known as the differential form of heat conservation equation (Jury et al., 1991; Scott, 2000)

$$\frac{\partial H}{\partial t} + \frac{\partial h}{\partial z} + r_r = 0 \qquad (17.42)$$

The H can also be expressed as a function of volumetric heat capacity of soil C_v and temperatures

$$H = C_v(T - T_{ref}) \qquad (17.43)$$

where T_{ref} is the reference temperature at which $H = 0$. Transferring Eq. (17.43) into (17.42), and assuming C_v a constant, we obtain

$$C_v \frac{\partial T}{\partial t} = \frac{\partial}{\partial z}\left(K_T \frac{\partial T}{\partial z}\right) \qquad (17.44)$$

if k_t is assumed to be independent of z then following heat flow equation is obtained.

$$\frac{\partial T}{\partial t} = D_T \frac{\partial^2 T}{\partial z^2}$$ (17.45)

where $D_T = K_T/C_V$, and is known as soil thermal diffusivity ($L^2 T^{-1}$ or $m^2 s^{-1}$). D_T can be expressed in three different ways for soil water diffusivity, diffusion in air, and soil thermal diffusivity as follows:

Definition	Equation	Unit
Soil water diffusivity	$D_\theta = K_\theta(d\Phi_m/d\theta)$	$cm^2\ s^{-1}$ or $L^2\ T^{-1}$
Diffusion coefficient in air	$D_s = D_0*0.66*f_a$	$cm^2\ s^{-1}$ or $L^2\ T^{-1}$ (see Chapter 18)
Soil thermal diffusivity	$D_T = K_T/C_V$	$cm^2\ s^{-1}$ or $L^2\ T^{-1}$

17.16 MEASUREMENT OF THERMAL CONDUCTIVITY OF SOIL

Similar to hydraulic conductivity (K_s), thermal conductivity (K_T) can be measured using the steady state method or transient methods. Let us assume a soil column of thickness L placed between two glass plates of thickness d. The outer surface of each of these two plates is at constant temperature. The temperature is measured at several different positions as shown in Fig. 17.8. Let us assume the datum at point A, i.e., $z = 0$ at A_1 and that heat flux through soil and glass plates is equal, glass and soil are in good contact, and lateral movement of heat is negligible. If the temperature measured at the plate at A is T_0, then according to Fourier's law, the heat flux across the plate and soil column can be written as

$$q_h = -K_T \frac{(T_2 - T_0)}{d}$$ (17.46)

$$q_h = -K_{Te} \frac{(T_3 - T_2)}{L}$$ (17.47)

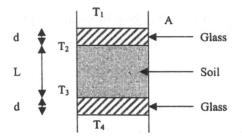

FIGURE 17.8 Apparatus for measurement of thermal conductivity under steady state.

where K_{Te} is the effective thermal conductivity of porous media. Equating these two equations [(17.46) and (17.47)] and rearranging them gives the expression for effective thermal conductivity as follows

$$K_{Te} = -\frac{K_T L(T_2 - T_0)}{d(T_3 - T_2)} \qquad (17.48)$$

The steady state methods are useful for measurement of K_T in the laboratory. The basic drawback of this method is that a nonuniform profile within the column is created due to the redistribution of water under the influence of a steady state temperature gradient (Jury and Miller, 1974). The transient method for the measurement of K_T can be used in situ and does have the drawback of the steady state method (De Vries and Peck, 1968). The method consists of a thin metal wire, which serves a heat source. The wire and the sensors for temperature measurement (for example thermocouples) are kept inside a cylindrical tube, which is inserted into the soil. The flow of heat takes place radially from the wire and the temperature of the thermocouple probe in contact with the soil is given by the following equation (Carslaw and Jaeger, 1959)

$$T = T_0 = \frac{q_h}{4\pi k_T}[d + \ln(t + t_0)] \qquad (17.49)$$

where T_0 is the temperature at $t = 0$, $T - T_0$ is rise in temperature, q is heat flowing per unit time and length of wire, d is a constant, which depends on the location of thermocouple, and t_0 is a correction constant, which depends upon the dimensions of the probe. The equation (17.49) can be rearranged for $t >>> t_0$ and the K_T can be calculated by the following relationship

$$K_T = \frac{q_h}{4\pi m} \qquad\qquad\qquad (17.50)$$

where m is the slope measured by plotting T versus $\ln(t)$. The q_h is calculated from the current (I) applied to the wire and the measured resistance per unit length of wire.

17.17 MANAGEMENT OF SOIL TEMPERATURE

Soil temperature can be managed in a number of ways, which includes mulching, tillage, irrigation, drainage, cover crop or shading, and application of dark or light powder. The management options depend whether the temperature of soil needs to be increased or reduced. The duff layer, which is the thatch of plant material on the surface, reduces the frequency of freeze–thaw cycles in the seed zone, maintains aggregates, and prevents crusting. Different types of mulches are used to either lower or raise the soil temperature, depending upon the need. Light-colored mulches (e.g., chopped straw, plastic mulches) reflect a portion of incoming solar radiation and reduce the amount of radiant flux reaching the soil surface, thus lowering the soil temperature (Fig. 17.9). On the other hand, application of thick and dark mulches (e.g., charcoal or bitumen) enhance the soil temperature. Dark plastic mulch absorbs most of the radiant solar energy but transmits very little to the soil, keeping the soil cool (Fig. 17.10). On the other hand, transparent plastic mulch transmits short-wave (visible

FIGURE 17.9 A light-colored mulch (aluminium foil) decreases soil temperature. (Lal, 1975, field experiments.)

FIGURE 17.10 A dark-colored plastic mulch decreases soil temperature. (Field experiments, IITA, Ibadan, Nigeria, 1975.)

light) radiation to the soil surface while preventing infrared (long-wave) radiation, creating a greenhouse effect and warming the soil (Fig. 17.11) (Lal, 1979). Soil temperature (in °C) at 5 cm depth under a maize crop one week after planting and under black plastic, clear plastic, straw mulch, ridges, bare flat, and aluminum foil shows that the soil temperature fluctuation is the minimum for straw mulch and the maximum for ridges (Fig. 17.12). The soil temperature for bare flat treatment under different crops was in the order cassava > soybean > maize = cowpea (Fig. 17.13) (Lal, 1979).

Ridge tillage increases surface soil temperature by increasing the area exposed to radiation and decreasing soil moisture (Fig. 17.14). The surface 5 cm, or the seed zone, of no-till soils, may warm more slowly in spring and cool more slowly in autumn than in cultivated soils. Below 5 cm, in the root zone, no-till soils may be warmer and wetter from fall through spring. The amplitude of temperature variation at the soil surface is greater in plowed than no-till soils. However, because of the lower thermal conductivity of plow-till soil, the amplitude decreases more rapidly in plow-till than in no-till soils (van Duin, 1956). The maximum and minimum temperatures at the surface of a plow-till soil are also much higher than a no-till soil. Soil temperature in early spring is significantly affected by tillage methods (Table 17.11) (Fausey and Lal, 1989). Similar to no-till, mulching with crop residue decreases the maximum soil temperature and increases the minimum soil temperature (Fig. 17.15).

FIGURE 17.11 A clear plastic creates a greenhouse effect and increases soil temperature. (Field experiments, IITA, Ibadan, Nigeria, 1975.)

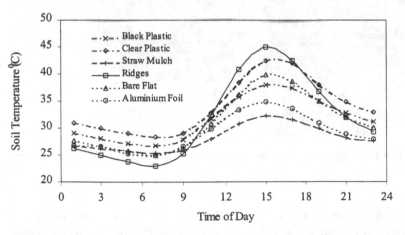

FIGURE 17.12 Soil temperature at a 5 cm depth under different mulches one week after planting crops during the first growing season in 1977 in Ibadan, Nigeria. (Redrawn from Lal, 1979.)

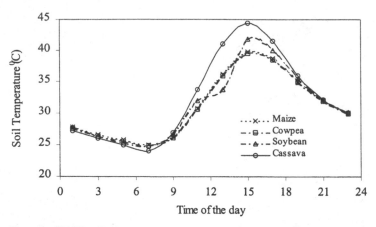

Figure 17.13 Soil temperature at a 5 cm depth under different crops for bare flat seedbed preparation during the first growing season in 1977 in Ibadan, Nigeria. (Redrawn from Lal, 1979.)

Figure 17.14 Ridge tillage may decrease the minimum and increase the maximum soil temperature.

Irrigation with cold water during the summer results in bringing down the temperature of surface soil (Fig. 17.16). Similarly, drainage has a strong influence on soil temperature. During spring, wet soils are cold at the soil surface, because of the increase in K_T of soil, which results in the conduction of heat in a downward direction reducing the temperature of the surface of

TABLE 17.11 The Tillage and Drainage Effects on Mean Daily Maximum Soil
Temperature in °C

	Distance from drain (m)		
Tillage	0	9	27
No-till	6.2	6.1	6
Ridge-till	5.4	5.3	5.1
Plow-till	6.2	5.8	5.5
Beds	6	5.7	5.7

Source: Modified from Fausey and Lal, 1989.

FIGURE 17.15 Mulching with crop residue decreases the maximum and increases
the minimum soil temperature.

soil. The high water content also increases heat capacity of soil, thus
reducing the temperature of the surface of soil. The evaporation of water
from wet soil also consumes the energy, which results in a reduction of
temperature of the soil surface. Therefore, removal of excess water by
surface or subsurface drainage increases soil aeration, which in turn warms
the soil surface and improves seed germination and root growth. Building
large mounds in a poorly drained (hydromorphic soil) increases soil
temperature (Fig. 17.17). The effects of tillage and drainage on soil
temperature are presented in Table 17.11, which show that as distance from

FIGURE 17.16   Irrigation lowers the soil temperature in summer and raises it in winter.

FIGURE 17.17   Farmers in West Africa construct large mounds in hydromorphic soils to create well-aerated root zones and raise soil temperature. (Field experiments, IITA, Ibadan, Nigeria.)

drain increases, the mean daily temperature of soil reduces for all the treatments namely, no-till, ridge-till, plow-till, and beds.

Example 17.1

If the Stefan–Boltzmann constant (σ) is $5.67 \ 10^{-8} \, \mathrm{Wm}^{-2} \mathrm{K}^{-4}$ and emisssivity (ε) is 0.94, calculate the long-wave thermal radiation energy flux R_{earth} for a temperature 300 K.

Solution

The energy flux density can be calculated as

$$R_{earth} = \varepsilon \sigma T^4 = 0.94 * 5.67 \times 10^{-8} * (300)^4$$
$$= 431.713 \, \mathrm{Wm}^{-2} = 0.61 \, \mathrm{calcm}^{-2} \mathrm{min}^{-1}$$

Example 17.2

A soil column contains 40 cm of dry sand over 20 cm of dry loam soil. Both ends are attached to a constant temperature bath with the top maintained at 25°C and the bottom at 4°C. If the thermal conductivity of sand (K_{Ts}) is $0.5 \, \mathrm{mcal}^{-1} \mathrm{s}^{-1} {}^\circ \mathrm{C}^{-1}$ and that of loam (K_{Tl}) is $0.25 \, \mathrm{mcal}^{-1} \mathrm{s}^{-1} {}^\circ \mathrm{C}^{-1}$, calculate the steady state heat flux through the two layers and the temperature at the sand–loam interface.

Solution

The equivalent thermal conductivity of the sand–loam system (k_{eq}) for the total thickness of the sand–loam system (i.e., 60 cm) can be calculated as below

$$\frac{60}{K_{eq}} = \frac{40}{K_{Ts}} + \frac{20}{K_{Tl}} \Rightarrow K_{Teq} = \frac{60}{(40/0.5) + (20/0.25)} = \frac{60}{(80 + 80)}$$
$$= 0.375 \, \mathrm{mcal}^{-1} \mathrm{s}^{-1} {}^\circ \mathrm{C}^{-1}$$

The heat flux equation across the entire soil column of sand and loam will provide the steady state heat flux across column as follows

$$q_h = -\frac{K_{Teq}}{(40 + 20)} (25 - 4) = -\frac{0.375 * 21}{60} = -0.131 \, \mathrm{calcm}^{-2} \mathrm{s}^{-1}$$

The temperature across the sand–loam interface (T) can be calculated as

$$q_h = -\frac{K_{Tl}}{20} (T - 4) \Leftrightarrow -0.131 = -\frac{0.25 * (T - 4)}{20} \Leftrightarrow T = 14.48^\circ \mathrm{C}$$

Example 17.3

If the bulk density of a soil 1.45 gcm^{-3} and is at two different volumetric water contents: (i) 0.50 and (ii) 0.25, what will be the ratio of volumetric heat capacity, C_v, for these two situations?

Solution

The C_v can be calculated from Eq. (17.26).

For $\theta = 0.50$, $\quad C_{v1} = (1.45 + 0.5) C_g$
And for $\theta = 0.25$, $\quad C_{v2} = (1.45 + 0.25) C_g$

$$\frac{C_{v1}}{C_{v2}} = \frac{1.95}{1.70} = 1.15$$

Example 17.4

If the particle density of a soil is $2.65 \times 10^3 \, kgm^{-3}$, and bulk density is $1.45 \times 10^3 \, kgm^{-3}$, assuming the soil is water-saturated, calculate the volumetric heat capacity (C_v) of the soil if volumetric organic matter content is 15% of solid mass. Assume volumetric heat capacity of mineral, organic matter, and water as 2×10^6, 2.5×10^6, and $4.2 \times 10^6 \, Jm^{-3}$ deg, respectively.

Solution

The total porosity of the soil can be calculated as

$$f_t = \frac{(\rho_s - \rho_b)}{\rho_s} = \frac{2.65 * 10^3 - 1.45 * 10^3}{2.65 * 10^3} = 0.453,$$

since soil is saturated volumetric fraction of water is equal to porosity
Therefore volumetric fraction of solids is $= 1 - 0.453 = 0.547$
Organic matter fraction $= 0.547 * 0.1 = 0.0547$
Mineral matter fraction $= 0.547 * 0.9 = 0.492$
Therefore volumetric heat capacity (C_v) can be calculated as

$$C_v = f_m C_m + f_o C_0 + f_w C_w$$
$$C_v = 0.492 * 2 * 10^6 + 0.0547 * 2.5 * 10^6 + 0.453 * 4.2 * 10^6$$
$$= 4.25 * 10^6 \, Jm^{-3} \text{degree}$$

Example 17.5

For a temperature difference of 20°C across a 30-cm thick soil sample, calculate the one-dimensional thermal flux and total heat transfer under steady state condition. Assume thermal conductivity of soil as $1.6\,\mathrm{J\,m^{-1}\,s^{-1}\,°C^{-1}}$.

Solution

The heat flux across a soil column is expressed as

$$q_h = -K_T \frac{(T_2 - T_0)}{L} = \frac{1.6 * 20}{0.3} = 106.67\,\mathrm{J\,m^{-2}\,s^{-1}}$$

Total heat transfer $= q_h * t = 106.67 * 3600 = 3.84 \times 10^5\,\mathrm{J\,m^{-2}}$

Example 17.6

Assuming that the diurnal temperature wave is symmetrical and mean temperature is equal throughout the soil profile with surface temperature equal to mean temperature at 6 A.M. and 6 P.M., calculate the temperatures at noon and midnight for depths 10 cm and 25 cm. Assume daily maximum and minimum soil surface temperature 36°C and 8°C, respectively, and damping depth as 10 cm.

Solution

The temperature T at any depth z and time t can be calculated as follows

$$T(z, t) = T_{ave} + \frac{A_0[\sin(\omega t - (z/d))]}{e^{z/d}}$$

where A_0 is minimum value above mean, ω is the radial frequency ($2\pi/24$), z is depth, and d is the damping depth. The average temperature $T_{ave} = (38 + 8)/2 = 23°$
 Temperature above mean $A_0 = 38 - 23 = 15°$
 At soil surface $z = 0$
 Temperature 6 h after mean temperature, i.e., noon temperature

$$T(0, 6) = 23 + \frac{15 * [\sin((\pi/2) - 0)]}{e^0} = 23 + 15 = 38°C$$

At midnight

$$T(0, 18) = 23 + \frac{15 * [\sin((3\pi/2) - 0)]}{e^0} = 23 - 15 = 8°C$$

At depth 10 cm

$$T(10,6) = 23 + \frac{15 * [\sin((\pi/2) - 0.1/0.1)]}{e^{0.1/0.1}} = 26°C$$

and at midnight $T(10,18) = 22°C$

PROBLEMS

1. If the particle density of a soil is $2.65 \times 10^3 \, \mathrm{kgm^{-3}}$, and bulk density is $1.45 \times 10^3 \, \mathrm{kgm^{-3}}$, assuming the soil is (a) dry and (b) volumetric water content is 30%, calculate the volumetric heat capacity (C_v) of the soil if volumetric organic matter content is 8% of solid mass. Assume volumetric heat capacity of mineral, organic matter, and water as 2×10^6, 2.5×10^6, and $4.2 \times 10^6 \, \mathrm{Jm^{-3}}$ deg, respectively.

2. Calculate one-dimensional thermal flux and total heat transfer under steady state condition for a temperature difference of 10°C across a 25-cm thick soil sample. Assume thermal conductivity of soil as $1.6 \, \mathrm{J \, m^{-1} \, s^{-1} \, {}°C^{-1}}$

3. How much heat is required to change 20 kg of ice at −8°C to steam at 100°C?

4. A soil column contains 35 cm of dry sand over 15 cm of dry loam soil. Both ends are attached to a constant temperature bath with top maintained at 28°C and bottom at 8°C. If the thermal conductivity of sand (k_s) is $0.48 \, \mathrm{mcal^{-1} \, s^{-1} \, {}°C^{-1}}$ and that of loam (k_L) is $0.23 \, \mathrm{mcal^{-1} \, s^{-1} \, {}°C^{-1}}$, calculate the steady state heat flux through the two layers and the temperature at the sand–loam interface.

5. Compute the amount of heat required to raise the temperature of a unit area of soil ($\rho_b = 1.25 \, \mathrm{g \, cm^{-3}}$, $w = 0.2 \, \mathrm{g \, g^{-1}}$) from an initial temperature of 10°C to 20°C to a depth of 50 cm.

6. Calculate the direction and quantity of heat per unit that will flow in one day, when soil temperature at the surface is 30°C and at 5 cm depth is 25°C. Assume thermal conductivity $= 3 \times 10^{-3} \, \mathrm{cal \, cm^{-1} \, s^{-1} \, c^{-1}}$.

7. Why is soil temperature more important than air temperature?

REFERENCES

ATSDR (1999). Mercury CAS#7439-97-6. Agency for Toxic Substance and Disease Registry, U.S. Dept. of Health and Human Services 1600 Clifton Road NE, Atlanta GA30333 (http://www.atsdr.cdc.gov).

Bohn H.L., McNeal B.L., and O'Connor G.A. (1934). Soil Chemistry. John Wiley & Sons, New York, p 52.

Carslaw H.S. and Jaeger J.C. (1959). Conduction of Heat in Solids. Oxford University Press, London.

Chang J.H. (1968). Microclimate of sugar cane. Hawaiian Planter's Rec. 56:195–225.

Childs P.R.N., Greenwood J.R. and Long C.A. (2000). Review of temperature measurement. Review of Scientific Instrument. 71(8):2959–2978.

Chudnovskii A.F. (1966). Plants and light. I. Radiant energy. Pp1-51. In: Fundamentals of Agrophysics, Isreal Prog. For Scientific Translations, Jerusalem.

de Vries D.A. (1958). Simultaneous transfer of heat and moisture in porous media. Trans. Am. Geophys. Un. 39:909–916.

de Vries D.A. (1975). Heat transfer in soils. In: de Vries, D.A. and Afgan N.H., eds. Heat and Mass Transfer in Biosphere. Scripta Book Co., Washington DC, p 5–28.

de Vries D.A. and Peck A.J. (1968). On the cylindrical probe method of measuring thermal conductivity with special reference to soils. Aust. J. Phys. 11:255–271.

Fausey N.R. and Lal R. (1989). Drainage–Tillage effects on Crosby-kokomo soil association in Ohio II. Soil temperature regime and infiltrability. Soil Technology 2:371–383, Cremlingen.

Geiger R. (1965). The climate near the ground. Harvard University Press, Cambridge, MA.

Ghuman B.S. and Lal R. (1985). Thermal conductivity, thermal diffusivity, and thermal capacity of some Nigerian soils. Soil Science 139(1):74–80.

Jury W.A. and Miller E.E. (1974). Measurement of transport coefficients for coupled flow of heat and moisture in a medium sand. Soil Sci. Soc. Am. Proc. 38:551–557.

Jury W.A., Gardner W.R. and Gardner W.H. (1991). Soil Physics. John Wiley and Sons, USA.

Lal R. (1979). Soil and micro-climate considerations for developing tillage systems in tropics. In Lal R., (ed.) Soil Tillage and Crop Production. International Institute of Tropical Agriculture, Ibadan, Nigeria, Proceedings Series number 2, p 52.

Nakshabandi G.A. and Kohnke H. (1965). Thermal conductivity and diffusivity of soils as related to moisture tension and other physical properties.

Scott H.D. (2000). Soil physics agricultural and environmental applications. Iowa State University Press, Ames, IA.

Soil Science Society of America (1987). Glossary of soil science terms. Madison, WI.

van Duin R.H.A. (1963). The influence of soil management on the temperature wave near the surface. Technical Bulletin 29, Institute of Land and Water Management Research, Wageningen.

Weast R.C. et al., eds. (1989). Handbook of Chemistry and Physics. Boca Raton, Florida, CRC Press, 69th edition.

WHO (1979). Environmental health criteria 14: Ultraviolet radiation. WHO, Geneva, Switzerland. 110pp.

18

Soil Air and Aeration

18.1 AIR

Earth is surrounded by a gaseous envelope of air about 80 km thick called the atmosphere. The origin of Earth's atmosphere is still a subject of speculation. One theory seems fairly certain that some five billion years ago when Earth was formed, it was extremely hot and did not have an atmosphere. It is generally accepted that the first atmosphere, created when Earth cooled down, consisted of helium (He), hydrogen (H_2), ammonia (NH_3), and methane (CH_4). Assuming that five billion years ago volcanoes emitted similar gasses as in the modern era, Earth's second atmosphere probably consisted of water vapor (H_2O), carbon dioxide (CO_2), and nitrogen (N_2), because these gasses are emitted from Earth's interior by a process known as "outgassing." With colonization by plants, which absorb CO_2 and emit O_2 during photosynthesis, the atmosphere eventually contained a large concentration of O_2, which now constitutes one-fifth of its volume.

In fact, the envelope of air is a mixture of many discrete gases. Each gas has a distinct physical and chemical property. The atmosphere comprises two types of gases: those whose concentration remains essentially constant or permanent (by percent), and those that are variable and have changing concentrations over a finite period of time. Among the permanent gases, nitrogen (78.1%) and oxygen (20.9%) constitute about 99% of the atmosphere. Other permanent gases are argon (Ar, 0.9%), neon (Ne, 0.002%), helium (He, 0.0005%), krypton (Kr, 0.0001%), and hydrogen (H_2, 0.00005%). The variable gases are water vapor (H_2O, 0 to 4%), carbon dioxide (CO_2, 0.037%), methane (CH_4, 0.0002%), ozone (O_3, 0.000004%), and nitrous oxide (N_2O, 0.00009%) (www.met.fsu.edu/explores/atm comp.html). A brief description on some of these gases is given in the following sections.

18.1.1 Nitrogen

Nitrogen gas (N_2) is composed of molecules of two nitrogen atoms, and occupies 78.1% of Earth's atmosphere. It is colorless, odorless, and tasteless. The atomic weight of N_2 is 14. Nitrogen is a principal nutrient. The low content of nitrogen in most soils exists in stark contrast to its abundance in the air. This is because gaseous N_2 molecules have very strong bonds, which make the gas chemically stable, but unusable by most biological organisms. Some species of bacteria absorb N_2 from the air and convert it to ammonium, which can be used by plants. This process is called "biological nitrogen fixation" and is the principal natural means by which atmospheric nitrogen is added to the soil by nitrogen-fixing bacteria living in nodules on the plant roots. An example of a leguminous nitrogen-fixing crop is soybean (*Glycine max*).

18.1.2 Oxygen

Oxygen gas (O_2) is composed of molecules of two oxygen atoms, and occupies 20.9% of Earth's atmosphere by volume. It is colorless, odorless, and tasteless, and constitutes 86% of the oceans and 60% of the human body. It is the third most abundant element found in the Sun. The atomic weight of oxygen is 16. Almost all plants and animals require oxygen for respiration to maintain life. Oxygen is flammable, reactive, and oxidizes most elements. A chemical reaction in which an oxide is formed is known as "oxidation." The rate at which oxidation occurs varies with the element with which oxygen is reacting, (e.g., burning involves a rapid oxidation, whereas rust, or iron oxide, forms slowly). Carbon in fossil fuels, for example, can be quickly oxidized to carbon monoxide (CO) and carbon dioxide (CO_2), with a considerable amount of heat being given off. Within the stratosphere (the second major layer of the atmosphere, which occupies the region of the atmosphere from about 12 to 50 km above Earth),

O_2 molecules combine with free oxygen atoms to form ozone (O_3). It absorbs ultraviolet (UV) radiation from the Sun.

18.1.3 Trace Gases

Oxygen and nitrogen together constitute about 99% of the atmosphere, and the remaining 1% is made up of trace gases whose concentrations are very small. The most abundant of the trace gases is the noble gas argon (atomic weight = 39.9). Noble gases, which also include neon (20.2), helium (4), krypton (83.8), and xenon (131.3), are very inert and do not generally involve any chemical transformation within the atmosphere. Hydrogen (1.008) is also present in trace quantities in the atmosphere. Although low in concentrations, the important trace gases in Earth's atmosphere are the so-called "greenhouse gases." These greenhouse gases include carbon dioxide (44), methane (16), nitrous oxide (44), water vapor (18), ozone (48), and sulfur hexafluoride (SF_6, 146.1). These gases allow sunlight, which is radiated in the visible and ultraviolet spectra, to enter the atmosphere unimpeded, but prevent most of the outgoing infrared radiation from the surface and lower atmosphere from escaping into outer space. The greenhouse gases absorb reflected infrared radiations (heat), thus trapping the heat in the atmosphere. Thus, these gases keep Earth warm through the so-called natural "greenhouse effect," which has raised Earth's temperature from $-18°C$ to $15°C$, an increase of $33°C$. (Refer to the footnote on p. 532.)

Variable greenhouse gases, can be divided into two categories: (i) those that occur naturally in the atmosphere (e.g., water vapor, CO_2, CH_4, and N_2O) and (ii) those that result from human activities (e.g., chlorofluoro-carbons (CFCs), hydrofluorocarbons (HFCs), perfluorocarbons (PFCs), and sulfur hexafluoride). Human activities can also enhance the concentration of naturally occurring greenhouse gases. Each greenhouse gas differs in its ability to absorb heat in the atmosphere, and HFCs and PFCs are the most heat-absorbent. The atmospheric lifetime of CH_4, a greenhouse gas 21 times more effective than CO_2 in trapping its long-wave radiation, is approximately ten years. Methane (CH_4) can trap 21 times more long wave radiation per molecule than CO_2, and N_2O can absorb 310 times more long wave radiation per molecule than CO_2 (IPCC, 2001). Methane, in contrast to CO_2 and other greenhouse gases, has the unique property of being partly converted to H_2O by cosmic radiation in the mesosphere.

The global mean surface air temperature has increased between approximately 0.3 and 0.6°C during twentieth century (IPCC, 2001). Globally, sea level has risen 10–20 cm over the past century. Worldwide precipitation over land has increased by about one percent. The frequency of extreme rainfall events has increased throughout much of the

TABLE 18.1 Concentration of Some of the Atmospheric Gases in $1\,cm^3$ Volume

Gas	Formula	Volume $(gmol^{-1})$	Concentration (% vol.)	Molar mass $(gmol^{-1})$	Concentration $(g\,cm^{-3})$
Nitrogen	N_2	22.4	78	28	9.75×10^{-4a}
Oxygen	O_2	22.4	21	32	3.0×10^{-4}
Carbon dioxide	CO_2	22.4	0.033	44	6.0×10^{-6}
Methane	CH_4	22.4	0.0002	18	1.6×10^{-9}

[a] $\frac{289}{mol} \cdot \frac{78}{100} \cdot \frac{1L}{10^3\,cm^3} \cdot \frac{mol}{24.4L} = 9.75 \times 10^{-4}\,g\,cm^{-3}$.

United States (IPCC, 2001). Some of the sinks, which absorb CO_2, are oceans, soils, and trees. Each year those sinks absorb hundreds of billions of tons of carbon in the form of CO_2. Concentration of trace/greenhouse gases in the atmosphere is also highly variable over time and space. Gaseous concentration is expressed on the basis of density or gL^{-1}, and can be calculated using Avogadro's law (see the footnote to Table 18.1).

Avogadro's law (1811) states, "Identical volumes of any gas at a standard identical temperature and pressure contain the equal number of molecules regardless of their chemical nature and physical properties." This number, known as "Avogadro's number" (N'), is 6.023×10^{23}. It is the number of molecules of any gas present in a volume of 22.41 L and is the same for a very light gas (e.g., H_2) as for a heavy gas (e.g., CO_2 or Bromine, Br). Avogadro's number is now considered to be the number of atoms present in 12 grams of the carbon-12 isotope (one mole of carbon is 12 g).

The concentration of atmospheric gases in a $1\,cm^3$ volume, can be calculated from the fact that a gram molecular weight of a gas occupies 22.4 L of volume at standard temperature and pressure (STP). Thus, the concentration of O_2 in the atmosphere is $3 \times 10^{-4}\,g\,cm^{-3}$. Similarly, the atmospheric concentration of other gases can be computed (Tables 18.1).

18.2 SOIL AIR

Soil air refers to air in the soil. It is located in the air porosity, whose volume is inversely proportional to that of the soil water $(f_a \propto \theta^{-1})$. Thus, as the volume of soil water (θ) increases, that of soil air (f_a) decreases, and vice versa. A compacted soil or an undrained soil has smaller amounts of soil air than a well-structured and drained soil. In a well-structured soil the soil air content is higher with soil air occupying most of the large or macropores. In general, soil air content (f_a) and water content (θ) are nearly equal at field moisture capacity for well-structured soils. The increase in bulk density (ρ_b) decreases the total porosity (f_t) and for given water content (θ) decreases the soil air content (f_a). Soil air content is also affected by drainage conditions in the field as poor or improper drainage increases the water content of soil thus lowering the air content. Composition of soil air is highly variable and depends on numerous factors (e.g., soil structure, bulk density, drainage conditions). In a well-aerated soil, the oxygen content of soil air is similar to that of the atmosphere because the consumed O_2 is readily replaced and CO_2 generated is readily removed from the soil-air system. In soils with restricted exchange, soil air differs from atmospheric air in several respects. The CO_2 concentration in soil air is much higher and O_2 concentration much lower than atmospheric air. Soil air is also relatively moister than atmospheric air, and it contains numerous trace gases (e.g., H_2S). The composition of soil air varies greatly from place to place in the soil, as plants consume some gases and microbial processes release others (Tables 18.2 and 18.3). The amount and composition of soil air is determined by the water content of soil unless the soil is very dry. The O_2 content in a well-aerated soil is higher than that of a poorly aerated soil. The latter has higher concentrations of CO_2, CH_4, and N_2O than atmospheric air. As the depth of soil profile increases, the concentration of CO_2 increases with a corresponding decrease in O_2 concentration; however, the sum of these two

TABLE 18.2 Measured O_2 and CO_2 Content in Soil Air (% by Volume) at Two Depths

Soil management	O_2 (%)		CO_2 (%)	
	15 cm	46 cm	15 cm	46 cm
Arable land manured	20.52	20.33	0.34	0.50
Arable land unmanured	20.32	20.35	0.34	0.45
Grassland	18.44	17.87	1.46	1.64

Source: Modified from Russel and Appleyard, 1915.

TABLE 18.3 Measured O_2 and CO_2 Content (% by Volume) in Soil Air Collected During Summer and Winter

Cropping systems		O_2 (%)	N_2 (%)	CO_2 (%)
Arable land manured and cropped	Summer	20.74	79.03	0.23
	Winter	20.31	79.32	0.37
Arable land unmanured and cropped	Summer	20.82	78.99	0.19
	Winter	20.42	79.37	0.21

Source: Modified from Russel and Appleyard, 1915.

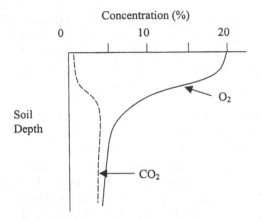

Figure 18.1 Schematic of variation of concentrations of O_2 and CO_2 in soil air with depth.

Table 18.4 CO_2, O_2 and N_2 Contents in Soil Air for Well-Drained Treatments with Constant Water Table Depths

Water table position	Date	CO_2 (%)	O_2 (%)	N_2 (%)
No water table	2 July	1.2	17	78
	30 July	2.0	18.5	76
	17 August	0.3	17.8	76.5
15 cm depth	2 July	6.8	15	75.8
	30 July	8.5	11	81
	16 August	8	7	80.2
30 cm depth	2 July	2.2	16.5	77
	30 July	6.2	11.5	73.5
	16 August	3	17	77

Source: Modified from Lal and Taylor, 1969.

concentrations never exceeds 21% (Fig. 18.1). A soil is considered healthy if the air filled pore spaces are about 50% of the total porosity, and composition of soil air is similar to that of atmospheric air. The reduced soil aeration results from excess water in the soil profile, which may be due to the poor drainage, a shallow groundwater table, soil compaction, swelling clays, or decomposition of organic matter by microorganisms with low O_2 replenishment. As the water table falls below the root zone, the CO_2 concentration in soil air decreases with a corresponding increase in O_2 (Table 18.4). Air permeability of soil, tillage practices (Table 18.5), soil

TABLE **18.5** Soil CO_2 Concentration Data for No-Till (NT) and Moldboard Plow (MB) Plots for Early (21 July), Mid (24 August), and Late (1 October) Season, 1998

| | Average CO_2 concentration (ppm) | | | | | |
| | no-till | | | moldboard plow | | |
Depth (cm)	early	mid	late	early	mid	late
5	2000	3000	1000			
10	8000	6000	2000			
20	28000	23000	4000			
30	34000	24000	5000	20000	9000	3000
50	36000	28000	9000	25000	18000	8000
70	35000	27000	13000	27000	16000	10000

Source: Modified from Reicosky et al., 2002.

TABLE **18.6** O_2 Consumption and CO_2 Release for a Cropped and Bare Soil in January (Soil Temperature 3°C) and July (Soil Temperature 17°C)

| | Cropped (g m^{-2} d^{-1}) | | Bare (g m^{-2} d^{-1}) | |
	January	July	January	July
O_2	2	24	0.7	12
CO_2	3	35	1.2	16

Source: Modified from Curry, 1970.

temperature, and microbial activities (Table 18.6) also affect concentration of CO_2 in soil air. Soil management practices, which improve soil structure, also improve soil aeration. These include no-till, residue mulch, application of manures, conversion of cropland to pasture, etc.

18.3 SOIL AERATION

Soil aeration, the process of the exchange of air (O_2 and CO_2) between soil (or plant roots and soil microorganisms) and the atmosphere is important to plant growth because it maintains O_2 concentration in the root zone at the level needed for root and microbial respiration. Soil aeration is a vital process for controlling the twin processes of respiration and photosynthesis. Plant roots absorb O_2 and release CO_2 during respiration. The O_2 in soil air also governs the chemical reactions, which provide the necessary conditions for oxidation of reduced elements (Fe^{+2}, Mn^{+2}), which may otherwise be toxic to plant growth. Respiration involves the oxidation of organic compounds (such as glucose), and can be represented as follows:

$$C_6H_{12}O_6 + 6O_2 \underset{\text{Photosynthesis}}{\overset{\text{Respiration}}{\rightleftharpoons}} 6CO_2 + 6H_2O + \text{energy} \qquad (18.1)$$

In photosynthesis, the above reaction is reversed (right to left). The total energy is 2883 kJ and biologically useful energy is 1270 kJ. The respiration process increases the concentration of CO_2 in the soil pores and at the same time reduces the O_2 concentration, which creates a concentration gradient, and O_2 flows in the soil profile through the process of diffusion and pushes the CO_2 out of the soil. The rate of O_2 diffusion into the soil profile is proportional to the aeration porosity. The aeration porosity has been defined as the pore space filled with air when the soil sample is placed on a porous plate and equilibrated at 50 cm of suction (Φ_m). The air circulation in and out of soil matrix also moderates the temperature of the soil. In addition to plant growth, soil air composition alters production and emission of trace gases (e.g., CH_4 and N_2O).

18.4 OXYGEN DEFICIENCY AND PLANT GROWTH

The influence of soil air on plant growth is a complex process and can be grouped into direct and indirect effects. The direct influences are related to the physiological effects of O_2 and CO_2 while the indirect influences affect the biological and chemical transformations in the soil. A decrease in soil O_2 concentration results in a decrease in aerobic microbial population and at the same time an increase in anaerobic microbial population, which is responsible for the changes in soil respiration, enzyme activity, and oxidation-reduction or redox potential. Among physiological influences, most of the effects are solely caused by the lack of O_2 for metabolic activities. The O_2 deficiency restricts the root respiration, growth of plant, water, and nutrient uptake, and changes root metabolism toward fermentation. The reliable index of O_2 availability to plant roots is termed the oxygen diffusion rate (ODR; Glinski and Stepniewski, 1985). The diffusion coefficient of O_2 increases with temperature as a result of decrease in O_2 solubility (Letey et al., 1961). After a certain value of ODR, the seedling emergence remains almost a constant, below this value the seedling emergence declines very rapidly with decrease in ODR. The limiting and critical values of ODR for some crops are presented in Table 18.7. At a critical value of ODR (20×10^{-8} g O_2 cm^{-2} min^{-1}) (Stolzy and Latey, 1964), the emergence falls to zero, i.e., no germination of seedling takes place. The

TABLE 18.7 Limiting and Critical Values of ODR for Some Crops

Crop	ODR ($\mu g\,m^{-2}\,s^{-1}$)	
	Limiting	Critical
Barley	25	8
Oats	30	12
Beans	33	12
Wheat	40	8
Flax	40	13
Maize	40	16
Tomato	40	25
Sugar beet	50	13
Rye	50	12

Source: Modified from Glinski and Stepniewski, 1985.

deficiency of O_2 results in restricted root respiration, which has adverse influences on plant growth, and nutrient and water uptake. The deficiency of O_2 for root metabolism also leads to increase in ethanol (C_2H_5OH) concentration, which decreases the emergence of seedlings. The adjustment of stomata aperture regulates the transpiration, heat balance, photosynthesis, and respiration in plants (Glinski and Stepniewski, 1985). The factors affecting stomata aperture are the partial pressures of CO_2, light, water stress, and temperature. The O_2 deficiency to roots results in stomata closure (Sojka and Stolzy, 1980). The wilting thus caused, despite inundation, is called "scalding."

18.5 OXYGEN DEFICIENCY AND SOIL PROPERTIES

Increase in the degree of saturation reduces O_2 content in the soil air. This scenario is very common in undrained or poorly drained soils, where waterlogging or inundation results in O_2 deficiency in soil. The high water content alters soil structural and water transmission properties such as air-filled porosity at a given suction, air permeability, saturated hydraulic conductivity, infiltration characteristic, and compressive strength (Hundal et al., 1976). Soil bulk density may be higher in undrained than drained soil (Table 18.8). The saturated hydraulic conductivity, air-filled porosity at 1 bar (100 kPa), and soil strength may increase with drainage or lowering of the water table (Table 18.8). Soil organic carbon concentration also decreases with drainage or lowering of the water table (Table 18.9). Increase

TABLE 18.8 Effect of Drainage on Soil Physical Property

Property	Undrained		Drained	
	0–15 cm	15–30 cm	0–15 cm	15–30 cm
ρ_b (g cm^{-3})	1.29	1.36	1.22	1.32
w (%)	30.4	29.2	30.1	29.6
K_s (cm h^{-1})	0.1	0.08	2	0.8
P_a	9	7	15	10
UCS (kg cm^{-2})	2.5	3.0	1.8	2.2

[a]Where ρ_b is bulk density; w is gravimetric moisture content at 1 bar (%); K_s is saturated hydraulic conductivity; P_a is air filled porosity at 0.5 bar; UCS is unconfined compressive strength.
Source: Modified from Hundal et al., 1976.

TABLE 18.9 Effect of Water Table Depth on SOC (Mgm^{-3})

Treatment	Sample number	Depth (cm)	
		8 to 16	16 to 24
Drained	7	2.37	2.3
	4	2.53	2.37
Undrained	7	2.58	2.34
	4	2.62	2.59

Source: Modified from Sullivan et al., 1997.

in soil water content also decreases soil temperature (see Table 17.11 in Chapter 17), which depending upon the prevalent climate of area, can increase the intensity of hot/cold, and freeze/thaw cycles, thereby causing a change in soil aggregation and overall structural properties.

18.6 SOIL RESPIRATION

Soil respiration is the amount of oxygen consumption or CO_2 evolution in the soil. The rate of soil respiration varies with space and time and depends upon soil water content, soil type, plant cover, and agriculture measures and amendments. Soil respiration can be measured both under field and laboratory conditions using various types of respirators or respirometers. The respiratory coefficient, which provides useful information on soil aeration, is the ratio of the volume of CO_2 produced to the volume of O_2 consumed. For a well-aerated soil, the respiratory quotient is equal to one.

The aerobic or anaerobic conditions of soil can be checked as follows (Monteith et al., 1964):

$$R = R_0 Q^{T/10} \tag{18.2}$$

where R is the flux at $T°C$ and R_0 at $0°C$ and Q is equal to 3 (Monteith et al., 1964). The concentration of O_2 consumed and CO_2 released in a cropped and bare soil is also presented in Table 18.6 as an example.

18.7 OXIDATION REDUCTION PROCESS IN SOIL

The chemical and biochemical reactions, which occur in soil under anaerobic condition, are dentrification and reduction of manganese (Mn), iron (Fe), and sulfate (SO_4). Nitrate (NO_3) is reduced to nitrite (NO_2), then to nitrous oxide (N_2O), and eventually to elemental nitrogen (N_2). The process of decrease in nitrate content with time in a flooded or saturated soil is known as denitrification. The rate of denitrification depends on soil saturation, pH, and temperature. Denitrification is an anaerobic process and an indicator of the absence of O_2 in at least a part of soil volume. The end products in a denitrification process are gaseous (N_2O, NO, and N_2) (Ponnamperuma, 1972).

$$3NO_3^- + 6H^+ + 6e^- \rightarrow 3NO_2^- + 3H_2O \rightarrow N_2O + NO_x + 3H_2O \tag{18.3}$$

$$2NO_2^- + 8H^+ + 6e^- \rightarrow N_2 + 4H_2O \tag{18.4}$$

Manganese reduces from a manganic (Mn^{+4}) to magnous (Mn^{+2}) state, iron from a ferric (Fe^{+3}) to ferrous (Fe^{+2}) state, and sulfate (SO_4) to hydrogen sulfide (H_2S).

$$Mn^{4+} + 2e^- \rightarrow Mn^{2+} \tag{18.5}$$

$$Fe^{3+} + e^- \rightarrow Fe^{2+} \tag{18.6}$$

$$SO_4^{2-} + 10H^+ + 8e^- \rightarrow H_2S + 4H_2O \tag{18.7}$$

Some of the toxic substances produced during anaerobic conditions are (H_2S), ethylene (C_2H_4), and acetic ($C_2H_4O_2$), butyric ($C_4H_8O_2$) and phenolic (C_6H_5OH) acids.

The process of production of CH_4 is known as methanogenesis. Methanogenic bacteria generate CH_4 biologically, largely from acetate (CH_3COOH) dissimilation and CO_2 reduction. The methanogens are capable of obtaining energy for growth by converting CO_2 and molecular hydrogen into CH_4 and H_2O.

$$CO_2 + 4H_2 \rightarrow CH_4 + 2H_2O \qquad (18.8)$$

Some methanogenic bacteria are also capable of transforming acetate into CH_4 and CO_2.

$$CH_3COOH \rightarrow CO_2 + CH_4 + 35.6\,kJmol^{-1} \qquad (18.9)$$

The H_2 is a product of anaerobic degradation of organic matter. The H_2 with acetate is one of the most important intermediates in the methanogenic degradation of organic matter and serves as a substrate for methanogenic process (Conrad, 1999). Methanogenesis is a major pathway for organic matter decay in sediments. The factors controlling methanogenesis are temperature, concentration of other electron acceptors, water table position, substrate (e.g., H_2) availability, and oxygen supply (Boon and Mitchell, 1995; Grunfeld and Brix, 1999). As temperature increases, water table in the root zone rises and other electron acceptors (e.g., NO_3, Fe_3, SO_4) reduce, methanogenesis increases (Kluber and Conrad, 1998). Methanogenesis occurs in flooded soils, as well as in soils at low water content incubated under anaerobic condition (Boon and Mitchell, 1995). Rice fields are estimated to contribute $100 \pm 50\,Tg\,yr^{-1}$ of the greenhouse gas CH_4 (Kluber and Conrad, 1998). Production of CH_4 occurs during fermentation process by anaerobic bacteria. In flooded soils, CH_4 appears from several days to weeks after flooding. The organic matter amendment stimulates CH_4 formation in alkaline soils, whereas it is suppressed in acid soils (Glinski and Stepniewski, 1985).

$$C_6H_{12}O_6 \rightarrow 2CH_3COOH + CO_2 + CH_4 + 346.8\,kJmol^{-1} \qquad (18.10)$$

The electron transfer is the primary source of energy needed by microorganisms for various processes. Glucose releases electron upon oxidation as follows

$$C_6H_{12}O_6 = 2CH_3COCOOH + 4H^+ + 4e^- \qquad (18.11)$$

In anaerobic conditions O_2, NO_3, H^+, and high valency iron and manganese accept electrons and are reduced to H_2O, N_2, H_2, lower valency Fe^{+2}, and Mn^{+2}, respectively.

$$O_2 + 4H^+ + 4e^- \rightarrow 2H_2O \qquad (18.12)$$

$$2H^+ + 2e^- \rightarrow H_2 \qquad (18.13)$$

This tendency of a substance to accept or donate electrons is measured in terms of the oxidation-reduction potential, commonly known as the oxidation-reduction potential or "redox potential." It is defined as "the potential in volts required in an electric cell to produce oxidation at the anode and reduction at the cathode." The redox potential is a relative term and is measured relative to a standard hydrogen electrode also known as reference electrode whose potential is assumed to be zero. The potential has an inverse relationship with the rate of reduction of substances. The redox potential of soil is closely linked to the availability of O_2, especially at low O_2 levels, and can identify the changes in availability of O_2. The redox potential can be represented as follows:

$$E_h = E_0 - \frac{RT}{nF} \ln\left(\frac{\text{Ox}}{\text{Red}}\right) \qquad (18.14)$$

where E_h is the potential difference between the reference electrode and inert platinum (Pt) electrode, E_0 is the potential of reference electrode, R is the gas constant, T is absolute temperature, n is the number of electrons transferred in the reaction, F is Faraday's constant, "Ox" is activity of oxidized specie, and "Red" is the activity of reduced specie. From Eq. (18.14), it is clear that the E_h is proportional to the natural log of reduced and oxidized products. In well-drained soils, a sufficient amount of O_2 is available, therefore, they can be called oxidized systems. The typical E_h values for oxidized systems are > 400 mV. The O_2 disappears at about 300 mV, NO_3^- is removed between 200 and 300 mV, and Mn_4^+, Fe_3^+, and SO_4^- are reduced sequentially with decrease in E_h value (Poonamperuma, 1972; Scott, 2000).

18.8 FLOW OF AIR IN SOIL

The gaseous exchange between soil and atmosphere occurs by two processes: convection and diffusion. The convective flow of air in soil occurs as a result of the total pressure difference between the soil air and

TABLE 18.10 Increase in CO_2 Content for Calcareous Silty Clay Loam (SCL) and Sandy Loam (SL) Near Field Capacity Under Tensions for Short Period of Time

Soil	Duration minute	Tension (kPa)	CO_2 (%)
SCL	35	35	6.6
SCL	40	29	8.7
SCL	67	41	4.8
SL	18	39	17.4
SL	20	28	4.6

Source: Modified from Boynton and Reuther, 1938.

outer atmosphere. The pressure difference is caused as a result of O_2 consumption by plant roots, CO_2 production in the soil (Table 18.10), change in the barometric pressure in the atmosphere, soil temperature, moisture content, or water table depth of soil due to evaporation, drainage, or water supply by rainfall or irrigation, etc. Various studies have pointed out that convection of air in soil is predominant for shallow depths and in soils with large pores (Rolston, 1986). The convective flow of air in the soil is similar to water flow and is proportional to the pressure gradient across the flow domain. However, since air is compressible, the density and viscosity are also the functions of pressure and temperature. Unlike water flow, gravity is not important for airflow. Air is not attracted to mineral particles and occupies the larger pores. Using Darcy's law for water flow [refer to Eq. (12.3)] the convective flux (q_a) for laminar airflow is given as follows

$$q_a \propto -\nabla P \tag{18.15}$$

where ∇ is the three-dimensional gradient of soil air pressure. If the permeability of air-filled pore space is k_a, and viscosity of soil air is η_a, then one-dimensional convective flow can be given as follows (Hillel, 1998)

$$q_a = -\frac{k_a}{\eta_a}\left(\frac{dP}{dx}\right) \tag{18.16}$$

If the density of soil air is ρ_a, then air flux (q_a) expressed as mass flow per unit area per unit time is

$$q_a = -\frac{k_a \rho_a}{\eta_a}\left(\frac{dP}{dx}\right) \tag{18.17}$$

If soil air is assumed to be an ideal gas at pressure, P, occupying a volume, V, then the ideal gas equation for soil air can be written as

$$PV = nRT \tag{18.18}$$

where n is number of moles of gas, R is the universal gas constant per mole, and T is absolute temperature. Substituting the density, $\rho_a = M/V$, and $M = nm$, in Eq. (18.18) where m is the molecular weight, and after rearranging, the following relationship for density is obtained.

$$\rho_a = \frac{m}{RT} P \tag{18.19}$$

For a one-dimensional compressible fluid, the rate of change of pressure with respect to time is equal to the rate of change of mass flux with respect to length of fluid mass and can be expressed as

$$\frac{\partial \rho}{\partial t} = -\frac{\partial q_a}{\partial x} \tag{18.20}$$

Substituting Eqs. (18.17) and (18.19) into Eq. (18.20) results in

$$\frac{m}{RT} \frac{\partial P}{\partial t} = \frac{\partial}{\partial x} \left(\frac{\rho_a k_a}{\eta_a} \frac{\partial P}{\partial x} \right) \tag{18.21}$$

For small pressure differences, $\rho_a k_a / \eta_a$ can be assumed a constant (Hillel, 1998).

$$\frac{\partial P}{\partial t} = \alpha \frac{\partial^2 P}{\partial x^2} \tag{18.22}$$

where $\alpha = (RT\rho_a k_a)/m\eta_a$. The above equation is an approximate equation for the transient-state convective flow of air in soil. Convective flow rarely meets more than 10% of the O_2 demand of plant roots (see Example). Thus, diffusion is the more important mechanism of soil aeration (Russell, 1952).

18.9 FICK'S LAW AND GASEOUS DIFFUSION IN SOIL

The gaseous transport of O_2 and CO_2 in the soil occurs both in the gaseous and liquid phases. The process of diffusion (random thermal molecular movement from high to low concentration; also refer to Chapter 16) maintains the air exchange between soil and surrounding atmosphere, whereas the supply of O_2 and removal of CO_2 from the plant roots or live tissues takes place by diffusion through water films. According to Fick's law, the mass rate of transfer of a diffusing gas through a unit area of bulk soil is proportional to the concentration gradient measured normal to the surface through which diffusion is taking place. If D is the diffusion rate $(cm^2\ s^{-1})$, C is the concentration of diffusing substances $(g\ cm^{-3})$, q_x is the rate of transfer of mass per unit area $(g\,cm^{-2}\,s^{-1})$, and x is the distance of diffusion (cm), the diffusion of gases in both phases can be represented by the following one-dimensional equation

$$q_x = -D\frac{\partial C}{\partial x} \tag{18.23}$$

The three-dimensional diffusion of gases according to Fick's law is represented as follows:

$$q_x + q_y + q_z = -\left(D_x\frac{\partial C}{\partial x} + D_y\frac{\partial C}{\partial y} + D_z\frac{\partial C}{\partial z}\right) \tag{18.24}$$

where q_x, q_y and q_z are the rate of transfer of mass per unit area, and D_x, D_y, and D_z are gaseous diffusivity, in x, y, and z directions. The partial differential equation of diffusion can be derived, similar to Laplace's equation, by equating the difference between the inflow and outflow of a diffusing substance in a volume element to the change in concentration with time.

$$\frac{\partial C}{\partial t}\Delta x\Delta y\Delta z = -\left(\frac{\partial q_x}{\partial x} + \frac{\partial q_y}{\partial y} + \frac{\partial q_z}{\partial z}\right)\Delta x\Delta y\Delta z \tag{18.25}$$

or

$$\frac{\partial C}{\partial t} = -\left(\frac{\partial q_x}{\partial x} + \frac{\partial q_y}{\partial y} + \frac{\partial q_z}{\partial z}\right) \tag{18.26}$$

From Eqs. (18.24) and (18.26), assuming the diffusion coefficient is independent of direction, the differential equation for three-dimensional gas flow is obtained as follows:

$$\frac{\partial C}{\partial t} = D\left(\frac{\partial C^2}{\partial x^2} + \frac{\partial C^2}{\partial y^2} + \frac{\partial C^2}{\partial z^2}\right) = \nabla D \tag{18.27}$$

TABLE 18.11 Diffusion Coefficient of Some Gases Under Standard Pressure and Temperature

Gas	Density ($kg\,m^{-3}$)	Diffusion coefficient ($m^2\,s^{-1}$)	
		In air	In water
O_2	1.429	1.78×10^{-5}	2.6×10^{-9}
CO_2	1.977	1.39×10^{-5}	1.91×10^{-9}
N_2	1.251	1.8×10^{-5}	1.9×10^{-9}
H_2	0.08	6.34×10^{-5}	5.85×10^{-9}
Water vapor	0.768	2.39×10^{-5}	—
NH_3	0.771	1.98×10^{-5}	2.0×10^{-9}
N_2O	1.978	1.43×10^{-5}	—
C_2H_4	1.261	1.37×10^{-5}	—

Source: Data from Weast et al., 1989.

The one-dimensional form of gaseous diffusion in a porous medium is given by

$$\frac{\partial C}{\partial t} = D \frac{\partial C^2}{\partial x^2}$$

(18.28)

which is similar to Eq. (16.17) when mass flow (second term on the right hand side) is zero. D varies inversely with the molecular weight of gas and is a direct function of temperature and pressure of the gaseous medium. Under standard pressure and temperature, the D in soil air is 10,000 times greater than in soil water (Table 18.11). Under normal atmospheric pressure and 25°C, the D ranges from 0.05 and $0.28\,cm^2\,s^{-1}$; the value depends on the volume of phase available for diffusion. The D is not affected by the shape of solid surfaces or by the particle size or pore size distribution of soil solids because mean free path of diffusing molecule is generally much smaller than the width of the pores.

Considering the diffusive path in the air phase of soil, the diffusion coefficient in soil D_s is much smaller than in air D_a. The ratio D_s/D_a is known as relative diffusion coefficient. The D_s and D_a are related by some function of air-filled porosity (f_a), which are presented in Table 18.12. The tortuosity coefficient of 0.66 (Table 18.12) (Penman, 1940) suggests that straight-line paths are only 66% of total average path of diffusion in soil. Van Bavel (1952) suggested the value of coefficient to be 0.61 rather than 0.66. The advantage of using the dimensionless coefficient or ratio is that the effects of state variables such as pressure, temperature, and type of gas are cancelled.

TABLE 18.12 Models of D_s/D_a as a Function of Volumetric Air Content

Relationship	Reference
$\dfrac{D_s}{D_a} = \kappa f_a^2$	(κ is a constant), Buckingham (1904)
$\dfrac{D_s}{D_a} = \dfrac{f_a}{[k - f_a(k - 1)]}$	Burger (1919)
$\dfrac{D_s}{D_a} = 0.66 f_a$	Penman (1940)
$\dfrac{D_s}{D_a} = 0.61 f_a$	Van Bavel (1952)
$\dfrac{D_s}{D_a} = f_a^{1.5}$	Marshall (1959)
$\dfrac{D_s}{D_a} = \alpha f_a^\beta$	Currie (1960)
$\dfrac{D_s}{D_a} = \dfrac{f_a^{10/3}}{\phi^2}$	Millington (1959), Millington and Quirk (1961)
$\dfrac{D_s}{D_a} = -0.12 + 0.9 f_a$	Wesseling and Van Wijk (1957)
$\dfrac{D_s}{D_a} = -0.1 + 0.9 f_a$	Wesseling (1962)

The O_2 and CO_2 can diffuse both in gaseous and aqueous systems, a diffusion constant K_a can be defined, which separates the contribution from these two phases. The diffusion constant in air (K_a) is given as follows:

$$K_a = f_a D_a \tag{18.29}$$

And diffusion constant in water K_w is

$$K_w = \theta D_w \tag{18.30}$$

The ratio of Eqs. (18.29) and (18.30) after rearrangement yields

$$K_W = \frac{\theta}{f_a} \frac{D_W}{D_a} K_a = a_b \frac{D_W}{D_a} K_a \qquad (18.31)$$

where a_b is Bunsen's solubility coefficient.

18.10 SOURCES AND SINKS OF GASES IN SOIL

The continuity equation states that the rate of change of concentration of a diffusing gas equals the rate of change of flux with distance. Mathematically, it is expressed as follows:

$$\frac{\partial C}{\partial t} = -\frac{\partial q_x}{\partial x} \qquad (18.32)$$

Equation (18.32) implies that a diffusing substance follows the law of conservation of matter. However, during the transport of CO_2 and O_2 through the soil system, the plant roots or anaerobic activities along diffusional path absorb O_2 and release CO_2. Considering S_g to be a source and sink term in time and space, Eq. (18.32) is modified as follows (Hillel, 1998; Scott, 2000):

$$\frac{\partial C}{\partial t} = -\frac{\partial q_x}{\partial x} \pm S_g(x, t) \qquad (18.33)$$

Substituting Eq. (18.23) into Eq. (18.33) and assuming D constant in diffusional path yields

$$\frac{\partial C}{\partial t} = -\frac{\partial}{\partial x}\left(-D\frac{\partial C}{\partial x}\right) \pm S_g(x, t) \qquad (18.34)$$

$$\frac{\partial C}{\partial t} = D\frac{\partial^2 C}{\partial x^2} \pm S_g(x, t) \qquad (18.35)$$

After a rainstorm or irrigation, the larger pores drain quickly and smaller pores or intraaggregate micropores drain slowly. The gaseous diffusion also takes place rather rapidly from interaggregate macropores. The plant roots are also confined to larger pores between aggregates but do not penetrate them. Therefore, larger pores remain well aerated whereas micropores remain anaerobic.

18.11   MEASUREMENT OF SOIL AERATION

Measurement of soil aeration involves assessing: (i) fractional pore space, (ii) composition of soil air, and (iii) rate of diffusion of O_2 from atmosphere into the soil. The aeration is measured by measuring the air-filled porosity at a standard value of soil suction or soil water content. This is done by collecting a core sample from a soil at field capacity (normally 24 to 48 h after a deep wetting or at soil water suction of about 50 cm of water) and measuring air-filled space with an air pycnometer. Alternately, first obtaining total porosity from bulk density (ρ_b) and particle density (ρ_s), and subtracting the water content of core can calculate the air-filled porosity (f_a). Measurement of the relative concentration of O_2, CO_2, and other gases in the soil air provides important information on the aeration and soil structure. Depletion of O_2 level content in soil air is a good indicator of the restricted gas exchange in the soil matrix. This method, although static, is better than the measurement of air volume alone. However, it requires extraction of a sample that is large enough to provide a measurement but at the same time small enough to be representative. Another drawback of this method is soil disturbance and contamination or mixing of air from the atmosphere. The repeated measurements of O_2 or CO_2 concentrations in soil air without extracting a sample can also be obtained by the electrode methods (McIntyre and Philip, 1964; Phene, 1986). The measurement of depletion of O_2 or increase in CO_2 can be made both in situ or in a laboratory by gas chromatography technique, which provides reliable measurements. The method allows rapid and precise measurement of N_2, O_2, Ar, CO_2, CH_4, Ne, H_2, CO, NO, C_2H_4, and C_2H_6 by employing a wide range of methods, detectors, and column packing (Blackmer and Bremner, 1977). The in situ method for measuring O_2 and CO_2 are based on detecting the thermal conductivity by paramagnetic oxygen analyzer and potable carbon analyzer (van Bavel, 1965), respectively.

An early approach to measure aeration involved the determination of the fractional air space or air filled porosity (f_a) at a standardized value of soil wetness. This was measured by either taking a core sample from the field two days after a deep wetting, or saturating the core sample with water and then subjecting it to a suction of 50 cm. All the pores with an effective diameter greater than 0.06 mm ($r = 0.147/50$ cm) are drained of water. The air space as a fraction of porosity now can be determined with an air pycnometer (Page, 1948; Vomocil, 1965). Alternatively, the air space can be determined by the difference of porosity and volumetric wetness ($f_a = f_t - \theta$). However, these two methods are not adequate as considerable uncertainties exist in the measurement and aeration dynamics remains almost untouched.

FIGURE 18.2 Soil air diffusion tube installed in a greenhouse water table management experiment. Similar diffusion tube is used under field conditions. Soil air sample is taken from the tube using a syringe.

The other traditional method involves the determination of the composition of soil air (Fig. 18.2). This method, although again static, is better than the measurement of air volume alone. The depletion of O_2 content in soil air can be a good indicator of the restricted gas exchange in the soil matrix and between soil and the atmosphere (Fig. 18.3). Still, the main concern here is how to extract a sample that is large enough to provide a measurement but at the same time small enough to represent the sample point and to avoid disturbances and mixing of soil air or contamination from the atmosphere. The gas chromatography technique can provide reliable measurements. An alternative method, which permits repeated measurements of oxygen concentrations in soil air without extracting a sample, is based on the use of membrane-covered electrodes (McIntyre and Philip, 1964).

Soil aeration can be characterized by the oxygen diffusion rate in the soil or ODR, (Erickson and van Doren, 1960). The method is based on the hypothesis that the moisture films around plants roots and organism limit the rate of O_2 diffusion. The ODR can be measured by a platinum electrometer under a constant electric potential (Lemon and Erickson, 1955). Once the O_2 present near electrode surface is depleted further depletion is a function of O_2 diffusion to electrode surface or current. The electric current (I, A) is proportional to the rate of O_2 flux at the electrode

FIGURE 18.3 A static chamber is used under field conditions to assess the gaseous emission over a short period of 10 to 15 minutes. (Waterman Farm, Columbus, OH, 1998.)

surface and can be expressed as follows:

$$ODR = \frac{60MI}{nFA} \qquad (18.36)$$

where ODR is oxygen diffusion rate (g m^{-2} s^{-1}), M is the molar mass of oxygen (32 g mol^{-1}), F is the Faraday's constant (96,500 coulombs equiv.$^{-1}$ mol^{-1}), A is the electrode surface area (m^2), and n is equal to four (equiv. mol^{-1}) and is the number of electrons required to reduce one molecule of O_2. The ODR values in soils vary from 0 to 200 μg m^{-2} s^{-1} and increase with suction and air-filled porosity of soil (Glinski and Stepniewski, 1985). The ODR method is satisfactory in soils having higher aeration and is less effective for poorly drained or flooded soils. The methods of soil aeration measurement are listed in Table 18.13. Another approach of characterizing soil aeration is to measure the air permeability.

18.12 AIR PERMEABILITY

The gaseous exchange between soil and the atmosphere and the transport of gases within the soil are complex phenomena. Characterizing soil aeration by measuring content and composition of soil air are inadequate because they do not take into account the process dynamics, directions, and rate of change. Air permeability of soils has been recognized as an important

TABLE 18.13 Methods of Measurement of Soil Aeration

Method	Reference
Air pycnometer	Page (1948), Vomocil (1965)
Membrane covered electrodes	McIntyre and Philip (1964)
Gas chromatography	Bremner and Blackmer (1982)
Closed chamber	Matthias et al. (1980)
Flow through chamber	Denmead (1979)

parameter for soil aeration and contaminant remediation techniques and is fundamental to our understanding of environmental problems in the vadose zone. The vadose zone comprises the region between the land surface and underlying groundwater aquifers varying in depth and composition. It is the geologic zone through which water, solutes, nutrients, and/or contaminants travel prior to reaching groundwater. In agricultural research, knowledge of air-filled pores, pore size distribution, tortuosity, air permeability, and their variation along the cross section or depth is important to describe aeration, structure, and compaction of the soil. Precise impact of these parameters on crop yield is not known. In general, poor structure, low air-filled porosity, and water permeability adversely affect crop yield (Moore and Attenborough, 1992).

Air permeability of porous media, including soils, is governed by the convective transport of air through the media under a pressure gradient. The gaseous flow as a consequence of the pressure head difference is often reported as the mass flow of gas. The other mechanism of gas transport is the diffusion, which occurs due to the change in concentration gradients or the partial pressures of the components of the gaseous mix. If the concentration and pressure gradients exist concurrently, both these processes can occur simultaneously. The mass flow of gas is important when differences in pressure are due to the change in barometric pressure, temperature, or soil water content. However, diffusion is considered the primary mechanism.

In general, a soil matrix consists of a mixture of fluid and gaseous phases. Since viscosity of air is small compared to that of water, soil air remains at most phases in the soil matrix at or near atmospheric pressure. A small pressure gradient is sufficient for soil air to move into or out of the soil system. As a result it has a negligible effect on flow of water and therefore most water transport analysis ignores the simultaneous movement of soil air. The negligible influence due to the low-pressure gradients in soil air is generally, but not necessarily always, true. In case of border irrigation, effects of air compression ahead of the wetting front during infiltration of water into the soil can occur (Dixon and Linden, 1972; Morel-Seytoux and Khanji, 1974). During drainage, air entry through the restrictions within the soil pore space causes surge of water in the draining soil columns (Corey and Brooks, 1975). Airflow through soils is essentially nondestructive and air permeability is sensitive to the changes in soil structure (Corey, 1986). Air permeability can be used as a soil quality indicator to characterize the changes in soil structure resulting from different soil management practices (Ball et al., 1988).

Air permeability is a function of pore characteristic and several soil hydrological properties, which are often more difficult to measure. Air permeability at −100 cm soil suction is a potential indicator for providing information about changes and differences of soil structure (Kirkham et al., 1958). Air permeability is related to air-filled macroporosity at different water contents to identify the changes in soil structure and soil water dynamics by soil management practices and biological activities (Blackwell et al., 1990), which are useful for studying the remediation of contaminated soils by modeling the soil-vapor extraction system (Moldrup et al., 1998). Tortuosity expresses a structural condition of soil and can be used as an index of soil structure (Moldrup et al., 2001). Soil structure has a strong influence on air permeability, and convective transport of air takes place through the larger pore networks in well-structured soils. The flow pattern in well-structured soils can be different for the air and water flow because of the differences in geometries and tortuosities of the two mediums. The saturated hydraulic conductivity is strongly correlated to air permeability at −100 cm of suction (Loll et al., 1999). The relationship between air permeability and saturated hydraulic conductivity in undisturbed soil media can be developed using pore scale network models (Fisher and Celia, 1999).

18.12.1 Governing Principles

According to Darcy's law for laminar flow, velocity of a given fluid is proportional to the pressure difference and inversely proportional to the length of flow path (Kirkham, 1946). Therefore, Darcy's law is applicable for the airflow through soils. The pore sizes and macropores or cracks greatly contribute to airflow in a soil. According to Poiseuille's equation, air flow through a single pore varies as the fourth power of the pore radius ($Q_a \propto r^4$). According to Darcy's law, air permeability (k_a) can be defined by the following relationship:

$$q_a = -\frac{k_a}{\eta_a} * \frac{dp}{dx} \qquad (18.37)$$

where q_a is the volume flux per unit area $(L^3 L^{-2} T^{-1})$; η_a is the dynamic viscosity of air $(ML^{-1} T^{-1})$; p is the pressure of air $(ML^{-1} T^{-2})$; and x is the distance in direction of flow (L). Air permeability of soil samples is calculated by modifying Eq. (18.37) as follows:

$$k_a = \frac{Q_a \eta_a L}{A \rho_a g \Delta H_a} \tag{18.38}$$

where Q_a is the volumetric flow rate $(cm^3 s^{-1})$, ΔH_a is difference in pressure head (cm), ρ_a is density of air $(g\,cm^{-3})$, g is acceleration due to gravity $(cm\,s^{-2})$, A is the cross-sectional area (cm^2) and L is the length of the sample (cm). Note the dimensions of air permeability coefficient, k_a as L^2, which are similar to the intrinsic permeability of soil and therefore k_a is also referred to as intrinsic permeability of air (Reeve, 1953). The cross-sectional area and the length of soil sample are replaced by a shape factor to measure in situ air permeability.

18.12.2 Air Permeability Measurement Methods

Wyckoff and Botson made air permeability measurements as early as 1936 by forcing a mixture of water and air through long tubes of unconsolidated sands. The experiments were repeated with different flow velocities and water and air permeability were measured simultaneously. The air permeability methods can be broadly divided into steady state and unsteady state methods. The steady state methods are based on establishing a steady airflow rate at the inlet through the soil sample and measuring the flow rate and pressure head difference across the sample. The transient methods are generally quicker, easy to use, and require less volume of air for the experiment (Smith et al., 1998). Another approach to air permeability measurement is known as the acoustic technique. The following sections describe these methods in more detail, and the merits and demerits of each method are presented in Table 18.14.

Steady State Methods

Steady state methods can also be described as constant pressure gradient methods or constant flux methods (Grover, 1955). In this method, air is pressed across a core sample at a constant pressure above the atmospheric pressure (Fig. 18.4). The flow rate of air at the inlet end of the core is measured over a given time interval. The constant pressure gradient method is suitable for highly permeable samples and those at higher water content.

TABLE 18.14 Merits and Demerits of Air Permeability Measurement Methods

Steady state laboratory method		Transient method		Acoustic method
Pressure gradient	Flux method	Core method	Field method	Field method
Constant pressure gradient	Constant flux	Drop in pressure in air tank	Drop in pressure in air tank	Reflection and transmission of audio frequency
Easy and simple	Constant flux and gradients are difficult to attain	Rapid and easy	Practical, rapid economical and easy	Rapid but requires skilled labor
Suitable for highly permeable soils	Suitable for less permeable soil	Suitable for both	Suitable for both	Suitable for homogeneous soils
Does not alter water content	Water content is altered	Does not alter water content	Does not alter water content	Does not alter water content
Disadvantage of air flow between soil and core	Disadvantage of air flow between soil and core	Disadvantage of air flow between soil and core	—	—
—	Soil shrinkage	—	—	—
Well developed	Well developed	Well developed	Well developed	Under development

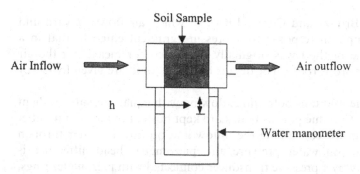

FIGURE 18.4 Schematic of the apparatus of measure air permeability (k_a).

Because the air pressure gradient is small for these soils, the measurements can be made easily without altering the water content or liquid phase of the samples significantly. The air permeability of soil using a constant pressure gradient method for air pressure less than 0.2 m of water can be calculated by Eq. (18.38).

The pressure difference across the core sample can be measured for an applied constant air flux at one end of the column (Blackwell et al., 1990). This method is relatively simple and requires a gas cylinder with a flow meter. The core sample can be placed inside a chamber and is connected to a water manometer for measuring the pressure difference across the sample, and the pressure gradient may be as high as 0.2 m of water. The air permeability can be calculated by the constant flux method as follows (Kirkham, 1946):

$$k_a = \frac{Q_a \eta_a L}{A \rho_a g \Delta H_a} \left[1 - \frac{\Delta H_a}{2 H_i} \right]^{-1} \tag{18.39}$$

If $\Delta H_a \ll 0.2$ m, the term in parentheses approaches unity and Eq. (18.39) reduces to Eq. (18.38). The constant flux method is simple and straightforward and is useful for soils of low permeability. However, it has two basic disadvantages: (i) the dry air changes the water content of core sample, and thus the air permeability and (ii) the constant air flux and gradient are difficult to achieve.

The conservation of moisture in the core sample is an important requirement for air permeability measurement. The water content of the core sample is a function of the pressure difference between both air and water; the pressure gradient between these two competing fluids should be equal in magnitude and direction. One of the methods is the stationary liquid method (Brooks and Corey, 1964), in which air flows upward and through the sample in response to a pressure gradient equal to that in a static liquid. The method was originally developed for measuring the air permeability of porous rocks. The details of this method are given by Corey (1986).

Another method uses both fluids flowing with equal pressure gradient in any direction. Ceramic porous plates are kept both at the inlet and outlet end of the system for controlling the inflow and outflow of water through the sample. The soil water pressure and piezometric head difference is measured directly by a pressure transducer connected with piezometer rings. The procedure allows the water to keep on flowing while permeability measurements are made. The inflow and outflow can be adjusted to obtain a steady state flow condition. A more accurate assessment can be made by simultaneously measuring the change of water content during permeability measurement mounting the soil sample on a scale and calculating the change in water content of soil sample by change in weight (Brooks and Corey, 1966). The merits and demerits of this method are presented in Table 18.14.

Transient Methods

Transient methods are more practical, quicker, less expensive, and easier to use (Table 18.14). They can be employed for air permeability measurements both in a lab on soil cores and in situ in the fields. They also require less volume of air to pass through the soil core or soil volume for a given permeability determination. The duration of the tests is shorter and soil desaturation or drying is less as compared to a steady state method. Kirkham (1946) first proposed the transient method as an in situ field method. In the transient method one end of a soil core is connected to a close pressurized air tank and the other end is kept open (Fig. 18.5). The rate of drop of air pressure in the tank is measured as air flows out through the other end and is used to calculate the air permeability of soil (Kirkham,

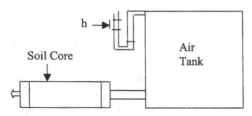

FIGURE 18.5 Schematic of an air permeability apparatus on soil cores for a transient method.

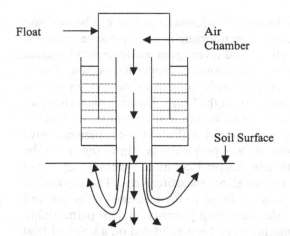

Float

Air
Chamber

Soil Surface

FIGURE 18.6 Schematic of an in situ air permeability apparatus for a transient method.

1946; Smith and Mullins, 1991; Stephens, 1996; Smith et al., 1997; Smith et al., 1998). Schematics of an in situ permeameter are given in Fig. 18.6 (Kirkham, 1946; Grover, 1955). Iversen et al. (2000) further modified the in situ air permeameter and replaced the float by a gas cylinder and a water manometer. Their apparatus is suitable for soil permeability measurements in situ, on site and in the laboratory on soil cores.

The in situ methods of air permeability measurements are preferred over the lab methods and have the advantage of assessing scale effects on permeability and anisotropy (Garbesi et al., 1996). However, these methods involve insertion of steel or plexiglass containers in the ground. Some of the disadvantages of these methods are the unknown sample length, disturbance of natural soil condition, and lack of or low reproducibility. Traditionally, air permeability calculations are made for isothermal conditions. Smith et al. (1997) showed that the traditional isothermal method is inaccurate because the pressure in the air tank cools the air inside. As a result, rate of change of air pressure in the tank does not represent the true mass flux out of the air tank and through the soil sample. Smith and colleagues proposed mathematical expressions for compensating the temperature effects for known temperature changes in the soil tank.

Acoustic Methods

Sound reflected from a soil surface interferes with the incident sound and causes an interference pattern in the total sound field. If the acoustic properties of porous media are known, interference patterns can be modeled from theory of sound propagation. Attenborough (1985) presented analytic approximations for calculating the sound levels from outdoor sound sources on or near the ground in terms of acoustic properties of soil. For a homogeneous porous media these acoustic properties depend on the air filled porosity of soil surface (Attenborough, 1987). The acoustic techniques involve the measurement of both the reflection and transmission of audio frequency and sound by the soil. Sound reflection measurements give qualitative indications of relative air permeability. Inserting a probe microphone at a given depth and keeping another vertically separated above ground makes sound transmission measurements. The theoretical predictions for homogeneous soils are fitted to the measured reflection and quantitative information on surface air filled porosity, and air permeability is obtained. The acoustic techniques have been validated on a series of trial plots, for a variety of soils and have been found within 10% of those obtained conventionally (Sabatier et al., 1990, Moore and Attenborough, 1992). They have shown a potential for the measurement and monitoring of management induced or seasonal changes in soil surface properties (Table 18.14).

18.13   MANAGEMENT OF SOIL AERATION

Poor aeration, due to inundation or compaction, affects crop growth by seedling mortality (Fig. 18.7) and creating poor soil physical conditions (Fig. 18.8). Therefore, management of soil aeration is important, which can be accomplished in a number of ways (e.g., tillage, drainage, mulches, etc.). Tillage is usually defined as the mechanical manipulation of soil to improve soil aeration conditions, control weeds, and incorporate organic matter in the soil, which directly affect crop production. Tillage practices on one hand open the soil and increase the porosity and soil aeration at least temporarily, and on the other hand compact the soil surface, increase bulk density, and reduce aeration for the soil below the plow layer. The use of conservation tillage and/or no-till improves the soil physical conditions and quality (Lal, 1989). A no-till system improves organic carbon concentration in soil, which in turn improves aggregation and aeration porosity of soil. The aeration porosity and ODR under different tillage practices are generally significantly correlated (Flowers and Lal, 1998). The data in Table 18.15 show that aeration from no-till treatment is much higher than moldboard or chisel treatment for 5 cm and 15 cm depths, however, corresponding ODR values are not necessarily a direct function of soil aeration (Flowers and Lal, 1998). This apparent

FIGURE 18.7 Poorly drained sites are inundated during spring leading to seedling mortality, patchy stand, and low yields. (Courtesy of Dr. N. R. Fausey, USDA Drainage Unit, Columbus, OH.)

FIGURE 18.8 Soil structure is adversely affected in poorly drained soils. (Courtesy of Dr. N. R. Fausey, USDA Drainage Unit, Columbus, OH.)

discrepancy may be due to the fact that intraaggregate ODR measurements are lower and represent anoxic conditions, whereas interaggregate measurements are higher and reflect higher aeration status between aggregates (McCoy and Cardina, 1997). The crop residue on soil surface

TABLE 18.15 Mean Aeration Porosity (f_a) and ODR at 5 cm Depth from Moldboard Plow (MP), Chisel Plow (CP), and No-Till (NT)

Moldboard plow		Chisel plow		No-till	
AP ($cm^3 cm^{-3}$)	ODR ($\mu g\, cm^{-2} s^{-1}$)	f_a ($cm^3 cm^{-3}$)	ODR ($\mu g\, cm^{-2} s^{-1}$)	f_a ($cm^3 cm^{-3}$)	ODR ($\mu g\, cm^{-2} s^{-1}$)
5 cm depth					
0.054	12.2	0.13	12	0.185	12.5
0.06	13.8	0.145	11.8	0.195	12
0.12	22	0.19	23	0.23	17.5
15 cm depth					
0.002	7.5	0.035	10	0.058	8.8
0.018	5.8	0.045	5	0.072	9
0.07	13.5	0.095	21	0.105	19

ORD = oxygen diffusion rate; f_a = air porosity.
Source: Modified from Flowers and Lal, 1998.

TABLE 18.16 The Dependence of Oxygen Diffusion Rates on Volumetric Water Content (θ)

θ ($cm^3 cm^{-3}$)	ODR ($\mu gm^{-2} s^{-1}$)	θ ($cm^3 cm^{-3}$)	ODR ($\mu gm^{-2} s^{-1}$)	θ ($cm^3 cm^{-3}$)	ODR ($\mu gm^{-2} s^{-1}$)
0.34	86	0.38	68	0.5	25
0.36	74	0.52	28	0.57	20

Source: Modified from Flowers and Lal, 1998.

maintains the soil temperature and soil water content and increases earthworm activities and macropore channel formations, etc., which result in increased porosity and soil aeration. Soil aeration can also be improved by maintaining the soil below saturation levels. This can be achieved by installing surface or tile drainage systems especially in areas where groundwater table is shallow. The soil ODR values are strongly affected by soil water content and fluctuate in response to rainfall (Sojka, 1997; Flowers and Lal, 1998). As the water content increases from 0.38 $cm^3 cm^{-3}$ to 0.5 $cm^3 cm^{-3}$, a sharp decline in ODR from 68 to 28 $\mu gm^{-2} s^{-1}$ (>100%) is observed (Table 18.16) (Flowers and Lal, 1998). The use of different kinds of mulches on soil surface protect the soil from hot and cold cycles, freeze and thaw processes, and help maintain aggregation in soil consequently maintaining higher soil aeration.

18.14 WATER TABLE MANAGEMENT IN POORLY DRAINED SOILS

Water table management is defined as "the management or regulation of amount or volume of soil water in the profile for a healthy environment for plants." Water table management also implies the management of drainage system for maintaining the temperature and aeration in the soil for sustainable agriculture. Water table management consists of: (i) conventional subsurface drainage, (ii) controlled drainage, and (iii) subirrigation (www.ohioline.ag.osu-state.edu).

Water table management in its simplest form is synonymous with conventional drainage. It is practiced commonly in the midwestern United States to remove excess water from the soil profile. A conventional drainage system essentially consists of an outlet (ditch or a stream) for discharging excess water and drainage pipes or tiles, which are made of corrugated plastic tubing, clay, or concrete tile. The excess water in the soil profile enters the perforated drainage pipes installed at certain depth from soil surface and flows out to the open ditch or stream through gravity. The controlled drainage system is very similar to a conventional drainage system except that in the former, drainage or outflow of water is intercepted by a control device, which manages the water table at any specified level below soil surface. The soil moisture status is better managed in a controlled drainage system. A subirrigation system is a combination of controlled drainage and an irrigation system and is used for drainage as well as

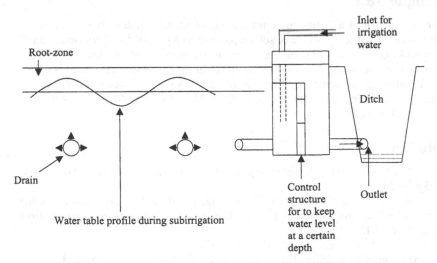

FIGURE 18.9 Schematic of a subirrigation method.

irrigation (Fig. 18.9). The drain spacing for a subirrigation is usually 30 to 50% less than conventional drainage system. Irrigation occurs below the ground surface and water table depth is maintained at a desired depth within the crop root zone.

The lowering of the water table during spring and fall facilitates field operations, and the raising of the water table during growing season by controlled drainage and/or subirrigation provides plants with much needed water. Drainage control strategies are used to improve the quality of surface water by keeping nitrate and other chemicals within the soil profile and not letting them flow out into surface drains. For reducing nitrate concentration in surface waters, three strategies are basically followed: (i) reduce air-filled porosity by maintaining a shallow water table within the root zone, thus creating anaerobic conditions to enhance denitrification; (ii) keep the water table shallow to reduce the volume of outflow from drainage; and (iii) decrease the leaching potential of soil nitrate by decreasing the depth of soil profile through which water infiltrates (Dinnes et al., 2002). Thus, in addition to improving quality of drainage water by reducing leaching of agrochemicals from soil profile, water table management can increase efficiency of crop production in three ways: (i) by retaining more nitrate in soil profile for plant, thus reducing fertilizer costs; (ii) increasing water availability during water demand periods; and (iii) keeping adequate soil air in the profile (Mejia et al., 2000).

Example 18.1

Corn was grown in a 10 ha agricultural field, which has an effective root zone of 75 cm. Assuming the daily rate of soil respiration as $8\,g\,O_2m^{-2}$ and transpiration as 5 mm, calculate the fraction of the O_2 requirement supplied by convection. Also calculate the volume of CO_2 drawn from the atmosphere. Note that the air is drawn from the atmosphere immediately by the pressure difference created in the soil by soil moisture extraction, molecular weight of O_2 is 34 and CO_2 is 44, and 22.4 liter = 1 mole of gas at STP.

Solution

The volume of water extracted by roots from $1\,m^2$ area $= 1\,m^2 * 0.005\,m = 0.005\,m^3 = 5\,L$.

The same volume of air from the atmosphere will replace the volume of water extracted or removed from the soil. Therefore, volume of air drawn from atmosphere $= 5\,L$.

The O_2 content of air is 21% and CO_2 is 0.037%.

Volume of CO_2 drawn from atmosphere $= (5*0.037)/100 = 0.00185\,L$

Mass of CO_2 drawn from atmosphere $= (0.00185/22.4)*44 = 3.63 \times 10^{-3}\,g$

Volume of O_2 drawn from atmosphere $= (5*21)/100 = 1.05\,L$
Mass of O_2 drawn from atmosphere $= (1.05/22.4)*32 = 1.5\,g$
Percentage of daily O_2 requirement supplied by convection $= (1.5/8)*100 =$
18.75%

Example 18.2

Determine the gaseous concentration of O_2, N_2, and CO_2 at standard temperatures and pressure (25°C and 101.3 kPa, respectively). The partial pressure for O_2 is $0.21*atm$ and for N_2 is $0.79*atm$, where atm is the atmospheric pressure $= 1.013 \times 105\,Pa$ at sea level, the gas constant, R is equal to 8.314 Jmol^{-1}K^{-1}.

Solution

Concentration can be calculated from the ideal gas law. Writing Eq. (18.18) in terms of partial pressures, the concentration on a mass basis is given as

$$C = \frac{M}{RT}p$$

where C is concentration (kg m^{-3}), M is molar mass (kgmol^{-1}), R is gas constant, and p is partial pressure of gas (Pa).

$$\text{Concentration of } O_2 = \frac{0.032}{8.314*298}0.21*1.013x10^5 = 0.275\,kg\ m^{-3}$$
$$\text{Concentration of } CO_2 = \frac{0.044}{8.314*298}0.0003*1.013x10^5 = 0.00054\,kg\ m^{-3}$$
$$\text{Concentration of } N_2 = \frac{0.028}{8.314*298}0.79*1.013x10^5 = 0.9044\,kg\ m^{-3}$$

Example 18.3

For a soil of bulk density 1.2 Mgm^{-3} and a constant water content of 0.2 m^{3}m^{-3}, calculate the amount of O_2 and CO_2 in the soil for (a) water content retained at 0.2 m^3 m^{-3} (b) if 1 cm depth of water is removed by evaporation or drainage, and (c) if 1 cm depth of water enters the soil matrix. Assume air replaces water instantaneously and concentration of dissolved air does not change.

Solution

(a) The total porosity of soil $= f_t = 1 - (\rho_b/\rho_s) = 1 - (1.2/2.65) = 0.55$
 The air-filled space in the soil $f_a = f_t - \theta = 0.55 - 0.2 = 0.35\,cm^3\,cm^{-3}$

In a well-aerated soil, air contains 21% of O_2 and 0.05% of CO_2.
Calculating the concentration on $1\,m^3$ basis

The concentration of $CO_2 = (44\,g/22.4\,L) * 0.05 * (1000\,L/1\,m^3) = 98.2\,g\;m^3$

The concentration of $O_2 = (32\,g/22.4\,L) * 0.21 * (1000\,L/1\,m^3) = 300\,g\;m^3$

Since air filled porosity was 35%, assuming $1\,m^3$ volume of soil, the total volume of air in the soil system $= 0.35\,m^3$

The amount of CO_2 in the soil $= 98.2 * 0.35 = 34.37\,g$

The amount of O_2 in the soil $= 300 * 0.35 = 105\,g$

(b) Total water in the soil profile $(1\,m^3) = 0.2 * 1 * 1 * 1 = 0.2\,m^3$

If 1 cm depth of water is removed $= 1/100 * 1 * 1 * 1 = 0.01\,m^3$

The total volume of water left in the soil profile after 1 cm of water is removed $= 0.19\,m^3$

Water content of the soil $= 0.19\,m^3\,m^{-3}$

The air-filled space in the soil $f_a = f_t - \theta = 0.55 - 0.19 = 0.36\,cm^3\,cm^{-3}$

The amount of CO_2 in the soil $= 98.2 * 0.36 = 35.35\,g$

The amount of O_2 in the soil $= 300 * 0.36 = 108\,g$

(c) If 1 cm water is added, the air-fiilled pore space changes to $= 0.55 - 0.21 = 0.34$

The amount of CO_2 in the soil $= 98.2*0.34 = 33.39\,g$

The amount of O_2 in the soil $= 300*0.34 = 102\,g$

Example 18.4

If the topsoil of a recently cleared and cultivated tropical forest with a topsoil depth of 35 cm and bulk density of $1.35\,g\;cm^{-3}$, contains 2.5% of readily decomposable organic residues having a carbon content of 30%, assuming constant O_2 consumption rate of $0.06\,kg\;m^{-2}\,d^{-1}$, calculate how much carbon is released to the atmosphere in three weeks. Note atomic weight of O_2 is 16 and C is 12.

Solution

Mass of soil in top layer $= 35 * 1.35 = 47.25\,g\;cm^{-2}$

The mass of decomposable organic matter $= 0.025 * 47.25 = 1.18\,g\;cm^{-2}$

The mass of carbon in top layer $= 0.3 * 1.18 = 0.35\,g\;cm^{-2}$

The daily Carbon release rate (from CO_2) $=$ atomic weight of C/atomic weight of $O_2 = 12/32$

O_2 consumption rate $= 0.375 * 0.06 = 0.0225\,kg\,m^{-2}$

Total mass of C released in 21 days $= 21 * 0.0225 = 0.4725\,kg\,m^{-2}$

Daily organic matter decomposed $= 0.0225/0.25 = 0.09\,kg\,m^{-2}$

Total organic matter decomposed in three weeks $= 0.09 * 21 = 1.89\,kg\,m^{-2}$

Example 18.5

40 cm deep homogeneous soil matrix is at uniform water content of 25%. If the density of soil air as $1.275 \, kg \, m^{-3}$, and k/η ratio equal to $60 \, \mu m^2 s^{-1}$, calculate the convective flow through soil if the pressure head difference is 10 cm.

Solution

The convective airflow through a unit cross section (q_v) can be calculated from Eq. (18.16).

$$q_v = 60 \times 10^{-6} * 0.1/0.4 = 1.5 \times 10^{-6} \, m^2 s^{-1}$$

The convective flow in terms of mass $= q_v * \rho_{sa} = 1.5 \times 10^{-6} * 1.275 = 1.91 \times 10^{-6} \, kg \, m^{-1} s^{-1}$.

Example 18.6

Assuming a homogeneous soil profile having a bulk density of $1.48 \, gcm^{-3}$, and water content of 30%, calculate (a) air filled porosity and (b) effective diffusion coefficient (D_S), and diffusion rate using at least three models from Table 18.12 when O_2 concentration diminishes linearly from 21% at soil surface to 12% at 80 cm. The bulk air diffusion coefficient (D_0) is given as $0.185 \, cm^2 s^{-1}$, particle density of soil as $2.65 \, g \, cm^{-3}$ and concentration of O_2 in atmosphere as $0.0003 \, g \, cm^{-3}$).

Solution

(a) The porosity of soil $= (1 - (\text{bulk density/particle density})) = 1 - (1.48/2.65) = 0.44$
    Therefore, air filled porosity $= 0.44 - 0.3 = 0.11$.
(b) The concentration of O_2 in atmosphere reduces to $0.12 \times 32/22.4 = 0.00017$ g cm^{-3}. We chose the following three models and calculated effective diffusion coefficient (D_s) and the steady state one-dimensional diffusive flux from Fick's law [Eq. 18.16)].

Change in O_2 concentration $= 0.0003 - 0.00017 = 0.00013 \, g \, cm^{-3}$
Change in elevation $= 80 - 0 = 80 \, cm$

Models	$D_s \, (cm^2 s^{-1})$	$q_x \, (g \, cm^{-2} s^{-1})$
Penman (1940)	0.0137	2.23×10^{-8}
Marshall (1959)	0.0067	1.09×10^{-8}
Millington and Quirk (1961)	0.0033	0.54×10^{-8}

PROBLEMS

1. Under normal standard pressure and temperature conditions (i.e., $101.3\,kPa$, and $25°C$, respectively), calculate the ratio of mobility of O_2 and CO_2 in air and water. Use the diffusion coefficient values for O_2 and CO_2 as given in Table 18.11. The solubility coefficient for O_2 and CO_2 is 0.0333 and 0.942, respectively.

2. Calculate the amount of oxygen contained in the top $20\,cm$ soil profile, if the soil volume is composed of 30% air, 25% of which is O_2. Soil consumes $6\,gm^{-2}\,d^{-1}$ of O_2. Assuming no replenishment, if the consumption rate is constant, how many days will this stored O_2 last?

3. A homogeneous soil profile has a porosity of 0.48, and moisture content of 32%. The O_2 concentration diminishes linearly from 21% at soil surface (concentration of O_2 in atmosphere as $0.0003\,gcm^{-3}$) to $1/3$ at a depth of $60\,cm$. Calculate the diffusion rate using any five models from Table 18.12. The bulk air diffusion coefficient (D_0) is given as $0.185\,cm^2\,s^{-1}$.

4. A static gas chamber has a diameter of $15\,cm$ and is $15\,cm$ high. Concentration of CO_2 in the chamber is increased by $200\,ppmv$ in 15 minutes. Calculate efflux of CO_2 from soil air to the atmosphere.

5. Assume an effective rooting depth of $50\,cm$ in corn at the initial tasselling stage of growth. The oxygen demand is $10\,gm^2\,d^{-1}$. Calculate the proportion of O_2 supplied by convection (displacement of water by air), if the rate of water intake is $5\,mmd^{-1}$.

6. While measuring CO_2 flux, a field experiment conducted using a diffusion chamber technique showed that the CO_2 concentration in the chamber was $500\,ppm$ in 10 minutes. If the chamber is $50\,cm^2$ in a cross-sectional area and $10\,cm$ high, calculate the flux assuming that CO_2 concentration in the atmosphere is 0.03%.

7. For a soil with bulk density of $1.2\,gcm^{-3}$ and gravimetric water content of 0.15, calculate the weight of O_2 and CO_2 (in gm^{-3}) in $1\,m^3$ of soil.

8. Calculate the effective diffusion coefficient (D_s) for the soil in problem 7, assuming that diffusion coefficient in air is $0.2\,cm^2\,sec^{-1}$.

9. Compute O_2 influx from atmosphere into soil using effective diffusion coefficient calculated in problem 8, when O_2 concentration in $50\,cm$ depth is 59% of that in the atmosphere.

10. The following data on methane emission was obtained from an experiment in Houston, Texas. Review these data and list soil physical properties and processes that will reduce CH_4 emissions.

Treatment	Seasonal methane emission (gm^{-2})
Straw	
No straw	27.4
With straw	35.6
Tillage	
Late tillage	15.8
No-tillage	16.5
Early tillage	14.2
Flooding	
Late flood	15
Normal flood	9.3
Midseason flood	4.9
Multiple aeration	1.2

REFERENCES

Attenborough K. (1985). Acoustic impedance models for outdoor ground surfaces. Journal of Sound and Vibration 99:521–544.

Attenborough, K. (1987). On the acoustic slow wave in air-filled granual media. Journal of Acoustic Soc. of Am. 81:93–102.

Ball B.C., M.F. O'Sullivan, and R. Hunter, (1988). Gas diffusion, fluid flow, and derived pore continuity indices in relation to vehicle traffic and tillage. J. Soil Sci. 39:327–339.

Blackmer A.M. and J.M. Bremner, (1977). Gas chromatographic analysis of soil atmospheres. Soil Sci. Soc. Am. J. 41:908–912.

Blackwell P.S., A.J. Ringros-Voase, N.S. Jayawardane, K.A. Olsson, D.C. McKenzie, and W.K. Mason, (1990). The use of air-filled porosity and intrinsic permeability to air to characteristic structure of macropore space and saturated hydraulic conductivity of clay soils. J. Soil Sci. 41:215–228.

Boon P.I. and A. Mitchell, (1995). Methanogenesis in the sediments of an Australian freshwater wetland: comparison with aerobic decay and factors controlling methanogenesis. FEMS Microbiol. Ecol. 62:143–150.

Boynton D. and W. Reuther, (1938). A way of sampling soil gases in dense subsoils and some of its advantages and limitations. Soil Sci. Soc. of Am. Proceedings 37–42.

Bremner J.M. and A.M. Blackmer, (1982). Composition of soil atmospheres. In A.L. Page et al. (ed.) Methods of soil analysis. Part 2, 2nd ed. Agronomy 9:873–902.

Brooks R.H. and A.T. Corey, (1964). Hydraulic properties of porous media. Hydrology Paper No. 3, Colorado State Univ., Fort Collins, Colorado. pp 27.

Brooks R.H. and A.T. Corey, (1966). Properties of porous media effecting fluid flow. J. Irrig. Drain. Div., Am. Soc. Civ. Eng. 92:455–467.

Buckingham E. (1904). Contributions to our knowledge of the aeration of soils. U.S. Bur. Soils Bulletin 25.

Burger H.C. (1919). Das Leitvermogen verdunnter mischkristall-freier Legierungen. Phyz. Z. 20:73–75.

Conrad R. (1999). Contribution of hydrogen to methane production and control of hydrogen concentrations in methanogenic soils and sediments. FEMS Microbiol. Ecol. 28:193–202.

Corey A.T. (1986). Air permeability. In A. Klute (ed.). Methods of soil analysis, Part I. Physical and Mineralogical methods. Agronomy Monograph 9 2nd edition, p 1121–1136.

Currie J.A. (1960). Gaseous diffusion in porous media. II. Dry granular materials. Br. J. Appl. Phys. 11:318–324.

Curry J.A. (1970). In: Sorption and Transport Processes in Soils. Chem. Ind. Monograph No. 37, 152–171, London.

Denmead O.T. (1979). Chamber systems for measuring nitrous oxide emission from soils in the field. Soil Sci. Soc. Am. J. 43:89–95.

Dinnes D.L., D.L. Karlen, D.B. Jaynes, T.C. Kasper, J.L. Hatfield, T.S. Colvin, and C.A. Cambardella, (2002). Nitrogen management strategies to reduce nitrate leaching in tile-drained midwestern soils. Agron. J. 94:153–171.

Dixon R. and D.R. Lindon, (1972). Soil-air pressure and water infiltration under border irrigation. Soil Sci. Soc. Am. J. 36:948–953.

Erickson A.E. and D.M. van Doren, (1960). The relation of plant growth and yield to soil oxygen availability. In Trans. 7th Int. Congr. Soil Sci., Madison, WI. 54:428.

Fisher U. and M.A. Celia, (1999). Prediction of relative and absolute permeabilities for gas and water from soil water retension using a pore scale network model. Water Resour. Res. 35:1089–1100.

Flower M.D. and R. Lal, (1998). Axle load and tillage effects on soil physical properties and soybean grain yield on a mollic ochraqualf in northwest Ohio. Soil Till. Res. 48:21–35.

Garbesi K., R.G. Sextro, A.L. Robinson, J.D. Wooley, J.A. Owens, and W.W. Nazaroff, (1996). Scale dependence of soil permeability to air: Measurement method and field investigation. Water Resour. Res. 32:547–560.

Grunfeld S. and H. Brix, (1999). Methanogenesis and methane emissions: effects of water table, substrate type and presence of Phragmites australis. Aquatic Botany 64:63–75.

Hillel D. (1998). Environmental Soil Physics. Academic Press. New York.

Hundal S.S., G.O. Schawab, and G.S. Taylor, (1976). Drainage system effects on physical properties of lakebed clay soil. Soil Sci. Soc. Am. J. 40:300–305.

IPCC (2001). Climate change 2000. Intergovernmental panel on climate change. Cambridge Univ. Press, Cambridge, UK.

Iversen B.V., P. Schjonning, T.G. Poulsen, and P. Moldrup, (2000). In situ, on-site, and laboratory measurements of soil air permeability: Boundary conditions and measurement scale. Soil Sci. 166(2):97–106.

Kirkham D., M. De Brodt, and L. De Leiheer (1958). Air permeability at field capacity as related to soil structure and yield. Overdruck Uit Mededelingen van de Landbouwhogeschool en de Opzoekingsstations van de staal te Gent deel XXXIV No. 1. Int. Symp. Soil Moisture, Gent Belgium p 337–391.

Kirkham D. (1946). Field methods for determination of air permeability of soil in its undisturbed state. Soil Sci. Soc. Am. Proc. 11:93–99.

Kluber H.D. and R. Conrad, (1998). Effects of nitrate, NO and N2O on methanogenesis and redox processes in anoxic field soil. FEMS Microbiol. Ecol. 25:301–318.

Lal R. (1989). Conservation tillage for sustainable agriculture: tropical vs temperate environments. Adv. Agron. 42:85–196.

Lal R. and G.S. Taylor, (1969). Drainage and nutrient effects in a field lysimeter study Part I: Crop yield and soil conditions. Soil Sci. Soc. Am. Proceedings 33:937–941.

Lemon E.R. and A.E. Erickson, (1955). Principles of the platinum microelectrode as a method of characterizing soil aeration. Soil Sci. 79:383–392.

Letey J., L.H. Stolzy, G.B. Blank, and O.R. Lunt, (1961). Effect of temperature on oxygen diffusion rates and subsequent shoot growth, root growth and mineral content of two plant species. Soil Sci. 92:314–321.

Loll P.P., P. Moldrup, P. Schjonning, and H. Rilley, (1999). Predicting saturated hydraulic conductivity from air: Application in stochastic water infiltration modeling. Water Resour. Res. 35:2387–2400.

Marshall T.J. (1959). The diffusion of gas through porous media. J. Soil Sci. 10:79–82.

Matthias A.D., A.M. Blackmer, and J.M. Bremner, (1980). A simple chamber technique for field measurement of emissions of nitrous oxide. J. Environ. Qual. 9:251–256.

McCoy E.L. and J. Cardina, (1997). Characterizing the structure of undisturbed soils. Soil Sci. Soc. Am. J. 61:280–286.

McIntyre D.S. and J.R. Philip, (1964). A field method for measurement of gas diffusion into soils. Aust. J. Soil Res. 2:133–145.

Mejia M.N., C.A. Madramootoo, and R.S. Broughton, (2000). Influence of water table management on corn and soybean yield. Agriculture Water Management 46:73–89.

Millington R.J. (1959). Gas diffusion in porous media. Science 130:100–102.

Millington R.J. and J.P. Quirk, (1961). Permeability of porous solids. Trans. Faraday Soc. 57:1200–1207.

Moldrup P., T.G. Poulsen, P. Schjonning, T. Olesen, and T. Yamaguchi, (1998). Gas permeability in undisturbed soils: measurements and predictive models. Soil Sci. 163:180–189.

Moldrup P., T.G. Poulsen, P. Schjonning, T. Olsen, and T. Yamaguchi, (2001). Gas permeability in undisturbed soils: measurements and predictive models. Soil Sci. 163(3):180–189.

Monteith J.L., G. Szeicz, and K. Yukubi, (1964). Crop photosynthesis and the flux of carbon dioxide below the canopy. J. Appl. Ecol. 6:321–337.

Moore and Attenborough, (1992). Acoustic determination of air-filled porosity and relative air permeability of soils. J. Soil Sci. 43:211–228.

Morel-Seytoux H.J. and J. Khanji, (1974). Derivation of an equation of infiltration. Water Resour. Res. J. 10:795–800.

Ohioline.ag.osu-state.edu (1998). In: L.A. Zucker and L.C. Brown (eds.), Agricultural Drainage: Water Quality Impacts and Subsurface Drainage Studies in the Midwest. The Ohio State University Extension Bulletin 871.

Page J.B. (1948). Advantages of the pressure pycnometer for measuring the pore space in soils. Soil Sci. Soc. Am. Proc. 12:81–84.

Penman H.L. (1940). Gas and Vapor movements in the soil: 1. The diffusion of vapors through porous solids. J. Agr. Sci. 30:437–461.

Phene C.J. (1986). Oxygen electrode measurement. In: A. Klute (ed.), Methods of soil analysis, part 1:physical and mineralogical methods. Monograph No. 9, Amer. Soc. Agron., Madison, WI.

Poonamperuma F.N. (1972). He chemistry of submerged soils. Adv. in Agron. 24:29–96.

Reeve R.C. (1953). A method of determining the stability of soil structure based upon air and water permeability measurements. Soil Sci. Soc. Am. J. 17:324–329.

Reicosky D.C. (2002). Long-term effects of moldboard plowing on tillage induced CO_2 loss. P87–97. In (ed.), J.M. Kimble et al. Agricultural practices and policies for carbon sequestration in soil. Lewis Publishers, CRC Press, Boca Raton, FL.

Rolston D.E. (1986). Gas Flux. In: A. Klute (ed.), Methods of soil analysis, part 1:physical and mineralogical methods. Monograph No. 9. Amer. Soc. Agron., Madison, WI.

Russell M.B. (1952). Soil aeration and plant growth. In: B.T. Shaw (ed.), Soil Physical Conditions and Plant Growth. Academia Press, New York, pp 253–301.

Sabatier J.M., H. Hess, W.P. Arnott, K.K. Attenborough, M.J.M. Romkens, and E.H. Grissinger, (1990). In-situ measurements of soil physical properties by acoustic techniques. Soil Sci. Soc. Am. J. 54:658–672.

Scott H.D. (2000). Soil physics agriculture and environment applications. Iowa University Press, Iowa.

Smith J.E. and C.E. Mullins, (1991). Soil analysis physical methods. Marcel Dekker, New York.

Smith J.E., M.J.L. Robin, and R.R. Elrick, (1998). Improved transient-flow air permeameter design: dampening the temperature effects. Soil Sci. Soc. Am. J. 62:1220–1227.

Smith J.E., M.J.L. Robin, and R.R. Elrick, (1997). A source of systematic error in transient flow permeameter measurements. Soil Sci. Soc. Am. J. 61:1563–1568.

Sojka R.E. and L.H. Stolzy, (1980). Soil oxygen effects on stomatal response. Soil Sci. 130:350–358.

Sojka R.E., D.J. Horne, C.W. Ross, and C.J. Baker, (1997). Subsoiling and surface tillage effects on soil physical properties and forage oat stand and yield. Soil Till. Res. 40:125–144.

Stephens D.B. (1996). Vadose zone hydrology. CRC Press, Boca Raton, FL.

Stolzy L.H. and J. Latey, (1964). Characterizing soil oxygen conditions with a platinum microelectrode. Adv. Agron. 16:249–279.

Sullivan M.D., N.R. Fausey, and R. Lal, (1997). Long-term effects of subsurface drainage on soil organic carbon content and infiltration in the surface horizons of a lakebed soil in northwest ohio. In: Ratan Lal et al. (eds.) Advances in soil science—management of Carbon Sequestration in soil. 1997. pp. 73–82.

Van Bavel C.H.M. (1952). Gaseous diffusion and porosity in porous media. Soil Sci. 73:91–104.

van Bavel C.H.M. (1965). Composition of soil atmosphere. In C.A. Black et al. (ed.) Methods of soil analysis. part I. Agronomy no. 9. American Society of Agronomy, Madison, WI, pp 315–346.

Vomocil J.A. (1965). Porosity. In: A. Klute (ed.), Methods of soil analysis, part 1: physical and mineralogical methods. Monograph No. 9. Amer. Soc. Agron. Madison, WI.

Weast et al., eds. (1989). CRC Handbook of Chemistry and Physics. 69th edition. CRC Press, Boca Raton, FL.

Wesseling J. (1962). Some solutions of the steady state diffusion of CO_2 through soils. Neth. J. Agr. Sci. 10:109–117.

Wesseling J. and W.R. van Wijk, (1957). Article in Luthin, Drainage of Agriculture Lands. Ist ed. Am. Soc. of Agron. Madison, WI.

Wyckoff R.D. and H.G. Botset, (1936). The flow of gas–liquid mixtures through unconsolidated sands. Physics 7:325–345.

19

Physical Properties of Gravelly Soils

19.1 SKELETAL SOIL: CLASSIFICATION

Skeletal soils are those that contain coarse fragments including gravels, stones, or rocks (Figs. 19.1 and 19.2). According to FAO (1977) classification, the particles in a skeletal soil are classified based on size as gravel (2–75 mm), stones (75–250 mm), and boulders (>250 mm). The Soil Survey Manual (Soil Survey Staff, 1951) classifies skeletal soils according to the diameter of round, subround, angular, or irregular fragments into gravel, pebbles, cobbles, stones, and boulders (Table 19.1a), as well as according to the length of flat fragments as channer, flagstone, stone, and boulder (Table 19.1a). The upper limit of sand content (2 mm) corresponds to the lower limit of coarse fragments. The pebbles, cobbles, and stones are all in the range from 76 mm to 250 mm. The Soil Survey Staff (1993) have classified the gravels in two major categories: (i) spherical, cubelike, or equiaxial, and (ii) flat (Table 19.1b).

According to soil taxonomy, soil fragments are classified at family and series level based on the particle size. The family particle size classes refer to whole soil including coarse fragments and are different than USDA textural classes (soil particles <2 mm). Soils with more than 35% of coarse fragments

FIGURE 19.1 A gravelly soil in western Nigeria with high gravel concentration in the subsoil horizon.

are known as "skeletal" soils. Soils having large coarse fragments but very little fine earth to fill interstices larger than 1 mm in diameter are known as "fragmental" soils. The volume percentage of rock fragments is used in identifying and naming map units during soil survey. If volume percentage of coarse fragments is less than 15% no special term is used. For volume percentage of coarse fraction between 15% and 35%, the map unit name includes the class name of rock as a modifier of textural class (for examples, gravelly loam or cobbly loam). For volume percentage of rock fraction between 15% and 35%, the term "very" precedes the textural classification (e.g., very gravelly loam) and for greater than 60% the term "extremely" (e.g., extremely cobbly loam).

Out of the recognized 12,620 soil series in United States, 2181 or 17% of the soil families are those, which contain 35% or more coarse fragments. The loamy skeletal particle size class contains about 1489, i.e., two-thirds of

FIGURE 19.2 A gravelly soil in Valencia, Spain, used for citrus orchards with fertigation technology.

these families, and about 1136, or three-fourths, of them are in the western United States (Miller and Guthrie, 1984). Coarse fragments are generally referred to as a combination of gravels and stones, such as the particles in the size range of 2 mm and 250 mm (Brakensiek and Rawls, 1994; Torri et al., 1994). This chapter specifically deals with only those coarse particles, which are larger than 2 mm and smaller than 75 mm, and are known as "gravels." Thus, soils containing large proportion of gravels are called "gravelly soils."

19.2   GRAVELLY SOILS

The gravels in a soil are generally physically and chemically inert and can be from a sedimentary, igneous, or metamorphic type of rock. Gravelly soils are wide spread in semiarid and arid regions (Fig. 19.1). Regardless of their inert nature, gravels have a strong impact on soil physical, mechanical, and hydrological properties. These properties must be corrected for gravel content, and special procedures are needed to quantify the physical properties of such soils.

19.2.1   Spatial Distribution of Gravels in Soils

Gravels are distributed in the soil profile including on the soil surface and deep inside the soil profile. Therefore, concentrations of gravels occur both

TABLE 19.1 Classification of Rock Fragments According to the Shape and Size of Fragments

Rock fragment[a]	Size (cm)
ACCORDING TO DIAMETER OF A REGULAR OR IRREGULAR FRAGMENT	
Fine gravel	0.2–0.5
Medium gravel	0.5–2
Coarse gravel	2–7.6
Cobble	7.6–25
Stone	25–60
Boulder	>60
ACCORDING TO THE LENGTH OF THE FLAT FRAGMENT	
Channer	0.2–15
Flagstone	15–38
Stone	38–60
Boulder	>60

Shape and size[b]	Noun	Adjective
SPHERICAL, CUBELIKE, OR EQUIAXIAL		
2–75 mm diameter	Pebbles	Gravelly
2–5 mm diameter	Fine	Fine gravelly
5–20 mm diameter	Medium	Medium gravelly
20–75 mm diameter	Coarse	Coarse gravelly
75–250 mm diameter	Cobbles	Cobbly
250–600 mm diameter	Stones	Stony
>600 mm diameter	Boulders	Bouldery
FLAT		
2–150 mm long	Channers	Channery
150–380 mm long	Flagstones	Flaggy
380–600 mm long	Stones	Stony
>600 mm long	Boulders	Bouldery

Source: [a]Modified from Soil Survey manual (1981).
[b]Adapted from USDA (1993).

vertically in soil profile and laterally in the surface. Concentration of gravels is generally higher in surface than subsoil horizons (Nettleton et al., 1989; Parsens et al., 1992). The high concentration of gravels at the surface can occur as (i) removal of fine earth by wind or water erosion, and (ii) upward migration of coarse fragment by freeze–thaw cycles, etc. (Cooke et al., 1993). The gravel content of soil can have a high spatial variability in a field (Childs and Flint, 1990; Webster, 1985).

19.2.2 Volumetric Gravel Content

The amount of gravels in the topsoil can be expressed by: (i) gravimetric content, (ii) volumetric content, and (iii) coverage of the soil surface. Sieving and weighing the gravel fraction provides the information on the gravimetric content. The sieved and air-dried gravels can be immersed in a known volume of water and the net volume change (total displacement) determines the volume of gravels to express the volumetric gravel content.

19.2.3 Effect of Gravels on Soil Physical and Hydrological Properties

Presence of gravels in the soil makes the determination of physical and hydrological properties difficult. The mechanical analysis for particle size distribution on <2 mm fraction and assessment of the gross bulk density could lead to erroneous conclusions for soils with high gravel content. Thus corrective methodologies for the determination of physical properties of gravelly soils are needed and are outlined in Table 19.2 (Lal, 1979). Large gravel concentration at the soil surface or even partly incorporated in the topsoil affect porosity, rainfall interception, moisture distribution, water infiltration, overland flow, evaporation, and land use and productivity. Presence of gravels below the soil surface also influences porosity, water infiltration, percolation, and runoff. Effects of gravels on some soil physical and hydrological properties are discussed briefly in the following section.

19.2.4 Structure

The gravels at the soil surface prevent sealing and crusting by reducing the impact of raindrops on soil surface. They protect soil aggregates, reduce dispersion of soil aggregates by raindrop impact, freeze–thaw cycles, etc., and improve soil structure. The gravels below the soil surface can reduce compaction and bulk density and can either support or improve the existing soil structure (Ravina and Magier, 1984).

19.2.5 Texture

The water and nutrient uptake zone in a soil is generally up to 60 cm, which is also the zone where most of the plant roots exist. The variability in soil texture, which influences the water holding capacity, root development, cation exchange capacity, and ease of harvesting below ground crops (e.g., potato, cassava, yam, sweet potato, turmeric, etc.), is also high in this depth range (i.e., 0 to 60 cm). Various textural classes are grouped together and

TABLE **19.2** Some Recommended Methods for Determination of Physical Properties of Gravelly Soils

No.	Soil property	Recommended methodology	Remarks
1	Texture	Hydrometer or pipette method (thorough dispersion both mechanical and chemical)	Determination of gravelly material (>2 mm) must be made
2	Bulk density	Soil excavation	Gravel correction and bulk density of fine earth material should be determined
3	Moisture content	Neutron probe tensiometer	Calibration before use for each horizon
4	pF	In situ measurement, measurements in lab on large cores	The rock soil system should be present in core
4	Aeration	Sampling by diffusion tubes	
5	Air permeability	Air permeameter	In situ
6	Erodibility	Unit runoff plot under natural rainfall conditions	Lab simulations are not recommended

Source: Modified from Lal, 1979.

assigned a code number based upon the coarseness or fineness of a soil (Table 19.3) (Mansfield, 1979).

19.2.6 Porosity

The porosity of soil varies spatially in general and when stones or fragments are present, the variability increases because of the range and tortuosity associated with gravel fragments (Tables 19.4 and 19.5). The bulk volume of gravelly soils is the sum of the volume of gravels and volume of soil. The porosity of each of these fractions can be measured. However, the total porosity of soil is the porosity of these two fractions plus the space between the soil particles and gravels, which is difficult to measure. The space between soil and large gravels often contains large pores or channels, which are also known as macropores (Bevan and Germann, 1982). The increase in total porosity associated with gravelly soils can be attributed to: (i) space between gravels is incompletely filled by fine earth, (ii) the smaller particles

TABLE 19.3 Grouping of Textural Classes According to the Coarseness of Material

Textural class[a]	Grade	Code No.
CL, SC, C, SiC, SiCL	Fine	1
L, SCL, SiL, Si	Medium	2
S, LS, SL	Coarse	3
CS, LcS, cSL	Very coarse	3c

[a]CLl is clayloam; SC is sandy clay; C is clay; SiC is silty clay; SiCL is silty clay loam; L is loam; SCL is sandy clay loam; SiL is silty loam; Si is silt; S is sand; LS is loamy sand; SL is sandy loam; CS is clayey sand; LcS is loam coarse sand; and cSL is coarse sandy loam.
Source: Modified from Mansfield, 1979.

TABLE 19.4 Soil Physical Properties: Total Bulk Density (ρ_b), Volumetric Rock Fragment (V_v), Particle Density of Rock Fragments (ρ_{pr}), Porosity of Rock Fragments (f_r), Bulk Density of Rock Fragments (ρ_r), and Available Water Content in Rock Fragment (AWC_r)

Parent material	ρ_b (Mg m^{-3})	V_v (%)	ρ_{pr} (Mg m^{-3})	f_r (%)	ρ_r (Mg m^{-3})	AWC_r (%)
Granite	0.98–1.32	5.5–10.0	2.62–2.78	16.5–17.3	2.17–2.35	1.6–4.1
Metasediment	1.05–1.53	10.5–40.2	2.50–2.87	11.8–34.5	1.64–2.46	5.3–27.2
Pumice and ash	0.69–0.72	12.3–32.2	2.13–2.33	52.2–60.3	0.84–1.11	16.2–24.7
Basalt	1.42–1.65	40.8–55.7	2.60–2.66	23.3–37.1	1.75–2.14	21.1–36.3

Source: Modified from Flint and Childs, 1984.

TABLE 19.5 Porosity of Parent Material

Material	Porosity (%)
Basalt	23–37
Granite	15–20
Metasediment	13–29
Pumice and ash	50–60
Tuff and breccia	40–60

Source: Modified from Flint and Childs, 1978.

cannot pack as closely to the larger particle as they can with each other, (iii) the differences in the behavior of fine earth and gravels during the process of wetting and drying or freezing and thawing, and (iv) change in the nature of fine soil fraction in presence of gravels (Poesen and Lavee, 1994; Stewart et al., 1970). Therefore, the total porosity of soils containing gravel fractions is larger than without them. With increasing gravel fraction, the decaying organic matter, water, and fertilizer inputs are all concentrated on a decreasing mass of fine soil. The increase in organic matter content of fine soil improves the soil structure and increases porosity (Childs and Flint, 1990; Poesen and Lavee, 1994). The porosity of soil gravel fraction (f_g) can be calculated as follows

$$f_g = 1 - \frac{\rho_b}{\rho_s} \tag{19.1}$$

$$\rho_b = \rho_{bf} + V_g(\rho_g - \rho_{bf}) \tag{19.2}$$

where ρ_b is the overall bulk density of soil (including gravels); ρ_{bf} is the bulk density of soil without gravels, also known as bulk density of fine earth material; V_g is volumetric gravel fragments (as a fraction of total volume); and ρ_s is the particle density of soil, including gravel fraction.

19.2.7 Bulk Density

Two types of bulk density values are generally required for a soil containing gravels: total bulk density of gravelly soil (ρ_b) and bulk density of fine earth (ρ_{bf}). The extra porosity associated with the gravels in soil profile decreases the bulk density of the soil. In general, as proportions of gravels increases the bulk density decreases (Stewart et al., 1970; Torri et al., 1994). Some typical values of bulk density of gravels and other particles are given in Tables 19.4 and 19.6. The bulk density of a gravelly soil is corrected for

TABLE **19.6** Bulk Density of Different Fragments

Fragment	Bulk density ($Mg\ m^{-3}$)	Reference
Basalt	1.95	Childs and Flint (1990)
Granite	2.17	Ingelmo et al. (1994)
Limestone	2.08	Alberto (1971)
Sandstone	2.56	Childs and Flint (1990)
Shale	2.07	Hanson and Blevins (1979)
Siltstone	1.97	Montagne et al. (1992)
Quartzite	2.43	Ingelmo et al. (1994)

stone content as follows:

$$\rho_{bf} = \frac{M_t - M_g}{V_t - V_g} = \frac{M_s}{V_t - V_g} \tag{19.3}$$

where ρ_{bf} is the bulk density of fine earth or rock free soil, M_t is the total mass of soil including gravels, M_s is the mass of fine earth or gravel-free soil, M_g is the mass of gravels, V_t is the total volume of soil plus gravels, and V_g is volume of gravels in V_t. The volume of soil (V_s) occupied by mass M_s and the volume of voids introduced by gravels (V_v) can replace the denominator ($V_t - V_g$) in Eq. (19.3) as follows:

$$V_t - V_g = V_s + V_v \tag{19.4}$$

Transferring Eq. (19.4) into (19.3) and diving both numerator and denominator by V_s yields

$$\rho_{bf} = \frac{M_s}{V_s} \frac{V_s}{(V_s + V_v)} \tag{19.5}$$

$$\rho_{bf} = \rho_b \frac{V_s}{(V_s + V_v)} \tag{19.6}$$

where ρ_b is the bulk density of soil in the absence of gravels. Adding and subtracting the volume of voids, V_v by gravels in Eq. (19.6) results in:

$$\rho_{bf} = \rho_b \left(\frac{V_s - V_v + V_v}{V_s + V_v} \right) = \rho_b \left(\frac{1 - V_v}{V_s + V_v} \right) \tag{19.7}$$

The ratio $(1 - V_v)/(V_s + V_v)$ largely depends on the gravel content, and its shape and size (Torri et al., 1994). The ratio $(1 - V_v)/(V_s + V_v)$ is replaced by a power function and an empirical relationship can be used for calculating the bulk density of soil as follows (Torri et al., 1994):

$$\rho_{bf} = \rho_b \left(1 - 1.67 M_g^{3.39} \right) \tag{19.8}$$

where M_g is the gravimetric gravel content.

19.2.8 Water Retention

Skeletal fractions can hold substantial quantities of water, which are available for the plant roots (Hanson and Blevins, 1979). The genetic or depositional layers increase water retention in gravelly soils (Clothier et al., 1977). If the gravels are porous they can further increase water retention depending upon their water holding capacity. The water is stored between the contact points for rocks up to 3 cm. For large gravels ($\sim$7.5 cm) and cobbles and rock fragments ($>$10 cm), water can also be held as puddles on the surface. However, the water release depends on the properties of fine earth in which they are embedded. Some skeletal soil particles, which are in the range 0.2–7.5 cm, also weather. They are more porous and can absorb larger quantities of water per unit mass (Childs and Flint, 1990). The bulk water content on a volume basis for gravelly soil (θ) can be calculated from gravimetric water content (w) (Table 19.7) as follows (Flint and Childs, 1984):

$$\theta_f = w_f * \rho_{bf} \tag{19.9}$$

$$\theta_g = w_g * \rho_g \tag{19.10}$$

$$\theta = \theta_f * (1 - V_v) + \theta_g * V_v \tag{19.11}$$

where θ_f and θ_g are the volumetric water contents of fine earth and gravels, respectively, w_f and w_g are the gravimetric water contents of fine earth and gravels, respectively, and ρ_{bf} and ρ_g are bulk densities of fine earth and gravels, respectively.

The effects of gravels on water retention are usually linked to the existence of large pores, which can develop as a consequence of packing or as cracks during drying and can be observed at small matric suction.

TABLE 19.7 Average Gravimetric Moisture Content at Saturation for Different Fragments

Fragment	W_r (%)	Reference
Basalt	0.4	Gras and Monnier (1963)
Granite	0.4	Gras and Monnier (1963)
Limestone	5.2	Gras and Monnier (1963)
Sandstone	13.1	Gras and Monnier (1963)
Shale	34.0	Hanson and Blevins (1979)
Siltstone	12.3	Montagne et al. (1992)

The increase in gravel content up to about 40% increases the volume extracted at low suction (Ravina and Magier, 1984). The water content at field capacity is directly related to the porosity of the gravel content. In skeletal soils, the total available water can range from 1.6 to 52.1% (Table 19.4) (Flint and Childs, 1984). The volumetric soil water content at −33 kPa is inversely related to percent coarse fragment by weight. However, volumetric water content at −1500 kPa is uncorrelated with gravimetric percent coarse fragment (Petersen, 1968). For a soil containing gravels, the available water capacity (AWC), which is the difference between water content at field capacity (FC) and wilting point (WP), can be calculated as follows (Moormann et al., 1975):

$$AWC = \frac{(FC - WP)\rho_b z}{\rho_w (1 + (M_g/M_s))} \tag{19.12}$$

where ρ_b is overall bulk density of soil and gravels, ρ_w is the density of water, M_g is the mass of gravel, M_s is the mass of fine earth, and z is thickness of soil horizon. The above equation implies that as gravel content of soil increases the available water capacity decreases (Fig. 19.3).

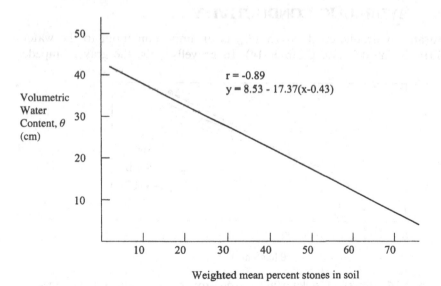

FIGURE 19.3 The influence of stones on water content of a 120 cm deep soil profile. Data presented here is from 40 soil profiles representing four soil series in Oregon. (Modified from Dryness, 1969, and Fisher and Binkley, 2000.)

19.3   EFFECT OF SOIL TEXTURE ON NEUTRON PROBE CALIBRATION

Soil water content determined by the neutron thermolization technique is also affected by the presence of gravels and other coarse fragments in the soil. Soil texture significantly altars the thermal neutron count (Gormat and Goldberg, 1972). The presence of gravels in the soil decreases the density of thermal neutrons by scatter or reflection (McHenry, 1963). The calibration curve relating neutron count ratio to moisture content can have similar slopes for sand and clayey soils. The intercept for a clayey soil is mostly positive compared to a sandy soil having a zero intercept (Fig. 19.4); however, for gravelly soils, the slope of the neutron probe calibration curve is greater than that for fine-textured soils (Fig. 19.4) (Lal, 1974). Therefore, developing a site-specific calibration of the neutron moisture probe is essential for gravelly soils. Similarly, use of time domain reflectometery (TDR) for field moisture content, gypsum block for electrical resistance (which can be converted to suction and moisture content by developing empirical relationships), and tensiometer (for soil suction) may also be problematic because of the poor contact between the probe and soil. The site-specific calibrations are also suggested for these devices.

19.4   HYDRAULIC CONDUCTIVITY

Saturated hydraulic conductivity (K_s) is an important input to the water infiltration models (see Chapter 14). In gravelly soils, the gravels impede

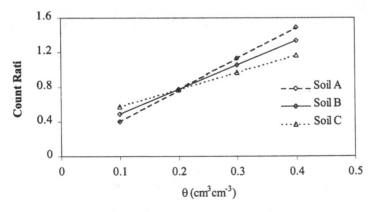

FIGURE 19.4   Effect of soil texture on neutron probe calibration for moisture content determinations for soil A (0.8% gravel, 41% sand, and 15.2% clay content), soil B (59.7%, 27.1%, 7.3%), and soil C (8.2%, 16.3%, 63.9%). (Recalculated and drawn from Lal, 1974.)

both excavation of the hole with controlled geometry and collection of a representative sample. When K_s values are expressed as a function of matric potential, the gravel content does not influence their values. However, expressing K_s as a function of volumetric moisture content, the apparent conductivities for given moisture content are higher when gravels are present. The unsaturated conductivity $[K(\theta)]$ for a gravelly soil can be estimated from measurements made on fine-earth fraction by applying a correction, which reduces the moisture content and area available for flow. The gravels present in the soil can decrease the overall $K(\theta)$ (Mehuys et al., 1975). The soil hydraulic conductivity for a gravelly soil can be calculated as follows (Peck and Watson, 1979):

$$\frac{K_s}{K_{sf}} = \frac{2(1 - V_g)}{(2 + V_g)} \tag{19.13}$$

where K_s is the saturated hydraulic conductivity of bulk soil (gravel + soil), K_{sf} is the saturated hydraulic conductivity of fine-earth, and V_g is the volume fraction of gravels. Equation (19.13) assumes that gravels have zero conductivity, which is a major drawback of this equation. At higher gravel contents, the equation can over predict or under predict the conductivity depending upon the porosity and moisture status of gravels. This drawback can be overcome by relating the soil conductivity to the void ratios for both bulk soil gravel system and pure sand as follows (Bouwer and Rice, 1984):

$$\frac{K_s}{K_{sf}} = \frac{e}{e_s} \tag{19.14}$$

where e is the bulk void ratio of a sand gravel mixture, which is the volume of voids divided by volume of solids, and e_s is the void ratio of sand. Equation (19.13) can be modified by taking into account the ratios of bulk density of fine-earth fraction and gravel fraction as follows (Brakensiek and Rawls, 1994):

$$V_g = \frac{aM_g}{1 - M_g * (1 - a)} \tag{19.15}$$

where M_g is the gravimetric gravel fraction, and a is the ratio of bulk density of fine-earth material and gravel fraction as follows:

$$a = \frac{\rho_{bf}}{\rho_g} \tag{19.16}$$

Substituting Eq. (19.15) into (19.13) yields

$$\frac{K_s}{K_{sf}} = \frac{(1 - M_g)}{1 - M_g(1 - (3a/2))}$$
(19.17)

19.5 WATER INFILTRATION

The influence of a pure gravel layer on water infiltration can be different based upon whether the gravel layer is located below a layer of fine earth material or above it. The water flow through the gravel soil interfaces depends upon the suction gradients across the interface. Let us consider two situations: (i) a gravel layer underlying a fine earth material, and (ii) gravel layer overlaying a fine-earth material. Let us apply water on both soil profiles for a long period. If a gravel layer exists below a layer of fine earth material (Fig. 19.5a), the infiltrating water will not move downward into the gravel layer unless the top fine earth layer is saturated. Thus, the flow is impeded or retarded. This is because the gravelly layer is at atmospheric

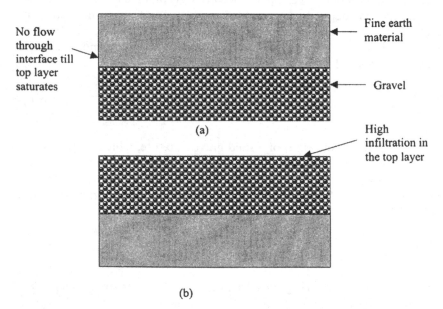

FIGURE 19.5 The schematic of gravel and fine earth material: (a) gravel layer underlying a fine earth material (impedes infiltration) and (b) gravel overlying a fine earth material (increases infiltration.)

pressure and therefore, at higher pressure than the overlaying fine soil. Once the overlaying soil layer gets saturated, the positive pressure (positive pressure gradient between fine earth and gravels) forces the water entry into the gravelly horizon. On the other hand, if gravels are overlaying a layer of fine earth material (Fig 19.5b), the high porosity associated with gravels will let water infiltrate in the profile rather quickly. Regardless, the rate of water infiltration into the layered profile is less than that into a homogeneous soil.

Quantifying factors affecting infiltration of water into soil is crucial for modeling soil erosion and water balance. Infiltration characteristics are significantly influenced by soil properties, soil sealing, and gravel content (Valentin, 1994). During rainfall, gravels at the soil surface intercept rain, which can be: (i) stored at surface, (ii) absorbed by the gravels, (iii) carried as overland flow, or (iv) evaporated on surface.

The gravel cover at the soil surface has an ambivalent effect on infiltration, percolation, and overland flow (Poesen and Lavee, 1994). The gravels in the top layer exert a significant control on infiltration and in some cases can produce flow, which increases the overland flow (Casenave and Valentin, 1992; Poesen et al., 1990). On the other hand, gravels protect the soil surface from rainfall impact and reduce surface crusting and soil sealing, which increases infiltration rate of soil (Valentin, 1994). In general, gravels increase or decrease the total amount of infiltration depending on their size, position, cover, and the structure of the fine earth around them (Poesen et al., 1990; Valentine, 1994). The size and position of gravels are important factors at the microscale, which controls overland flow near the edge. The cover percentage of gravels is important at macroscale, which controls the continuity of overland flow along a hill slope.

19.6 SOIL EROSION

Gravels at the soil surface or in the top layer can significantly influence soil erosion by water and wind. Depending upon the temporal and spatial scales, gravels can have a favorable or adverse effect on water erosion. Some of the favorable effects can be grouped into two classes: direct and indirect influences. The direct effects include: (i) reduction in soil detachment by raindrop splash, (ii) interception of splashed sediment, and (iii) reduction in sediment loaded runoff. The indirect influences are: (i) reduction in surface sealing and compaction, (ii) increase in aggregation, infiltration, and percolation, and (iii) reduction in loss of nutrient and dissolved organic carbon (DOC). Similar effects apply to wind erosion.

On a temporal scale, the soil erosion by water can expose large sized gravels, which act as an erosion pavement and reduce soil erosion in the long

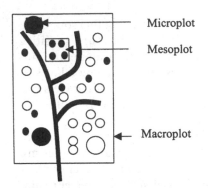

Microplot

Mesoplot

Macroplot

FIGURE 19.6   The schematic of micro-, meso-, and macroplots for studying soil erosion by water. (Modified from Poesen et al., 1994.)

run. On a spatial scale (micro-, meso-, and macroplots) (Fig. 19.6), single large gravel, which corresponds to a microplot, protects the soil under it from erosion by rainfall or runoff. The mass of sediment detached by raindrop impact on a bare interrill soil surface partly covered by gravels per unit area per unit time (SD) can be estimated by:

$$SD = (1 - C_g)(KE)R^{-1} \tag{19.18}$$

where C_g is gravel cover (fraction), KE is the kinetic energy of rainfall per unit area and per unit time ($J\,m^{-2}\,h^{-1}$), and R is the resistance of bare soil detachment ($J\,kg^{-1}$) (Poesen et al., 1994).

Moving on to a mesoplot scale, the interrill area around the large gravels is also included. On a macroplot scale a combination of interrill, rill, and in some cases gullied areas are all included. Therefore, in a natural diversity of an entire slope, a combination of several erosion processes take place simultaneously (Poesen et al., 1994). In a mesoscale, i.e., on interrills, effects of gravel on soil erosion can be ambivalent and largely depend on their size, land slope, and type of soil. On macroscale, the soil erosion is the result of combined subprocesses taking place at the micro and mesoscales. An overall less sediment yield on a macroplot can overshadow the higher sediment yield from a mesoplot. It has been generally observed that mean decrease in relative interrill and rill sediments yield with gravel cover can be expressed by an exponential decay function as follows (Poesen and Ingelmo-Sanchez, 1992):

$$IR = e^{-b(C_g)} \tag{19.19}$$

where IR is relative interrill sediment yield and b is a coefficient indicating the effectiveness of gravel cover. For cultivated topsoils, b values range from 0.02 to 0.06 (Poesen and Ingelmo-Sanchez, 1992).

19.7 TEMPERATURE AND EVAPORATION

A surface layer of gravel has a profound influence on temperature regimes of surface horizon. In arid regions especially, the diurnal and seasonal temperature fluctuation are large, and the maximum temperature is obtained in the afternoon. The temperature fluctuations decrease very rapidly with depth, and at 50 cm and deeper, soil temperatures are nearly constant although considerable lower in winter than summer. The thermal properties of gravels are different than soil, therefore, gravels in soil alter the temperature profile in the soil–gravel system.

The amplitude of temperature below large gravel is smaller than bare soil at a given depth (Fig. 19.7). This is mainly because: (i) it acts as an insulator during hot hours of the day and help retains soil heat during night, and (ii) heat flows from under it toward surrounding soil (Jury and Bellantuoni, 1976). Soil water conditions are generally better under gravels, and soil temperature fluctuations in a gravelly soil are also dependent on soil water content. The temperature fluctuations in a moist gravelly soil are

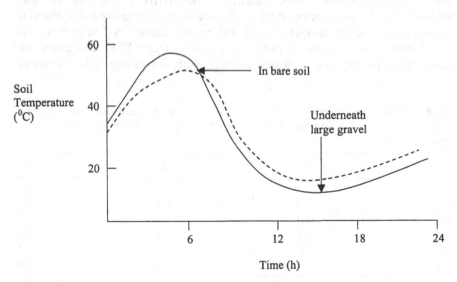

FIGURE 19.7 Schematic of a soil temperature profile at a given depth below and close to large gravel in the bare soil.

lower than those in a dry one (Mehuys et al., 1975). The decrease in the amplitude for moist soil at or near the soil surface occurs as evaporating water cools the soil surface. In a dry soil surface with large gravel content, the daytime temperatures directly beneath the soil are lower than away from it. However, for a similar situation in a moist soil, the temperature beneath a surface gravel are higher than in the surrounding soil because of the loss of latent heat to atmosphere from a gravel-free soil and increased conduction of heat downward in moist areas.

The temperature is also dependent on the dimensions and color of the gravels. A lighter colored gravel is cooler on its undersurface than a gray one (Larmuth, 1978). A layer of gravels (embedded or lying on surface) at the soil surface can also act as mulch. This gravel layer changes the radiation balance and temperature. Porous gravels have generally low unsaturated hydraulic conductivity [$K(\theta)$] at small suctions, which lets only a small volume of water to be transported to the soil surface via capillary rise. Therefore, evaporation losses are low (Unger 1971).

19.8 THERMAL PROPERTIES OF GRAVELLY SOILS

Heat transfer in a soil is characterized by the thermal diffusivity (D_T, cm^2s^{-1}), which is the ratio of thermal conductivity (k_T, $Jm^{-1}s^{-1}K^{-1}$) to heat capacity (C_v, $Jm^{-3}K^{-1}$ or C_g, $Jkg^{-1}K^{-1}$) (refer to Chapter 17 on temperature and heat flow). Thermal conductivity is the rate of heat transfer along a unit temperature gradient and heat capacity is the amount required to raise the temperature of soil by one degree. As the gravel content increases so does the thermal conductivity (Table 19.8). For a dry soil the heat capacity also increases linearly with increasing gravel content

TABLE 19.8 The Volumetric Rock Fragment Content and Relative Thermal Conductivity (K_t), Relative Heat Storage Capacity (H_C), and Thermal Diffusivity (D_f) in a Dry Soil

V_v	K_t	H_C	D_T
0	1	1	1
0.2	1.5	1.2	1.2
0.4	2.7	1.5	1.7
0.5	3.6	1.6	1.9
0.8	7.0	1.8	3.1
1.0	9.2	2.1	4.8

Source: Modified from Poeson and Lavee, 1994.

(Table 19.8). This is the reason why dry gravelly soils get warmer more rapidly than gravel-free soils. Since the thermal diffusivity of dry gravelly soils is high, the total heat flux into the soil is also high. The heat also penetrates deeper in a gravelly soil than a gravel-free soil (Childs and Flint, 1990). Gravels can also induce lateral movement of heat, which can cause water movement as well. Therefore, both heat and water vapor move, mainly because of the horizontal temperature gradients induced by gravels, from the adjacent soil towards the underside of the cooler fragments and causes moisture condensation (Jury and Bellantuoni, 1976a,b).

19.9 ROOT GROWTH

Gravels influence soil's edaphic environments. Gravelly soils may have adverse effects on root growth if presence of gravels lead to a closed packing arrangement. In that situation, smaller-sized gravels can have a more pronounced adverse effect on crop root development than large sized gravels (Babalola and Lal, 1977b). The gravels at shallow depths have inhibitory effects on root development. Total root length and root penetration decrease with increasing gravel content (Table 19.9). The literature suggests as much as a 40% to 75% decline in root growth in gravelly soils, and as much as 70% in coarse sandy soils (Takijima and Sauma, 1967; Babalola and Lal, 1977a). The penetration depth of root declines rapidly with increase in gravel content. As depth to gravel layer increases the depth of penetration of roots also increases. The shoot growth is also significantly affected by the depth to gravel layer. In the leaf area, fresh and dry shoot weights increase with increasing depth to gravel layer (Babalola and Lal, 1977a). For open packing, as is the case in clayey soils,

TABLE 19.9 Effect of Gravel Concentration on Relative Depth and Length of Root Penetration

Gravel concentration (%)	Relative root depth (%)	Relative root length (%)
0	100	100
10	75	70
20	30	38
30	22	20
50	10	18
70	5	10

Source: Modified from Babalola and Lal, 1977a.

effect on root growth may be positive. In this case, gravels increase the porosity. Water retention and availability for clayey soils, therefore, cause no adverse effects on root development or productivity (Unger, 1971).

Example 19.1

If the gravimetric gravel content of a soil is 25 g, the total soil bulk density is 1.5 Mgm^{-3}, bulk density of fine earth is 1.35 Mgm^{-3}, particle density of soil gravel is 2.7 Mgm^{-3}, and the bulk density of gravel is 1.2 Mgm^{-3}, calculate the porosity of gravelly soil.

Solution

The porosity of gravel can be calculated from Eq. (19.1) as follows:

$$f_g = 1 - \frac{1.5}{2.7} = 0.44$$

The volumetric gravel fraction (V_g) of soil can be calculated by Eq. (19.2):

$$V_g = \frac{(1.5 - 1.35)}{(2.7 - 1.35)} = 0.11$$

If the total volume of soil is 1 m^3, the volumetric gravel content $= 0.11$ m^3.

Example 19.2

If the gravimetric mass of a soil–gravel fraction is 520 g and volume 442 cm^3, calculate the bulk density of the soil–gravel system. If the gravimetric mass of gravels is 25 g and the volume displaced when immersed in pure water is 25 mL, calculate the bulk density of gravel and gravel-free soil.

Solution

The bulk density of a soil–gravel system can be directly calculated from weight and volume of the soil–gravel system $= 520/442 = 1.18$ Mgm^{-3}.

The bulk density of gravels $= 25/25 = 1$ Mgm^{-3}.

The bulk density of fine earth material (gravel-free soil) $= (520-25)/(442-25) = 1.19$ Mgm^{-3}.

PROBLEMS

1. If the gravimetric mass of a gravel soil fraction is $550\,g$ and volume $442\,cm^3$, calculate the bulk density of the soil–gravel system. If the gravimetric mass of gravel is $35\,g$ and the volume displaced when immersed in pure water is $30\,mL$, calculate the bulk density of gravels and gravel-free soil.
2. Calculate the porosity of gravel from data in Problem 1.
3. If the saturated hydraulic conductivity of a gravelly soil is $2\,cmh^{-1}$, calculate the saturated hydraulic conductivity of bulk soil for the volumetric gravel fraction from Problem 1.

REFERENCES

Alberto F. (1971). Considerations sur la pierrosite des sols bruns a croute calcaire du basin de l'Ebre. Bull. Rech. Agron. Gembloux, 6:180–185.

Babalola O. and R. Lal (1977a). Subsoil gravel horizon and maize root growth I. Gravel concentratiom and bulk density effects. Plant and Soil, IITA Journal Series 70, 46:337–346.

Babalola O. and R. Lal (1977b). Subsoil gravel horizon and maize root growth II. Effects of gravel size, intergravel texture, and natural gravel horizon. Plant and Soil, IITA Journal Series 70, 46:347–357.

Beven, K. and P. Germann (1982). Macropores and water flow in soils. Water Resour. Res. 18:1311–1325.

Bouwer H. and R.C. Rice (1984). Hydraulic properties of stony vadose zones. Ground Water 22(6):696–705.

Brakenslek D.L and W.J. Rawls (1994). Soil containing rock fragments: effects on infiltration. In: J. Poesen and H. Lavee (ed.) Rock fragments in soil: surface dynamics. Catena, 23:99–110.

Casenave A. and C. Valentin (1992). A runoff capability classification system based on surface features criteria in the arid and semi-arid area of West Africa. J. Hydrol. 130:231–249.

Childs S.W. and A.L. Flint (1990). Physical properties of forest soils containing rock fragments. In: S.P. Gessel, D.S. Lacate, G.S. Weetman, and R.F. Powers (eds.), sustained productivity of forest soils. University of British Columbia, Faculty of Forestry Publ., Vancouver, B.C., 95–121.

Clothier B.E., D.R. Scotter, and J.P. Kerr (1977). Water retension in soil underlain by a coarse textured layer: Theory and field application. Soil Sci. 123:392–399.

Cooke R.U., A. Warren, and A.S. Goudie (1993). Desert Geomorphology. UCL Press.

Dryness C.T. (1969). Hydrological properties of soils on three small watersheds in the western cascades. USDA Forest Service Research Note PNW-111.

FAO (1977). Guidelines for soil profile description. Food and Agriculture Organization of the United Nations (F.A.O.), Rome, p 66.

Fisher R.F. and D. Binkley (2000). Ecology and management of forest soils. John Wiley & Sons, New York.

Flint A.L. and S. Childs (1984). Physical properties of rock fragments and their effect on available water in skeletal soils. In: J.D. Nichols, P.L. Brown and W.J. Grant (eds.), Erosion and Productivity of soils containing rock fragments. Soil Science Society America, Madison, WI, Special publication no. 13 p 91–101.

Gornat B. and D. Goldberg (1972). The relation between moisture measurements with a neutron probe and soil texture. Soil Sci. 114:254–258.

Gras R. and G. Monnier (1963). Contribution de certains elements grossiers a l'alimentation en eau des vegetaux. Soil Sol. 1:13–20.

Hanson C.T. and Blevins R.L. (1979). Soil water in coarse fragments. Soil Sci. Soc. Am. J. 43:813–820.

Ingelmo F., S. Cuadrado, A. Ibanez, and J. Hernandez (1994). Hydric properties of some Spanish soils in relation to their rock fragment content: implication for runoff and vegetation. In: J. Poesen and H. Lavee (ed.) Rock fragments in soil: surface dynamics. Catena 23:87–97.

Jury W.A. and B. Bellantuoni (1976a). Heat and water movement under surface rocks in afield soil: I. Thermal effects, Soil Sci. Soc. of Am. J. 40:505–509.

Jury W.A. and B. Bellantuoni (1976b). Heat and water movement under surface rocks in afield soil: II. Moisture effects, Soil Sci. Soc. of Am. J. 40:509–513.

Lal R. (1974). The effect of soil texture and density of the neutron and density probe calibration for some tropical soils. Soil Sci. 117(4):183–190.

Lal R. (1979). Physical characteristics of soils of the tropics: determination and management. In: R. Lal and D.J. Greenland (eds.) Soil Physical Properties and Crop Production in tropics. John Wiley & Sons, New York, p 7–44.

Larmuth L. (1978). Temperature beneath stones used as daytime retreats by desert animals. J. Arid Environ. 1:35–40.

Mansfield J.E. (1979). Land capability for annual rainfed arable crops in Northern Nigeria based on soil physical limitation. In: R. Lal and D.J. Greenland (eds.) Soil Physical Properties and Crop Production in tropics. John Wiley & Sons, New York, p 408–437.

McHenry F.R. (1963). Theory and application of neutron scattering in the measurement of soil moisture. Soil Sci. 95:294–307.

Mehuys G.R., L.H. Stolzey, and J. Letey (1975). Temperature distributions under stones submitted to a diurnal heat wave. Soil Sci. 120:437–441.

Miller F.T. and R.L. Guthrie (1984). Classification and distribution of soils containing rock fragments in the United States. In: J.D. Nichols, P.L. Brown, and W.J. Grant (eds.), Erosion and Productivity of soils containing rock fragments. Soil Science Society America, Madison, WI, Special publication no. 13 p 91–101.

Montagne C., J. Ruddell, and H. Ferguson (1992). Water retention of soft siltstone fragments in a Ustic torriorthent, Central Montana. Soil Sci. Soc. Am.J. 56: 555–557.

Moormann F.R., R. Lal, and A.S.R. Juo (1975). The soils of IITA.International Institute of Tropical Agriculture, Tech. Bull. No. 3, Ibadan, Nigeria.

Nettleton W.E. Gamble, B. Allen, G. Borst, and F. Peterson (1989). Relict soils of subtropical regions of the United States, Catena Suppl. 16:59–93.

Parsens A.J., A.D. Abrahams, and J.R. Simanton (1992). Microtopography and soil surface materials on semi-arid piedmont hillscope, Southern Arizona, J. Arid Environ. 22:107–115.

Peck A.J. and J.D. Watson (1979). Hydraulic conductivity and flow in non uniform soil, In: Workshop on Soil Physics an Field Heterogeneity. CSIRO Div. Of Environmental Mechanics, Canberra, p 31–39.

Poesen J.W. and F. Ingelmo-Sanchez (1992). Runoff and sediment yield from topsoils with different porosity as effected by rock fragment cover and position. Catena 19:451–474.

Poesen J.W. and H. Lavee (1994). Rock fragments in top soils: significance and processes. In: J. Poesen and H. Lavee (ed.) Rock fragments in soil: surface dynamics. Catena 23:1–28.

Poesen J.W., D. Torri, and K. Bunte (1994). Effects of rock fragments on soil erosion by water at different spatial scale: a review. In: J. Poesen and H. Lavee (ed.) Rock fragments in soil: surface dynamics. Catena 23:141–166.

Poesen J.W., F. Ingelmo-Sanchez, and H. Muler (1990). The hydrological response of soil surfaces to rainfall as effected by cover and position of rock fragments in the top layer. Earth Surf. Process, Landforms 15:653–671.

Ravina I. and J. Magier (1984). Hydraulic conductivity and water retention of clay soils containing rock fragments. Soil Sci. Soc. Am. J., 48:736–740.

Soil Survey Staff (1951). Soil Survey Manual. Agric. Handb. No. 18, USDA, U.S. Government Printing Office, Washington, D.C.

Soil Survey Staff (1993). Soil Survey Manual. USDA-SCS Agric. Handb. No. 18, U.S. Government Printing Office, Washington, D.C.

Stewart V.I., W.A. Adams, and H.H. Abdullah (1970). Quantitative pedological studies on soils derived from Silurian mudstones II. The relationship between stone content and apparent density of the fine earth. J. Soil Sci., 21:248–255.

Takijima Y. and H. Sauma (1967). The classification of paddy soils based on their suitability for reclamation. 2. Effect of sand and gravel content of the soil on the development of the root system of rice plants. J. Sci. Soil Manure, Tokyo 38:313–318.

Torri D., J. Poesen, F. Monaci, and E. Busoni (1994). Rock fragment content and fine soil bulk density: In: J. Poesen and H. Lavee (ed.) Rock fragments in soil: surface dynamics. Catena 23:65–71.

Unger P. (1971). Soil profile gravel layers: I. Effect on water storage, distribution, and evaporation. Soil Sci. Soc. Am. Proc. 35:631–634.

Unger P.W. (1971). Soil profile gravel layers. Effect on water storage, distribution and evaporation. Soil Sci. Soc. Am. Proc. 35:631–654.

Valentin C. (1994). Surface sealing as effected by various rock fragment covers in West Africa: In: J. Poesen and H. Lavee (ed.) Rock fragments in soil: surface dynamics. Catena 23:87–97.

Webster R. (1985). Quantitative spatial analysis of soil in the field. Adv. Soil Sci. 3:1–70.

20

Freezing and Thawing Effects, Swelling Soils, and Other Special Problems

The theory and experimental verification of isothermal moisture storage and movement in soils are well developed. Some of the processes such as freezing and thawing, presence of macropore flow channels, swelling and shrinking of soils, and salinity complicate the flow and transport processes through soils by modifying the physical and water transmission properties of porous media. It is generally recognized that during freezing and thawing processes, soil moisture and thermal states of a soil system are coupled. Inadequate understanding of this complex problem of phase changing processes during freezing and thawing and sparsity of development in regions where these processes are significant are some of the reasons why enough progress has not been made in this field. This chapter discusses some of the mathematics of coupled heat and moisture transport models as well the effects of freezing and thawing on soil's physical properties. The modeling of macropore flow and its effect on infiltration and solute transport through soil profile, characteristics, and water flow in swelling soils, and pressure potentials in saline soils, are also discussed. Another important topic covered in this chapter is water-repellent soils. This is an important research topic to which an increasing attention has been paid in view of its. This chapter describes hydrological processes in water-repellent soils.

20.1 FROZEN SOILS

Freezing and thawing can have a profound influence on the stability, hydrology, chemistry, biology, and ecology of soils. Under frozen conditions the soils become less permeable to rainfall/snowmelt, water within the frozen part of the soil profile becomes immobile and therefore unavailable for leaching, chemicals within the soil profile are redistributed due to the presence of temperature gradients and nonuniform freezing, and microbiological activity is reduced due to lowered temperature. The mass and energy balance conditions within the soil profile are also greatly affected by the existing surface cover conditions. For instance, the presence of a snow pack can increase the intensity of microbiological activity within the soil profile due to the insulation of the soil by the snow.

Laboratory and field experiments can provide useful information about the winter processes within the soil profile. However, experimentation, especially under in situ conditions, is very difficult. Freezing processes in soil are also complicated, and use of simulation models can be advantageous. The processes of coupled water flow, thermal energy transport, and solute transport within a variably saturated, variably frozen soil profile can be described by a set of coupled conservation (mass and energy) equations, appropriate constitutive laws (Darcy's, Fick's, Fourier's), and equilibrium thermodynamic relations. A number of numerical solution models have been developed to solve these systems of equations, with some methods imposing numerous simplifying assumptions, while others solving the full set of equations.

20.1.1 Frozen Soil Composition

Soil moisture conditions influence water and contaminant migration in frozen soils. Frozen soils consist of four main phases including air, unfrozen water, ice, and soil particles. The four volumetric components of frozen soil are illustrated in Fig. 20.1. The conversion of water into ice results in increase in volume. The state of water in soil water systems for ice-free soil can be expressed as the matrix pressure of water and air (Φ_{mwa}) as follows (Black, 1990)

$$\Phi_{\mathrm{mwa}} = P_w - P_a \tag{20.1}$$

where P_w and P_a are soil water and air pressures. Similarly the state of water in an air-free frozen soil can be explained as

$$\Phi_{\mathrm{miw}} = P_i - P_w \tag{20.2}$$

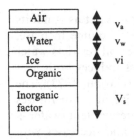

FIGURE 20.1 Schematic of volumetric distributions of frozen soil components where V_a, V_w, V_i, and V_s represent volumes of air, unfrozen water, ice, and soil particles, respectively.

where P_i is pressure of ice. The state of water in air-free frozen soils is often expressed in terms of temperature (T). For a solute-free soil, the Clapeyron equation relates the ice and water pressures to temperature, specific gravity ($\rho'_s = (\rho_s/\rho_w)$), and volumetric latent heat of fusion (L_0) as follows (Black, 1990)

$$\Phi_{\text{miw}} = (\rho'_s - 1)P_w - \frac{\rho'_s L_0}{273}T \tag{20.3}$$

20.1.2 Soil Freezing Process

Freezing process in soil can be explained as follows:

1. Water begins freezing when soil temperatures are below the pore-water freezing point.
2. Soil temperatures drop below pore waters freezing temperatures until enough energy exists to instigate pore-water nucleation.
3. Pore-water temperature then increases to its freezing point.
4. Soil temperatures remain constant at this temperature until all latent heat of fusion is released.
5. Soil temperatures then decrease if ambient temperatures are below the pore-water freezing point.

Soil-particle surface tension and capillary forces cause depression of pore-water freezing temperatures adjacent to the soil grains in saturated soils. This phenomenon is most evident in fine-grain soils such as clays and silts. The unfrozen water content decreases as soil temperature decreases. The unfrozen water layer surrounding soil particles acts as a conduit for contaminant transport. Unsaturated frozen soils contain little or no unfrozen water content. Therefore, air voids in pore spaces act as conduits for contaminant transport. This is in part due to a 9% volume increase, which occurs when water freezes to form ice. Ice in the soil restricts water and contaminant pathways. In general, freezing process induces both heat and mass transfer from warm regions to cold regions (Harlan, 1973). The water film on soil surface exists in equilibrium with ice at temperatures below 0°C. The film thickness is independent of total water content (water + ice) but dependent on temperature (Low et al., 1968) and decreases with temperature falling below the freezing point of water.

20.1.3 Water Flow in Frozen Soils

If soil water characteristic curve is expressed as per Eq. (20.4)

$$\Phi_m = \Phi_0(\theta/\theta_s)^{-b} \tag{20.4}$$

then the hydraulic conductivity ($K(\theta)$) can be represented by

$$K(\theta) = K_s(\theta/\theta_s)^{(2b+2)} \tag{20.5}$$

where Φ_m is the soil water matric potential (cm of water), Φ_0 is the air entry value (cm of water), K_s is the saturated hydraulic conductivity (cmh^{-1}), θ is unsaturated liquid water content (cm^3 cm^{-3}), θ_s is the saturated water content (cm^3 cm^{-3}), and b is a constant (Campbell, 1974). When ice is present in the soil pores at atmospheric pressure, the water potential can be approximated by the vapor pressure of pure ice as follows (Cary and Mayland, 1972):

$$1.2 * 10^4 T = \Phi_m + \Phi_\pi \tag{20.6}$$

where Φ_π is the osmotic potential of the soil solution (cm of water), and $1.2 * 10^4$ is a factor that approximately converts the temperature of ice T (°C) to total water potential (cm of water) (Cary et al., 1979). Combining and rearranging Eqs. (20.4), (20.5), and (20.6) results in the following equation, which describes the unsaturated soil water flow in a frozen soil system

$$
\begin{aligned}
J_l &= -K\frac{d(\Phi_m + \Phi_z)}{dz} \\
&= -K_s\left[\frac{1.2 * 10^4 - \Phi_\pi}{\Phi_0}\right]^{-2[1+(1/b)]}\left[1.2 * 10^4\frac{dT}{dz} - \frac{d\Phi_\pi}{dz} + 1\right]
\end{aligned} \tag{20.7}
$$

where J_l is flow in the liquid phase and z is soil depth. The above derivation assumes that the water content, including ice, is less than saturation, and that water transport due to anion exclusion, liquid ice interface phenomenon, vapor diffusion, and plastic flow of ice are all negligible. The value of Φ_π can be approximated by osmotic pressure and temperature relationship for the cases where solubility limits are not exceeded as follows

$$\Phi_\pi \sim \Phi_{\pi 0}(\theta_s/\theta) \tag{20.8}$$

Combining Eq. (20.8) with Eqs. (20.4) and (20.6) provides the value of Φ_t though iteration for T less than freezing point of soil solution (Cary et al., 1979)

$$\Phi_\pi = \Phi_{\pi 0}\left[\frac{1.2 * 10^4 - \Phi_t}{\Phi_o}\right]^{1/b} \tag{20.9}$$

For $T <$ the freezing point of soil solution, Eq. (20.9) converges rapidly to π for an initial π_0. The osmotic pressure effects in Eq. (20.4) do not correspond to those observed in clay soils and biological systems with selective semipermeable membranes. In these types of systems, water flows from regions of low osmotic pressure to high. When ice is present, the direction of flow is reversed. Solute increases the amount of liquid phase water, which flows towards the thinner films away from the regions of high osmotic pressure (Cary et al., 1979).

20.1.4 Simultaneous Heat and Fluid Transport

The simultaneous transport of heat and fluid in a partially frozen porous medium requires the knowledge of interrelationships among the laws of heat and fluid flow, equations of continuity for mass and energy, and the characteristics of the fluids and medium involved. The energy state of liquid water at temperatures less than 0°C in equilibrium with ice is a function of temperature but not of water content (except for very dry conditions). Hence, ice phase can be assumed to be behaving as a sink (water is added), or source (water removed from storage). The mass transport equation for one-dimensional steady or unsteady flow in a partially or fully saturated heterogeneous porous medium with freezing or thawing can be written as (Harlan, 1973)

$$\frac{\partial}{\partial x}\left[\rho_l K(x, T, \Phi_m)\frac{\partial \Phi_t}{\partial x}\right] = \frac{\partial(\rho_l \theta)}{\partial t}\Delta S_i \tag{20.10}$$

where x is the depth of soil under consideration (g), ρ_l is density of liquid fraction (g cm^{-3}), θ is volumetric water content (cm^3 cm^{-3}), K is hydraulic conductivity of soil, T is temperature (°C), Φ_t is total head (cm), Φ_m is matrix or capillary head (cm), and ΔS_i is change in ice content (g cm^{-3} min^{-1}). The total head or Φ_t is the sum of Gibbs free energy ($\Phi_m + \Phi_\pi$), or soil water potential, pneumatic pressure head (Φ_p), and elevation head (Φ_z).

The assumptions made to derive the heat transfer equation are: (i) soil is homogeneous, (ii) free thermal convection is negligible, (iii) the temperature distribution is smooth and continuous, (iv) the temperature of the fluid entering the medium is equal to the temperature along appropriate boundary, and (v) thermal resistance between fluids and soil matrix is small, and matrix temperatures are equal. Based upon these assumptions, the one-dimensional steady or nonsteady convection–conduction heat transport equation may be written as (Harlan, 1973)

$$\frac{\partial}{\partial x}\left[K_T(x,T,t)\frac{\partial T}{\partial x}\right] - C_w \rho_l \frac{\partial(v_x T)}{\partial x} = \frac{\partial(C_v T)}{\partial t} \tag{20.11}$$

where K_T is the thermal conductivity (cal cm^{-1}°C^{-1} min^{-1}), T is temperature (°C), C_w is bulk specific heat of water (cal g^{-1} cm^{-1} C^{-1} min^{-1}), v_x is the fluid flow velocity in x direction, C_v is apparent volumetric specific heat (cal cm^{-1}°C^{-1}), and is defined as follows:

$$C_v = C_v'(x,T,t) - L_0 \rho_{\text{ice}}\left(\frac{\partial\theta_s}{\partial T}\right) \tag{20.12}$$

where C_v' is the volumetric specific heat capacity (cal cm^{-1}°C^{-1}) and is defined as the sum of specific heat of the soil material, liquid water fraction, and ice fraction; L_0 is latent heat of fusion (cal g^{-1}); θ_s is volumetric ice fraction (cm^3 cm^{-3}); and ρ_{ice} is the ice density (g cm^{-3}).

20.1.5 Effect of Freezing and Thawing on Soil Physical Properties

The complexity of the hydrologic cycle is increased during winter conditions when the temperature of the soil reaches the freezing point. Winter can be visualized as a static period regarding the movement of water from precipitation through the soil profile and into ground water recharge. Factors that affect infiltration are hydraulic conductivity, saturation, soil makeup, and porosity. However, under frozen conditions, temperature becomes the primary factor in determining a soil's permeability. The contradictory accounts on effect of freezing and thawing on soil's physical properties are available. Frost causes a breakdown of soil's physical properties, i.e., decreased soil aggregation and increased bulk density (Bisel and Nielsen, 1964, 1967; Leo, 1963). Frost action can also improve soil's physical properties, e.g., increase aggregation, and reduce bulk density (Sillanpaa and Weber, 1961). In addition, changes in saturated water holding capacity, hydraulic conductivity, and soil bulk density as a result of freeze–thaw activity are significantly related to the type of soil, initial water content, degree of aggregation, initial bulk density, degree and rate of freezing, and number of cycles. (Mostaghimi et al., 1988; Benoit and Voorhees, 1990) Specific effects of freeze–thaw on soil properties are described below.

Aggregation

Aggregate stability, a measure of a soil aggregate's resistance to breakdown, influences many soil physical and hydraulic characteristics, such as surface sealing, infiltration, and hydraulic conductivity. Freeze–thaw cycles influence soil aggregation. The water contained in the pores of the soil expands on changing to ice, and freezing of water inside the soil pores affects soil aggregation. In coarse-textured soils, water freezes in situ, whereas in fine-textured soils, water moves towards the freezing sites and forms ice lenses. Because of the suction, the water is withdrawn from the areas where it is not frozen and compression and heaving of the soil occurs as ice lenses grow (Young and Warkentin, 1975). The compression of drier soil near the lenses leads to aggregation in fine-textured soils. The freezing and thawing can also cause crumbling of cloddy soils. It can produce cracks in the soils, which may open further during subsequent freezing and thawing cycles (Richardson, 1976). Aggregate stability is inversely proportional to soil water content at the time of freezing, constraint to expanding, and number of thaw–wet cycles (Bullock et al., 1988; Mostaghimi et al., 1988; Lehrsch et al., 1990). Aggregates in otherwise poorly aggregated soils become stable when frozen at intermediate water contents.

Infiltration and Hydraulic Conductivity

The knowledge of water infiltration in frozen soils is important for water conservation, runoff, or flooding and erosion in areas where freezing temperatures occur. The ice content of soil and water infiltration rate is inversely related (Granger et al., 1984; Thunholm and Ludin, 1990). The ice lenses on soil surface or at shallow depth impede water infiltration. Vertical freezing and thawing decreases soil hydraulic conductivity and is related to water content, initial aggregate size, and freezing temperature. The greatest decrease in hydraulic conductivity values in repacked soil columns is associated with larger aggregate size and higher water content upon freezing (Benoit, 1973).

20.2 MACROPORE FLOW

The biological, chemical, and physical processes and their interactions in soils are fundamental to the formation of preferential or macropore flow paths. In contrast to flow through homogeneous soil profile, in which water penetrates the entire porous network of soil, preferential flow occurs through distinct pathways, which constitute a small fraction of total soil pore space but can conduct as much as 80% of water percolating through soil profile. The macropore channels may be randomly or systematically distributed, reflecting the influence of land use practices. Preferential flow paths have a significant influence on water and solute transport in the soil. This influence is usually reflected in reduced travel time, increased volumes of water and solute concentrations in drainage water, or deeper penetration of chemicals into the soil profile, as compared to corresponding predictions from conventional flow theory (Beven and Germann, 1982). The mechanics of flow and transport through capillary pores in the soil matrix is well documented, as this phenomenon has always been considered to be the dominant flow process. However, since the mid 1970s, there has been an increase in the number of research studies confirming the importance of macropore flow in water and solute transport (McCoy et al., 1994), and several simple and complicated (able to account for the spatial variability of macropores into the description of water and solute movement in soils) models have been developed.

20.2.1 Preferential Flow Paths

Some of the terminologies used to describe macropore flow are confusing. Some commonly used terms are subsurface storm flow short-circuiting, macropore flow, bypass flow, channeling, fracture flow, preferential flow, or fingering (Bouma and Dekker, 1978). The preferential flow paths are often termed as macropores, which suggests that size is a sufficient criterion to define them. In practice, however, it is difficult to put a size limit on soil voids that function as macropore channels. The sizes ranging from 0.03 mm to 3 mm are being used as the lower limit for the equivalent diameter of a macropore channel (Beven and Germann, 1982). The ability of a soil void to contribute to short circuiting is dependent not only on size, but on pore structure and continuity, initial soil water content, and water application flux (Beven, 1982). However, the flow dynamics through macropore channels are more important than their size, and any size limit used in a particular application needs to be explicitly stated. Preferential flow paths have been traditionally separated in classes based on how closely they conform to a circle in cross section (Bouma et al., 1978). A more functional system of classification is based on the mode of formation and the persistence of a preferential flow path. According the morphology, macropores are classified into four groups: (i) pores formed by burrowing animals, (ii) pores formed by plant roots, (iii) cracks and fissures, and (iv) natural soil pipes (Beven and Germann, 1982). The first two groups are collectively called biopores, and the last two, cracks. Brief description of these two categories follows.

Biopores

Biopores are formed as a result of the interactions between the soil and the soil biota. This interaction is usually limited to the upper soil layers (the A and B horizons) in mineral soils. Biopores consist of decayed root channels, living roots, wormholes, insect burrows, and the burrows of small rodents, with the smaller holes being more prevalent. They tend to be roughly cylindrical in shape and can extend up to 1/m depth (Edwards et al., 1988). The void forming animals are classified as fossers who are equipped with excavating limbs; miners who use their jaws to bite soil particles off of aggregates, leaving the aggregates with jagged and pitted surfaces; and tunnelers that push or eat their way through the soil (Hole, 1981). On an average, small soil animals form holes from 10 mm in diameter and up to 1 m to 2 m deep, and these channels tend to be very stable (Hole, 1981).) Wormholes have no stress concentrations on their edges and as such do not change shape in response to moisture content changes in the soil matrix (Dexter, 1978). In general, biopores are very persistent or stable and have relatively high transport capacities. The size and frequency of biopores, which result from the action of soil flora, are dependent mainly on plant species, on soil water status during the plant-growing season, and on agrotechnical practices. Roots also increase macropore densities by growing into microcracks and enlarging them.

Cracks

Cracks in a soil system are formed by the shrinkage in clayey soils or by chemical weathering of bedrock material (Fig. 20.2). They can be formed by methods of cultivation as well as freeze–thaw cycling. When clay soils, particularly those rich in montmorillonite or other smectitic clays, dry out, they usually form desiccation cracks. These soils are so widespread that soil cracking has been used as a differentia at the highest level of soil

FIGURE 20.2 (a) Cracks in a vertisol and (b) in a structurally unstable soil conducting water rapidly until the cracks are sealed. (Southern India, 2001.)

classification, the order level, in most classification systems. Although crack volume decreases on rewetting, cracking is not a fully reversible phenomenon. Once a desiccation crack has been formed in an uncompacted soil, it cannot be removed by rewetting, since the particles on the edge of the crack change their orientation. The same phenomenon was also observed in compacted clays and in field soils, respectively (Boynton and Daniel, 1985; Jalalifarahani et al., 1993). Modifications in soil hydraulic behavior resulting from the formation of desiccation cracks in soils become significant enough to warrant the development of a new infiltration equation (Mailhol and Gonzalez, 1993). The new infiltration equation has two parameters to account for crack formation and persistence. One of these parameters is related to the intrinsic properties of the soil, while the other accounts for the prevailing conditions at the onset of infiltration.

The cracks that appear when montmorillonitic soils are saturated with seawater are known as synerisis cracks (Burst, 1965). Synrisis is defined as the spontaneous separation of an initially homogeneous colloid system into both a coherent gel and a liquid. Some organic fluids also have the same effect on clay soils (Anderson et al., 1981). The leachate from municipal solid waste also produces significant cracking in compacted clays with high concentrations of expansive clays (D'Appolonia, 1980; Hettiaratchi et al., 1988). The divalent cations, especially calcium ions, are the most important in the formation of synerisis cracks. The free swell potential, a measure of volume change, can be doubled by increasing the concentration in leachate from $0.009 \, molesL^{-1}$ to $0.09 \, molesL^{-1}$ (D'Appolonia, 1980; Hettiaratchi et al., 1988). Calcium ion concentrations ranging from $0.002 \, molesL^{-1}$ to $0.04 \, molesL^{-1}$ have been detected in municipal sludge (Fuller and Warwick, 1985).

20.2.2 Soil Management Practices

The formation and persistence of preferential flow paths is influenced by soil management practices (e.g., tillage, planting technique, drainage conditions). Cracks tend to originate from sites where microcracks are present (Briones and Uehara, 1977). Thus any operation that alters the soil surface should affect soil cracking and preferential flow path density. The cracking pattern can be altered in a clayey soil by altering planting techniques (Swartz, 1966). The runoff from an agricultural field can be controlled or reduced by controlling soil-cracking patterns by using a variable row spacing and skip seeding technique (Johnson, 1962). The subsurface drainage systems also influence the soil crack formation and have a totally different pattern (Godwin et al, 1981) than formed by different tillage methods (Culley et al., 1987). The traffic on the soil surface may compact the soil and reduce the number or density of macropore channels (Ankeny et al., 1990).

20.2.3 Water Infiltration

During the initial period of infiltration, the water intake rate in a soil having open cracks can be extremely high (Mitchell and van Genuchten, 1991). The infiltration rate is much higher than the saturated hydraulic conductivity of soil or the infiltration through a homogeneous soil profile without macropores (Fig. 20.3) mainly as the short-circuiting of water through open cracks. It is clear from Fig. 20.3 that if the infiltration process is long enough, the influence of cracks or macropore flow starts to diminish and the two infiltration curves from homogeneous soil and cracked soil approach the same constant infiltration rate. This phenomenon is possible mainly due to two factors. One of them is that the wetting front has penetrated well below the cracks and the second is the closure of the cracks due to the swelling of clay under prolonged imbibition (Mitchell and van Genuchten, 1991).

20.2.4 Solute Transport

The effect of preferential flow channels on solute transport is also well documented in the literature (Beven and Germann, 1982). A large number of field studies have been performed in which the effects of preferential flow paths on solute transport were examined (Steenhuis et al., 1990; Mohanty et al., 1998). Various tracers including different types of pesticides, and other tracers such as bromide, chloride, and nitrate have been applied to field plots under different land use and management systems. The significant

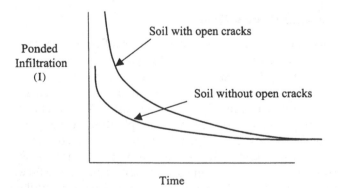

FIGURE 20.3 Schematic of infiltration-time curves for a soil with and without cracks.

differences in the transport behavior of these tracers have been observed as a result of preferential flow paths in general, and biopores in particular. Preferential flow paths are higher under no-till as macropore or biopore channels are better preserved than in plow-till soils. More water flow and movement of chemicals at a much faster rate through no-till soils are also generally observed, which results in less attenuation and higher leachate concentrations (Isensee et al., 1990).

20.2.5 Models of Macropore Flow

The analysis of water and solute flow through porous media is based on a continuum approach, which assumes that each variable of interest can be expressed as a continuous function of time and space. In reality, especially at pore scale, no porous medium is a continuum, rather, pore spaces and solids exist as separate regions. Therefore, for most analyses of water and solute flow, soil is considered a continuum at a macroscopic scale or at representative elementary volume (REV). The REV of a soil is a volume that is large enough so that the inclusion and/or exclusion of a few pores at the edge of REV does not change the soil properties, and at the same time small enough so that any variable measured at this scale is a continuous function of space (McCoy et al., 1994).

For a soil with macropore, the two physically based continuum approaches are available. One of them is the disjoint-volume approach, which treats macropores as not being part of the porous system. In this approach flow in each macropore or a set of macropores is described separately and the remaining soil system is dealt with as a separate continuum. This type of an approach requires three sets of equations to describe the flow process: (i) through macropores, (ii) through soil matrix, and (iii) exchange of water and solutes at the interface. The other approach is called a multicontinuum (multiporosity or multiregion) approach, where different types of pores are treated separately (or grouped in various categories) and soil mass assumed to be made up of more than one continuum. If only two continua are considered, the approach is called dual porosity or bicontinuum or two-region (Gerke and van Genuchten, 1993).

20.3 SWELLING SOILS

Soils that swell when water is added are known as swelling or expansive soils. Swelling soils typically contain clay minerals that attract and absorb water. When water is added to these expansive clays, the water molecules are pulled into gaps between the clay plates. As more water is absorbed, the plates are forced further apart, leading to an increase in soil pressure or an expansion of the soil's volume. Soils containing expansive clays become very sticky when wet and usually are characterized by surface cracks or a "popcorn" texture when dry. The presence of surface cracks is usually an indication of an expansive soil. Swelling bedrock contains a special type of mineral called claystone.

Vertisols are dark montmorillonite-rich clays with characteristic shrinking/swelling properties. This group of soils with a high clay content (>30% to at least 50 cm from the surface) and in dry state with typical cracks which are at least 1 cm wide and reach a depth of 50 cm or more, are often also called heavy cracking clay soils. They cover an area of about 340 Mha worldwide and 9.12 Mha in North America (Table 20.1). The shrinking of a soil on drying and swelling on wetting has a strong influence on soil structural and water movement, as well as on stability of pavement and buildings.

Several soils change their volume when their water content (especially close to saturation) is altered. The increase in volume of such soils with increase in water content is known as swelling. The porosity of such types of soil depends on water content (Fig. 20.4). The swelling index is the ratio of volume change for a soil saturated with water to that saturated with a nonpolar liquid. The water content at saturation does not limit the process of swelling but swelling can continue further, provided more water is available. The pressure needed to prevent a soil from swelling is called "swelling pressure." The swelling pressures can be calculated using double layer theory, and for monovalent cations the swelling pressure (P) resulting from osmotic pressure differential between the interparticle midplane can be calculated as follows (Kutilek

TABLE 20.1   The Distribution of Vertisols Around the World

Region	Area (Mha)
Africa	105
India (including some parts Asia and Far East)	58
Australia	48
Latin America	27
United States (including other parts of North America)	10
Middle East	5.7
Europe	5.4

Source: Modified from Hubble, 1981.

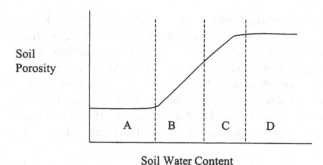

FIGURE 20.4 Schematic of water concent vs. porosity in swelling soils for (A) zero change in porosity with water content change, (B) change within zero and one, (C) change equal to one, and (D) zero change.

and Nielsen, 1991):

$$P = RT(n_c - 2n_0) \tag{20.13}$$

where R is universal gas constant, T is absolute temperature, n_c is ionic concentration, and n_0 is the concentration of ambient intermicellar solution. The above model assumes parallel orientation of clay platelates, which is seldom true in real situations, where a large number of factors, i.e., silt and sand particles, cations, clay minerals, random arrangement of clay particles, tactoid formation, and deflocculating can influence soil behavior during swelling.

20.3.1 Nature of Swelling Clay Soils

Swelling depends upon the clay content, mineralogical composition of clay, and exchangeable cations. The magnitude of shrinkage is dependent on the amount and type of clay minerals present. The montmorillonite type clay swells more than chlorite or illite type clays because monovalent exchangeable cations induce greater swelling than divalent cations. The swelling of kaolinitic soils is negligible. The physical dimensions, or volume, of coarse-grained soils, such as sand, are governed solely by loading stresses. In contrast to this, the volume of a clay soil is governed not only by external stresses but also by internal stresses. In humid areas natural clay soils with high initial water contents and which have not previously been subjected to drying or consolidation by loading will tend to shrink on drying or loading more than they will tend to swell on wetting or unloading. Soil shrinkage data is presented as specific volume change of the soil as a function of water content, which is also the reciprocal of bulk density. Since water content is not evenly distributed in soil profile, the soil drying is considered water lost from entire soil profile and expressed as a volume–change ratio. From the volume–change ratio, which is the ratio of change in soil bulk volume and change in volume of water between saturated and air-dried value, the shrinkage characteristic (S_c) can be presented as follows (Mitchell and van Genuchten, 1992)

$$S_c = \frac{\partial V_s}{\partial V_w} \qquad\qquad (20.14)$$

where V_s is the volume of soil and V_w the volume of soil water. Under other climatic conditions, clays that have been subjected to cyclic moisture change or previously subjected to higher loading may tend to swell greatly when allowed access to water under light loading. Their rate of swelling is governed by the rate at which water can move into the clay, i.e., the permeability of the clay. Because of small pore size and thus their low permeability, clays may take years to reach new moisture equilibrium conditions. In saturated clay soils, swelling is given by the change in soil water content:

$$dq = A_m\,dx \qquad\qquad (20.15)$$

where q is the volume of water in unit mass of clay, A_m is the surface area per unit mass of clay and x is the mean thickness of the water over surface, i.e., half of water thickness between two surfaces.

20.3.2 Pressure Potentials in Swelling Soils

The actual swelling in a clayey soil depends on the depth of each layer. The surface layer of soil swells freely upon wetting, however, deeper layers are prevented by the confinement of overlying soil, which is known as overburden or envelope pressure. The total soil water potential (Φ_t) for an unsaturated swelling soil can be given as

$$\Phi_t = \Phi_z + \Phi_s + \Phi_a + \Phi_m \qquad\qquad (20.16)$$

and for a saturated swelling soil, it is expressed as

$$\Phi_t = \Phi_z + \Phi_s + \Phi_p + \Phi_b \qquad\qquad (20.17)$$

where Φ_z, is gravitational potential, Φ_s is osmatic potential, Φ_a is air pressure potential, Φ_m is matrix potential, Φ_p hydrostatic pressure potential, and Φ_b is overburden pressure potential (Jury et al., 1991).

20.3.3 Models for Swelling

A two parallel plate model can describe the swelling behavior of montmorillonite sheets of clay. These two plates are subjected to repulsive and attractive forces (Iwata et al., 1988). The energy for attraction is caused by adhesive and van der Waals forces, which depend upon the orientation of the plates and pH of soil solution. When pH decreases the charges at the edge of the plate becomes positive and edge bonding develops, which is more like a plate-like arrangement than parallel arrangement of clay particles (montmorillonites). The repulsive forces between two parallel plates with negatively charged surfaces consists of forces owing to double layer (short distance) and forces owing to the differences in electrolyte concentration between the plates and outer solution (long distance). Since the chemical potential of·interlayer is greater than outer solution, addition of water causes the plates to drift apart till potential difference reaches equilibrium. The total potential energy is the sum of attractive and repulsive potential energies, which is not necessarily a monotonic function with distance.

20.3.4 Stages of Swelling

The swelling of plate like clay particles can be described by three stages of swelling (Norrish, 1972; Kutilek and Nielsen, 1999). The first stage is when the initial distance between the two plates is less than two nanometers (nm). During this stage, swelling is opposed by the electrostatic attraction between cations and negatively charged layers. Swelling beyond 2 nm is possible provided the hydration energy of cation is more than the energy of attraction. With the addition of water the distance between plates increases in discrete steps associated with each molecular water layer formed between the sheets. The swelling continues to the second stage if monovalent cations are present, however, if divalent and trivalent cations are present, swelling ends at the first stage. In the beginning of the second stage, the bonding of molecules to solid surface continues as swelling process continues. The distance between neighboring sheets rises smoothly up to tens of nm and edge-to-face forces are important for holding sheets together. In the third stage, the sheets are totally separated and form an arrangement caused by edge-to-face and edge-to-edge forces. During drying at first the decrease in volume is equal to volume of water drained (Fig. 20.5) and the degree of

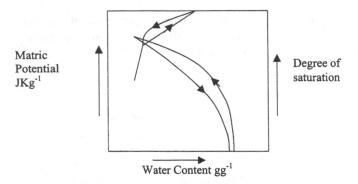

FIGURE 20.5 The stages of drying during the drying process for a block containing 64% clay. (Modified from Holmes, 1955.)

saturation remains fairly constant. However, in the second stage, volume decreases less rapidly than water content as air starts entering the soil.

20.3.5 Flow in Swelling Soils

Darcy's law, Darcy–Buckingham's law, and Richards' equation describe the flow in nonswelling soils. These equations along with continuity and conservation equations are described in Chapters 12 and 13. The theory of flow in swelling soils for a one-dimensional deformation using Darcy's equation needs to be modified and hydraulic conductivity redefined to relate the rate of water flow to the solid phase (Smiles and Rosenthal, 1968; Philips, 1969). Some of the possible reasons, which limit their application on swelling soil are (i) three-dimensional macroscopic volume change, (ii) swelling soils are structured and contain aggregates and voids and are highly permeable when dry and impermeable when wet depending upon the rate at which soil swells and voids close and (iii) profiles are structural and due to self weight, bulk density increases with depth (Smiles, 1981).

In swelling soils during unsteady vertical water flow, the soil solids are in motion, therefore Darcy's equation essentially describes the volume flux of water relative to solid framework for a water content (θ) dependent hydraulic conductivity (K) as follows

$$q = -K(\theta)\frac{\partial \Phi_t}{\partial x} \tag{20.18}$$

The total potential (Φ_t) of vertical system for one-dimensional flow for solute-free water can be written as follows:

$$\Phi_t = \Phi_b + \Phi_{mp} - \Phi_z \tag{20.19}$$

where Φ_b is overburden potential and Φ_{mp} is unloaded moisture potential, $(\Phi_b + \Phi_{mp})$ can be measured by a tensiometer.

The unsteady flow problems can be considered in a framework of physical space or material coordinate. Therefore, instead of Euler's coordinate system, Lagrange's coordinates need to be considered, which is arithmetically more succinct where material coordinate is defined as follows (Smiles and Rosenthal, 1968)

$$\frac{dm}{dx} = \frac{1}{1+e} \tag{20.20}$$

or

$$m = \int_{-\infty}^{x} \frac{dx}{1+e} = \int_{-\infty}^{x} (1 - f_t)\,dx \tag{20.21}$$

where m is cumulative volume of the solid component defined in length scales. Equation (20.21) states that the ratio of material coordinate m to Eulerian coordinate x equals the ratio of the volume of solid phase to total volume of soil. The e is void ratio, which is the ratio of volume of void and volume of solid. For saturated soils, $e = \theta_s$ and the equation of continuity of water can be written as follows:

$$\left[\frac{\partial \theta_s}{\partial t}\right]_m = -\left[\frac{\partial q}{\partial m}\right]_t \tag{20.22}$$

Equations (20.18), (20.19), (20.20), and (20.22) can be combined to yield a flow equation for swelling soils. The combined equation can be solved for known $K(\theta)$, $\Phi(\theta)$ relationship and e and θ values for a given f_t.

20.3.6 Measurement of Swelling

Assuming that soil in the field is homogeneous, saturated, and has one-dimensional drainage and compression, the test soil core collected from the field is trimmed and its weight and height are determined. This information is used to obtain the initial volume, initial density, void ratio, water content, and degree of saturation. To measure expansion characteristics, the soil sample is saturated under full load then allowed to expand after that seating load of $0.025\,\mathrm{kg\,cm^{-2}}$ is applied and the initial dial reading on a consolidometer are recorded. Then fill the pan in which the consolidometer stands with water and let the sample saturate. As the sample expands, increase the load as required holding the sample at its original height. Then gradually (once every 24 hours or longer) reduce the load to 1/2, 1/4, and 1/8 of maximum load, and finally, to the seating load. Measure the height of the soil sample with each load.

20.4 SALT-AFFECTED SOILS

Salt-affected soils have a high concentration of soluble salts. Such soils are of three types: saline soils, alkaline soils, or saline–alkaline soils. Saline soils are those which have an electrical conductivity (EC) of the saturation soil extract of more than $4\,dSm^{-1}$ at 25°C, exchangeable sodium percentage (ESP) less than 15, and pH about 8.5 (Richards, 1954). The ESP is the characteristic of alkalinity or sodic soils. Saline–alkali soils have EC greater than $4\,dSm^{-1}$ and ESP greater than 15. Electrical conductivity is measured in units of Siemens m^{-1} (Sm^{-1}). One $S\,m^{-1}$ equals one mho m^{-1}. These soils require treatment with gypsum to reduce high ESP before leaching and prevent dispersion. Nonsaline alkali soils have ESP greater than 15, salinity less than $4\,dSm^{-1}$, and high pH (8.5–10). These soils are also known as solonetz or black alkali. The EC value of $4\,dSm^{-1}$ is used worldwide for defining salinity although the terminology committee of the Soil Science Society of America has lowered the boundary between saline and nonsaline soils to $2\,dSm^{-1}$ in the saturation extract (Abrol et al., 1988). Excess salts keep the clay in a flocculated state and the soils have better structure and permeability to water than nonsaline soils (Abrol et al., 1988). Salt-affected soils may be primary or secondary. Primary salinization is caused by natural factors (e.g., parent material, drainage, ground water quality, etc.). In contrast, secondary salinization is caused by anthropogenic factors such as excessive irrigation with poor quality water and inadequate drainage (Fig. 20.6).

Saline soils cover about 190 Mha areas worldwide and about 0.13 Mha in North America (Oldeman, 1994· http://www.fao.org) (Fig. 20.7). Soluble salts present in saline soils are the chlorides and sulfates of sodium, calcium, magnesium, and nitrates, with sodium and chloride being the dominant ions, calcium and magnesium basically enough to meet the nutritional needs of crops. Saline soils show a considerable diversity in hydrological, physical, and chemical properties and can be calcium, sodium, or magnesium dominated, with a tendency toward structural degradation (which depends on the presence or absence of calcium). Under low rainfall and high evaporation conditions, salts present in the soil solution precipitates as white efflorescence, salt crusts, nonaggregated brown powder, black salt deposits, evaporative salt crystals, etc. Many saline soils contain high amount of gypsum $(CaSO_4, 2H_2O)$ and soluble carbonates (Abrol et al., 1988).

FIGURE 20.6 (a) Flood irrigation without adequate drainage. (b) Irrigation used for cotton-wheat rotation in central Asia may lead to secondary salinization if subsoil has high salts, water is of poor quality, and drainage is inadequate.

Soil salinity is used to designate a condition in which soluble salt concentration of soil reaches a level that is harmful to crops. Moderate salinity can often go undetected because it causes no apparent injuries other than restricted growth. Leaves of plants growing in salt infested areas may be smaller and darker blue to green in color than the normal leaves. Salinity causes increased succulence, especially for a high concentration of chloride ions in the soil solution. The appearance of plants in salt-affected soils and moisture stress (drought) conditions is almost similar. The wilting of plants

FIGURE 20.7   (a) A saline soil with a high water table. (b) Salt accumulation in a poorly drained depressional land in Haryana, India.

is far less prevalent because the osmotic potential of the soil solution usually changes gradually and plants adjust their internal salt content sufficiently to maintain turgor and avoid wilting.

Symptoms of specific element toxicities, such as marginal or tip burn of leaves, occur as a rule only in woody plants. Chloride and sodium ions and boron are the elements most usually associated with toxic symptoms. Nonwoody species may often accumulate as much or more of these elements in their leaves without showing apparent damage, as do the woody species.

20.4.1 Osmotic Pressure

The water molecules are dipole and other ions in the solution are attracted to them by the electric field to form clusters. The presence of solutes affects the thermodynamic properties of water and lowers the potential energy. Consider two compartments, one containing pure water and the other a solution, that are separated by a membrane permeable to pure water and impermeable to solute. The pure water will continue to cross over into the solution side, unless stopped by an opposing force. If the compartment on the solution side is a flexible diaphragm type, then the pure water entry will expand it. This will result in a rise in hydrostatic pressure that will eventually stop the flow of pure water into the solution compartment. The hydrostatic pressure at equilibrium is known as osmotic pressure (Π, erg cm^{-3}) of a solution and for a dilute solution it is expressed as follows

$$\Pi = C_s RT \tag{20.23}$$

where C_s is concentration of solution (moles cm^{-3}), R is universal gas constant (8.32×10^7 erg moldeg^{-1}), and T is absolute temperature (K). The solute potential differences tend to become uniform through the system owing to the process of diffusion and do not affect the value of any other soil water potential components at equilibrium. However, when membrane or diffusion barriers are present within the soil-water system and solute system, the solute potentials need to be included in the analysis of potentials. In order to further explain the osmotic pressure potentials and other components of soil water potentials, let us consider a vessel filled (Fig. 20.8) with a solution of osmotic pressure (or π) and a capillary tube filled with pure water. One end of the capillary has a perfect semiinfinite membrane, that restricts flow of solute into a capillary but allows flow of water into a capillary. The capillary risestops at equilibrium at height h.

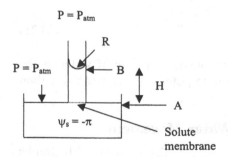

FIGURE 20.8 Schematic of a capillary tube at equilibrium with a solution of osmotic pressure (π).

At point A

The datum $z = 0$ and $P_0 = P_{atm}$

Evaluating the components of total soil water potential (Φ_t)

Gravitational pressure potential: $\Phi_z = 0$ because $z = z_0 = 0$

Air pressure potential: $\Phi_a = 0$ because $P = P_0$

Matric potential: $\Phi_m = 0$ because no soil is present

Hydrostatic pressure potential: $\Phi_p = 0$ because no hydrostatic pressure

Solute potential: $\Phi_\pi = -\pi$ (by definition)

Therefore, $\Phi_\tau = \Phi_z + \Phi_a + \Phi_m + \Phi_p - \Phi_\pi$

Or, $\Phi_t = -\pi$ at point A   (20.24)

At point B

$\Phi_a = 0$ because $P = P_0 = P_{atm}$

$\Phi_s = 0$ because pure water

$\Phi_z = \rho_w g h$

$\Phi_m = P_1 - P_a = -2\sigma/R$ because the contact

angle is zero, σ is surface tension

Therefore, $\Phi_t = \rho_w g h - 2\sigma/R$ (20.25)

Since the soil water system in the vessel and capillary is in equilibrium, Eqs. (20.24) and (20.25) are equal

$$\rho_w g h - 2\sigma/R = -\pi \tag{20.26}$$

or

$$h = 2\sigma/\rho_w g R - \pi/\rho_w g \tag{20.27}$$

The equation shows that due to solute potential the rise of water in the capillary is smaller than without it. If solute potential becomes larger than $2\sigma/R$, there will be no capillary rise.

20.4.2 Effects of Salinity on Water Movement

Darcy's law states that the soil water flux is the product of hydraulic conductivity and the driving force. The driving force consists of gravitational and pressure potentials for solute free soils. For soils containing salts, the osmotic potential gradient is the additional driving force for water movement through a semipermeable membrane by restricting the flow of solutes and at the same time allowing the flow of water. For the situations where solute flow is totally restricted the total hydraulic head is sum of all the three driving forces (i.e., gravitational, pressure, and osmotic) and flux of water (q) for a soil of hydraulic conductivity K can be given as follows

$$q = -K\left(\frac{d\Phi_t}{dx} + \frac{1}{\rho_l g}\frac{d\Phi_\pi}{dx}\right) \tag{20.28}$$

where $\Phi_t = \Phi_m + \Phi_z$, ρ_l is the density of solution, g is acceleration due to gravity, and x is the distance along the direction of flow. When solutes are restricted to movement relative to the water solvent, such a phenomenon is known as salt sieving. For field situations, a total restriction of solute particles from flow seems unrealistic; therefore, an osmotic efficiency factor (F_0) is introduced in Eq. (20.28), which changes to

$$q = -K\left(\frac{d\Phi_t}{dx} + \frac{F_0}{\rho_l g}\frac{d\Phi_\pi}{dx}\right) \tag{20.29}$$

Experimental studies have demonstrated that F_0 is close to zero under saturated conditions. However, for unsaturated conditions at high suction values, F_0 becomes significant and is reported as 0.03 for suction of 0.25 to 1 bar (Letey, 1968). The solutes have a profound influence on the saturated soil hydraulic conductivity (K_s) because aggregates tend to collapse by the dispersion of clay, which also blocks the interaggregate pores, and high exchangeable sodium percentage and low salt concentrations cause swelling and dispersion of clay—both of which ultimately reduce the K_s of soil. The negatively charged clay particles form a diffuse double layer by attracting cations. When the solution concentration is less than 200–400 meql^{-1}, this process of imbibition causes swelling in soils, which reduces the osmotic pressure difference between the soil solution (or more appropriately ambient solution, which is the soil solution away from soil particles) and clay particle, and weakens interparticle bond (McNeal, 1974). This results in dispersion of clay and reduction in K_s of soil.

20.4.3 Leaching Requirement

In arid regions where irrigation with water containing salts is applied to crops, the twin processes of evaporation and transpiration results in rise in salt concentration in the root zone. On the other hand, if a shallow groundwater table exists in the area, then salt is brought in the root zone by the process of capillary rise. The excess salt present in the soil is removed by leaching, which is a process in which the optimal quantity of water equal to the leaching requirement is applied to the field and allowed to flow through and past the root zone so that excess salts are removed (Richards, 1954). Leaching may result in a slight increase in soil pH by lowering of salt concentration, but saline soils rarely become strongly sodic upon leaching. Unless the water table is very deep and lateral movement of water fast, the process of leaching can cause water table buildup. Therefore, an adequate drainage system is a necessity for leaching. Leaching requirement (LR) is defined as the fraction of irrigation water that must be leached out from the bottom of root zone to keep soil salinity level within a specific limit (usually $4\,dSm^{-1}$). LR depends on the evapotraspiration, salt tolerance of crops, and salt content of soil profile and irrigation water. The LR can be obtained by first making a salt balance, which is the total salt input and output for a given volume or depth of soil as follows:

$$\rho_w\left(V_r c_r + V_i c_i + V_g c_g\right) + M_s + M_a - \left(M_p + M_c + \rho_w V_s c_s\right) = \Delta M$$

$$(20.30)$$

where V and c are the volume of water entering or leaving the soil root zone (per unit surface area or equivalent depth) and concentration (EC), respectively, subscript r, i, and g are for rainfall, irrigation, and groundwater, respectively. M_s and M_a are the mass of salts from soil and soil amendment or fertilizers, M_p and M_c are mass of salt precipitated and removed by crop, V_s and c_s is the volume of water drained from soil and concentration, respectively, and ΔM is the total change in mass of salt. Disregarding the changes in salt balance in soil profile by precipitation, agricultural inputs, evapotraspiration, drainage, and groundwater or capillary rise, Eq. (20.22) is simplified as follows:

$$V_i c_i = V_s c_s \tag{20.31}$$

Equation (20.31) is for the steady state conditions where water content and salinity of soil profile is constant, and no precipitation or dissolution of salt is taking place.

$$V_s = V_i - V_{ET} \tag{20.32}$$

where V_{ET} is volume of evapotranspiration. Transferring Eq. (20.24) into (20.23)

$$V_i c_i = (V_i - V_{ET})c_s \tag{20.33}$$

or

$$V_i = \left[\frac{c_s}{c_s - c_i}\right] V_{ET} \tag{20.34}$$

or in terms of depth of irrigation water (d_i), equivalent depth of evapotranspiration by crop (d_{ET}), EC of drainage (s) and irrigation water (i), the equation can be written as follows (Richards, 1954)

$$d_i = \left[\frac{EC_s}{EC_s - EC_i}\right] d_{ET} \tag{20.35}$$

Equation (20.35) suggests that by varying the amount of water for leaching the concentration of salts in root zone can be reduced to the desired level.

20.5 SOIL WATER REPELLENCY

Water repellency is defined as a phenomenon of repulsion of water by soil particles. Soil hydrophobicity, also called "water repellency or non-wetting," reduces the affinity of soil for water. Hydrophobicity can reduce the infiltration capacity of a soil to the extent that the soil does not wet up even after weeks of being in contact with water. This can lead to inhibited plant growth, increased overland flow and accelerated soil erosion, uneven wetting patterns, and preferential flow generation. Hydrophobicity is known to vary temporally, being generally most extreme after long dry periods and reduced or absent after long wet spells (DeBano, 2003).

Water repellency is mostly associated with organic matter and its decomposition, particularly where fungi growth is involved. Exudates and biomass produced by plant roots and soil microbes can alter the surface characteristics of soil particles and lead to the development of hydrophobic particle surfaces that may reduce water transport and retention. This type of organic coating does not necessarily require covering the entire soil particle; just a partial covering can render it water-repellent. The degree of soil hydrophobicity is most severe at the soil surface and within the top 5 cm of soil profile, but can be as deep as 15 cm or have patches of a hydrophobic layer within the soil profile (DeBano, 2003).

Water repellency has been a concern for both land managers and researchers since the early part of the twentieth century. It is a soil property with important repercussions for plant growth, surface and subsurface hydrology, and soil erosion. It is generaly confined to coarse-textured soils in regions with specific vegetation types and seasonally dries climate and/or areas affected by fire. However, research conducted during the 1980s and early 1990s showed that its occurrence is far more widespread. Water repellency can occur at much lower levels or a localized scale in soil profile and can contribute to preferential flow of water and nutrients. At low levels, repellency may not have a deleterious impact on water retention and may even enhance microbial diversity through the preferential alteration of soil pores by organisms. The hydrophobic substances causing water repellency are also beneficial for conserving water by reducing the capillary rise of water and the attendant evaporation, and leaching of nutrients. The water repellency is often characterized in terms of wetting coefficients ($C_w = \cos\theta$; where C_w is wetting coefficient and θ is contact angle) (Bahrani et al., 1970), surface roughness (Bond and Hammond, 1970), and water surface tension and water–solid contact angles (refer to Chapter 9; Watson et al., 1971).

20.5.1 Wetting Pattern in Water-Repellent Soils

Dry soils are wetted when water is applied to them. A drop of water disappears and wets soil because the force of attraction between soil particles and water results in loss of cohesion in the latter, which lets it flow along the surfaces of particles. Once the attractive forces between soil and water droplet are nonexistent, water remains as a droplet and does not wet the soil. Before water starts infiltrating uniformly or percolating inside the soil matrix, the presence of a continuous film of water over soil particle surface is a prerequisite (Fig. 20.9).

The fundamental principle underlying the process of wetting shows that a reduction in the surface tension of a solid (to be wetted) reduces its

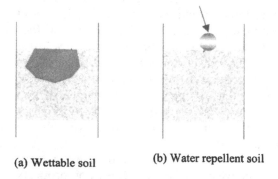

(a) Wettable soil   (b) Water repellent soil

FIGURE 20.9 Applied water droplets make a film of water percolating in the soil matrix due to suction gradients in (a) wettable soil. It retains its shape as a droplet in a (b) water-repellent soil.

TABLE 20.2 Water Repellency Classes

WDPT (seconds)	Repellency class	Water repellency
<5	0	Non-repellent
5–60	1	Slightly repellent
60–600	2	Strongly repellent
600–3600	3	Severely repellent
>3600	4	Extremely repellent

Source: Modified from Dekker and Ritsema, 2003.

wettability, or a reduction in the surface tension of applied liquid increases the wettability. The common method of classifying the water repellency is the empirical water drop penetration time (WDPT). In this method, three drops of deionized water are placed on a smoothened soil surface and the time over which drops are completely absorbed is recorded. The time required for the drops to be absorbed depends on the temperature of water and relative humidity of air. The increase in water temperature reduces the surface tension and the time required for wetting. The increase in the relative humidity of air increases the time for which the drops remain on the soil surface. The water repellency classification given in Table 20.2 shows that soil is considered water-repellent for WDPT > 5 seconds, (Dekker and Jungerius, 1990; Dekker and Ritsema, 2003).

20.5.2 Effects of Water Repellency on Soil Processes

Water Infiltration

A water repellent soil does not get wet when water is applied under zero or negative potential because contact angle is greater than 90°. Thus, a positive pressure must be applied to force the entry of water into a soil. The value of the positive pressure depends on the contact angle as well as pore dimension, and it increases with the contact angle and decreases with the pore radius (Feng et al., 2001). The water content and the attendant water pressure potential diagram (Fig. 20.10) with respect to time show that non-water-repellent sand has a stable Richards-type imbibing front, slightly less saturated than the total porosity of soil. The matric potential at the imbibing front for non-water-repellent sand is negative. For a water-repellent soil, the matric potential behind the imbibing front is slightly positive (Fig. 20.10b).

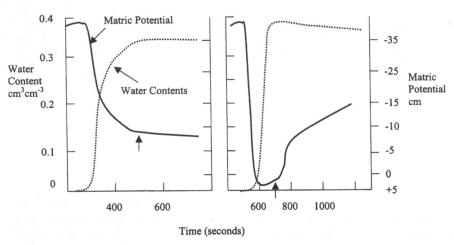

FIGURE 20.10 Schematic of matric potential and moisture content for (a) wettable and (b) water-repellent soil. The arrow points out to a negative matric potential for wettable soil and positive potential for repellent soil at imbibing front. (Redrawn from Bauters et al., 2003.)

In water-repellent soils, water movement is severely limited and the infiltration rates are low. Most of the rainfall falling in a dry water-repellent soil may be lost as runoff. However, as the dry water-repellent soil becomes wetter, the infiltration and water movement gradually increases. The main difference between a hydrophilic and hydrophobic soil is the shape of the wetting front. Infiltrating water in hydrophilic soils forms an unconditionally stable horizontal Richards-type wetting front (refer to Chapter 13), whereas a hydrophobic soil forms an unstable front with fingers (Fig. 20.11). Therefore, water distribution in the soil can have large variability with high water content in the ectorganic layer (also known as "humus," which protects the soil from erosion, while enhancing aggregation) beneath which there can be a dry water-repellent layer, which may be underlaid by a moist, less hydrophobic layer. The hydraulic conductivity of water-repellent soil increases with depth of ponding (Carrillo et al., 2000).

Preferential Flow

The preferential transport of water and solutes can take place through soil matrix via cracks formed in well-structured soils (such as clay or peat) due to shrink–swell mechanism or biopores formed by soil fauna or the channels left behind by decayed roots. In nonstructured sandy soils, the preferential

FIGURE 20.11   Irregular wetting in a water-repellent soil.

flow can occur due to the formation of unstable wetting fronts, which can grow into fingers because of the lateral diffusion (Ritsema et al., 1998).

Soil Erosion

Water repellence may exacerbate soil erosion on sloping land by contributing to surface runoff. In crop or pasture situations it causes an uneven distribution of water through the surface soil horizons, which may result in patchy growth and reduced yield as seeds fail to germinate in dry spots. This poor seedling emergence and crop growth exposes the soil to wind and water, accentuating the risks of soil erosion.

20.5.3   Effect of Wild Fire on Water Repellency

A thin layer of soil at or below the mineral soil surface can become hydrophobic because of high temperature following an intense heating. The hydrophobic layer is formed as the result of a waxy substance that is derived from plant material burned during an intense fire. The waxy substance penetrates into the soil as a gas and solidifies after cooling, forms a coating around soil particles, and appears similar to nonhydrophobic layers. Plant leaves, twigs, branches, and needles form a layer of litter and duff on the forest floor and shrubs accumulating the hydrophobic substances in this layer. During an intense fire, these substances move into the mineral soil. Some soil fungi also excrete substances, which render both litter and surface layer water-repellent. However, not all wildfires create a water-repellent layer. Four factors that commonly influence the formation of this layer include: (i) a thick layer of plant litter prior to the fire, (ii) high-intensity surface and crown fires, (iii) prolonged periods of intense heat, and (iv) coarse soil texture.

Very high temperatures are required to produce the gas that penetrates the surface. The gas is forced into the soil by the heat of the fire. Soils that have large pores, such as sandy soils, are more susceptible to the formation of hydrophobic layers because they transmit heat more readily than heavy-textured soils, such as clay. The coarse-textured soils also have larger pores that allow deeper penetration of the gas. The hydrophobic layer is generally 2 to 8 cm beneath the soil surface and can be 2 to 3 cm thick. The continuity and thickness of the layer vary across the landscape (USDA NRCS Soil Quality Institute, 2000).

20.5.4 Management of Water-Repellent Soil

Management options for water-repellent soils can be broadly classified as adaptation strategies and avoidance strategies (Blackwell, 2003). Some plant species and cultivars are naturally adapted to water-repellent soils. Conversion to a no-till method of seedbed preparation can reduce the drying of soil surface under the mulch, increase porosity by the root network from previous crops, and enhance infiltration of rainwater and water redistribution in the soil profile. Water repellency in soils can be also be reduced or masked by the addition of materials with a high surface area, e.g., clay. Dispersed clay due to wetting exposes a greater surface area of hydrophilic clay surface and the drying process binds sand and clay particles due to the surface tension masking the hydrophobic organic coated surfaces and reducing water repellency (Ward and Oats, 1993). Wetting agents, such as common household detergents or surfactants, can also be used for amelioration and management of soil water repellency. The performance of a wetting agent is influenced by the degree of water repellency. Wetting agents have a strong affinity for the hydrophobic soil and are strongly absorbed on the surface of hydrophobic soil particles, increasing water infiltration and ameliorating the soil (Dekker et al., 2003). Water-repellent soils can be avoided by either digging up the surface layer and planting seed on nonrepellent subsoil or totally removing water-repellent topsoil and exporting them to other areas.

20.6 SCALING METHODS IN SOIL PHYSICS

Knowledge of soil's physical properties is important to understand and manage soil resources at scales ranging from pedon to plot to global. Most of the sampling for measuring soil properties is done using cores and monoliths on a small scale with an objective to reconstructing soil physical properties across a wider scale such as a field or watershed. The multiscale characterization of soil physical properties and processes is useful as a research topic and as a practical tool for data assimilation. Scaling of soil physical properties is a useful tool to integrate chemical, biological, and physical properties and processes affecting soil quality and the environment. "Scaling" refers to developing/formulating/identifying relationships between soil physical properties (data) at different scales. The term "upscaling" is used when soil properties are integrated at a larger scale than the one for which data is available, and "downscaling" is the opposite of upscaling. The scaling theories can be applied using similar media concepts, which imply that the two or more soil systems are similar and the properties of one can be predicted from those of the other by using a single scaling factor. The single scale factor describes either system exactly relative to the other.

20.6.1 Methods

Application of scaling to the characterization of soil properties is done by several different ways, such as dimensional and similitude analysis, and regression analysis. In the regression analysis scale factors are obtained by minimizing the sum of squares between scaled and measured data points (Warrick et al., 1977).

Ed and Bob Miller, who introduced the concepts of scaling during the 1950s, showed that for constant water content, the matric potential (Φ_m) and unsaturated hydraulic conductivities ($K\theta$) for two similar soil systems (1 and 2) can be explained by a single scale factor as follows (Miller and Miller, 1956).

$$\lambda_1 \Phi_{m1} = \lambda_2 \Phi_{m2} = \lambda_{ave} \Phi_{mave} \tag{20.36}$$

$$K_{\theta 1}/\lambda_1 = K_{\theta 2}/\lambda_2 = K_{\theta ave}/\lambda_{ave} \tag{20.37}$$

The similar media concept is often used to scale the field data for soil water characteristic relationships and unsaturated hydraulic conductivity. Warrick et al. (1977) extended this concept of Miller and Miller for a single scaling factor and introduced another variable known as degree of saturation (s) to scale the volumetric water content. The technique employed was regression analysis, also known as functional normalization, where scale factors were obtained by minimizing the sum of squares between scaled and measured data points. This method is effective in coalescing large sets of data into a very narrow range. Some of the advantages of the regression method are the possibility to include all data points without prior smoothing and to approximate the soil heterogeneity in terms of a single stochastic parameter, i.e., the normalized scaling factor (Russo and Bresler, 1980).

Another method used in soil science to derive scale factors for soil properties is the "dimensional method" also known as "similitude analysis," which is based on physical characteristics of a soil system, and involves reducing the number and complexity of the physical process or a phenomenon. If a phenomenon depends on n dimensional variables, dimensional analysis will reduce the problem to only k dimensionless variables where $n-k = 1$, 2, 3, or 4 depends on the complexity of the problem involved. The similitude analysis is used in experimental research and design and analysis by the means of a model or the correlation of field data (Tillotson and Nielsen, 1984; Shukla et al., 2002.) The physically significant scale factors can be determined by dimensional and inspectional analysis, and empirical scale factors can be obtained by functional normalization (Tillotson and Nielsen, 1984; Sposito and Jury, 1985; Shukla et al., 2002). Scale factors obtained through functional normalization are not necessarily related to those obtained from dimensional techniques. The scaling parameters and similarity groups developed in an inspectional analysis also depend on the boundary and initial conditions imposed, as well as on the special physical hypothesis invoked. Vodel et al. (1991) proposed that instead of assuming a heterogeneous field to be an ensemble of mutually similar homogeneous zones, the spatial variability of soil properties can be assumed to have two components: (i) linear and (ii) nonlinear. Scaling the soil properties with respect to the dominant component can do the functional normalization for such a system.

Macroscopic Miller similarity is based on the physical postulate that viscous flow and capillary forces govern water transport through unsaturated soil. It differs from the classical Miller similar media concept in which a scaling parameter is required for the volumetric water content and makes no direct reference to the geometric structure of a soil at the pore scale. Warrick extended the use of Miller's single scaling factor, introduced the degree of saturation, and eliminated the assumption of identical porosity (Kutilek and Nielsen, 1994). Warrick similarity analysis is formulated independently of the initial and boundary conditions while adopting three scaling parameters to define reduced forms of water content, matric potential, and hydraulic conductivity (Warrick et al., 1977). Nielsen similarity analysis is based on a zero-flux boundary condition and the analysis develops from the physical assumption that the water diffusivity, hydraulic conductivity, and matric potential in unsaturated soil are exponential functions of the volumetric water content (Warrick et al., 1977).

20.6.2 Applications of Scaling Theory

Some of the applications of scaling methods are

1. Description of hierarchical heterogeneities over a broad range of scales

2. Reduction of soil water characterization, unsaturated hydraulic conductivities, and infiltration rates
3. Steady state moisture distribution in heterogeneous fields
4. Diffusion and dispersion in porous media
5. Quantification of variability in hydraulic functions of soils and upscaling
6. Stochastic analysis of soil water regime
7. Water budget modeling

Detailed descriptions on various scaling procedures are given by Simmons et al. (1979), Hillel and Elrick (1990), Sposito (1998) and Papachepsky et al. (2003) among others.

Example 20.1

Calculate the net annual salt balance in the root zone of soil for an area where total annual amount of water was 40 cm by rainfall, 80 cm by irrigation, 15 cm by the total drainage, and 12 cm capillary rise. The concentration of salt was 30 ppm in precipitation, 500 ppm in irrigation, 750 ppm in drainage water, and 900 ppm in capillary water. The total salt input from fertilizer and amendments was 100 gm^{-2}, and the salt content removed by harvested crop was 60 gm^{-2}. Precipitation and dissolution of salt can be neglected.

Solution

Assuming a unit field area (1 cm^2) and bulk density of water as 1 g cm^{-3}, the net salt balance for the soil root zone can be calculated by Eq. (20.30):

$$\Delta M = 1 * (40 * 30 \times 10^{-6} + 80 * 500 \times 10^{-6} + 12 * 900 \times 10^{-6})$$
$$+ 100 \times 10^{-4} - 60 \times 10^{-4} - 1 * 15 * 750 \times 10^{-6} = 4.48 \times 10^{-2} g$$

Since DM is positive, it indicates that soil is accumulating salts at a rate of 4.48 g cm^{-2}yr^{-1}.

Example 20.2

Calculate the leaching requirement and depth of leaching for a field where evapotraspiration is 1000 mm, if EC of irrigation water is 1.2 mmhocm^{-1} and that of drainage water can be as high as 4 mmhocm^{-1}.

Solution

The leaching requirement (d_i) can be calculated by Eq. (20.35):

$$d_i = \frac{4}{4 - 1.2} * 1000 = 1428.6 \, \text{mm}$$

The leaching depth $= d_i - d_{ET} = 1428.6 - 1000 = 428.6 \, \text{mm}$

Example 20.3

Calculate the total osmotic pressure of a $0.01 \, M$ (or $10^{-5} \, \text{mol cm}^{-3}$) solution of chloride and KCl at 27°C.

Solution

The total osmotic potential for chloride solution can be calculated by Eq. (20.23):
$\Pi = 10^{-5} * 8.32 * 10^7 * 300 = 2.5104 \, \text{erg cm}^{-3} = 0.25 \, \text{atm}.$
For KCl it will be $= 2 * 0.25 = 0.5 \, \text{atm}.$

PROBLEMS

1. Calculate the net annual salt balance in the root zone of soil for an area where total annual amount of water was 50 cm by rainfall, 90 cm by irrigation, 25 cm by the total drainage, and 15 cm capillary rise. The concentration of salt was 40 ppm in precipitation, 600 ppm in irrigation, 850 ppm in drainage water, and 1000 ppm in capillary water. The total salt input from fertilizer and amendments was $150 \, \text{gm}^{-2}$, and the salt content removed by harvested crop was $70 \, \text{gm}^{-2}$. Precipitation and dissolution of salt can be neglected. Also indicate whether salt accumulation or release is taking place.

2. Calculate the leaching requirement (LR) and depth of leaching for a field where evapotraspiration is 1200 mm, if EC of irrigation water is $1.5 \, \text{mmhocm}^{-1}$ and that of drainage water can be as high as $4 \, \text{mmhocm}^{-1}$. If EC of drainage water can be doubled, calculate the LR.

References Abrol I.P., Yadav J.S. and Massoud F.I. (1988). Salt effected soils and their management. Food and Agriculture Organization of the United Nations (FAO), Rome.

Anderson D., Brown K.W. and Green J. 1981. Organic Leachate effects on the permeability of clay liners. In: Land disposal of hazardous waste. (ed. EPA) EPA-600/9-81-0021, 119–130.

Ankeny M.D., Kaspar T.C. and Horton R. (1990). Characterization of tillage and traffic effects on unconfined infiltration measurements. Soil Sci. Soc. Am. J. 54:837–840.

Bahrani B., Mansell R.S. and Hammond L.C. (1970). Wetting coefficients for water repellent sand. Soil and Crop Sci. Soc. Florida Proc. 30:270–274.

Benoit G.R. (1973). Effect of freeze–thaw cycles on aggregate stability and hydraulic conductivity of three soil aggregate sizes. Soil Sci. Soc. Am. J. 37:3–5.

Benoit G.R. and Voorhees W.B. (1990). Effect of freeze–thaw activity on water retension, hydraulic conductivity, density, and surface strength of two soils frozen at high water content. In: (ed.) Cooley K.R., Proceedings International Symposium on Frozen Soil Impacts on Agricultural, Range, and Forest Lands. March 21–22, Spokane, US Army Corps of Engineers, CRREL Special Report 90-1, Washington, pp. 45–53.

Beven K. and Germann P. (1982). Macropores and water flow in soils. Water Resour. Res. 18:1311–1325.

Bisel L.D. and Nielsen K.F. (1964). Soil aggregates do not necessarily breakdown over winter. Soil Sci. 98:345.

Bisel L.D. and Nielsen K.F. (1967). Effect of frost action on size of soil aggregates. Soil Sci. 104:268–272.

Black P.B. (1990). Three functions that model empirically measured unfrozen water content data and predict relative hydraulic conductivity. U.S. Army Corps of Engineers, CRREL report 90–5.

Blackwell P.S. (2003). Management of water repellency in Australia. In: (eds.) Ritsema C.J. and Dekker L.W., Soil Water Repellency—Occurrence, Consequences, and Amelioration. Elsevier, Amsterdam, Netherlands, pp. 291–301.

Bond B.D. and Hammond L.C. (1970). Effect of surface roughness and pore shape on water repellency of sandy soils. Wetting coefficients for water repellent sand. Soil and Crop Sci. Soc. Florida Proc. 30:308–315.

Bouma J. and Dekker L.W. (1978). A case study on infiltration into dry clay soil, I. Morphological observations. Geoderma 20:27–40.

Bouma J., Dekker L.W. and Wosten J.H.M. (1978). A case study on infiltration into dry clay soil, II. Physicsl measurements. Geoderma 20:41–51.

Boynton, A.M. and Daniel D.E. (1985). Hydraulic conductivity tests on compacted clay. Journal of Geotechnical Engineering 111:465–478.

Briones A.A. and Uehara G. (1977). Soil elastic constants: II. Application to analysis of soil cracking. Soil Sci. Soc. Am. J. 41:26–29.

Bullock M.S., Kemper W.D. and Nelson S.D. (1988). Soil cohesion as effected by freezing, water content, time and tillage. Soil Sci. Soc. Am. J. 52(3):770–776.

Burst J.F. (1965). Subaqueously formed shrinkage cracks in clays. Journal of Sedimentary Petrology 35:891–901. D'Appolonia (1980).

Campbell G.S. (1974). A simple method for determining unsaturated conductivity from moisture retension data. Soil Sci. 117:311–314.

Carrillo M.L.K., Letey J. and Yates S.R. (2000). Unstable water flow in a layered soil: I. The effects of stable water repellent layer. Soil Science Soc. Am. J. 64: 456–459.

Cary J.W. and Mayland H.F. (1972). Salt and water movement in unsaturated frozen soil. Soil Sci. Soc. Am. J. 36:549–555.

Cary J.W., Papendick R.I. and Campbell G.S. (1979). Water and salt movement in unsaturated frozen soil: principles and field observations. Soil Sci. Soc. Am. J. 43:3–8.

Culley, J.L.B., Larson W.E. and Randall G.W. (1987). Physical properties of a Typic Haplaquoll under conventional and no tillage. Soil Sci. Soc. Am. J. 51:1587–1593.

D'Appolonia D. (1980). Soil bentonie slurry trench cut-offs. Journal of Geotechnical Engineering 106:399–418.

Dasog G.S. and Shashidhara G.B. (1993). Dimension and volume of cracks in a vertisol under different crop covers. Soil Science 156:424–428.

DeBano L.F. (2003). Historical overview of soil water repellency. In: (eds.) Ritsema C.J. and Dekker L.W., Soil Water Repellency—Occurrence, Consequences, and Amelioration. Elsevier, Amsterdam, Netherlands, pp. 3–24.

Dekker L.W. and Ritsema C.J. (2003). Wetting patterns in water repellent Dutch soils. In: (eds.) Ritsema C.J. and Dekker L.W., Soil Water Repellency—Occurrence, Consequences, and Amelioration Elsevier, Amsterdam, Netherlands, pp. 151–166.

Dekker L.W. and Jungerius P.D. (1990). Water repellency in the dunes with special reference to Netherlands. Catena Suppl. 18:173–183.

Dekker L.W., Oostindie K., Kostka S.J. and Ritsema C.J. (2003). Treating water repellent surface layer with surfactant. In: (eds.) Ritsema C.J. and Dekker L.W., Soil Water Repellency—Occurrence, Consequences, and Amelioration. Elsevier, Amsterdam, Netherlands, pp. 281–289.

Dexter A.R. 1978. Tunnelling in soil by earthworms. Soil Biol. Biochem. 10:447–449.

Edwards et al. (1988). Edwards W.M., Shipitalo M.J. and Norton L.D. (1988). Contribution of macroporosity to infiltration into a continuous corn no-tilled watersheds: Implications for contaminant movement. J. Contam. Hydrol. 3: 193–205.

Feng G.L., Letey J. and Wu L. (2001). Water ponding depths affect temporal infiltration rates in a water repellent sand. Soil Sci. Soc. Am. J. 65:315:320.

Fuller W.H. and Warrick A.W. (1985). Soils in Waste Treatment and Utilization. Volume II, Boca Raton, CRC Press Inc.

Gardner R. (1945). Some effects of freezing and thawing on aggregation and permeability of dispersed soils. Soil Sci. 60:437–443.

Gerke H.H. and van Genuchten M.Th. (1993). A dual-porosity model for simulating the preferential movement of water and solutes in structured porous media. Water Resour. Res. 29(2):305–319.

Gish T.J. and Jury W.A. (1983). Effect of plant roots and root channels on solute transport. Transactions of the ASAE 26:440–444, 451.

Godwin R.J., Spoor G. and Leeds-Harrison P. (1981). An experimental investigation into the force, mechanics and resulting soil disturbance of mole ploughs. J. Agr. Eng. Res. 26:477–497.

Granger R.J., Gray D.M. and Dyck G.E. (1984). Snowmelt infiltration to frozen Prairie soils. Can. J. Earth Sci. 21:669–677.

Harlan R.L. (1973). Analysis of coupled heat-fluid transport in partially frozen soil. Water Resour. Res. 9(5):1314–1323.

Harris, G.L., Nicholls P.H., Bailey S.W., Howse K.R. and Mason D.J. (1994). Factors influencing the loss of pesticides in drainage from a cracking clay soil. J. Hydrol. 159:235–253.

Hettiaratchi J.P.A., Hrudey S.E., Smith D.W. and Sego D.C.C. (1988). Shrinkage behavior of clay liner material exposed to simulated municipal waste landfill leachate. Can. J. Civ. Eng. 15:500–508.

Hettiaratchi J.P.A., Hrudey S.E., Smith D.W. and Sego D.C.C. (1987). A procedure for evaluating municipal solid waste leachate components capable of causing volume shrinkage in compacted clay soils. Envir. Tech. Lett. 9:23–34.

Hillel D. and Elrick D.E. (ed.) (1990). Scaling in Soil Physics: Principles and Applications. SSSA Spec. Publ. 25. SSSA, Madison, WI.

Hole F.D. (1981). The effects of animals on soil. Geoderma 25:75–112.

Isensee A.R., Nash R.G. and Helling C.S. (1990). Effect of conventional vs. no-tillage on pesticide leaching to shallow groundwater. J. Environ. Qual. 19: 434–440.

Iwata S.T.T. and Warkentin B.P. (1988). Soil Water Interactions. M. Dekker, New York and Basel.

Jalalifarahani H.R., Heerman D.F. and Duke H.R. (1993). Physics of surge irrigation II. Relationship between soil physical and hydraulic parameters. Transactions of the ASAE 36:45–50.

Johnson W.C. (1962). Controlled soil cracking as a possible means of moisture conservation on wheatlands of the southwestern Great Plains. Agron. J. 54: 323–325.

Jury A.W., Gardner W.R. and Gardner W.H. (1991). Soil Physics. 5th ed., John Wiley, New York.

Kutílek M. and Nielsen D.R. (1991). Soil Hydrology. Catena Verlag, 38162 Cremlingen-Destedt, Germany.

Kutilek M. and Nielsen D.R. (1994). Soil hydrology. GeoEcology Textbook. Catena Verlag, Cremlingen-Destedt, Germany.

Lehrsch G.A., Sojka R.E., Carter D.I. and Jolley P.M. (1990). Effects of freezing on aggregate stability of soils differing in texture, mineralogy, and organic matter. In: (ed.) Cooley K.R., Proceedings International Symposium on Frozen Soil Impacts On Agricultural, Range, and Forest Lands. March 21–22, Spokane, US Army Corps of Engineers, CRREL Special report 90–1, Washington, pp. 61–69.

Leo M.W.M. (1963). Effect of freezing and thawing on some physical properties of soils as related to tomato and barley plants. Soil Sci. 96:267–274.

Letey J. (1968). Movement of water through soil as influenced by osmotic pressure and temperature gradients. Hilgardia 39:405–418.

Low P.F., Hoekstra P. and Anderson D.M. (1968). Some thermo-dynamic relationships for soils at or below the freezing point. 2. Effects of temperature and pressure on unfrozen soil water. Water Resour. Res. 4(3):541–544.

Mailhol J.C. and Gonzalez J.M. 1993. Furrow irrigation model for real-time applications on cracking soils. Journal of Irrigation and Drainage Engineering 119:768–783.

McCoy E.L., Boast C.W., Stehouwer R.C. and Kladivko E.J. (1994). Macropore hydraulics: Taking a sledgehammer to classical theory. In: (ed.) Lal R. and Steward B.A., Soil Processes and Water Quality. Boca Raton, Fla.: Lewis Publishers, pp. 303–348.

McNeal B.L. (1974). Soil salts and their effects on water movement. In: (ed.) van Schilfgaarde J., Drainage for Agriculture, Monograph 17 Am. Soc. Agron. Madison, Wisconsin.

Miller E.E. and Miller R.D. (1956). Physical theory for capillary flow phenomenon. J. Appl. Phys. 27:324–332.

Mitchell A.R. and van Genuchten M.Th. (1992). Shrinkage of bare and cultivated soil. Soil Sci. Soc. Am. J. 56:1036–1042.

Mohanty B.P. Bowman R.S., Hendrickx J.M.H., Simunek J. and van Genuchten M.T. (1998). Preferential transport of nitrate to a tile drain in an intermittent-flood-irrigated field: Model development and experimental evaluation. Water Resour. Res. 34(5):1061–1076.

Mostaghimi S., Young R.A., Wilts A.R. and Kenimer A.L. (1988). Effects of frost action on soil aggregate stability. Transactions of the ASAE 31:435–439.

Norrish K. (1972). Forces between clat particles. Proc. Intl. Clay Conf., Madrid, pp. 375–383.

Papachepsky Y., Radcliffe D.E. and Selim H.M. (ed.) (2003). Scaling Methods in Soil Physics. CRC Press, New York.

Philips J.R. (1969). Hydrostatics and hydrodynamics in swelling soils. Water Resour. Res. 5:1070–1077.

Richards L.A. (ed.) (1954). Diagnosis and Improvement of Saline and Alkali Soils. U.S. Dept. Agr. Handbook, p. 60.

Richardson S.J. (1976). Effect of artificial weathering cycles on the structural stability of a dispersed silt soil. J. of Soil Sci. 27(3):287–294.

Ritsema C.J., Nieber J.L., Dekker L.W. and Steenhuis T.S. (1998). Modelling and field evidence of finger formation and finger recurrence in a water repellent sandy soil. Water Resourc. Res. 34:555–567.

Shukla M.K., Kastanek F. and Nielsen D.R. (2002). Inspectional analysis of convective-dispersion equation and application on measured BTCs. Soil Sci. Soc. Am. J. 66(4):1087–1094.

Sillanpaa M. and Webber L.R. (1961). The effect of freezing-thawing and wetting-drying cycles on soil aggregation. Can. J. Soil Sci. 41:182–187.

Simmons C.S., Nielsen D.R. and Biggar J.W. (1979). Scaling of field-measured soil water properties. Hilgardia, J. Agri. Sci. UCDAVIS, 47(4):77–173.

Smiles D.E. (1981). Water relations of cracking soils. In: (ed.) McGarity J.W. et al. Proceedings of Symposium on: The properties and Utilization of Cracking Clay Soils. August 24–28, pp. 143–149.

Smiles D.E. and Rosenthal M.J. (1968). The movement of water in swelling material. Aust. J. Soil Res. 6:237–248.

Sposito G. (1998). Scale dependence and scale invariance in hydrology. Cambridge University Press, Cambridge, New York.

Sposito G. and Jury W.A. (1985). Inspectional analysis in the theory of water flow through unsaturated soil. SSSAJ, 49(4):791–798.

Steenhuis, T.S, Staubitz W., Andreni M.S., Surface J., Richard T.L., Paulsen R., Pickering N.B., Hagerman J.R. and Geohring L.D. 1990. Preferential movement of pesticides and tracers in agricultural soils. J. Irrig. Drain. Engg. 116, 50–66

Swartz, G.L. 1966. Modification of the cracking pattern on a black earth of the Darling Downs, Queensland. Qd. Agr. J. 23:279–285.

Thunholm B. and Ludin L.C. (1990). Infiltration into a seasonally frozen clay soil. In: (ed.) Cooley K.R., Proceedings International Symposium on Frozen Soil Impacts on Agricultural, Range, and Forest Lands. March 21–22, Spokane, US Army Corps of Engineers, CRREL Special report 90–1, Washington, pp. 156–160.

Tillotson P.M. and Nielsen D.R. (1984). Scale factors in soil science. SSSAJ 48(5):953–959.

USDA NRCS Soil Quality Institute (2000). Soil Quality Resource Concerns: Hydrophobicity. Available at http://www.statlab.iastate.edu/survey/SQI/. The site was visited on 29 July 2003.

Vogel T., Cislerova M. and Hopman J.W. (1991). Porous media with linearly variable hydraulic properties. Water Res. Res. 27:2735–2741.

Warrick A.W., Mullen G.J. and Nielsen D.R. (1977). Scaling field-measured soil hydraulic properties using a similar media concept. Water Resources Research, 13(2):355–362.

Watson C.L., Letey J. and Mustafa M.A. (1971). The influence of liquid surface tension and liquid solid contact angle on liquid entry into porous media. Soil Sci. 178–183.

Yong R.N. and Warkentin B.P., (1975). Soil Properties and Behaviours. Elsevier, Amsterdam.

Appendix A

The Greek Alphabet

A	α	alpha
B	β	beta
Γ	γ	gamma
Δ	δ	delta
E	ε	epsilon
Z	ζ	zeta
H	η	eta
Θ	θ	theta
I	ι	iota
K	κ	kappa
Λ	λ	lambda
M	μ	mu
N	ν	nu
Ξ	ξ	xi
O	o	omicron
Π	π	pi
P	ρ	rho
Σ	σ	sigma
T	τ	tau
Y	υ	upsilon
Φ	ϕ	phi
X	χ	chi
Ψ	ψ	psi
Ω	ω	omega

Appendix B

Mathematical Signs and Symbols

$\propto$	is proportional to
$=$	is equal to
$\approx$	is approximately equal to
$\neq$	is not equal to
$>$	is greater than
$\gg$	is much greater than
$<$	is less than
$\ll$	is much less than
$\leq$	is less than or equal to
$\geq$	is greater than or equal to
Σ	sum of
$\bar{x}$	average value of x
Δx	change in x
$\Delta x \to 0$	Δx approaches zero
$n!$	$n(n-1)(n-2)$

Appendix C

Prefixes for SI Units

Factor	Prefix	Abbreviation
10^{-1}	deci	d
10^{-2}	centi	c
10^{-3}	milli	m
10^{-6}	micro	μ
10^{-9}	nano	n
10^{-12}	pico	p
10^{-15}	femto	f
10^{-18}	atto	a
10	deca	da
10^{2}	hecto	h
10^{3}	kilo	k
10^{6}	mega	M
10^{9}	giga	G
10^{12}	tera	T
10^{15}	peta	P
10^{18}	exa	E

Appendix D

Values of Some Numbers

$\pi = 3.1415927$
$e = 2.7182818$
$\sqrt{2} = 1.4142136$
$\sqrt{3} = 1.7320508$
$\ln 2 = 0.6931472$
$\ln 10 = 2.3025851$
$\log_{10} e = 0.4342945$
$\text{rad} = 57.2957795$

Appendix E

SI Derived Units and Their Abbreviations

Quantity	Unit	Abbreviation	In terms of base units[a]
Force	newton	N	$kg\,m/s^2$
Energy and work	joule	J	$kg\,m^2/s^2$
Power	watt	W	$kg\,m^2/s^3$
Pressure	pascal	Pa	$kg/m\,s^2$
Frequency	hertz	Hz	s^{-1}
Electric charge	coulomb	C	$A\,s$
Electric potential	volt	V	$kg\,m^2/A\,s^3$
Electric resistance	ohm	Ω	$kg\,m^2/(A\,s^3)$
Capacitance	farad	F	$A^2\,s^4/(kg\,m^2)$
Magnetic field	tesla	T	$kg/(A\,s^2)$
Magnetic flux	weber	Wb	$kg\,m^2/(a\,s^2)$
Inductance	henry	H	$kg\,m^2/(s^2\,A^2)$

[a]kg = kilogram (mass), m = meter (length), s = second (time), A = ampere (electric current).

Appendix F

Unit Conversion Factors

| Multiply the U.S. customary unit | | | To obtain the SI unit | |
Name	Symbol	By	Symbol	Name
		ACCELERATION		
feet per second squared	ft/s²	0.3048	m/s²	meters per second squared
inches per second squared	in./s²	0.0254	m/s²	meters per second squared
		AREA		
acre	acre	0.4047	ha	hectare
acre	acre	4.0469×10^{-3}	km²	square kilometer
Square foot	ft²	9.2903×10^{-2}	m²	square meter
Square inch	in.²	6.4516	cm²	square centimeter
Square mile	mi²	2.5900	km²	square kilometer
Square yard	yd²	0.8361	m²	square meter
		ENERGY		
British thermal unit	Btu	1.0551	kJ	joule
foot-pound (force)	ft-lb	1.3558	J	joule
horsepower-hour	hp-h	2.6845	MJ	megajoule
kilowatt-hour	kW-h	3600	kJ	kilojoule
kilowatt-hour	kW-h	3.600×10^6	J	joule
watt-hour	W-h	3.600	kJ	kilojoule
watt-second	W-s	1.000	J	joule
		FORCE		
pound force	lb	4.4482	N	Newton
		FLOW RATE		
cubic foot per second	ft³/s	2.8317×10^{-2}	m³/s	cubic meters per second
gallons per day	gal/d	4.3813×10^{-2}	L/s	liters per second

(continued)

Multiply the U.S. customary unit		By	To obtain the SI unit	
Name	Symbol		Symbol	Name
gallons per day	gal/d	3.7854×10^{-3}	m^3/d	cubic meters per day
gallons per minute	gal/min	6.3090×10^{-3}	m^3/s	cubic meters per second
gallons per minute	gal/min	6.3090×10^{-2}	L/s	liters per second
million gallons per day	Mgal/d	43.8126	L/s	liters per second
million gallons per day	Mgal/d	3.7854×10^3	m^3/d	cubic meters per day
million gallons per day	Mgal/d	4.3813×10^{-2}	m^3/s	cubic meters per second
		LENGTH		
foot	ft	0.3048	m	meter
inch	in.	2.54	cm	centimeter
inch	in.	0.0254	m	meter
inch	in.	25.4	mm	millimeter
mile	mi	1.6093	km	kilometer
yard	yd	0.9144	m	meter
		MASS		
ounce	oz	28.3495	g	gram
pound	lb	4.5359×10^2	g	gram
pound	lb	0.4536	kg	kilogram
ton (short: 2000 lb)	ton	0.9072	Mg (Metric ton)	megagram (10^3 kilogram)
ton (long: 2240 lb)	ton	1.0160	Mg (Metric ton)	megagram (10^3 kilogram)
		POWER		
British thermal units per second	Btu/s	1.0551	kW	Kilowatt
foot-pounds (force) per second	ft.lb/s	1.3558	W	watt
horsepower	hp	0.7457	kW	Kilowatt

PRESSURE (FORCE/AREA)

atmosphere (standard)	atm	1.0133×10^2	kPa (kN/m²)	kilopascal (kilonewtons per square meter)
inches of mercury (60° F)	in Hg	3.3768×10^3	Pa (N/m²)	pascal (newtons per square meter)
inches of water (60° F)	(60° F) in H_2O	2.4884×10^2	Pa (N/m²)	pascal (newtons per square meter)
pounds (force) per square foot	(60° F) lb/ft²	47.8803	Pa (N/m²)	pascal (newtons per square meter)
pounds (force) per square inch	lb/in.²	6.8948×10^3	Pa (N/m²)	pascal (newtons per square meter)
pounds (force) per square inch	lb/in.²	6.8948	kPa (kN/m²)	Kilopascal (kilonewtons per square meter)

TEMPERATURE

degrees Fahrenheit	°F	$0.555 \, (°F - 32)$	°C	degrees Celsius (centigrade)
degrees Fahrenheit	°F	$0.555 \, (°F + 459.67)$	°K	degrees Kelvin

VELOCITY

feet per second	ft/s	0.3048	m/s	meters per second
miles per hour	mi/h	4.4704×10^{-1}	km/s	kilometers per second

VOLUME

acre-foot	acre-ft	1.2335×10^3	m³	cubic meter
cubic foot	ft³	28.3168	L	liter
cubic foot	ft³	2.8317×10^{-2}	m³	cubic meter
cubic inch	in.³	16.3871	cm³	cubic centimeter
cubic yard	yd³	0.7646	m³	cubic meter
gallon	gal	3.7854×10^{-3}	m³	cubic meter
gallon	gal	3.7854	L	liter
ounce (U.S. Fluid)	oz (U.S. fluid)	2.9573×10^{-2}	L	liter

Source: A. D. Ward and W. J. Elliott (1995) Environmental Hydrology. CRC/Lewis publisher, Boca Raton, FL.

Appendix G

Unit Conversions (Equivalents)

LENGTH

1 in = 2.54 cm
1 cm = 0.394 in
1 ft = 30.5 cm
1 m = 39.37 in = 3.28 ft
1 mi = 5280 ft = 1.61 km
1 km = 0.621 mi
1 nautical mile (U.S.) = 1.15 mi = 6076 ft = 1.852 km
1 fermi = 1 femtometer (fm) = 10^{-15} m
1 angstrom (Å) = 10^{-15} m
1 light-year = 9.46×10^{15} m
1 parsec = 3.26 light years

TIME

1 day = 8.64×10^4 s
1 year = 3.156×10^7 s

SPEED

1 mi/h = 1.47 ft/s = 1.609 km/h = 0.447 m/s
1 km/h = 0.278 m/s = 0.621 mi/h
1 ft/s = 0.305 m/s = 0.682 mi/h
1 m/s = 3.28 ft/s = 3.60 km/h
1 knot = 1.151 mi/h = 0.5144 m/s

ANGLE

1 radian (rad) = 7.30 = 57°18'
1 = 0.01745 rad
1 rev/min (rpm) = 0.1047°rad/s

MASS

1 atomic mass unit (u) = 1.6605×10^{-27} kg
1 kg = 14.6 kg
1 kg = 0.0685 slug
(1 kg has a weight of 2.20 lb where g = 9.81 m/s^2)

FORCE

1 lb = 4.45 N
1 N = 10^5 dyne = 0.225 lb

ENERGY AND WORK

1 J = 10^7 ergs = 0.738 ft.lb
1 ft.lb = 1.36 J = 1.29×10^{-3} Btu = 3.24×10^{-4} kcal
1 kcal = 4.18×10^3 J = 3.97 Btu
1 Btu = 2.52 cal = 278 ft lb = 1054 J
1 eV = 1.602×10^{-19} J
1 kWh = 3.60×10^6 J = 860 kcal

POWER

1 W = 1 J/s = 0.738 ft.lb/s = 3.42 Btu/h
1 hp = 550 ft.lb/s = 746 W

PRESSURE

1 atm = 1.013 bar = 1.013×10^5 N/m^2 = 14.7 lb/in.2 = 760 torr
1 lb/in.^{2} = 6.90×10^3 N/m^2
1 Pa = 1 N/m^2 = 1.45×10^{-4} lb/in.^{2}

Appendix H

Conversion Factors for Non-SI Units

Unit	Abbreviation	Value
atmosphere	atm	$101325\,\text{Pa}$ (definition)
torr	torr	$133.322\,\text{Pa} = 1/760\,\text{atm}$
atomic mass unit	amu	$1.66054 \times 10^{-27}\,\text{kg}$
bar	bar	$1 \times 10^{5}\,\text{Pa}$
electron volt	eV	$1.602178 \times 10^{-19}\,\text{J}$
poise	P	$0.1\,\text{kg}\,\text{m}^{-1}\text{s}^{-1}$
liter	L	$1 \times 10^{-3}\,\text{m}^3 = 1\,\text{dm}^3$
angstrom	Å	$1 \times 10^{-10}\,\text{m}$
debye	D	$3.335641 \times 10^{-30}\,\text{Cm}$
calorie	cal	$4.184\,\text{J}$ (definition)
inch	in	$0.0254\,\text{m}$ (definition)
pound	lb	$0.4536\,\text{kg}$

Appendix I

Conversion Among Units of Soil Water Potential

1 atmosphere	$= 1.013 \times 10^5 \, \text{N} \text{m}^{-2} = 1.013 \times 10^5 \, \text{Pa} = 101.3 \, \text{kPa}$
	$= 1.013 \, \text{bar}$
	$= 1.013 \times 10^6 \, \text{dynes} \, \text{cm}^{-2}$
	$= 14.7 \, \text{PSI}$ or $2.12 \times 10^3 \, \text{lb} \, \text{ft}^{-2}$
	$= 76 \, \text{cm}$ of Hg or $760 \, \text{mm}$ of Hg or $760 \, \text{torr}$
	$1.03 \times 10^4 \, \text{mm-H}_2\text{O}$ at $4°\text{C}$
1 bar	$= 10^5 \, \text{N} \text{m}^{-2}$
1 dyne cm^{-2}	$= 0.1 \, \text{N} \text{m}^{-2}$
1 Kg cm^{-2}	$= 9.85 \times 10^4 \, \text{N} \text{m}^{-2}$
1 PSI (lb m^{-2})	$= 6.90 \times 10^3 \, \text{N} \text{m}^{-2}$
1 lb m^{-2}	$= 47.9 \, \text{N} \text{m}^{-2}$
1 lb ft^{-2}	$= 47.9 \, \text{N} \text{m}^{-2}$
1 cm-H$_g$	$= 1.33 \times 10^3 \, \text{N} \text{m}^{-2}$
1 mm-Hg	$= 133 \, \text{N} \text{m}^{-2} = 1 \, \text{torr}$
1 mm H$_2$O at $4°\text{C}$	$= 9.81 \, \text{N} \text{m}^{-2}$

Appendix J

Surface Tension of Water Against Air

Temperature (°C)	Surface tension (g, dynes cm^{-1})
−8	77.0
−5	76.4
0	75.6
5	74.9
10	74.22
15	73.49
18	73.05
20	72.75
25	71.97
30	71.18
44	69.56
50	67.91
60	66.18
70	64.40
80	62.60
100	58.90

Appendix K

Density of Water from Air

Temperature (°C)	Density (g cm^{-3})
0	0.99987
3.98	1.0000
5	0.99999
10	0.99973
15	0.99913
18	0.99862
20	0.99823
25	0.99707
30	0.99567
35	0.99406
38	0.99299
40	0.99224
45	0.99025
50	0.98807
55	0.98573
60	0.98324
65	0.98059
70	0.97781
75	0.97489
80	0.97183
85	0.96865
90	0.96534
95	0.96192
100	0.95838

Appendix L

The Viscosity of Water 0°C to 100°C

°C	π (cp)	°C	π (cp)
0	1.787	29	0.8148
1	1.728	30	0.7975
2	1.671	31	0.7808
3	1.618	32	0.7647
4	1.567	33	0.7491
5	1.519	34	0.7340
6	1.472	35	0.7194
7	1.428	36	0.7052
8	1.386	37	0.6915
9	1.346	38	0.6783
10	1.307	39	0.6654
11	1.271	40	0.6529
12	1.235	41	0.6408
13	1.202	42	0.6291
14	1.169	43	0.6178
15	1.139	44	0.6067
16	1.109	45	0.5960
17	1.081	46	0.5856
18	1.053	47	0.5755
19	1.027	48	0.5658
20	1.002	49	0.5561
21	0.9779	50	0.5468
22	0.9548	51	0.5378
23	0.9325	52	0.5290
24	0.9111	53	0.5204
25	0.8904	54	0.5121
26	0.8705	55	0.5040
27	0.8513	56	0.4961
28	0.8327	57	0.4884

(continued)

°C	π (cp)	°C	π (cp)
58	0.4809	80	0.3547
59	0.4736	81	0.3503
60	0.4665	82	0.3460
61	0.4596	83	0.3418
62	0.4528	84	0.3377
63	0.4462	85	0.3337
64	0.4398	86	0.3297
65	0.4335	87	0.3259
66	0.4273	88	0.3221
67	0.4213	89	0.3184
68	0.4155	90	0.3147
69	0.4098	91	0.3111
70	0.4042	92	0.3076
71	0.3987	93	0.3042
72	0.3934	94	0.3008
73	0.3882	95	0.2975
74	0.3831	96	0.2942
75	0.3781	97	0.2911
76	0.3732	98	0.2879
77	0.3684	99	0.2848
78	0.3638	100	0.2818
79	0.3592		

The above table was calculated from the following empirical relationships derived from measurements in viscometers calibrated with water at 20°C (and one atmosphere), modified to agree with the currently accepted value for the viscosity at 20°C of 1.002 cp.

$$0° \text{ to } 20°C: \log_{10}\pi r = \frac{1301}{998.333 + 8.1855(T - 20) + 0.00585(T - 20)^2} - 1.30233$$

(R. C. Hardy and R. L. Cottington, J. Res.NBS 42.573 (1949).)

$$20° \text{to} 100°C: \log_{10}\frac{\pi r}{\pi_{20}} = \frac{1.3272(20 - T) - 0.001053(T - 20)^2}{T + 105}$$

(J. F. Swindells, NBS, unpublished results.)
Source: Handbook of Chemistry & Physics (1988) CRC Press, Boca Raton, FL.

Appendix M

Effect of Temperature of Vapor Pressure, Density of Water Vapor in Saturated Air, and Surface Tension of Water

Temperature (°C)	Vapor pressure of water (mm of Hg)	Mass of water vapor in saturated air (g/m)^3	Surface tension of water (dynes/cm)
−20	0.776	0.892	—
−10	1.950	2.154	—
0	4.579	4.835	75.6
4	6.101	6.330	75.0
5	6.543	6.761	74.9
10	9.209	9.330	74.2
15	12.788	12.712	73.5
20	17.535	17.118	72.7
25	23.756	22.796	72.0
30	31.824	30.039	71.2
40	55.324	50.500	69.6
50	92.510		67.9
75	289.100		63.5
100	760.000		58.9

Appendix N

Osmotic Pressure of Solutions of Sucrose in Water at 20°C

Molality (n)	Molar concentration (c)	Observed osmotic pressure (atm)	Calculated osmotic pressure (dyne cm^{-2})	Calculated osmotic potential (ω) (erg g^{-1})
0.1	0.098	2.59	2.36	−2.364
0.2	0.192	5.06	4.63	−4.638
0.3	0.282	7.61	6.80	−6.812
0.4	0.370	10.14	8.90	−8.915
0.5	0.453	12.75	10.90	−10.919
0.6	0.533	15.39	12.80	−12.822
0.7	0.610	18.13	14.70	−14.726
0.8	0.685	20.91	16.50	−16.529
0.9	0.757	23.72	18.20	−18.232
1.0	0.825	26.64	19.80	−19.835

Appendix O

Constant Humidity

The following table shows % humidity and the aqueous tension at the given temperature within a closed space when an excess of the substance indicated is in contact with a saturted aqueous solution of the given solid phase.

Solid phase	t°C	% humidity	Aq. tension mm Hg
$H_3PO-1/2H_2O$	24	9	1.99
$KC_2H_2O_2$	168	13	738
$LiCl \cdot H_2O$	20	15	2.60
$KC_2H_4O_2$	20	20	3.47
KF	100	22.9	174
$NaBr$	100	22.9	174
$NaCl$, KNO_3 and $NaNO_3$	16.39	30.49	4.23
$CaCl_2 \cdot 6H_2O$	24.5	31	7.08
$CaCl_3 \cdot 6H_2O$	20	32.3	5.61
$CaCl_2 \cdot 6H_2O$	18.5	35	5.54
CrO_3	20	35	6.08
$CaCl_2 \cdot 6H_2O$	10	38	3.47
$CaCl_3 \cdot 6H_2O$	5	39.8	2.59
$Zn(NO_4)_2 \cdot 6H_2O$	20	42	7.29
$K_2CO_3 \cdot 2H_2O$	24.5	43	9.82
$K_2CO_3 \cdot 2H_2O$	18.5	44	6.96
KNO_3	20	45	7.81
$KCNS$	20	47	8.16
NaI	100	50.4	383
$Ca(NO_3)_2 \cdot 4H_2O$	24.5	51	11.6
$NaHSO_4 \cdot H_2O$	20	52	9.03
$Na_2Cr_2O_3 \cdot 2H_2O$	20	52	9.03

(*continued*)

Constant Humidity (Continued)

Solid phase	t°C	% humidity	Aq. tension mm Hg
$Mg(NO_3)_2 \cdot 6H_4O$	24.5	52	11.9
$NaClO_3$	100	54	410
$Ca(NO_3)_2 \cdot 4H_2O$	18.5	56	8.86
$Mg(NO_3)_2 \cdot 6H_2O$	18.5	56	8.86
KI	100	56.2	427
$NaBr \cdot 2H_2O$	20	58	10.1
$Mg(C_2H_2O_2)_2 \cdot 4H_2O$	20	65	11.3
$NaNO_3$	20	66	11.5
NH_4Cl and KNO_3	30	68.6	21.6
KBr	100	69.2	526
NH_4Cl and KNO_3	25	71.2	16.7
NH_4Cl and KNO_3	20	72.6	12.6
$NaClO_3$	20	75	13.0
$[(NH_4)_2SO_4]$	108	75	754
$NaC_2H_2O_2 \cdot 3H_2O$	20	76	13.2
$H_2C_2O_4 \cdot 2H_2O$	20	76	13.2
$Na_2S_2O_3 \cdot 5H_2O$	20	78	13.5
NH_4Cl	20	79.5	13.8
NH_4Cl	25	79.3	18.6
NH_4Cl	30	77.5	24.4
$(NH_4)_2SO_4$	20	81	14.1
$(NH_4)_2SO_4$	25	81.1	19.1
$(NH_4)_2SO_4$	30	81.1	25.6
KBr	20	84	14.6
Tl_3SO_4	104.7	84.8	768
$KHSO_4$	20	86	14.9
$Na_2CO_3 \cdot 10H_2O$	24.5	87	20.9
$BaCl_2 \cdot 2H_2O$	24.5	88	20.1
K_2CrO_4	20	88	15.3
$Pb(NO_3)_2$	103.5	88.4	760
$ZnSO_4 \cdot 7H_2O$	20	90	15.6
$Na_2CO_3 \cdot 10H_2O$	18.5	92	14.6
$NaBrO_4$	20	92	16.0
K_3HPO_4	20	92	16.0
$NH_4H_3PO_4$	30	92.9	29.3
$NH_4H_3PO_4$	25	93	21.9
$Na_2SO_4 \cdot 10H_2O$	20	93	16.1
$NH_4H_3PO_4$	20	93.1	16.2
$ZnSO_4 \cdot 7H_2O$	5	94.7	6.10
$Na_3SO_3 . 7H_2O$	20	95	16.5

(*continued*)

Constant Humidity (Continued)

Solid phase	t°C	% humidity	Aq. tension mm Hg
$Na_3HPO_4 \cdot 12H_2O$	20	95	16.5
NaF	100	96.6	734
$Pb(NO_3)_2$	20	98	17.0
$CuSO_4 \cdot 5H_2O$	20	98	17.0
$TlNO_3$	100.3	98.7	759
TlCl	100.1	99.7	761

For concentrations of sulfuric acid solution refer to tables relating density to percent composition.

CONSTANT HUMIDITY WITH SULFURIC ACID SOLUTIONS

The relative humidity an pressure of aqueous vapor of air in equilibrium conditions above aqueous solutions of sulfuric acid are given below.

Density of acid solution	Relative humidity	Vapor pressure at 20°C	Density of acid solution	Relative humidity	Vapor pressure at 20°C
1.00	100.0	17.4	1.30	58.3	10.1
1.05	97.5	17.0	1.35	47.2	8.3
1.10	93.9	16.3	1.40	37.1	6.5
1.15	88.8	15.4	1.50	18.8	3.3
1.20	80.5	14.0	1.60	8.5	1.5
1.25	70.4	12.2	1.70	3.2	0.6

For concentrations of sulfuric acid solution refer to tables relating density to percent composition.

Appendix P

Some Common Algebraic Functions

Linear $\qquad$ $y = ax + b$
Quadratic $\qquad$ $y = a + bx + cx^2$
Polynomial $\qquad$ $y = a_1 + a_2x + a_3x^2 + a_4x^3 + \cdots + a_nx^n$
Exponential $\qquad$ $y = ae^{bx}$
Power $\qquad$ $y = ax^b$
Logarithmic $\qquad$ $y = a + b(\ln x) = a + b2.303\log x$
$\qquad$ (ln has a base of e and log has 10)
Sine $\qquad$ $y = a \sin bx$ (a sine curve has zeros
at the beginning, middle and end
of a cycle. It reaches its maximum
and minimum values at the $\frac{1}{4}$ and
$\frac{3}{4}$ mark, respectively)
Cosine $\qquad$ $y = a \cos bx$ (a cosine graph begins and
end at its maximum point. In the middle,
it is at its minimum value, and has zeros
at the $\frac{1}{4}$ and $\frac{3}{4}$ mark)

For $y = f(x)$

(i) 1st derivative $\qquad$ $\dfrac{dy}{dx} = \lim_{\Delta \to 0} \dfrac{\Delta y}{\Delta x} = y' = \lim_{\Delta \to 0} \dfrac{f(x + \Delta x) - f(x)}{\Delta x}$

(ii) 2nd derivative $\qquad$ $\dfrac{d^2y}{dx^2} = \lim_{\Delta \to 0} \dfrac{\Delta y'}{\Delta x} = y'' = \lim_{\Delta \to 0} \dfrac{[df(x + \Delta x)/dx] - [df(x)/dx]}{\Delta x}$

(iii) For $y = ax + b$ $\qquad$ $\dfrac{dy}{dx} = a$ and $\dfrac{d^2y}{dx^2} = 0$

(iv) For $y = ax^n$ $\qquad$ $\dfrac{dy}{dx} = anx^{(n-1)}$ and $\dfrac{d^2y}{dx^2} = an(n-1)x^{(n-2)}$

Integration

(i) For $y = f(x)$ $\qquad$ $\displaystyle\int_a^b f(x)dx = \lim_{\Delta x \to 0} \sum_0^{n-1} f(x_i)dx = F(x)|_a^b + C(\text{Constant})$

(ii) For y = af(x) $\int af(x)dx = a \int f(x)dx$

(iii) For y = x^n $\int x^n dx = \dfrac{x^{n+1}}{n+1} + C$

∇ $\dfrac{\partial}{\partial x} + \dfrac{\partial}{\partial y} + \dfrac{\partial}{\partial z}$

Arithmetic Progression $x, (x+a), (x+2a), (x+3a), \ldots\ldots (e.g.\ 1,3,5,7,9,11,\ldots\ldots)$

Geometric Progression $ax, a^2x, a^3x, a^4x, a^5x, \ldots\ldots (e.g.\ 1,2,4,8,16,32,\ldots\ldots)$

Index

Absolute pressure (P), 335
Absorbed water, 288
Absorption, 9
  calculating amount of heat needed for, 72
Active pool, 78, 79
Activity ratio (AR), 240
Adhesion, 232
Adhesive forces, 260–261
Adiabatic process, 522
Adsorption, 9, 468
  batch experiments for, 500
  equation for, 491
  at equilibrium, 493
  Freundlich model of, 485
  Langmuir model of, 485
  opposite of, 484
  of polymers on clay surfaces, 138
  of soil moisture content, 347
Adsorption isotherms
  defined, 50
  defining shape of, 50
  schematic of, 485
Aeration, 9
  governed by soil water temperature, 527
  indices for measuring, 129

[Aeration]
  management of, 586–588
  measurement of, 576–578, 579
Aeration porosity, 563
Aerobic condition of soil, 567
Aggregate
  continuity of pores within. See Porosity
  defined, 97
  hydrophobicity of, 115
  ideal breaking of, 204
  integrity of, 150
  measuring stability of, 124
  measuring strength of, 124
  properties of, 114–115
  schematic of, 381
  strength of, 115
  weakest state of, 180
Aggregate analysis
  expression of results of, 126, 127, 128
  indices to express results of, 128
Aggregate coalescence, 175
Aggregated (silt + clay) index, 129
Aggregate disruptions, 174
Aggregate geometry models, 507
Aggregate hierarchy model, 107
Aggregate strength method, 127

Aggregation, 65, 96, 99–100
  aggregate hierarchy model, 107
  biotic factors in, 118
  bonding agents for, 110–114
  calcium-linkage theory, 101, 102
  clay-domain theory, 104, 105
  clay-water structure, 101–102
  correlation with organic matter,
      113–114
  distinction between flocculation and,
      108–109
  edge-surface proximity concept, 102,
      103
  effect of added organic matter on,
      120–121
  effect of decline in, 165
  effect of drying and wetting on,
      116–117
  effect of fertility management on, 137
  effect of freezing on, 117–118
  effect of no-till farming on, 433
  effect of soil conditioners on, 138
  effect of soil tillage on, 118–120
  effect of thawing on, 118
  effect of tillage on, 137
  effect of water management on, 137
  effect of weather on, 169
  Emerson's model, 102–104
  factors affecting, 115–121
  from farming systems, 136–137
  and frozen solids, 631
  laboratory methods for assessing,
      123–124
  methods for improving, 178
  methods of assessment for, 121–126
  microaggregate theory, 105–107
  organic bond theory, 104
  pedological methods for assessing,
      122–123
  POM nucleus model, 107–108
  quasi-crystal theory, 104–105
  Russell's theory of crumb formation,
      100–101
  stages of, 107
  and structural formation,
      108–114
  and structural resiliency,
      132–133
Agricultural lands, management of soil
    compaction in, 217–221

Agricultural machinery
  and soil compaction, 210–212
  and soil compaction prevention,
      218–219, 220
Agricultural productivity
  and fertilizer, 7
  importance of soil physics to, 6–9
Agricultural sustainability, 6–9
Agriculture
  application of soil strength in, 197
  importance of soil solids to, 80, 81
  importance of understanding solute
      transport to, 465–466
  relevance of soil compaction to, 206
Agroecology, 5
Agronomic capability, 4, 5
Agronomic operations, 82, 83
Agronomic yield, 85
Air entry point, 342
Airflow in soil, 569–571
Air permeability, 578–586
  measurement methods for, 581–586
Air porosity (f_a), 560
  range of, 25
Air pressure potential, 326
Air ratio (α), 25
Air temperature, global mean surface,
      559–560
Albedo, 524–525, 533
Algebraic functions, 689–690
Alkali soils, 644
Alumino-silicates, 35
Ammonia (NH_3), 557
Anabolic reactions, 535
Anaerobic condition
  of soil, 567
  toxic substances produced during, 567
Anaerobiosis, 16
Analogous laws, 540
Anion exchange capacity (AEC), 55
Anion exclusion, 477, 478, 484
  defined, 486
  equilibrium model of, 486, 488
  two-region model of, 491–492
Anisotropic soil, 372
Anisotropy, 372
Annual cycle, the, 529–530
Apparent diffusion coefficient, 502
Apparent dispersion coefficient (D),
      502–503

Applied liquid, 474
Arable land area, global shrinkage of, 8
Archimedes' principle, 27
Arhennius equation, 526–527
Atmospheric gases
  concentration of, 560
  pressure of, 269
Atterberg constants, 232–233
  applications of, 243–246
  applied to soil shrinkage, 246
  cohesion limit, 238
  factors affecting, 240–241
  lower plastic limit, 238
  measurement of, 241–243
  shrinkage limit, 237–238
  soil indices based on, 238–240
  sticky limit, 238
  upper limit of viscous flow, 238
  upper plastic limit, 238, 239
Available water capacity (AWC),
      293–297, 611
  calculating, 296
Avogadro's law, 560

Backscatter technique, 215
BC equation, 396–397, 398
BET method, 51
Bimetallic, 518
  development of temperature scales
      and, 516–517
  electric resistance, 518
  liquid-in-glass, 517–518
  remote sensing thermometer, 519–520
  thermocouple, 519
  thermoelectric, 518–519
Binding agents
  persistent, 113–114
  temporary, 111
  transient, 111
  types of, 109–110, 111
Bingham model, 191
Bingham plastic fluids, 266
Biochemical reactions, under anaerobic
      conditions, 567
Biological crusts, 166
Biological nitrogen fixation, 558
Biomass productivity, 9
Biopores, 633
Biota, effect on soil structure, 118
Bivalent cations, 64

Blackbody, 523
Black plastic mulch, 456
Blue water, 255
Boltzmann constant, 63
Boltzmann transform method, 393–394
Bond dipole moment, 57
Bonding agents, 110–114
Bonding mechanisms, 107
Bonding pores, 154
Bottleneck effect, 348
Boundary condition, 495
  at exit, 495
  for pulse input, 495
  for step-type input, 496
Boussinesq equation, 212
Breakthrough curves (BTC)
  defined, 474
  interpretations of, 476–479
  solute input in 474–476
Brownian movement, 41, 467
Buffering capacity, 33
Bulk density (ρ_b), 20, 608–609
Bulk modulus, 195
Burger model, 191
Bypass flow, 632–633

Calcium-linkage theory, 101, 102
Caloric, 521
Canopy surfaces, albedo for, 524
Capacitance method, 306–307
Capillarity, 261–264
Capillary bundle concept, 356–357
Capillary depression, 263
Capillary potential, 325–326
Capillary rise, 263
Carbon dioxide (CO_2), 557
  release into atmosphere, 16
Casagrande test, 242
Catabolic reactions, 535
Cation exchange capacity (CEC),
      55
Celsius (°C), 515
  conversion from one scale to another,
      516
Cementing agents
  soil structure index based on, 131
  types of, 109–110
Channeling, 632–633
Charge distribution, 171
Chemical crusts, 166

Chemical reactions
under anaerobic conditions, 567
endothermic, 536
exothermic, 536
and heat evolution, 535–536
Childs and Collis-George method, 390
Chisel plow, 588
Clausius-Clapayron equation, 268, 411
Clay
adsorption of polymers on, 138
charge properties of, 59–60
evaluating transport from surface to
subsoil, 278–279
field moisture capacity (FC) of, 290
heat of wetting, 71
hydraulic gradients in, 360–361
minerals in. *See* Clay minerals
relation with soil properties and
processed, 82
secondary minerals found in, 53
size fractions of, 35–37
and soil plasticity, 241
sources of charge in, 53–55
stability of suspension of, 63–66
strength of, 200
Clay-domain-theory, 104, 105
Clay minerals
classification of, 54
electrical double layer and zeta
potential of, 60–63
primary bonds in, 55–58
secondary bonds of, 58–59
silicate, 53
specific surface area of, 60, 72
structural units in, 52–54
Clay particles
swelling and shrinkage of, 66–68
theory for bonding of, 100–101
water adsorption on, 69–71
Clay ratio, 129
Clay-water structure, 101–102
Clear plastic mulch, 455
Coagulation, 63
Coefficient of linear extensibility
(COLE), 249–250
Cohesion, 232
Cohesion limit, 238
Cohesive forces, 237
Collision efficiency, 63–64
Colloidal hydration, 67

Colloidal suspension, 64–65
Column displacement experiments, 494
Combined nondimensional transport
equations, 496, 497–499
Compaction. *See* Soil compaction
Compensation method, 519
Compression, 208
Condensation, 533–534, 535
Conduction, 533
Cone penetrometer, 215–217
Conservation tillage, 458
Conservative solutes, 466
Consolidation, 208
Constant flux method, 583
Constant head method, 362–363
Constant-volume hydrogen scale, 517
Consumptive water use, computing,
276
Contact angle, 259–261
Contact angle effect, 348–349
Continuity equation, 380, 384–385
Convection, 467, 533–534
equation for, 470–471
Convective diffusion transport, 490
Convective dispersion process, 477, 478
Convective-dispersive equation (CDE),
483–484
nondimensional, 496, 498–499
simplest form of, 482–483
Convective transport, 492
Coshocton wheel sampler, 278, 281
Coulomb's law, 55–56, 198
Covalent bonds, 57–58
Cracking, 183–184
assessing degree of, 184
effect on soil water evaporation,
452
in vicinity of porous blocks, 337
Cracks, 634–635
Creep, 194
Critical soil organic matter content index,
131
Critical state theory, 237
Cropping systems, influence on soil
structures, 136–137
Crop residue mulch, 120
Crops, albedo from, 525
Crop yields, effect of soil compaction on,
206, 207, 208, 217–218
Crumb structure, 104

Crusting
conceptual model of, 175
defined, 165
effect of rough seedbed on, 171
general model for, 174–176
versus hardsetting, 181
impacts on plant growth, 176
mechanisms of, 171–173
and modulus of rupture, 203
soil and crop management options for
reducing, 178
soil properties susceptible to, 170
Crusts
characterization of, 176
figures of, 169, 170
formation of. *See* Crusting
hydraulic resistance of, 423
management of, 176–179
measuring strength of, 176, 177
methods of determining properties of,
177
multiple layer, 174
properties of, 173–174
types of, 166–167
and water infiltration, 422–423
Crust-topped method, 395
Cryic soils, 530–531
Crystalline minerals, 54
Cubic particles, 48–49
Cumulative evaporation, 451
Cylindrical shearing test, 202

Darcian flow, 467
Darcy-Buckingham equations, 382–383
applied to unsaturated flow, 384
diffusivity form of, 426
Darcy's law, 357–359, 580
calculating intrinsic permeability with,
362
calculating saturated hydraulic
conductivity of layered soil
with, 373–374
compared to Laplace equation, 372
equating fluxes with, 423
for hydraulic head difference, 376
limitations of, 360–361
one-dimensional flow according to,
370–371
for predicting infiltration rates,
412–413

[Darcy's law]
for saturated hydraulic conductivity,
361–362
for steady upward flow of water, 443
using permeability (k) in place of
conductivity, 366
validity of, 359–360
Dark-colored soil, albedo for, 524
Deflocculation, 63, 168–171
Deformation stresses, 180–181
Degree of saturation (s), 24, 25
Delayed meniscus formation effect, 349
Denitrification, 567
Density, 527
Density of water, 528, 684
Deposition, 535
Depositional crust, 167
Depth of penetration (X_p)
Desiccation, 173
Desorption, 347
Dew point temperature, 269
Dielectric constant (E), 306
Dielectric properties, measuring in soil,
305–308
Diffuse emitter, 523
Diffusion, 9, 467
equation for, 471
of gases, 571–575
Diffusion coefficient, 543, 573
Diffusion constant, 574
Digital elevation models (DEMs),
309–310
Dilatant fluids, 267
Dimensional method, 657–658
Dipole moment, 98
Dirac pulse, 476
Direct method
for measuring soil moisture content,
298–300
pros and cons of, 217
Direct shear test, 201
Dirichlet's boundary condition, 425
Disjoint-volume approach, 637
Dispersion, 37, 63
classifying soil according to, 124
factors leading to, 171–173
process of, 479–480
Dispersion agents, 38
increasing zeta potential with, 64–65
Dispersion coefficient, 481–482

Dispersion forces, 98
Dispersion ratio, 126, 129
Dispersity, 63
Dispersive transport
  causes of, 472
  equations for, 471–474
Dispersivity, 480
Displacing liquid, 474
Distribution percent by (DPW), 128
Diurnal temperature range (DTR), 528
DLVO theory of colloid stability, 63–64
Double layer repulsion
  at different electrolyte concentrations
    or valencies of counterions, 173
  equation for, 172
Double ring infiltrometer, 425, 428, 430
Downscaling, 657
Drainage, 548–549
  effects on soil physical property, 566
    Dry bulk density (ρ_b)
  range of, 25
Drainage to groundwater, 426
Drop-Cone test, 242
Dry sieving, 124
Dry specific volume (V_b), 21, 25
Duff layer, 545
Dynamic penetration test, 215

Earth's atmosphere
  first, 557
  nitrogen, 558
  origin of, 557
  oxygen, 558–559
  permanent gases, 558
  second, 557
  trace gases, 559–560
  variable gases, 558
Ecological approach, 95–96
Ecology, 2–3
Ecosystems
  agricultural. See Agroecology
  categorizations of solutes in, 465
  components of, 2–3
  effects of soil structure on, 165, 168
  natural versus managed, 3
  soil quality in, 3
Edaphological approach, 94
Edaphology, 5
Edge-surface proximity concept, 102, 103
Effective diameter, 44

Elasticity, 194–195, 196
Elastoplastic soil, 196
Electrical conductance, 313
Electrical conductivity (EC), 644
  and capacitance, 300
Electrical resistance, measuring,
    336–337
Electric double layer, 171
Electrokinetic potential, 61–63
Electron transfer, 568–569
Electrostatic bonds, 55–57
Electrostatic forces, 99
Elovich model, 487
Emerson's model, 102–104
Emissivity coefficient, 531
Endogenous factors, 115
Endothermic reactions, 535–536
Energy balance of soil, 536–537
Energy hill diagram, 535–536
Engineering
  application of soil strength in, 197
  importance of soil solids to, 80, 81
Engineering approach, 95
Enthalpy, 138
Entrapped air effect, 349
Entropy
  for change reversible system, 522
  defined, 411
  for physical process, 522
  of water, 257
Entropy effect, 349–350
Environment quality
  applications of soil physics to,
    10–11
  importance of soil solids to, 80, 81
Equation of continuity, 473
Equilibrium adsorption (S_s), 485
  equations for, 491
Equilibrium solute transport equation,
    496
Erosion, 9, 615–617
Evaporation, 535
  cumulative, 451
  defined, 439
  governed by soil water temperature,
    527
  and gravelly soils, 617–618
  nonisothermal, 452
  potential, 441–442
  rate as affected by texture of soil, 446

[Evaporation]
of soil water. *See* Soil water evaporation
transient, 446–447
Evaporation rate (*e*) estimation, 444
Evaporemeter, 274
Exchangeable cations, 241
Exchangeable sodium percentage (ESP), 644
Exergonic reactions, 535
Exogenous factos, 115–116
Exothermic processes, 536
Exothermic reactions, 535–536
Expansive soils. *See* Swelling soils

Fahrenheit (°F), 515
conversion from one scale to another, 516
Falling head method, 365–366
Farming systems, influence on soil structures, 136–137
Fick's laws, 469–470, 571–575
for calculating flux density, 508–509
second law of diffusion, 388
Field moisture capacity, 288–291, 427–428
assessing with neutron moisture meter, 302
effect of clay content on, 291
effect of organic matter on, 291
Field moisture content, 170
Filtering capacity, 33
Filter paper method, 337
Finger flow, 422
Fingering, 632–633
First atmosphere, 557
First law of heat conduction, 539–540
First law of thermodynamics, 521–522
First order model, 487
Fitting flow velocity, 501
Flocculation, 63
cause of, 171
causing, 64, 65
dispersion forces in, 98
distinction between aggregation and, 108–109
edge-to-face type of, 102
electrostatic forces in, 99
forces involved in, 97–99
gravitational forces in, 99

[Flocculation]
incomplete, 65
intermolecular and intramolecular forces in, 98
preventing, 64–65
random, 65
van der Waals forces in, 98
Floccules, 63
card-house structure of, 103
formation of, 65
Fluidity, 267–268
Flux controlled infiltration, 408
Flux, equating with Darcy's law, 423
Flux density, 358–359
calculating, 508–509
in unsaturated flow, 382
Fokkar-Plank equations, 388
Force of attraction, 56, 57
Force of buoyancy (*F_b*), 39
Fortuosity, 580
Fourier law, 533, 539–540, 543
Fractals
defined, 133
and soil structure, 133–134
using for pore size distribution, 157
Fractionation, 37–38
Fracture flow, 632–633
Fragmental soils, 602
Free surface energy, 259
Free water, 70–71
Freezing, 535, 630–632
Freundlich adsorption model, 485, 487
Friable consistence, 233, 236
Friction force, 39
Frigid soils, 530–531
Frozen soils, 626–632
composition of, 626–627
freezing process, 627–628
simultaneous heat and fluid transport in, 629–630
water flow in, 628–634
Functional normalization, 657
Fungi, as binding agent, 112, 113
Furrow irrigation, 432, 433

Gamma radiation, 310
Gamma ray attenuation, 304–305
merits and limitations of, 313
Gaseous components, 15
Gaseous diffusion in soil, 571–575

Gas thermometry, 520
Gas tortuosity (ξ_g), 368
General Freudlinch model, 487
General Langmuir-Freundlich model, 487
Geometric mean diameter (GMD), 128
Geometric mean particle diameter (GMD_p), 369
Gibbs equation, 520–521
Gouy-Chapman model, 60
Gravelly soils, 601–623
  bulk density, 608–609
  thermal properties of, 618–619
Gravel mulching, 456
Gravels
  effect of gravel on soil physical and hydrological properties, 605
  and soil structure, 605
  and soil texture, 605–606
  spatial distribution of, 603–604
  volumetric content, 605
Gravimetric gravel content, 609
Gravimetric heat capacity (C_g), 523
Gravimetric soil moisture content (w), 23
  converting into volumetric moisture content (Θ), 294
  range of, 25
Gravitational force (F_g), 39, 99
Gravitational potential (Φ_z), 330
  of soil moisture, 322
Gray water, 255
Greek alphabet, 667
Green-Ampt model, 412–415
  calculating parameters of, 434
  compared to other methods, 421
Greenhouse effect, 559
  effect of soil properties and processes on, 10
Greenhouse gases, 559
Green water, 255
Griffith's tensile failure theory, 200
Groundwater, 255

Hardsetting, 181–183
Hardsetting soils
  figure of, 182
  management of, 181, 183
  structure of, 181
Hazen's coefficient, 44

Heat
  diurnal and seasonal variations of, 529–530
  intensity, 521
  quantity, 521
Heat capacity, 618
  of gases, 538
  gravimetric (C_g), 523, 527, 538
  of Nigerian soils, 539
  of soil, 537–538
  volumetric (C_y), 523, 527
Heat conservation equation, 541–543
Heat dissipation technique, 334, 337
Heat evolution, 535–536
Heat flow in soils, 539–541
Heat of hydration, 72–73
Heat of wetting, 71
  causes of, 72
  measuring, 73
Heat transfer, 9
  conduction, 533
  convection, 533
  radiation, 531–533
Helium (He), 557
Helmholtz model, 60
Henry's law, 265
H-flume, 277
Holtan model, 419–420
  compared to other methods, 421
Homogenous soil, 371
Hookean model, 190
Horizontal flow, calculating, 374–375
Horizontal infiltration, 415–416
Horton model, 419
  calculating parameters of equation of, 434
  compared to other methods, 421
Humic substances, 77
Humidity, constant, 686–688
Humus, 78
Hydration energy, role in swelling process, 67
Hydraulic conductivity, 243, 355, 628
  and frozen solids, 631–632
  in gravelly soils, 612–614
  models for, 397–399
  reducing, 179
  relative, 397–399
  in saturated zones. *See* Saturated hydraulic conductivity

[Hydraulic conductivity]
and soil drying, 442
in unsaturated zones. *See* Unsaturated hydraulic conductivity
Hydraulic diffusivity, 448–450
Hydraulic gradients (ΔH), 357–358
calculating in absence of high conductance porous plate, 364–365
in clayey soils, 360–361
in coarse-textured soils, 360
rate of change of, 396
Hydraulic head, 331
in Darcy's law, 370–371
at inlet and outlet, 364
Hydraulic potential (H), 356
Hydraulic resistance, of crusted soil, 423
Hydraulic weighting device, 282
Hydrodynamic dispersion, 467, 468, 469
Hydrodynamic dispersion coefficient (D), 479–480
Hydrogen (H_2), 557
Hydrogen bonds, 58–59, 70
Hydrological process, 321
Hydrologic cycle, 269–270
components of, 271–281
Hydrophilic soil, 260–261
Hydrophobicity, 115
Hydrophobic soil, 260–261
Hydroscopic coefficient, 68
Hydrosphere
defined, 255
soil physical properties important to, 11
Hyperthermic soils, 531
Hyphae, as binding agent, 112
Hysteretic soil, 347

Ideal gas, 517, 520
Infrared thermometry, 519
Immiscible displacement, 474
Immobile (σ_{im}) water contents, 504–506
Immobile water zones, 468
Index based on texture and cementing agents, 131
Index of crusting, 131
Index of erodibility (I_e), 130
Index of resistance (I_r), 130
Index of structural stability (I_s), 130–131
Indirect methods for measuring soil moisture content, 300–312

Industrial building, importance of soil solids to, 80, 81
Infiltration. *See* Water infiltration
Infiltration rate, 406–407
Infiltrometers, 425, 428, 430
Initial conditions
at exit, 495
for pulse application, 495
for step-type input, 496
Inlet condition, 494
Inorganic bonding agents, 111
Inorganic components, 15
clay minerals, 52–66
packing arrangement of, 73–77
particle shapes of, 44–47
particle size distribution of. *See* Particle size distribution
primary particles of, 34
properties of, 34
secondary particles of, 34
specific surface area of, 47–52
swelling and shrinkage of, 66–68
water absorption in, 68–69
water adsorption in, 69–73
Inorganic soils
composition of, 26
particle density of, 20
In situ methods for measuring unsaturated hydraulic conductivity, 394–396
Insulation, factors affecting, 523–524
Interaggregate pores, 94, 95
Interlattice swelling, 66
Intermolecular attractions, 98
Internal drainage method, 395–396
Interparticla swelling, 66
Intraaggregate pores, 94, 95
Intramolecular attractions, 98
Intrinsic charge density, 59
Intrinsic permeability, 362
of air, 581
Inverse modeling technique, 503–504
Ionic bonds, 55–57
Irreversible (sink/source) model, 487
Irrigation, 548
Isobars, for soil-moisture potential, 331
Isomorphic substitution, 53
Isotropic soil, 372
Isotropism, 195

Kaolinite, 72
Kelvin (K) equation, 350, 515, 520
    conversion from one scale to another,
        516
Kelvin model, 191
Kinematic viscosity (η_k), 265–266
Kostiakov model, 418–419
    calculating parameters of, 434
    compared to other methods, 421
Kozeny-Carman equation, 368
Krilium, 121

Labile pool, 78, 79
Labile solutes, 466
Laminar flow, 359–360
    equation for relationship to pore
        radius, 366–367
Langmuir adsorption model, 485
Langmuir kinetic model, 487
Laplace equation, 370–372
Latitude, 524
Layered soil
    calculating Darcy's flux through,
        374–375
    calculating effective conductivity of,
        376
    calculating hydraulic and pressure
        heads in, 375–376
    evaporation of water in, 447–448
    saturated flow through, 372–376
    water infiltration in, 420, 422
Leaching, 3
    calculating amount of, 509
    method for, 299–300
    requirement for, 649–651
Least limiting water range, 294, 296–297
Least square fitting method, 503–504
Light-colored soil, albedo for, 524
Linear model, 191, 487
Liquid components, 15
Liquidity Index (LI), 239–240
Liquid ratio (σ_ρ), 24, 25
Liquid water, 288
    field moisture capacity (FC) of,
        288–291
Lithosphere, 3
Log normal statistical distribution, 128
London forces, 98
Longitudinal dispersion, 472
Longitudinal strain, 192

Lower plastic limit, 238
    in Casagrande test, 242
    in Drop-Cone test, 242
    indirect methods for measuring,
        242–243
    and subsoiling, 243
Lunch-time soils, 181
Lysimeters, 272–273, 274, 276
    drainage, 283
    for evaluating components of
        hydrologis cycle, 281
    figures of, 278–280
    under plastic shelter, 284
    water-filled pillows beneath, 282
    zero-tension, 508
Lysimetric analysis, 272–274, 276–281
Lysimetric measurements, 284

Macroaggregates
    reduction of, 179
    stabilization of, 114
Macrofauna, 224
Macropore flow, 632–637
    models of, 637
    and soil management practices,
        635–636
    and solute transport, 636–637
    and water infiltration, 636
    and water-repellent soil, 654–655
Macropores. *See* Non-matrix pores
Macroscopic miller similarity, 658
Macroscopic mixing, 468
    causes of, 479
Management of soil temperature, 457,
        458
    with drainage, 548–549
    with irrigation, 548–549
    with mulches, 545–546
    with tillage methods, 546
Mariotte bottle technique, 428–429
Mass flow, 467
Mass transfer model, 487
Mass transport equation, 470–471
Mathematical signs and symbols, 668
Matric potential (Φ_m), 325–329
    measurement of, 333–337
    and relative hydraulic conductivity,
        383
    using vapor pressure to compute, 346
Matric suction (Φ_m), 380–382

Matrix pores, 151
Maxwell model, 191
Mean weight diameter (MWD), 126, 128
Measurements, common units for, 88–89
Mechanical analysis
  dispersive agents needed to remove
    binding agents prior to, 38
  forces in, 39
  fractionation, 37–38
  methods by sedimentation technique,
    42
Mechanical puddling, 180
Melting, 535
Mercury intrusion method, 160
Mercury manometer tensiometers,
  326–328
Mesic soils, 531
Mesopores, 155
Metabolism, 535
Metallic bonds, 59
Methane (CH_4), 557, 567
Methanogenesis, 568
Microaggregates, 104, 105–107
  formation of, 109
Microaggregate theory, 105–107
Microbial processes, 535
Micropores, 155
Microrelief, 171
Minerals, particle density of, 21
Miscible displacement, 474
Mobile (σ_m) water contents, 504
Modulus of rupture, 202–203
Modulus of shearing, 195
Mohr-Coulomb maximum shear
  strength, 200
Mohr theory, 198
Moldboard plow, 588
Molecular diffusion, 468
Mole drainage channels, 245
Monovalent cations, 64
Mounded seed beds, 460
Mulches, 453–456, 545–546
Mulching
  benefit to soil water evaporation,
    453–456
  effect on soil temperature, 457, 458
  techniques for, 221, 224
  and water infiltration, 432–433
Multicolumn approach, 637
Multidivider tank, 276

Negative adsorption, 486
Nernst heat theorem, 522
  net flux of, 540–541
Nernst's potential, 61
Net radiation (R_N), 533
Neumann's boundary condition,
  426
Neutron moisture meter, 313
Neutron moisture readings, 301–304
Neutron probe calibration, 612
Neutron thermalization, 301–304
Newtonian fluids, 266, 267
Newtonian model, 190
Newton's first law, 534
Newton's law of viscosity, 266, 357
Nielsen similarity analysis, 658
Nitrate reduction, 590
Nitrogen, 558
Nitrogen sorption, 160–161
Nonconservative solutes, 466
Nonequilibrium transport,
    488–489
  analytical solutions for equations of,
    498
  reduced to nondimensional equations,
    496, 497
  two-region model of, 489–491
  two-site model of, 491–492
Nonisothermal evaporation, 452
Non-limiting water range (LLR), 294
Non-matrix pores,
  describing by diameter size, 152
  determining number of, 151
  formation of, 155
  pore size distribution method for, 156
  shape and continuity of, 154–155
Non-Newtonian fluids, 266–267
Nonpolar bonds, 57, 98
Non-SI units conversion factors, 678
Normal stress (τ), 192
  relationship to tangential stress (τ),
    198
No-till farming
  with crop residue mulch, 461
  effect on aggregation, 433
No-till treatment, 586, 588
nth order model, 487

Organic bonding agents, 111
Organic bond theory, 104

Organic components, 15
  in soil, 77–80
  soil properties and processes affected
    by, 84
Organic fraction, 83–86
Organic matter
  classifications of, 78–79
  correlation of aggregation with,
    113–114
  effect on available water capacity
    (AWC), 293
  effect on plasticity, 241
  and pF curve of soils, 343
  relation with soil properties, 85
  role in soil fertility management, 78
Organic soils
  composition of, 26
  particle density of, 20
Osmotic potential (Φ_o), 330–331
Osmotic pressure, 264, 647–648
  of solutions of sucrose in water, 685
Outlet condition, 494
Overburden potential (Φ_o), 331
Oxidation, 558
Oxidation reactions, 535–536
Oxidation reduction process, 567–569
Oxidation-reduction potential, 569
Oxygen, 558–559
Oxygen deficiency
  and plant growth, 564–565
  and soil properties, 565–566
Oxygen diffusion rate (ODR), 564, 565
Oxygen diffusion rate, 577–578

Packing arrangements
  composite form, 76
  cubic form, 74, 75
  and grade of soil material, 77
  open versus closed, 76
  orthorhombic configuration, 75
  orthorhombic form, 74
  rhombohedral configuration, 75
  rhombohedral form, 74
  and soil porosity, 73, 75–76
Para plow, 431
Particle density (ρ_s), 19–20
  of inorganic versus organic compo-
    nents, 24, 25
  range of, 25
  of soil minerals, 21

Particle geometry, 48–50
Particle repulsion, 172
Particles
  collision efficiency of, 63–64
  compression of, 206
  packing arrangement of, 73–77
  rearrangement of, 172–173
  shapes of, 41, 48
  sizes of, 48
  specific surface area of, 47–52
Particle shape
  angularity of, 45
  elements determining, 44–45, 47
  figures of, 46
  indices of, 89
  roundness of, 47
  sphericity of, 47
Particle size
  importance of, 34
  uniformity of, 44
Particle size analysis, 42
Particle size distribution
  assessment of particle size fractions,
    37–42
  defined, 34
  size fractions, 34–37
Particle size fractions, 37–42
Partition coefficient, 506
Passive pool, 78–79
Pedological approach, 94
Pedosphere, 3
Pedospheric processes, 11
Pedotransfer functions, 242–243
Penetration resistance, 215–217,
    218
Penetrometers, 215–217
Peptization, 63
Percent clay aggregated, 128
Percent silt plus clay aggregated,
    128
Pergelic soils, 530–531
Permanent charge, 59
Permanent wilting point (PWP),
    291–292
  relation between clay and volumetric
    water content at, 292
Persistent binding agents,
    113–114
pF curves, 242–243. *See also* Soil
    moisture characteristics

Philip equation, 434
Philip's model
  compared to other methods, 421
  for predicting water infiltration,
    415–418
Photons, 531
Photosynthesis, 563
Physical crusts, 166
Physical edaphology
  defined, 5
  role in sustaining agricultural
    production, 6–9
Physical nonequilibrium model, 503
Piezometer tube, 324–325
Piezometric head, 324
Piston flow, 477
Planck's law, 532
Plant available water capacity index, 132
Plant growth
  and available water capacity (AWC),
    293
  critical limit of air-filled porosity for,
    23
  effects of organic fraction on, 85
  importance of pores for, 155
  influence of soil air on, 564
  normal range of soil physical proper-
    ties in relation to, 25
  and soil structure, 94–95
Plant residues, 458
Plasticity, 195–196
  defined, 236
  dependence on clay content, 241
  effect of organic matter on, 24
  impact of soil tilth, 231
  index for range of, 23, 240
  influence of exchangeable cations on,
    241
  lower plastic limit, 238
  methods for measurement of, 241–243
  necessary conditions for, 236
  theories that explain, 236–237
  upper plastic limit, 238
Plasticity Index (PI), 239, 240, 241
Plastic Range, 239, 240, 241
Plate condensation, 65–66
Plate-shaped particles, specific surface
    area of, 50
Platinum resistance thermometer, 518,
    519

Pneumatic potential (Φ_a), 326
Poiseuille's equation
  for calculating water flow, 357
  for finding volumetric flow rate, 389
  for relationship between pore radius
    and laminar flow, 366–367
Poiseuille's law, 471–472
Poisson's ratio (v), 194
  vand tensile strength, 204
Polar bond, 57
Polymers, 121
Polyvalent cations, 64
POM nucleus model, 107–108
Ponding, 424
Pore geometry, 366–370
Pores
  assessing size distribution of, 155–161
  classification systems of, 152–154
  connectivity of, 468
  estimating permeability of, 368
  functional characteristics of, 154
  matrix, 151
  non-matrix, 151
  origin and formation of, 155
  proportion of textural versus
    structural, 199
  shape and continuity of, 154–155
  size classifications for, 152–154
  size distribution of, 151–154
  tortuosity of, 467
Pore size distribution
  field methods for, 156–157
  fractal analyses of, 157
  laboratory methods for, 157–161
  mercury intrusion method for, 160
  microscopic measurements for, 157
  nitrogen sorption for, 160–161
  water desorption method for, 158–160
Pore volumes, 476
Pore water velocity, 467, 501
  calculating, 509
  relation to dispersion coefficient, 481
Porosity (f_t), 22
  air-filled porosity (f_a), 23
  air ratio (α), 23
  assessment of, 155–161
  defined, 22, 149–150
  expressing degree of soil compaction
    in, 205
  of gravelly soil, 606–608

[Porosity]
  indices for measuring, 129
  methods of expression of, 151–155
  and packing arrangements of soil, 73,
    75–76
  pore size distribution, 156–161
  range of, 25
  structural, 150
  textural, 150
  total porosity (f_t), 22, 155
  void ratio (e), 23, 155
Porous blocks, 300
Porous material sensors, 334
  limitations of, 336–337
  measurements of, 336, 337
Potential evaporation, 441–442
Power model, 487
Precipitation, measurement of, 272, 273,
    275
Precipitation infiltration, 424–425
Preferential flow paths. See Macropore
    flow
Pressure potential (Φ_t), 324–325
Proctor compaction test, 209–210, 242
Profile-controlled infiltration, 408
Pseudoplastic fluids, 266
Psychrometers
  figure of, 336
  limitations of, 334, 335
  measurements of, 335
  types of, 335
Puddlability (P), 179
Puddling, 179–181
Pulse application, 495–496
Push-in electrodes, 307

Quanta, 531
Quasi crystals, 105
Quasi-crystal theory, 104
Radiation, 531–533
  emitted by the Sun, 532
  net, 536–537
Radiation methods, 213–215
  pros and cons of, 217
Radiation techniques
  gamma ray attenuation, 304–305
  measuring soil moisture content with,
    300–305
  neutron thermalization, 301–304
Rainfall, transmission of water after, 290

Rainfall simulators, 431
Rain gauge, 273
Reactive solutes, 466
Recalcitrant pool, 78–79
Redox potential, 569
Reduction reactions, 535–536
Red water, 255
Regression analysis, 657
Relative diffusion coefficient, 573
Relative hydraulic conductivity (K_s)
  model for predicting, 397–399
Relative solute concentration, 474, 475
Remote sensing, 309–310
Remote sensing method, 313
Remote Sensing Thermometer, 519–520
Representative elementary volume
    (REV), 637
Repulsion, 477, 478
Resident concentration, 467–468
Residual pores, 152, 154
  formation of, 155
Residual shrinkage, 247–248, 249
Resistance blocks, 336
Respiration, 563
Respiration rate (R_T), 525–526
Respiratory coefficient, 566
Retardation factor (R), 500–501
Retention pores
  formation of, 155
  increasing, 180
Reversible potential, 61
Reversible process, 522
Reynolds number (N_{Re}), 360
Rheology
  defined, 189, 231
  determining soil elasticity, 194
  models of, 189–191
  stress-strain relationship in, 192–193
Rheopectic fluids, 267
Richards' equation, 386–389
  for calculating decrease of moisture
    content with respect to time, 428
  capacitance form of, 425
  diffusivity form of, 425, 451
Rock fragment classifications, 604
Root channels, 221, 223
Root growth, 619–620
Roots, as binding agents, 111–112
Rothmund-Kornfeld ion exchange
    model, 487

Roundness, 47
Runoff, 272, 276–278
  example for calculation of, 281
Russell's theory of crumb formation,
  100–101

St. Vincent model, 190
Saline soils, 644
Saline-alkali soils, 644
Salinity, 645–646
  effects on water movement, 648–649
Salinization, 644
Salt-affected soils, 644–651
Salt sieving, 356
Sand, 34–35, 36
Sandy soils, strength of, 200
Saturated flow. *See also* Saturated
    hydraulic conductivity
  calculating with Laplace equation,
    370–372
  importance of knowing, 355
  through layered soil, 372–376
  versus unsaturated flow, 381
Saturated hydraulic conductivity (K_s),
    358, 580
  alternative system for determining,
    363–364
  calculating with Darcy's law, 361–362
  and constant head method, 362–363
  defined, 361
  ensuring saturation of soil for, 363–364
  and errors in volumetric flow rate
    measurement, 364
  estimating from pore geometry,
    366–370
  estimating with Kozeny-Carman
    equation, 368–369
  and falling head method, 365–366
  homogeneity and heterogeneity of, 372
  and intrinsic permeability, 362
  in layered soil, 372–376
  schematic of apparatus for, 363
Saturated uncompressible soils, 371
Saturated vapor pressure, 268
Saturation deficit, 269
Scalding, 565
Scale factors, 658
Scaling methods in soil physics, 656–659
Second atmosphere, 557
Second law of thermodynamics, 522–523

Second-order irreversible model,
    487
Sedimentation, 9, 38
Settling equation, 40, 41
Settling velocity, 41–42
Shearing stress, 192
  elastic relation for, 195
Shear strain (γ), 193
Shear strength, 197, 243, 244
Shear stress (τ), 266
Shear tests
  cylindrical, 202
  direct, 201
  triaxial, 201–202
  Vane, 203
Shrinkage characteristic, 640
Shrinkage limit, 237–238
  defined, 246
Shrinking, 116
Sieves, 140
Sieving, 37–38, 124
  indices to express results of aggregate
    analysis by, 128
  salt, 356
Silica:sesquioxide ratio, 130
Silicon tetrahedron, 52–54
Silt, size fractions of, 35, 36
SI units
  abbreviations, 671
  prefixes for, 669
Similitude analysis, 657–658
Sinks, 575
Size fractions
  of clay, 35
  of sand, 34–35, 36
  of silt, 35
Skeletal soil, 601–603
Skin seal, 173–174
Slaking, 134
  cause of, 168
  defined, 407
  effects of, 170
  factors affecting, 168
  field moisture content and, 170
  in hardsetting process, 181
Slichter equation, 359
Slump test, 127
Snow gauge, 273
Sodium adsorption ratio (SAR),
    422–423

Soil
   adhesive versus cohesive forces in, 232
   agricultural functions of, 9
   agricultural sustainability of, 7–8
   agronomic capability of, 4
   albedo from, 525
   assessing stress-strain behavior in,
      190–191
   buffering capacity of, 33
   cementing agents in, 109–110
   cohesion limit of, 238
   cohesive strength, 202
   color changes as indicator of moisture
      content, 310
   compaction of. *See* Soil compaction.
   as component of ecosystem, 2–3
   components of, 5, 15–16, 18
   compression of, 208
   computing relative density of, 210
   computing uniformity coefficient of, 46
   consolidation of, 208
   for construction purposes, 77
   contact angle in, 260–261
   crack initiation in, 183
   in critical state, 237
   degradation of, 4–5
   degradative processes of, 86
   dense verse porous, 22
   determining erodibility of, 129, 130
   diffusion coefficient in, 471
   dispersivity values in, 480
   dry, 17
   drying process of, 441–443
   effect of drying and wetting on,
      116–117
   effect of freezing on, 117–118
   effect of organic component on prop-
      erties of, 84
   effect of puddling on, 179–180
   elasticity, 194–195
   elastoplastic, 196
   engineering functions of, 9
   ensuring saturation of, 363–364
   environmental functions of, 9
   environmental purification functions
      of, 10–11
   estimating hydraulic functions of,
      389–402
   factors affecting compactibility of,
      208–212

[Soil]
   field moisture capacity of, 427–428
   filtering capacity of, 33
   finite resources of, 1
   friable consistence of, 232, 233, 236
   friction between metal and, 244–245
   frictional forces in, 199
   functions of, 4
   general description of, 1
   general physical properties of, 25
   general properties of phases and
      components of, 26
   as geomembrane, 10
   hardsetting of, 181–183
   harsh consistence of, 232, 233
   hydrophilic versus hydrophobic,
      260–261
   hysteresis of, 347–350
   hysteretic, 347
   impact of decline in soil structure on,
      166
   importance of porosity to, 149
   importance of studying science of, 1
   indices for measuring properties of,
      129
   infiltration rate of, 406
   influence on air quality, 11
   influence on water quality, 11
   initial condition for, 494
   inorganic components of.
      See Inorganic components
   interaction with environment. *See*
      Pedosphere
   interparticle bond forces in, 199
   interrelationship among properties of,
      25–26
   isotropic versus anisotropic, 372
   isotropism in, 195
   liquid consistence of, 233
   lower plastice limit of, 238
   major stabilizing agents for, 108
   measuring dielectric properties of,
      305–308
   measuring electrical resistance of,
      336–337
   measuring hydraulic function of,
      391
   minimizing pore space of, 15
   models for sorption, 485–486, 487
   moist, 17

[Soil]
moisture content of, 200. *See* Soil moisture content
movement of water in, 322
non-homogeneity in, 372
organic compounds in, 77–80
penetration depth of, 310
penetration resistance of, 215–217
phases of, 5, 15–16, 18
plastic consistence of, 233
plasticity of, 195–196, 231, 236–237
porosity of. *See* Porosity (f_t)
properties of, 18
quality of. *See* Soil quality
relation of organic matter with properties of, 85
in relation to plant growth. *See* Edaphology
removing water from, 299
as reservoir of freshwater, 255
as reservoir of water, 271
saturated, 17
sensitivity of, 95
shape of particles in, 44–47
shear strength of, 197, 198
shrinkage limit of, 237–238
soft consistence of, 232
sticky consistence of, 233
sticky limit of, 238
stress-strain relationship in, 192–193
structural stability of, 110
structureless versus structured, 122
susceptibility to puddling, 179
tensile strength of, 184, 200–202
thermal conductivity of, 308–309
as three-phase system, 259–260
total porosity (f_t) of, 155
upper plastic limit of, 238
viscoelastic soils, 197
viscosity of. *See* Viscosity
viscous flow of, 237, 238
void ratio (e) of, 155–156
water absorption capacity of, 68
water infiltration in. *See* Water infiltration
Soil aeration, 563–564
Soil aggregates, 203–204. *See also* Aggregate

Soil air, 560–563
composition of, 26
influence on plant growth, 564
O_2 and CO_2 content in, 561
Soil architecture. *See* Porosity
Soil bulk density, 199, 206–207
methods of measuring, 212–215
pros and cons of methods for measuring, 217
using cold method for, 211
wet versus dry, 213
Soil cohesion (C), 198
Soil colloids
charge distribution, 171
water absorption on, 68–69
Soil compaction, 9, 119
alleviation of, 219, 431
biological measures for preventing, 221, 222–224
causes of, 207–208
defined, 205
in dynamic situation
effect on crop yields, 206, 207, 208, 217–218
expressing degree of, 205
factors affecting, 208–212
farm equipment causing, 208, 209, 210–212
versus hardsetting, 181
influence of texture on, 82
managing in agricultural lands, 217–221
measurement of, 212–217
prevention of, 218–219
relationship between moisture content and, 210
relevance to agriculture, 206
and soil bulk density, 212–215
and soil moisture content, 245–246
in static situation, 205
strategies for management of, 218–221
using penetration resistance to measure, 215
and wheel traffic, 210–212
Soil compressibility, 208
Soil conditioners, 138
Soil consolidation, 221
Soil consistence
attributes of, 232
define, 231–232

[Soil consistence]
  forms of, 232–233
  impact on soil tilth, 233, 236
  Soil Survey Division Staff's levels of, 233, 234–235
Soil consistency, liquidity index (LI) for, 239–240
Soil core, saturating, 363–364
Soil cracking, 183–184
Soil deformation, 199
Soil degradation, 4–5
Soil density (ρ_s)
  bulk density (ρ_b), 20
  dry specific volume (V_b), 21
  particle density, 19–20
  relative density, 21
  specific gravity (G_s), 21
  units of measurement for, 19
Soil dispersion, 134
Soil drying stages, 441–443
Soil energy balance, 524
Soil erosion, 655
Soil fabric, 96–97
Soil fertility, 7. *See also* Soil quality
Soil fertility management, effects on soil structure, 137–138
Soil friability, 199
Soil infiltration capacity, 406
Soil management, 24
Soil matrix, 24
  defined, 33
  entry of water into, 406
  hydraulic gradient (ΔH) across, 357–358
  movement of solutes inside. *See* Solute transport
  pore water velocity through, 359
  ratio of smaller to larger pores in, 389–390
  total solute resident concentraction $(C, g\ cm^{-3})$
  unsaturated, 379–380
  volumetric flow rate through, 358
Soil minerals, particle density of, 21
Soil moisture
  ability to suck water from pure water reservoir, 330
  characteristics of. *See* Soil moisture characteristics
  defined, 23

[Soil moisture]
  energy status of, 321–322
  gravitational potential energy of, 322
  presence of solutes in, 330
  retention curves for, 344
Soil-moisture capacity function, 387–388
Soil-moisture characteristic curve (SMCC), 390
Soil moisture characteristics
  computing from relative humidity, 346
  defined, 341
  factors that affect, 342–344
  methods of determining, 345
  for soils of contrasting texture, 342
Soil moisture content, 23–24
  adsorption and desorption of, 347
  available water capacity (AWC) of, 293–297
  choosing method for measurement of, 312–313
  and compaction, 245–246
  defined, 287
  difficulties in assessing, 297
  discontinuity of, 422
  in Drop-Cone test, 242
  effect of clay minerals on, 241
  effect of water infiltration on, 413
  expressing rate of change of, 387
  expression of measurements of, 312, 314
  field capacity (FC) of, 288, 427–428
  forces acting on, 287
  importance of measuring, 297
  least limiting water range of, 294, 296–297
  measuring by chemical properties, 311
  measuring by dielectric properties of soil, 305–308
  measuring by electrical conductivity and capacitance, 300
  measuring by evaporation method, 299
  measuring by leaching method, 299–300
  measuring by low-energy ultrasonic waves, 311
  measuring by radiation technique, 300–305
  measuring by remote sensing, 309
  measuring by thermal conductivity, 308–309

[Soil moisture content]
  measuring by volume displacement method, 311–312
  methods of measurement of, 297–312
  mobile (σ_m) and immobile (σ_{im}), 504–506
  permanent wilting point (PWP) of, 291–292
  principles underlying methods of assessment of, 298
  redistribution of, 426–427
  relationship to soil-matric potential (Φ_m). *See* Soil moisture characteristics
  relationship to soil volume, shrinkage behavior, and soil consistency, 241
  relationship to soil water diffusivity, 389
  in Richards equation, 386–387
  shrinkage limit of, 237–238
  in terms of Atterberg's constants, 237–240
Soil-moisture hysteresis
  bottleneck effect for, 348
  contact angle effect for, 348–349
  defined, 347
  delayed meniscus formation effect for, 349
  entrapped air effect for, 349
  entropy effect for, 349–350
  importance of, 350
Soil-moisture potential
  applications of, 350–351
  characteristics of, 323
  computing components of, 338–341
  conversion units for, 352
  defined, 322, 323
  under field conditions, 331–333
  gravitational potential (Φ_z), 330
  matric potential (Φ_m), 325–329
  measuring matric potential (Φ_m), 333–337
  osmotic potential (Φ_o), 330–331
  overburden potential (Φ_o), 331
  plotting profile of, 332
  pressure potential (Φ_t), 324–325
  total. *See* Total soil-moisture potential (Φ_t)
  units of measurement of, 337–341

Soil moisture redistribution, 426–427
Soil-moisture retention, 342, 343
  equations for description of, 397
Soil organic carbon (SOC), 79
  calculating rate of change of, 80
  relationship to agronomic yield, 85
Soil particle density, 27–29
Soil particles. *See* Particles
Soil physical quality, 24–25
Soil physics
  and agricultural sustainability, 6–9
  applications of, 7, 8
  defined, 5, 19
  and environmental quality, 10–11
  global importance of, 11
  interaction with basic and applied sciences, 7
  principles of, 5–6
  scope of, 16, 19
Soil porosity. *See* Porosity (f_t)
Soil properties
  oxygen deficiency and, 565–566
  surface charge on, 171
  susceptible to crusting, 170
Soil quality
  and agronomic productivity, 5
  defined, 4
  elements dependent on, 5
Soil respiration, 566–567
Soil science
  and agrosystems, 5
  and ecology, 2–3
  importance of, 1
Soil scientists, systems used by, 34
Soil separates, 34–37
Soil shrinkage
  application of, 250
  causes of, 246
  defined, 246
  methods for determining, 248–250
  normal, 246–247, 249
  residual, 247–248, 249
Soil slaking. *See* Slaking
Soil solids
  importance of, 80–82
  organic fraction and soil processes of, 83
  processes in, 33
  properties of, 199–200
  texture and soil processes of, 82–83

Soil-solute interactions, 477
Soil solution, 26
Soil strength
    applications of, 197
    cohesiveness factor (C) in, 201
    and compaction, 245–246
    defined, 192, 197
    factors affecting, 199–200
    indices for measuring, 129
    in situ determination of, 203
    measurement of, 200–204
    measuring modulus of rupture to
        determine, 202–203
    mohr theory of, 198
    and penetration resistance, 215
    types of, 197
Soil structure, 199
    aggregation and structural formation
        in, 108–114
    bonding mechanisms in, 107
    classifications according to shape, 123,
        125
    complexity of, 93–94
    crusting and surface seal formation on,
        165–167
    defined, 93
    deformation stresses on, 180–181
    degradation of, 134–135
    ecological approach to, 95
    edaphological approach to, 94–95
    effect of added organic matter on,
        120–121
    effect of decline in, 134
    effect of fertility management on, 137
    effects of soil conditioners on, 138
    effects of structural degradation in,
        165, 166, 167, 168
    effect of tillage on, 137
    effect of water management on, 137
    effect of wheel traffic on, 119
    elasticity of, 196
    endogenous versus exogenous factors
        in, 115–116
    engineering approach to, 95
    environmental impacts of, 135–136
    factors affecting aggregation in,
        115–121
    and forces involved in flocculation,
        97–99
    fractal analyses, 133–134

[Soil structure]
    functional entity of. See Porosity
    indices of, 126, 129–132
    influence of farming systems on,
        136–137
    laboratory methods for assessing,
        123–124
    management of, 135–138
    mechanisms of aggregation in, 99–108
    methods of assessment for, 121–134
    multidisciplinary approach to, 135
    pedological approach to, 94
    pedological methods for assessing,
        122–123
    permanent deformation of, 196
    pore size distribution of, 156–161
    properties of aggregates in, 114
    schematic of bonds in, 103
    shapes and size classes of, 125
    versus soil fabric, 96–97
    and soil moisture characteristics,
        342–344, 345
    structural resiliency of, 132
    types of pores in, 94–95
    using puddling to improve, 179–181
    Zakhrov system of classification of,
        122
Soil suction, 386–387
Soil surface, transient evaporation from,
        446–447
Soil Survey Division Staff, nine levels of
        consistence, 233, 234–235
Soil temperature
    affecting plant growth, 524–525
    effect of mulching on, 457, 458
    influence on evaporation, 527
    management of, 545–554
    mean annual, 530–531
    optimum range, 525
    and properties of water, 527
    regimes, 528
    variations in, 529–530
    varies as a result of, 527–528
Soil temperature classes, 530–531
Soil temperature regimes
    and the annual cycle, 529–530
    categories, 530–531
    variations in, 528
Soil texture. See Particle size distribution
Soil thermal diffusivity, 543

Soil thermal regime, 453
Soil tillage. *See* Tillage
Soil tilth. *See* Tilth
Soil type, identifying, 122
Soil water
  field moisture capacity (FC) of,
    288–291
  plotting characteristic curve of,
    401–402
  redistribution of, 426–427
Soil-water availability, 527
Soil-water diffusivity, 388–389, 543
  equation for, 394
  mathematical expression of, 449
  of wet soil region, 414
Soil water evaporation
  conditions for, 440
  drying of soil during, 441–443
  effect of cracks on, 452
  example for calculating rates of,
    461–462
  introduction to, 439–440
  from layered soils, 447–448
  management of, 452–461
  mathematical modeling of stages of
    drying, 448–452
  in presence of water table, 443–446
  process of, 440–441
  reducing with conservation tillage,
    458
  reduction of, 452–457, 458, 461
  theory of, 443–452
Soil-water movement, 527
Soil water potential, 679
Soil-water profile, 408
Soil water retention models, 396–397
Soil-water storage, 274
Solonetz, 644
Solute flux, 473
Solute flux density, 469–470
Solute input, 474–476
Solutes
  categories of, 466
  piston flow of, 477
  presence in soil, 466
  pulse application, 476
Solute transport
  breakthrough curves for, 474–479
  combined nondimensional equations
    for, 496, 497

[Solute transport]
  conditions required for solving
    equations of, 493–495
  dimensional conditions for pulse
    applications for, 495–496
  dispersion processes of, 479–480
  displacement of solutes in, 477
  equations for, 470–474
  equilibrium anion exclusion model for,
    486–488
  estimating apparent dispersion coeffi-
    cient for, 502–503
  estimating parameters of, 496, 500–507
  estimating parameters of two-region
    model (TRM) for, 503–507
  estimating retardation factor (R) for,
    500–501
  example for calculating, 508–510
  Fick's laws for describing, 469–170
  in heterogeneous soils, 488–489
  immiscible displacement, 474
  influence of displacement length on,
    477, 479
  introduction to, 465–466
  land use effects on, 508
  mathematical representation of,
    482–484
  miscible displacement, 474
  nondimensional variables for equa-
    tions of, 497
  nonequilibrium, 488
  one-dimensional, 490–491
  processes of, 466–468
  relations between dispersion coefficient
    and pore water velocity in, 481
  sorption phenomenon in, 484–486
  step input experiments for, 493–495
  two-region anion exclusion model for,
    491–492
  two-region nonequilibrium model for,
    489–491
  two-site nonequilibrium model for,
    491–492
Solute transport equation, 483
Sorption, 477, 478
  models of, 485–486, 487
  in solute transport, 484–486
  in two-site nonequilibrium transport
    model, 493
Sorptivity, 416

Sources, 575
Spatial fractals, 133
Specific gravity (G_s), 21
Specific humidity, 269
Specific surface area, 60
  of clay minerals, 72
  importance of knowing, 47
  indices of, 47–48
  particle geometry for, 48–50
  using adsorption isotherms to determine, 50–52
Specific water capacity, 344
Spherical particles, specific surface area of, 49–50
Sphericity, 47
Spider gauge, 275
"Spirit" thermometer, 516
Sprinkler method, 394–395
  for testing water infiltration limits, 406
Stability against water or wind method, 127
Static penetration test, 215
Stationary liquid method, 583–584
Steady flow (v) equation, 373–374
Steady state methods
  for measurement of K_T, 544
  for measuring soil's hydraulic function, 391
Stefan-Boltzmann equation, 519–520
Stefan-Boltzmann law, 531
Step input experiments, 493–495
Stern model, 60–61
Sticky limit, 238
Strain (ε)
  longitudinal (ε), 192
  in Poisson's ratio (v), 194
  relationship to stress in soil, 196–197
  rhelogical models for, 189–191
  shear (γ), 192
  time-dependent, 193
  in Young's modulus, 194
Strain hardening, 195–196
Stress (σ)
  acting inside soil body, 202
  normal (σ), 192
  in Poisson's ratio (v), 194
  relationship to strain in soil, 196–197
  rhelogical models for, 189–191
  tangential (τ), 192

[Stress (σ)]
  time-dependent, 193
  in Young's modulus, 194
Stokes law, 38, 39, 40
  assumptions of, 41
Storage pores, 152, 154
Structural crusts, 167
Structural degradation, environmental effects of, 165, 166, 167, 168
Structural form, 97
Structural pores, 94
Structural porosity, 150
Structural resiliency, 132–133
Structural stability
  bonding agents for, 110–114
  defined, 110
  methods of determining, 127
Structured soils, 122
Structured water, 70
Structureless soils, 122
Subirrigation, 589–590
Sublimation, 535
Submergence potential, 324
Subsidence, 9
Subsoil alleviation, 219, 221
Subsoiling, 243
Subsurface flow short-circuiting, 632–633
Suction head, 445
Sulfuric acid solutions, 688
Summation curve, 42, 44, 128
Surface aggregation ratio, 130
Surface area per unit bulk volume (a_b), 47
Surface area per unit mass (a_m), 47
Surface area per unit volume (a_v), 47
Surface runoff, 406
Surface seal, 169, 170
Surface tension, 158–159, 258–259, 527, 680, 684
Swelling, 116, 637–644
  clay and, 639–640
  cross-linking particles and, 67
  curtailing diffused double-layer repulsion in, 68
  defined, 66
  effect of exchangeable cations on, 66–67
  flow in, 642–643
  interlattice, 66
  interparticle, 66
  measurement of, 643–644

[Swelling]
  models for, 641
  pressure potentials in, 640–641
  process of, 67
  ratio of, 66
  stages of, 641–642
Swelling index, 66
Swelling pressure, 638–639
Synthesis cracks, 634–635

Tangential stress (τ), 192, 193
  relationship to normal stress (σ),
    198
Temperature, 515, 684
  conversion from one scale to another,
    516
  and gravelly soils, 617–618
  Remote Sensing Thermometer,
    519–520
    scales for, 516–517
    thermocouple, 519
    as thermodynamic property,
      520–521
Temporary binding agents,
    111
Tensile deformation, 204
Tensile strength, 203–204
Tensiometers
  absolute pressure (P) of, 335
  components of, 326
  defined, 326
  in internal drainage method, 395
  limitations of, 329, 333–335
  for measuring matric potential (Φ_m),
    333–335
  mercury manometer, 326–328
  types of, 327
  vacuum gauge, 327, 328–329
Tension infiltrometer, 428
Terminal velocity, 39
Terzaghi's effective stress equation,
    202
Textural classes, 42, 43
Textural pores, 94
Textural porosity, 150
Texture
  agricultural applications of, 84
  effect on field moisture capacity (FC),
    290
  impact on soil, 82–83

[Texture]
  soil properties and processes affected
    by, 83
  soil structure index based on, 131
Thawing, 625. See also Frozen soils
  and soil physical properties, 630–632
Thermal conductivity (K_T), 308–309,
    533, 618
  of certain metals, 534
  defined, 538
  effective, 544
  of gases, 538
  measurement of, 543–545
  merits and limitations of, 313
  of soil particles, 539
Thermal diffusivity, 539
Thermal expansion, 516
Thermal infrared radiation, 310
Thermalization, 301
Thermal radiation monitoring, 519
Thermic soils, 531
Thermistor, 519
Thermocouples, 519
  for the measurement of K_T, 544
Thermodynamic potential, 61
Thermodynamic properties, 520–521
Thermodynamics
  first law of, 521–522
  second law of, 522–523
  third law of, 522–523
  zeroth law of, 521–522
Thermodynamic temperature, 520
Thermoelectric effect, 519
Thermogravimetric method, 299
  merits and limitations of, 313
Thermometric liquid, 516
Thermoscope, 515–516
Third law of thermodynamics,
    522–523
Thixotropic fluids, 267
Thorburn subsoiling test, 127
Tillage, 118–120, 586
  conservation, 458
  defined, 456
  effect on aggregation, 118–120
  effect on soil temperature, 546
  effects on soil structure, 137
  methods of, 459, 546
  reduction of soil water evaporation,
    456–457

[Tillage]
and soil plasticity, 243–245
and water infiltration, 429, 432–433
Tilth
defined, 236
making objective classifications for, 236
production of, 243
and soil consistence, 233, 236
Time domain reflectometry (TDR), 307–308
merits and limitations of, 313
Time moment analysis, 501
Tortuosity, 468
gas (ξ_g), 368
of soil pores, 467
water (ξ_θ), 368
Total porosity (f_t), 22, 155
Total soil-moisture potential (Φ_t)
components of, 323–331
under field conditions, 331–333
units of measurement for, 323
Trace gases, importance in Earth's atmosphere, 559
Transient binding agents, 111–113
Transient evaporation, 446–447
Transient methods
for measurement of K_T, 544
for measuring soil's hydraulic function, 391–392
Transmission pores, 152, 154
formation of, 155
Transmission technique, 215
Transverse dispersion, 472
Travel time analysis, 501
Triaxial shearing test, 201–202
Tropical soils, measuring structural stability of, 131
Turbidity/slaking test, 127
Turbulence, 468
Two parallel plate model, 641
Two-region anion exclusion model, 491–492
Two-region model (TRM) parameters, 503–507
Two-site nonequilibrium transport model, 491–492

Unconfined compression test, 202
Uniformity coefficient, 44

Unit conversion factors, 672
Unit conversions, 676
for soil and water potential, 679
Unsaturated flow
application of continuity equation to, 384
application of Richards equation to, 386
flux density in, 382
functions of, 379
mechanisms of, 380–382
versus saturated flow, 381
across soil core, 382
total outflow rate of, 385
using Darcy-Buckingham equation to determine, 382–383
zone of, 379–380
Unsaturated hydraulic conductivity, 383–384
calculating parameters of, 400–401
calculating with VG equation, 401–402
empirical approaches to estimating, 399–402
measuring with Boltzmann transform method, 393–394
measuring with in situ methods, 394–396
measuring with laboratory methods, 391–393
measuring with soil water retention methods, 396–397
model for predicting, 397–399
relationship to soil porosity, 389–391
Upper plastic limit, 238
in Casagrande test, 242
in Drop-Cone test, 242
indirect methods for measuring, 242–243
Upscaling, 657
Unsteady infiltration, 425–426

Vacuum gauge tensiometer, 327, 328–329
Vadose zone, 379–380, 579
van der Waals forces, 59
colloidal stability and, 172
equation for attractive energy due to, 172
types of, 98
Vane shear test, 203
Vapor density, 269

Vapor in saturated air, 684
Vapor pressure, 268–269, 684
  estimating, 411
Vapor-wetting front, 409
Vegetation, 523–524
Vegetative mulch, 453, 454, 456
Vertical flow, 374–375
Vertical infiltration, 416–417
Vertisols, 638
Vesicular-arbuscular (VA) mycorrhizal
  fungi, 112–113
Viscoelastic fluids, 267
Viscoelastic soil, 197
Viscosity, 527
  of fluids, 265–266
  kinematic, 265–266
  Newton's law of, 266, 357
  of water, 528
Viscosity coefficient, 266
Viscous flow, 237
  versus plastic flow, 247
  upper limit of, 238
Visible and near infrared spectrum, 310
Void ratio (e), 155, 205
  range of, 25
Volume displacement method, 311–312
Volumetric compression, measuring
  resistance to, 197
Volumetric flow rate, 385
  equations for, 358
  using Poiseulle's law for, 389
Volumetric heat capacity (C_y), 523
Volumetric inflow rate, 385
Volumetric soil moisture content (Θ),
  23–24, 25

Warrick similarity analysis, 658
Water. *See also* Soil water
  boiling point of, 268–269, 516
  change from liquid to vapor, 268
  density and viscosity of, 528
  density of, 681
  entropy of, 257
  evaporation from soil. *See* Soil water
    evaporation
  flux density of, 471
  freezing point of, 516
  global distribution of, 255, 256
  global transfer rates of, 270
  hydrologic cycle of, 269–270

[Water]
  large specific heat of, 98
  lysimetric analysis of, 272
  measuring in unit quantities, 337
  measuring of runoff, 272, 276–278
  precipitation of, 272
  properties of, 256–269
  redistribution into soil, 426–427
  six phase changes of, 534–535
  soil as reservoir of, 271
  solute-free, 466
  as soil moisture content. *See* Soil
    moisture content
  total potential of, 356
  transport to soil surface, 441
  viscosity of, 682
  types of, 255, 256
Water absorption, 68–69
Water absorption isotherms, 68, 69
Water adsorption, 69–73
Water deficit, 296
Water desorption method, 158–160
Water film theory, 236–237
Water flow, 9
  capillary bundle concept of, 356–357
  through homogeneous soil profile,
    410–411
  horizontal, 371
  importance of knowing, 355
  laminar, 359–360
  land use effects on, 508
  and matric suction (Φ_m), 380–382
  principles of, 356–361
  in saturated soils. *See* Saturated flow
  steady, 373
  through 3-dimensional space, 370–372,
    384
  upward, 445
  in unsaturated soils. *See* Unsaturated
    flow
  using Darcy's law to determine,
    357–361
  vertical, 371
  at wetting front, 407–412
Water infiltration
  calculating depth of, 433
  calculating rate of, 433
  comparisons of equations for, 421
  conceptual models for predicting rates
    of, 412–418

[Water infiltration]
into crusted soils, 422–423
cumulative, 418, 419
description of, 405–407
empirical models for predicting rates
of, 418–420
flux controlled process of,
407–408
and frozen solids, 631–632
and gravelly soil, 614–615
horizontal versus vertical, 415–417
instantaneous rate of, 407–408
into layered soil, 420, 422
management of, 429, 431–434
measurement of, 428–429
from precipitation, 424–425
profile-controlled process of,
408
relation between surface flux and time
during, 424
unsteady, 425–426
for water-repellent soil, 653–654
Water management, 137
Water manometer, 327
Water molecules
description of, 256–257
electric field of, 257
hydrogen bonding of, 257, 258
surface tension of, 258–259
Water movement. *See* Water flow
Water properties
at atmospheric pressure, 285
boiling point, 268–269
capillarity, 261–264
contact angle, 259–261
fluidity, 267–268
Newtonian and non-Newtonian fluids,
266–267
osmotic pressure, 264
relevant to soil physical properties and
processes, 257
solubility of, 265
surface tension, 258–289
vapor pressure, 268–269

[Water properties]
viscosity, 266–266
water molecule, 256–258
Water repellency, 651–656
classes, 653
effect of wildfire on, 655
effects on soil processes, 653–655
management of, 656
Water retention
in gravelly soils, 610–611
indices for measuring, 129
Water-stable aggregates, 101
Water stage recorder, 276, 277, 430
Water surfaces, albedo for, 524
Water table, management of, 589–590
Water tortuosity (ξ_θ), 368
Water transmission, 129
Water vapor (H_2O), 557
Wet bulk density (ρ'_b), 25
Wetting, in water-repellent soils, 652–653
Wetting front
defined, 408
on layered soil profiles, 420, 422
movement of vapor ahead of, 411
schematic of depth of, 410
schematic of movement through soils,
410
water movement at, 407–412
Wien's law, 532
Wier, 277
Wildfires, 655–656
Work-energy principle, 322
Worm holes, 221, 222–223

Yield stress model, 190
Young's equation, 260
Young's modulus, 194

Zakhrov system of classification of soil
structure, 122
Zeta potential, 61–63
effect of increasing, 64–65
effect of lowering, 64, 65

Printed in the United States
by Baker & Taylor Publisher Services